Decolonizing Medical Education, Research and Practice

Decolonizing Medical Education, Research and Practice

Edited by

Faadiel Essop
Centre for Cardio-Metabolic Research in Africa (CARMA), Division of Medical Physiology, Faculty of Medicine and Health Sciences, Stellenbosch University, Cape Town, South Africa

ELSEVIER

ACADEMIC PRESS
An imprint of Elsevier

Academic Press is an imprint of Elsevier
125 London Wall, London EC2Y 5AS, United Kingdom
50 Hampshire Street, 5th Floor, Cambridge, MA 02139, United States

ISBN: 978-0-443-44998-7

For information on all Academic Press publications visit our website at https://www.elsevier.com/books-and-journals

Publisher: Megan Ball
Acquisitions Editor: Elizabeth Brown
Editorial Project Manager: Suman Deshwal
Production Project Manager: Vijayaraj Purushothaman
Cover Designer: Mark Rogers

Typeset by Integra Software Services

Contents

Contributors

Adedamola Adetiba
Centre for History of Science, Technology and Medicine, Faculty of Biology, Medicine, and Health, University of Manchester, Manchester, England, United Kingdom

Amanda Alblas
Biological Anthropology Research Unit (BARU); Victim Identification Service Unit (ViSUN), Division of Clinical Anatomy, Department of Biomedical Sciences, Faculty of Medicine and Health Sciences, Stellenbosch University, Cape Town, South Africa

Jaya Aysola
Perelman School of Medicine, University of Pennsylvania, Philadelphia, PA, United States

Juan Bornman
Division of Speech-Language and Hearing Therapy, Department of Health and Rehabilitation Sciences, Faculty of Medicine and Health Sciences, Stellenbosch University, Cape Town, South Africa

Faadiel Essop
Centre for Cardio-Metabolic Research in Africa (CARMA), Division of Medical Physiology, Faculty of Medicine and Health Sciences, Stellenbosch University, Cape Town, South Africa

Joe R. Feagin
Texas A&M University, College Station, TX, United States

Pedro W. Fernandez
Division of Urology, Faculty of Medicine and Health Sciences, Stellenbosch University, Cape Town, South Africa

Wandile Ganya
Division of Cardiology, Department of Medicine, Faculty of Medicine and Health Sciences, Stellenbosch University, Cape Town, South Africa

Wanele Ganya
Division of Nephrology, Department of Medicine, Faculty of Medicine and Health Sciences, Stellenbosch University, Cape Town, South Africa

Berna Gerber
Division of Speech-Language and Hearing Therapy, Department of Health and Rehabilitation Sciences, Faculty of Medicine and Health Sciences, Stellenbosch University, Cape Town, South Africa

Ramya Kumar
School of Public Health, University of Washington, Seattle, WA; University of California, Davis, CA, United States; Centre for Infectious Disease Research Zambia (CIDRZ), Lusaka, Zambia

Sefora Makuse
Department of Human Nutrition and Dietetics, School of Health Care Sciences, Faculty of Health Sciences, University of Limpopo, Polokwane, South Africa

Patricia Maritim
School of Public Health, University of Zambia, Lusaka, Zambia

Vanessa Mbhatsani
Department of Nutrition, Faculty of Health Sciences, University of Venda, Thohoyandou, South Africa

Xikombiso Mbhenyane
Division of Human Nutrition, Department of Global Health, Faculty of Medicine and Health Sciences, Stellenbosch University, Cape Town, South Africa

Keegan O. Meiring
Biological Anthropology Research Unit (BARU); Victim Identification Service Unit (ViSUN), Division of Clinical Anatomy, Department of Biomedical Sciences, Faculty of Medicine and Health Sciences, Stellenbosch University, Cape Town, South Africa

Rhoda Meyer
Department of Health Professions Education, Faculty of Medicine and Health Sciences, Stellenbosch University, Cape Town, South Africa

Mabitja Moeta
Faculty of Health Sciences, University of Johannesburg, Johannesburg, South Africa

Phila M. Msimang
Department of Philosophy, Stellenbosch University, Stellenbosch, South Africa

Michelle Munyikwa
Perelman School of Medicine, University of Pennsylvania, Philadelphia, PA, United States

Alida Naudé
Division of Speech-Language and Hearing Therapy, Department of Health and Rehabilitation Sciences, Faculty of Medicine and Health Sciences, Stellenbosch University, Cape Town, South Africa

Rachel Niehuus
Departments of Surgery and Social Medicine, University of North Carolina School of Medicine, Chapel Hill, NC, United States

Nomakhushe Nxusani
Division of Human Nutrition, Department of Global Health, Faculty of Medicine and Health Sciences, Stellenbosch University, Cape Town, South Africa

Sheila Patel
University of California San Diego School of Medicine; Gaja Collective, San Diego, CA, United States

Desiree C. Petersen
South African Medical Research Council Centre for Tuberculosis Research, Division of Molecular Biology and Human Genetics, Faculty of Medicine and Health Sciences, Stellenbosch University, Cape Town, South Africa

Muneera Rasheed
Independent Professional, Karachi, Pakistan

Kathryn Smith
VIZ.Lab, Department of Visual Arts, Faculty of Arts and Social Sciences, Stellenbosch University, Stellenbosch, South Africa

Saleha Suleman
Department of Public Health, University of Cape Town, Cape Town, South Africa

Caitlin Uren
South African Medical Research Council Centre for Tuberculosis Research, Division of Molecular Biology and Human Genetics, Faculty of Medicine and Health Sciences, Stellenbosch University, Cape Town, South Africa

Mthokozisi Zuma
Division of Human Nutrition, Department of Global Health, Faculty of Medicine and Health Sciences, Stellenbosch University, Cape Town; Research Support and Coordination Division, Agricultural Research Council, Hatfield, Pretoria, South Africa

About the editor

Prof. Faadiel Essop is the Director of the Centre for Cardio-metabolic Research in Africa (CARMA) at Stellenbosch University, where he investigates the effects of chronic stress on the onset of cardio-metabolic diseases. He completed his PhD at the University of Cape Town and worked as a Fulbright Fellow at the University of Texas-Houston Medical School. He served as a former President of the African Association of Physiological Sciences (AAPS) and a council member of the International Union of Physiological Sciences (IUPS). He is a Fellow of the IUPS, the American Physiological Society, the Royal Society of South Africa, and the Academy of Science of South Africa. He recently received the prestigious Teaching Advancement at University (TAU) Fellows Award for his outstanding contributions to transformative education, particularly for challenging scientific elitism within the medical field, and the 2025 IUPS' Julie Chan Award, which recognizes physiologists who have made outstanding contributions to international mentoring and/or career development.

Preface

Modern medicine has made major advances over the last century, resulting in improved diagnostic tools and therapeutic interventions and increased life expectancies. This remarkable progress is largely driven by sophisticated research studies, randomized clinical trials, and implementation science. However, such technical mastery in modern medicine is not a panacea as several underlying issues and contexts (e.g., social justice, equity, access to resources and medical care, lingering Eurocentric epistemic frameworks) continue to shape contemporary medical practice, research, and education.

This book is not about replacing existing medical practices; instead, it aims to extend contemporary ideas by focusing on the colonial roots of modern medicine and their ongoing effects. It is crucial to increase such awareness, especially with an increased emphasis on more inclusive health care systems and holistically trained clinical practitioners and biomedical scientists. A rhetorical question may arise, i.e., is it necessary to delve into such contentious issues at this time? We argue that it is now, more than ever, vital to evaluate the colonial roots of modern medicine and thereafter consider the best ways to decolonize our contemporary practices. Here, we argue that although the colonizers have departed, the structural remnants of colonialism remain, i.e., the phenomenon of coloniality. This book therefore addresses the tension between modern medicine's advances and the structural legacies of colonialism that continue to impact especially the Global South and marginalized communities in the Global North.

Although there are other texts dealing with some of these aspects, this is done in a relatively limited and fragmented way, and hence, we are of the opinion that there is a lack of a single volume covering *all* these dimensions (research, teaching, and clinical practice) in a comprehensive manner. This book bridges this gap by providing an in-depth analysis of such issues by initially focusing on the challenges posed by colonialism and scientific racism, thereafter providing some solutions to equip practitioners to better manage such challenges in their respective knowledge domains and practices. The uniqueness of this book is further strengthened by the inclusion of several authors from the Global South who are now provided with an opportunity to share their views and experiences. Moreover, it includes contributions from several authors within the medical field, thereby offering unique discipline-type insights to better relate to the target audience. This book is aimed at academics in medical schools/colleges, undergraduate and postgraduate students pursuing medical, allied health, and biomedical sciences, postdoctoral fellows, doctors/clinicians, health care professionals and workers, leaders in universities, leaders in research and health care, policymakers, and the broader public.

As editor, my task was to conceptualize the overall vision and layout for the book to create a unified theme that would be covered in a comprehensive manner. This endeavor was facilitated and made easier by the rigorous and impressive contributions provided by the array of authors included. For the review process of all submitted chapters and subsequent editing, my approach was a curatorial versus a prescriptive one to best capture the honest (and at times raw) reflections of the authors and to allow their intellectual honesty and awareness regarding the complexity of the topics to be showcased. I wish to extend my gratitude to all the authors for their excellent contributions to this compelling book.

I also wish to thank my loved ones who supported me over the last year and a half in bringing this book project to its fruition—my wife Rehana, my children and their spouses (Ziyaad, Aaliyah, Yasin, Tasmiya, Nawaaz and Harun Faadiel), my mother Fareda, and my good friend Mansoer Abrahams (now sadly deceased) for always encouraging me to write about such thorny issues.

Lastly, the book is an invitation for readers to critically engage with its content as part of an ongoing dialog, to trigger self-reflection and self-critique regarding current medical practices. Only by recognizing the hurtful past of colonial medicine and scientific racism can more inclusive medical practices be implemented to eventually result in meaningful and transformative changes. Moreover, this volume should also help ensure the training of a cohort of medical and allied health students that display epistemic humility and increased awareness regarding indigenous knowledge systems and are better equipped how to engage with diverse settings and populations with different needs and values.

Prof. Faadiel Essop

PART

I

Challenges

CHAPTER 1

From slave ships to contemporary clinics: The US neo-slavery medical complex

Joe R. Feagin
Texas A&M University, College Station, TX, United States

Introduction

For over four centuries, antiblackness has been a major structuring logic of United States (US) society, shaping all important institutions from slavery's beginnings to the present. This oppressive antiblack reality began in 1619, when the first enslaved Africans were imported on a European slave ship. African Americans were racially enslaved for more than half of US history, and once freed in 1865, they suffered legal racial segregation for another fifth of that history. This racial segregation lasted until the 1960s, when civil rights laws officially ended it. Black enslavement, followed by a century of Jim Crow segregation, and now many decades of informal discrimination, has ensured that Black Americans have had to endure systemic oppression, impoverishment, and poor living conditions over most of the history of the United States.

As detailed in my *Racist America* (2025), this systemic racism is not reducible to just individual racial prejudices and acts but is a foundational societal system—one rooted in centuries-old, white-controlled institutions, resource dominance, and a legitimating racial framing. This white racial frame constitutes the cognitive, emotional, and moral scaffolding that normalizes these racist societal arrangements, ensuring that antiblackness is not temporary but regularly reproduced across major institutions, including those of health care.

As Black critical scholars have often argued, antiblackness operates as a permanent, structuring white logic that positions Blackness as the paradigmatic site of economic dispossession and social death. That is, it renders Black people hypervisible as objects of racial exploitation but invisible as bearers of human rights and full US citizenship. This dominant white logic is not reducible just to the category of "racism" or to the everyday discrimination faced by Black Americans and other people of color. Rather, antiblackness is a broad, enduring white societal force undergirding the architecture of US systemic racism. In this way, it shapes not only the major US racial hierarchy but also the very terms of life and death in this society.

Yet, this antiblackness is not simply about white racial exploitation, but also about the resulting exposure of Black communities to extensive premature suffering,

Decolonizing Medical Education, Research and Practice. https://doi.org/10.1016/B978-0-443-44998-7.00013-2

neglect, and death—a white-imposed pattern that social analyst Achille Mbembe (2019) theorizes as "necropolitics". This white Western necropolitics determines who is allowed to live and who is consigned to die, and it dictates the timing and circumstances of their deaths. As a result, the US health care system is not just a site of major health resource inequality, but also a necropolitical institution—one long normalizing Black vulnerability to poor health and death as routine features of their individual and social lives.

Within this discriminatory framework, the US health care system emerged as a paradigmatic site of "medical sharecropping"—the centuries-long process in which Black labor, bodies, and knowledge have been essential to building up a US medicine greatly benefiting white Americans. Unfortunately, however, these medical benefits have often been systematically restricted or denied to Black communities. Furthermore, to fully grasp the persistence and depth of US racial health inequalities, we need to theorize Black health and health care as a site where not only this medical sharecropping but also antiblackness and necropolitics intersect. This intersectional reality involves a centuries-long legacy of racial injustice, one still demanding radical dismantling in a multiracial democratic society. To that end, in this chapter I extend my theory of systemic white racism by foregrounding these intersecting logics as central to understanding the societal origin and operation of US health and health care injustices.

Over several decades, my colleagues and I have conducted extensive research showing that authentic US health equality requires not just better African American inclusion in the contemporary "slave ship" that is still US medicine, but ultimately a radical dismantling of that ship's centuries-old racist health care structure. This societal deconstruction will necessitate an aggressive implementation of major reparative justice, not just slow reform, to finally abolish US health care's centuries-old patterns of highly racialized inequality.

The world's largest and costliest health care system

In a recent year, total US health care expenditures were about $4.9 trillion, more than $14,500 per American, and much more than in any other country. This stunningly high cost is mainly due to higher charges for health care services—including routine treatments, drugs, and administrative costs—and not from greater health care utilization than in other countries. This large-scale health care system is unmatched in employee size, namely about 17 million employees, including about 934,000 doctors and four million nurses, more than for any other US employer or any other country. This massive workforce includes health care employees distributed across numerous medical, nursing, dental, and other health care occupations (Centers for Medicare and Medicaid Services, 2024; Hartman et al., 2025).

Significant antiblack discrimination has long been *fundamental* to this large-scale health care system. One infamous contemporary example is the medical use of a specific Black person's deceased body—that of Henrietta Lacks, a poor Black Virginian. After her death from cancer in 1951, Johns Hopkins Hospital officials took her body's cells for long-term medical use without consent, an action typical of white medical practices during the long US racial segregation era. Her famous ("HeLa") cells became the first immortal human cell line, one used for significant biomedical research for three-quarters of a century, indeed to the present day. For decades, neither Lacks nor her relatives received adequate recognition or payment for this use of the cells, which has led to important scientific advances. Only since 2013, when her family reached an agreement with the US National Institutes of Health, has the family gained some control over the medical use of her groundbreaking cells (Summerside, 2017).

In addition, in the early decades of the 21st century, the common white practice of using unclaimed Black and other poor people's bodies (cadavers) for medical research continues, often still without consent. This medical practice is not a relic of some distant past but signals the persistence of antiblack practices in the present. For example, starting in 2019, more than 2,300 unclaimed bodies, disproportionately Black, were given to the University of North Texas Health Science Center by Texas county officials. Many were used for medical education. Initially, Center officials explained that this use of unclaimed Black bodies was critical for contemporary medical education. However, after significant media coverage, the Center suspended its body donation and dissection program (NBC News, 2024a; Hixenbaugh, 2024). These media investigations have not only demonstrated that such medical practices persist but also raised ethical questions about them. These highly questionable practices also help explain the commonplace distrust Black Americans have regarding the contemporary US medical system (NBC News, 2024b).

Unmistakably, the fact that such Black bodily exploitation remains routinized shows the continuing power of the white racial frame in US medical fields. This example is indeed a textbook illustration of the operation of that white frame: a collective worldview viewing Black American exploitation as invisible or justified and centering white medical dominance as legitimate. It is also a societally embedded *interpretive* framework encompassing common white narratives, emotions, and inclinations to discriminate that, for centuries, has justified historical and contemporary forms of medical and other forms of Black exploitation (Feagin, 2020).

Historical foundations: Systemic racism in US health care

Slave trade era: Commodification of Black bodies and emerging Western medicine

Lasting for at least three centuries, the transatlantic African slave trade involved several European nations and profoundly shaped the medical, professional, and ethical foundations of Western medicine. The trade embedded the extensive exploitation of Black bodies firmly within it. Still largely unknown to most whites in Western countries, this centuries-long African slave trade and subsequent African American enslavement greatly contributed to whites' shaping of Western medicine into a highly racialized and capitalistic profit-driven endeavor. Indeed, a few contemporary scholars of color have foregrounded the *slave ship* as a literal and symbolic crucible for the emergence of much Western medical knowledge. For instance, African American historian Carolyn Roberts (Roberts, 2017) has conceptualized this transatlantic slave trade as regularly being *medically managed*, highlighting how slave ships actually functioned as sites of Western medical experimentation on enslaved bodies. This was not only a matter of Black exploitation but also of routine Black exposure to unnecessary suffering and death—an early instance of the aforementioned necropolitical logic becoming foundational to US health care. White European physicians played a central role in this medical exploitation of enslaved African bodies on and off these many slave ships.

Roberts draws a direct line from African slavery to contemporary African American health and health care inequalities, including systemically racist realities such as the routine *lack* of adequate medical treatment for Black physical pain (Roberts, 2017; see also Kyere, 2020). Similarly, African American scholar Sowande Mustakeem has advanced this critical analysis by showing that thousands of slave ships were the birthplace of much global medical practice and emphasizing how the medical traumas regularly inflicted on enslaved bodies were fundamental to the professionalization of Western medicine (Mustakeem, 2016).

Across the centuries, from the origin of African enslavement in North America to the present, the white-controlled medical professions and their organizations have routinely been pillars in the systemic racism architecture. As Roberts and Mustakeem have shown, the hegemonic white racial frame has shaped not only the everyday practices of these medical personnel but also the development of medical professionalization more broadly. Moreover, on those thousands of enslaving ships, in the onshore slave markets, and at numerous other onshore settings, white European and North American doctors were operating not just as medical wage earners receiving banknotes, for they were often paid in enslaved Africans. Their medical discoveries, derived from the direct exploitation of Black bodies, became part of a brutally racist capitalistic system rather than mere commitments to expanding ethical medical care (Roberts, 2017; Mustakeem, 2016; see also Nong et al., 2023).

Indeed, the transatlantic slave trade substantially fostered the development of the *global* capitalistic system itself by blending traditional African and new European medical knowledge into the relentless pursuit of capitalistic profit in Europe and the Americas. This globalizing slave economy brought numerous innovations in Western disease care, surgical techniques, and drug development by forcefully extracting the conceptual and physical labor of African health care healers. This "borrowed" medical knowledge was early on incorporated into Western medical texts without credit to these non-European sources. In this way, the Western modernization of medicine was built, in part, on the exploitation of Black medical labor *and* knowledge—a foundational reality omitted from standard historical narratives about Western medicine (Roberts, 2017; Mustakeem, 2016).

In addition, the centuries of Western slave trade did far more than shape the knowledge and technical developments of modern medicine. These exploitative white actions also helped in forging numerous medical and ethical standards that continue to govern Western health care professions. From its earliest days, white physicians and other white health care workers constructed an increasingly professionalized medical system in which Black bodies were—repeatedly and often unreflectively—rendered as objects for medical experimentation and profit (Feagin and Bennefield, 2014; Bowleg, 2023). Ever since, the dominant white medical frame has continued to define what counts as legitimate medical knowledge, whose bodily pain is taken seriously, and whose lives are most valued.

US slavery: The rise of the medical–industrial complex

Over the centuries of North American enslavement, a thoroughly racist medical and health care system developed within the country soon called the United States of America, one that extended the horrific racist conditions of slave ships to this racially oppressive society. Eventually, this new white-run country enslaved many millions of Africans and their African American descendants. After their arrival in the US the enslaved Africans usually received little medical care that reflected their own choices. What care they did receive was often marked by outright brutality. Just as on the slave ships, during the US antebellum slavery period, numerous white medical innovations were frequently tested on enslaved Black bodies without consent (Cooper Owens, 2017). In addition, these early white physicians and other medical workers spread widely racist theories of Black biological inferiority, which justified their brutal medical experimentation and the larger society's extensive Black dehumanization. An infamous example is the dehumanized experimentation of the famous white physician J. Marion Sims, often named the "father" of modern medical gynecology. His medical advancements were often achieved through experimenting on enslaved Black women, again without their consent and while denying them their basic human rights. For example, in the 1840s, Sims subjected Anarcha, a young enslaved woman, to surgeries without anesthesia to develop a medical procedure to repair *white* women's vesicovaginal fistulas. His surgical methodology is still celebrated today as a major medical breakthrough. Yet, from the enslaved women's perspective, this was nothing more than dehumanizing gynecological torture (Cooper Owens, 2017, p. 45ff).

Repeatedly, these early white US medical practices were sustained by the prevailing white frame which, as noted previously, operates as a rationalizing worldview that provides whites and their medical knowledge with authority, while viewing Black bodies as inferior and exploitable without consent. The white medical experiments and data collections are examples of the *epistemic* violence central to the long era of US slavery, as well as to the oppression of the country's Indigenous inhabitants. Indeed, this epistemic violence is a *constitutive* pillar of the broader and systemic US racism as I theorize it: the forceful, systematic appropriation or devaluation of Black and Indigenous knowledges by white-controlled institutions, all justified by the dominant white racial frame. The long-term result has not only been the extensive material exploitation of Black labor and lives, but also the extensive exploitation or delegitimation of Black medical expertise.

The scholar Gayatri Chakravorty Spivak clearly demonstrated just how white racial power generally operates not only through physical force but also through shaping society's prevailing ordinary and scientific knowledge of the surrounding natural world (Spivak, 1988). Strikingly, Spivak and other scholars' colonial and postcolonial studies are exposing just how the dominant Western intellectual practices have assisted in perpetuating and rationalizing the exploitation of racially colonized peoples of color, indeed to the present day. White Western colonialism has not just dominated peoples of color materially, for it has also systematically constructed them as *intellectually* inferior, including in large-scale attempts to erase their established ways of human knowing. Thus, while this white intellectual process frequently steals from the colonized's health care knowledge, it more often delegitimizes these knowledges in favor of white Western medical and other health care knowledge. In this process, powerful whites impose a particular way of white-racist knowing on nonwhite others, including, over time, through the Western educational system and mainstream media (Smith, 2012).

Drawing on my systemic racism theory, I consider it important to recognize that these oppressive material and epistemic health care practices are not isolated but integral to the broader US and other Western societies' major resource transfers and imposed racial hierarchies. The extensive commodification of Black bodies in medical development is a paradigmatic example of Western societies' much larger and racialized processes of unjust enrichment and unjust impoverishment. This capitalistic process exemplifies the core mechanism of systemic racism, as extensively outlined in *Racist America* (2025): the transfer of major resources from Black communities to white-controlled institutions, resulting in the intergenerational accumulation of unjust enrichment for whites and unjust impoverishment for Blacks.

This transfer process is not incidental but *central* to the reproduction of historical and contemporary US racism. For instance, important in understanding the long enslavement period of African American oppression is the great wealth stolen by white generations through many decades of enslavement. In contemporary dollars, estimates of the value of stolen Black labor and its compounded effects range from $14 trillion to over $100 trillion, directly shaping today's huge US white-Black wealth gap (Craemer, 2015; Robinson, 2023; Wrigley-Field, 2024). In addition, at the peak of this US Black slavery system, white slaveholders secured about $3.5 billion (in today's dollars) in life insurance policies on many enslaved Black workers. Major insurers such as Aetna and New York Life issued hundreds of policies on these enslaved Black people, with individual policies ranging from $300 to $1000 in 19th-century dollars but often covering only a portion of the enslaved person's appraised value. Indisputably, this specific white commodification of enslaved Black lives was well connected to the country's ever-growing financial and other capitalistic institutions (Aetna Insurance Company, 1853; Patton, 2024).

Ironically, the enslaved Black workers who managed to flee their enslavement were defined by elite whites as engaged in *pathological* resistance and given racist diagnoses like "drapetomania" (flight-from-slavery "disease"). The white antebellum physician Samuel Cartwright wrote about Black runaways not only having this drapetomania disease but also being susceptible to "dysaethesia aethiopis," yet another phony disease caused by "Black blood" that he alleged had created their "barbarism" (Edwards-Grossi and Willoughby, 2024). Repeatedly, Black Americans were racially framed by elite slaveholders and their political allies as requiring the strict lessons of the white-run slavery system to "improve" their supposedly diseased Black mentality. Cartwright's mainstream medical work was influential, published in leading medical journals and cited by slavery's elite white defenders as scientific evidence for persisting Black inferiority. Here is another instance of the white racial frame at work, one where pseudoscientific racist narratives are mobilized to justify Black dehumanization and white institutionalized power. Cartwright's racist theories were integral to the dramatic rise in this era of "scientific racism," a white perspective that early on buttressed Black enslavement (Opara et al., 2022, pp. 225–226). This "scientific" racist framing of those enslaved as having imaginary diseases laid the foundation for many later decades of extensive racist stereotyping. Unsurprisingly, such racist stereotyping has supported the persisting white dehumanization of Black Americans while ignoring the white-imposed oppression shaping their very negative socioeconomic conditions.

More medical discrimination: "Jim Crow" segregation and eugenics

Over the long legal segregation ("Jim Crow") era, after the slavery era ended, as numerous African American scholars have emphasized, the white-imposed segregation perpetuated this racial dehumanization and associated impoverishment. The leading African American intellectual of this era, W.E.B. Du Bois, offered incisive analyses of what would soon be termed "institutional" racism, including in health care institutions. In his early 20th-century book, *The Health and Physique of the Negro American*, Du Bois (1906) pioneered in documenting that the persisting health inequalities between Black and white Americans were *not* due to white-alleged biological "race" differences, but rather to vastly different societal conditions historically forced by whites on Black Americans. The causes of these substantial health differences were thus primarily systemic—including poor nutrition, health care, education, and socioeconomic opportunities—all shaped by a white-imposed racist hierarchy. This pathbreaking research by a top Black scholar on Black health early on aggressively challenged the dominant white "race" paradigm, demonstrating that the roots of US racial health inequality lie in well-institutionalized white racism.

Decades before the US Civil War and the end of slavery, racial segregation laws had already been developed by Northern whites. They enforced the segregation of Black Americans in workplaces, hospitals, and public schools. These discriminatory laws set up a legal framework for the later imposition of Southern "Jim Crow" racial segregation, including that affecting Black health care availability, after the US Civil War (see Feagin, 2025).

After the 1865 emancipation of formerly enslaved African Americans, the US Congress created the Freedmen's Bureau, the *first*-ever federal health care effort. Intended to improve post-war health for Black and other poor Americans, these were

Jim Crow facilities, with new hospitals and clinics usually racially segregated and thus featuring separate wards for Black and white patients (NCpedia, n.d.). Sadly, serious underfunding of this program during the new postwar President Andrew Johnson's administration, along with vigorous opposition from former Confederates, forced this Bureau to prematurely close most health care facilities (North Carolina Civil War and Reconstruction History Center, n.d.; Cox, 2021).

In this postwar era, significant racial segregation in southern and border state areas extended beyond housing and public accommodations to local health care institutions. For nearly a century after the Civil War, this local segregation perpetuated the old Southern slave plantations' racist reality by placing Black patients in poorly funded medical facilities, in stark contrast to better-funded white facilities. Southern Black communities were systematically denied access to quality hospitalization, with white facilities refusing Black patients or relegating them to inferior segregated hospital wings. This systematic denial of Black health care, and the forcing of Black patients into sites of medical neglect, demonstrates yet again the *necropolitical* operations of US health care—constantly one where white institutional decisions routinely determine which lives are protected and which are not. Indeed, the term "pesthouse" segregation was sometimes used to describe this racially segregated Black care. To make matters worse, local housing segregation produced hospital "deserts," where Black residents had little geographical access to hospitals. This large-scale racial segregation continued during the long Jim Crow era (circa 1870s–1960s), including federal legislation creating new hospitals (e.g., the 1946 Hill-Burton Act). This legislation prioritized racially segregated facilities and excluded Black neighborhoods from new white-run hospital service areas (Newkirk II, 2016).

Sadly too, after the Civil War, the white development of scientific-racist framing in health care accelerated, including aggressive, white-run medical research endeavors. One infamous example is the Tuskegee syphilis research project, which lasted for decades (circa 1932–1972). In this racist project, white doctors and officials of the US Public Health Service intentionally withheld life-saving penicillin from Black men suffering terribly from this disease. Unsurprisingly, this racially discriminatory research was part of centuries of medical exploitation, constantly motivated by the prevailing white racial frame, one still riddled with pseudoscientific notions about racial matters (Centers for Disease Control and Prevention, 2025; Northington Gamble, 1997).

The US pharmaceutical industry has long played a complicit role in racializing medical experiments on African Americans. During the long Jim Crow era, poor Black communities were sites for yet other drug trials, again frequently without informed consent. For instance, the company Merck's 1950s trials of the drug ivermectin for blindness were conducted in Black communities (Merck and Co. Inc., 2025).

Relatedly, the Jim Crow segregation era saw an increasing medical emphasis on the scientific germ theory of disease. Led by the pioneering microbiological research of white scientists like Louis Pasteur and Robert Koch, this new germ theory helped to generate drugs involving an array of high-tech medical engineering. This, in turn, created an increasingly profitable US corporate drug industry. In addition, these engineered drugs, coupled with the germ theory, helped to establish a "more scientific" medicine increasingly focused on specific disease cures rather than on holistic health and disease prevention. Unlike Black folk medicine practitioners and traditional family physicians, most of the new white scientist-specialist physicians paid less attention to the major contributions of family, community, and workplace in generating human illness and healthy human wellness (Curtin, 2010; Adler, 2023).

Another chilling aspect of the Jim Crow era's brutal oppression is seen in the many forced sterilizations, the "surgical lynchings" disproportionately targeting Black women from the late 1800s to the 1960s. Many US states' racist eugenics laws encouraged the sterilization of poor women, including thousands of Black women, who were considered societally "unfit" to be mothers in the racist framing of white medical professionals. These mostly unnecessary sterilizations were often performed without consent and defended, yet again, in the racist pseudoscientific terms of the dominant white frame (Texas Woman's University, n.d.). Both state welfare agencies and federally funded clinics were medical sites for this racist coercion, with Black women sterilized at rates far exceeding those of white women (Stern, 2020; Minna Stern, 2020). Yet again, the white racist subordination of Black women was *bodily*, involving violent control over these women and their families' development.

Black midwives and the professionalization of white-dominated medicine

Racial oppression during the legal segregation era included the racist subjugation of many Black midwives. Several social scientists, including this author, have conducted research on the increasing exclusion of these Black, female, and working-class health care practitioners during the rise of white "scientific" medicine and white-imposed health care segregation. This research collectively documents how the increased professionalization (e.g., physician licensing) of modern medicine, over time and under white-male control, gradually excluded Black and other working-class midwives. Unsurprisingly, the resulting reduction in health care access increased its cost. During the centuries of US slavery, and well into the legal segregation era, Black midwives were essential health care providers in Black and white communities, especially in southern states. For the most part, they were poorly paid but still frequently relied upon by whites for health care. Yet, they were often denied appropriate white respect and professional status (Shead, 2022; Randall, 1993). Metaphorically, they were thus just "medical sharecroppers" because, like Black agricultural sharecroppers whose low-wage labor greatly enriched whites in the Jim Crow segregation era, these hardworking midwives provided major health care for Black and white women in a racially oppressive system that exploited them.

As the Southern medical system became more professionalized, it was increasingly dominated by white male physicians. Long-serving Black midwives were restricted or eliminated by white segregation laws and licensing requirements, as well as by white public campaigns falsely stereotyping them as incompetent. This racist exclusion was similar to the economic oppression of Black male farmers enduring the white agricultural sharecropping era. Black midwives lost the ability to practice independently, and the economic benefits of their hard work were transferred to white professionals and white-run medical institutions. In this era and subsequently, the major increase in white hospital-based, physician-attended births was frequently motivated by this white professional monopoly and profitmaking rather than genuine concern for Black, white, and other maternal health (Shead, 2022; see also Feagin and Bennefield, 2014).

The Flexner Report and medical school closures: Institutionalized racial exclusion

In addition, during the Jim Crow era, the new medical faculties at segregated white universities became dominant in US medical education. A 1910 white-run Carnegie Commission study (the Flexner Report) evaluated educational deficiencies in existing medical schools, and soon many of these schools were closed to "protect" the public (Flexner, 1910). Entrance requirements for white-run medical schools increased, usually excluding students from modest-income families and nonwhite applicants. Implementation of this Flexner Report resulted in the closure of most historically Black medical schools under the slogan of necessary medical "standardization." In addition, established as all-white and all-male, the new American Medical Association (AMA) not only advanced much biological pseudoscience, including racist eugenics, but also prioritized medical education that privileged white doctors instead of fostering authentic racial desegregation in the new medical schools. In this way, the AMA further institutionalized the racially skewed hierarchy within the US medical profession, a hierarchy that persists to the present.

One intended result of closing the existing medical schools was to increase the national importance of the white-run, racially segregated medical schools and thereby cut the total number of trained doctors, in turn reducing competition and raising incomes. By about 1920, only 3.3% of US physicians were Black (Sidel and Sidel, 2006, pp. 236–237). Medical care became more expensive and often geographically inaccessible to Americans of modest means, especially Americans of color. Nonwhite communities suffered significant health care losses from this newly professionalized medicine. Existing working-class practitioners, including Black midwives and herbalists in these communities, were driven out of practice by new state licensing requirements under the control of white medical societies. Indeed, much scholarship makes clear that the professionalization of US and European medicine has long been deeply entangled with the suppression of alternative healing traditions, including those of traditional Black health care practitioners—even as the medical knowledge of the latter has been repeatedly appropriated in the historical process of the new white-run health care (Brown, 1979, pp. 79–80, 96; Sidel and Sidel, 2006).

US health care: Contemporary manifestations of systemic racism

Contemporary legacies of centuries of racial exploitation

As we move beyond the long US slavery and Jim Crow eras—altogether *about 80%* of US history—into the shorter contemporary era of US history, the broad societal metaphor suggested earlier again helps summarize and assess centuries-long US health care development. As previously discussed, this *medical sharecropping* metaphor captures the centuries-long dynamic of exploitative Black labor and knowledge extraction within US health care.

Consider the interlocking societal processes of this medical sharecropping, some of which still operate today. First, millions of Black bodies and much Black labor have been central to the development of US health care, from the aforementioned use of enslaved Black people for white anatomical research and medical experimentation to the foundational roles played by Black midwives and herbalist caregivers in developing US health care. However, the resulting health care benefits, such as those from medical breakthroughs and their economic gains, have come disproportionately to white professionals, white patients, and white health care organizations. Second, as with agricultural sharecropping's cycle of Black debt and white profitability, Black communities have, in the past and present, frequently been dispossessed of some benefits of their own health resources—from their midwifery, community hospitals, and traditional healing practices—through white exploitation, exclusion, marginalization, and underfunding. Third, this medical sharecropping metaphor underscores how white medical institutions have not just neglected Black health matters but have actively controlled Black bodies by means of forced sterilization, exploitative drug trials, and racialized denial of effective medical treatments. These are not isolated discrimination issues but part of broader white societal domination justified by racist "scientific" theories and institutionalized in everyday medical practices.

Fourth, the medical sharecropping metaphor highlights how contemporary racist patterns persist into the present. Here, the concept of historical "path dependency" applies. As I argue in my systemic racism theory (Feagin, 2006), this path dependency is not just a matter of historical inertia but the ongoing reproduction of white advantage and Black disadvantage through the long *intergenerational* transmission of white-controlled institutional health care resources and structures. Past discriminatory decisions are not left behind; rather, they are embedded in the architecture of contemporary health care, shaping present-day outcomes and opportunities. Contemporary health care racial inequalities—unequal pain management, longer medical wait times, lower hospital admission rates for Black patients—are the historical legacy of a white-run health care system built on much earlier racial discrimination. Indeed, the implicit and explicit racist framing of many white clinicians, and their related racist practices in their health care institutions, still conceptualize Black patients as less medically deserving, thereby regularly perpetuating antiblack discrimination (Ezenwa et al., 2013; Anderson et al., 2009). These discriminatory practices are predictable outcomes of a centuries-old health care system regularly developed within the white racial frame and other institutional logics of US systemic racism.

Fifth, just as white agricultural sharecropping narratives blame continuing Black poverty on Black individual and family failings rather than institutionalized white discrimination and other forms of exploitation, the contemporary US medical system commonly locates the causes of Black poor health in their individual and family cultural weaknesses rather than in the institutionalized white racism routinely shaping those health problems.

Clearly, this extended medical sharecropping metaphor underscores the reality that the inferior medical and other health care received by Black Americans in the past and present is not accidental or merely the result of individual whites' racist attitudes but is instead woven into the institutionally discriminatory fabric of the dominant US health care institutions.

Contemporary discrimination against Black medical personnel

African American physicians have long faced institutionalized racial barriers. Excluded from the Jim Crow-era AMA until the mid-20th century, they developed their own organizations, such as the National Medical Association. Moreover, today, US medical care remains significantly inegalitarian and often racially segregated (Zephyrin et al., 2024). Currently, Black physicians are seriously underrepresented in the US medical workforce, comprising about 5.7% of physicians compared to approximately 14% of Black Americans in the US population. This percentage is lower than it was even in 1910—a discriminatory reality that again highlights major path dependency stemming from earlier Jim Crow segregation's impacts on racial discrimination in current medical education and practice. Even today, white doctors rarely establish their major practices in or near Black communities, while Black doctors still disproportionately serve Black patients in Black communities (Zippia, 2025; Chartis, 2025). In addition, Black faculty underrepresentation is pronounced in most US medical schools, and the relatively few Black faculty members there frequently face racial barriers to professional advancement (Simbo.ai, 2024).

The hardy persistence of these discriminatory patterns illustrates again how the hegemonic white racial frame continues to shape the social structure and culture of US health care institutions. Over numerous generations, this white framing and discrimination have not simply reflected white individuals' prejudices or isolated racial abuses. Instead, they have become firmly institutionalized in health care institutions, embedding a white racist logic into the core reality of these institutions and their "advancements." The professionalization of US and European medicine has, from its origins, developed within a broad, centuries-old white worldview systematically privileging white Western medical knowledge and practices while devaluing and suppressing those of people of color. The result is an ongoing white racist legacy: the contemporary health care system continues to be shaped by the institutionalized white-over-Black medical hierarchy, one routinely producing persisting major inequalities in health care access and outcomes for African Americans and other Americans of color.

Persistence of scientific racism in medical research and practice

Unfortunately, the old white ideology of "scientific racism" remains deeply embedded in US medical institutions and everyday practices. For instance, contemporary white medical professionals and researchers continue to discuss disease rates and physiological differences among the "races" as if these centuries-old categories are biologically fixed and meaningful, despite much countering evidence. As scholars Jessie Daniels and Amy Schulz have observed, numerous contemporary genome scientists persist in using these old racist categories as proxies for genetic variation, arguing that it is more convenient than examining individual genetic profiles. In this way, they routinely perpetuate the debunked "race" typology of "Caucasian," "African," "Asian," and "Indian," which actually originated in 18th-century racist science (Daniels and Schulz, 2008, p. 89ff).

A century of research in US physical and social sciences has shown that these "race" categories are not valid for *authentic* understandings of human biological ancestry and health. Yet, the ideological power of this white racial framing endures, and many scientists, mostly white, default to it in everyday research and clinical settings. This white tendency is reinforced by the fact that the broader white racial framing routinely normalizes linking these "race" groups to white discriminatory practices. Manifestations of this are evident in how racial inequalities in health continue to be attributed to genetic differences rather than to the discriminatory consequences of the country's persisting institutional racism. Recent scholarship, such as that of biological anthropologist Jonathan Marks, shows that even within serious scientific circles, the assertion of these ideas is still tolerated, further entrenching problematic "race" theory in US science (Marks, 2017). The persistence of this racist framing means the use of "race" as a biological variable continues to shape much medical research design and clinical decision-making, again relieving US health care institutions of responsibility for the continuing reality of racial health inequalities. To move beyond this racialized legacy, we must reject "race" as an actual biological reality and instead focus on dismantling the racist ideologies and institutions still perpetuating oppressive racism in US medicine (see Roberts, 2011).

White framing and epistemological ignorance: Clinical discrimination

Unsurprisingly, this hegemonic white racial frame is observed in much of US medical decision-making in the form of "epistemological ignorance"—the thought-structured failure to recognize white racism's commonplace clinical impacts. That is, this framing involves more than just a lack of knowledge; it is actively created by whites to center and implement white racial domination (Tuana, 2007). Unsurprisingly, numerous studies reveal pervasive racial biases in the everyday actions of white health care personnel. Sadly, this white clinical failure is frequently seen when Black patients' pain complaints are dismissed much more often than those of white patients—another path legacy from the early Sims-era Black "pain tolerance" stereotype. Black patients are thus less likely to receive pain medication compared to white patients for similar medical conditions (Hoffman et al., 2016).

This white racist framing continues to negatively shape medical treatment for Black patients in many contemporary areas. For example, Black youth are overdiagnosed with "conduct disorders" but underdiagnosed with attention-deficit/hyperactivity disorder (ADHD) compared to white youth (Swanson et al., 2020).

One medical review highlights the reality of implicit antiblack bias in the actions of many white health care providers—a white framing contributing significantly to inequalities in medical treatment and health outcomes (FitzGerald and Hurst, 2017). Moreover, in another study of emergency room patients, Black patients had to wait longer times, and their medical emergencies were not evaluated as accurately as those of white patients, again reflecting the institutionalized inequality in this type of health care (Hagiwara et al., 2023).

The persistence of such biased framing is also evident in white physicians' attitudes toward racial inequality in US health care. One recent study found that while three-quarters of white doctors believe patients are *rarely* treated unfairly based on their racial group, only a fifth of Black doctors share this optimistic view. In addition, nearly three-quarters of the mostly white physicians attributed unequal medical treatment to *individual* physician shortcomings rather than to institutionally racist systems (Burgess et al., 2007).

This disconnect with societal reality reflects broader white ignorance within US medicine about systemic racism's well-demonstrated role in shaping Black patients' frequently negative medical outcomes. As one study on medical treatments recently noted, the "efficacy of these (medical) interventions may be hampered by health care providers' work and learning environments, which are rife with discriminatory practices that sustain the very biases US health care professions are seeking to diminish" (Vela et al., 2022, p. 477). These savvy health care researchers add that their approach to dealing with this institutional racism in health care demonstrates that provider-level racial bias should be addressed by efforts "that systemically change structures inside and outside the health care system if the country is to succeed in influencing biases and reducing health inequities" (Vela et al., 2022, pp. 477-478). Unfortunately, most US health care institutions ignore or obscure their institutionalized racism through what I term "colorblind technocracy." For instance, a recent O'Neill Institute report details how white colorblind reasoning regularly shapes US health policy, legitimizing the persisting racial "disparities" by refusing to recognize their systemic determinants. Even the commonplace research term racial health disparities, as used here, tends to emphasize racial oppression's *individual* aspects rather than its systemically racist causes (Twinamatsiko, 2024).

Algorithmic racism and digital colonialism in health care

What is currently termed the problem of "algorithmic colonialism" in US institutions represents a recent variation in the white racial frame's adaptive abilities. As I have argued elsewhere (Feagin, 2020), the white frame regularly evolves to include new societal realities, ensuring the persistence of white advantage and Black disadvantage even in the supposedly neutral digital world. This algorithmic colonialism phrase names the new ways in which digital technologies exploit Black Americans and other people of color, but its inherent racial logic actually *perpetuates* earlier patterns of Western racial framing and exploitation.

Algorithmic colonialism describes how dominant white-run organizations, especially corporations, use old and new computerized technologies to extract value (e.g., huge amounts of online data) from, and thus control, the less powerful communities of color. Their exploitable human resources involve bodily and other personal data. However, these algorithmic technologies often misinterpret nonwhite cultures substantially because they are developed without significant input from communities of color (Benjamin, 2019).

In addition, current artificial intelligence (AI) medical diagnostic systems are frequently trained on racially skewed data from previous and current decades, while medical insurance algorithms often generate insurance redlining in the form of high medical "risk scores" for racially segregated areas. Algorithmic diagnostic bias can negatively impact the health care experiences of Americans of color. For example, there is an overrepresentation of *white* patients in AI health datasets used for disease diagnosis, which means these datasets are *less* relevant for patients of color. To cite a recent example, government data show that Black Americans are significantly underrepresented in important cancer drug experiments (American Cancer Society, 2025). This Cancer Society report further shows that Black Americans continue to have the highest US mortality rates for most cancer types, as a result of their drug trial underrepresentation and unequal access to necessary cancer care in the racist health care system.

Unsurprisingly, many contemporary algorithmic disease calculators have been created by white (and sometimes Asian American) analysts based on such racially biased statistics, which affect the accuracy of medical information for treating African Americans and other Americans of color (The Greenlining Institute, 2021; Obermeyer et al., 2019). Central to this negative algorithmic impact on medical decision-making is also the embedding of white health care providers' racial biases in AI health decision-making programs. Their commonplace white racist framing legitimizes the dominance of white health care authorities, yet tends to marginalize nonwhite critics. Unsurprisingly too, this algorithmic colonialism for Black and other Americans of color extends centuries-long historic white exploitation, yet more path dependency by advanced algorithmic means.

Contrasting white and Black framing of institutionalized health care

The societal power of white racial framing is not limited to shaping these technological logics. It is also vividly reflected in how Black and white Americans experience and perceive health care inequalities. Given their dramatic differences in contemporary societal conditions, it is unsurprising that their views of the current health care system differ dramatically. Thus, US health care institutions are largely evaluated by whites from their pervasive white framing, one usually grounded in white denial and ignorance regarding important aspects of US society's well-institutionalized racism.

For example, data from the Kaiser Family Foundation's recent national survey reveal that a substantial majority of white Americans still mistakenly believe Black

Americans have equal or superior access to US health care compared to whites, a faulty assessment contradicted by well-documented data on the actual health realities cited previously (Kaiser Family Foundation, 2024). Such white framing extends well beyond health care, shaping views of other major socioeconomic areas like education and employment. For example, roughly half of white respondents in that Kaiser survey felt that Black educational attainment equals or surpasses that of whites, despite census data showing that this is inaccurate. In addition, Black families continue to experience poverty at more than twice the white rate, and Black workers remain very disproportionately concentrated in lower-wage jobs, with the Black unemployment rate usually double or more than that of whites. Here, too, white Americans are much less likely than Black Americans to view these as *significant* racial problems. These data underscore the commonplace disconnect between centuries-old white racial framing and the harsh discriminatory realities still shaping Black lives (Kaiser Family Foundation, 2024).

In great contrast, the lived experience and historical memory of Black Americans motivate deep skepticism about the contemporary health care system. The Kaiser Family Foundation survey also found that Black respondents were far more likely than whites to see significant societal racism as a major health problem. One finding was that the level of trust in (white) health care providers to mostly do what is right for them and their communities was significantly lower among Black respondents who have experienced discriminatory health care (Kaiser Family Foundation, 2024). Moreover, both Black and Latin Americans' distrust of mainstream health care is reinforced by the often poor health care communications they receive from white health care workers (Armstrong et al., 2007).

More institutional racism in health care delivery: Nursing homes, hospitals, insurance companies

Consider, too, the well-institutionalized racial inequalities in other areas of US health care delivery. For instance, white-run nursing homes serving predominantly Black residents frequently provide poorer quality care than those serving mostly white residents. Homes with higher proportions of Black residents experience more hospitalizations, as well as more staff shortages and discriminatory financing (Travers et al., 2023). Unsurprisingly, there is continued hospital care discrimination in many US states. Indeed, an informal Jim Crow hospital legacy has aggressively persisted, with public hospitals serving Black areas often remaining underfunded and understaffed, while major hospital centers, usually located in predominantly white areas, have remained relatively well supported. In addition, urban hospitals tend to be located outside low-income communities of color, creating yet more barriers to health care availability (Braveman et al., 2022).

The relocation or closing of nonprofit hospitals in predominantly Black communities has also increased since the late 20th century, deepening racial inequality in health care and contributing to higher rates of life-threatening health complications for Black Americans. Hospital closures have created at least 439 maternity "deserts" in Black-majority counties. In addition, since the 1960s desegregation efforts, many of the once racially segregated hospitals have engaged in yet more covert forms of institutionalized discrimination. Black patients, especially those on public assistance, still often suffer discrimination by white hospital staff (Brown et al., 2023; LaVeist et al., 2013). Multiple research studies also confirm that hospitals serving historically redlined Black communities have significantly fewer intensive care unit (ICU) resources and higher disease (e.g., COVID-19) mortality rates (Zhilkova et al., 2023). Such discriminatory patterns are not just a result of white neglect but rather of ongoing white necropolitical views of Black lives being more disposable than white lives within this health care system.

Unsurprisingly, there are clear remedies for the hospital crises facing US communities of color. Hospital community-benefit mandates, such as those in the 2010 Affordable Care Act, can fund much of the large medical debt created in these impoverished communities of color. Community-benefit mandates require nonprofit hospitals to provide significant services addressing local community health, including public health efforts, in exchange for their tax-exempt status. Expanding these mandates for impoverished nonwhite and other communities reduces US health care inequality and helps to guarantee that federal resources are directed to those most affected by systemic health care racism (Figueroa et al., 2025). However, recent federal policy shifts have severely undermined these efforts. The summer 2025 federal funding legislation, signed by President Donald Trump, enacted historic Medicaid cuts of nearly $1 trillion over the next decade or so. It is projected to cause about 12 million people, especially in low-income areas, to lose their health insurance. Such severe cuts weaken the financial stability of their existing hospitals and will likely force significant hospital closures. In addition, funding cuts for community health services and stricter work rules will disproportionately harm nonwhite and other low-income Americans who depend heavily on these aid programs (Dow, 2025). This combination of reduced federal funding and associated medical neglect will worsen the current necropolitical conditions underlying their health care, a reality increasingly challenged by grassroots protests such as those led by the North Carolina Poor People's Campaign (Crouch, 2024).

Unfortunately, institutional racism is evident in yet other US health care policies, as is seen in significant inequalities in insurance coverage. Employment-based health insurance systems intersect with racially segmented US labor markets. Those employment areas with larger contingents of workers of color tend to have weaker or nonexistent health insurance coverage. Generally, Black Americans continue to face significantly higher uninsured rates—11% for Black adults but just 5% for white adults, as reported in a recent major study (Agency for Health care Research and Quality, 2023; see also Ortega and Roby, 2021).

Pervasive geographical and environmental racism

Consider, too, that environmental racism routinely intrudes on Black working-class communities, often greatly affecting their health and wellbeing. This includes significant water pollution. For example, decades of white-imposed racism set the stage for the infamous Flint Water Crisis, well documented in a Michigan Civil Rights Commission (2017) pathbreaking report. Here, researchers showed that serious discrimination was a central factor in this prolonged water crisis, including the local white government's delayed response to the predominantly Black, low-income residents' complaints about their dangerously contaminated water. Empirical analysis by Hanna-Attisha et al. (2016) revealed the serious health impacts of this large-scale government failure. After the city's major water source was switched to the Flint River, the percentage of Black children with elevated blood lead levels doubled. In fact, this water pollution crisis deepened preexisting racial inequalities in environmental health and in government remedies for environmental racism (Hanna-Attisha et al., 2016).

In addition, federal government policies governing major US health care programs, such as Medicare, initiated in the 1960s, give discretion to local white authorities—many of whom are racially biased—to restrict the programs' eligibility. This then disproportionately and negatively affects Black Americans and other Americans of color. Unfortunately, the government-subsidized health care assistance is often characterized by racially segregated *geographical* access. This substantial geographical segregation, a legacy of the earlier Jim Crow era, still hampers contemporary health care access for numerous predominantly Black and Latin communities (Smith, 2016). These communities endure fewer medical care resources and more hospital closures, again reinforcing the racially bifurcated health care system (Caldwell et al., 2017).

In addition, a major study in the *Journal of Health Politics, Policy and Law* reports on state decisions about Medicaid health care expansion, revealing that it is often strongly racialized (Grogan and Park, 2017). In fact, states with large Black populations are much less likely to expand this health care program. In these states, white officials' views of Medicaid expansion are frequently shaped by their racist framing and thus have negatively affected the allocation of such health programs, yet again contributing to racially bifurcated health care (Grogan and Park, 2017).

In addition, research on racialized US political barriers, such as the unnecessarily strict voter identification laws affecting voters of color, has revealed links to increased health care inequalities. In numerous Republican-dominated ("red") states, these and other discriminatory laws are used to disproportionately block nonwhite voters from the polls. As a result, these laws are often associated with less per capita government health care spending for disproportionately Black areas and, thus, with greater Black health problems (e.g., infant mortality) there than in the Democratically controlled "blue" states without such restrictive voting laws (Hing, 2019).

Predictably, several research studies find that life expectancy is linked to racial and geographical segregation. Graetz et al. (2023) have demonstrated a strong connection between neighborhood racial redlining and lower life expectancy, with differences of at least 5 years between the most and least racially redlined neighborhoods. Current data also show that Black Americans die at a much higher rate than whites because of the still omnipresent environmental hazards and health care deserts caused by residential segregation, as well as the everyday life stress stemming from such realities, e.g., local racist policing. During the recent COVID-19 pandemic, the US saw this racial inequality in mortality highlighted. Black Americans experienced mortality rates double those of whites due to factors like health care deserts and chronic stress resulting from dealing with the pandemic and other racially linked health care challenges. Routine antiblack discrimination regularly amplifies these harsh health realities, including providing the numerous US counties with majority Black populations limited access to critical medical treatments (e.g., COVID-19 treatments) (American Public Media Research Lab, 2023; Patient Engagement HIT, 2022).

The lifetime burden: Black stress, allostatic load, and health outcomes

The bottom line on the serious human effects of institutionalized health care racism, as demonstrated in numerous examples previously discussed, is that there are lifetime *cumulative* negative individual and collective impacts on Black Americans as they live out their lives. This everyday impact never ends until death. Research by David Williams and Selina Mohammed (2009, 2013) has emphasized the chronic stress and allostatic load that build up on Black Americans from these accumulating life issues. The chronic stress imposed by everyday racism increases Black susceptibility to hypertension, diabetes, and numerous other negative health conditions.

Unsurprisingly, the US Centers for Disease Control and Prevention (CDC) has documented that Black Americans disproportionately suffer higher mortality rates than white Americans—often due to commonplace chronic diseases such as heart disease and cancers. Inferior medical care is frequently involved in generating or accelerating these diseases (Centers for Disease Control and Prevention, 2017; Vann 2022). Clearly, Black Americans and other people of color are less likely than demographically comparable white patients to receive evidence-based health care for these chronic diseases (Smith, 2016). To take a major example, Black patients are significantly overrepresented among Americans with end-stage renal disease. Yet, they are much less likely to receive high-quality renal dialysis and are more likely to be treated at medical facilities with worse survival outcomes. On average, they are also less likely than white patients to receive kidney transplants. In addition, a recent report by the Agency for Health Care Research and Quality (2023) found that Black patients received worse care than white patients on *half* of their medical quality measures, including treatments for heart disease, cancer, and diabetes. Well-institutionalized white racism remains a major determinant of these significant inegalitarian patterns.

Moreover, the national Black maternal mortality rate is *far higher* than the white rate (circa 39.9 versus 14.1 per 100,000 live births recently), and increasingly so (Gender Equity Policy Institute, 2025; Rushovich et al., 2024). Similarly, the Black infant mortality rate remains more than twice that of whites. Black maternal and infant mortality rates in southern states with discriminatory abortion bans and limited

Medicaid services are disproportionately high. Unsurprisingly, too, Black voter suppression in these Southern states is linked to numerous persisting racial health inequalities, including higher Black infant and maternal mortality rates (Gender Equity Policy Institute, 2025; Rushovich et al., 2024). These persistent differences are not just indicators of US racial inequality but are yet more evidence of a *necropolitical* health care system that regularly normalizes Black vulnerability to premature disease and death.

Toward medical decolonization and reparative health justice

Aggressive decolonization of Western health care

In this chapter I have periodically referenced important reparative health care policies that can effectively "decolonize" health care for Americans of color. These include significant Medicaid expansion, rejection of the political disenfranchisement and gerrymandering associated with health care, and new community hospital mandates. Taken together, these are necessary forms of significant reparative social justice. Adding in significant income redistribution, these reparative reforms can result in an important rebuilding of the very troubled US health care system. The aggressively for-profit nature of white-run capitalism, together with white socioeconomic exploitation of communities of color, regularly generates Black and other nonwhite political calls for major *systemic* change. As Alang et al. (2023) argue, this well-institutionalized white racism needs to be dismantled by numerous strategic means. These include substantial government investments in more antiracist research and research that can draw on extant Black health care knowledge. These government investments should include supporting Black health care scholars and greatly expanding Black communities' preventive and other everyday health care. As we have seen previously, various aspects of systemic white racism still underlie contemporary US health care realities, and all of these require much dismantling and rebuilding to achieve authentic health and health care equality (Feagin, 2020).

This rebuilding of the racist health care system will require a forceful national confrontation of its inegalitarian capitalistic foundation, one created over centuries of elite whites stealing the labor, wealth, knowledge, and bodies of many ordinary individuals and communities, especially those of color. Achieving authentic health care justice requires not only major health care institutional changes but first a significant societal frame transformation, accenting the importance of more democratic political-economic institutions. This major democratic frame change should include more respect for, and use of, Black American counter-frames, Afrocentric epistemologies, and intersectional Black (e.g., Black feminist) thought as legitimate sources of new democratic knowledge and practice. The extensive Black American lived experience with the discriminatory health care system offers a reparative vantage point for dismantling the controlling white racist frame and implementing a much more democratic and just health care system. Major democratic innovations in this troubled system should include integrating important Black and other nonwhite health counter-frames, such as those of the African-origin Sankofa care (see below), linking ancestral and contemporary nonwhite healing knowledge to radically reshape contemporary white-run health care institutions.

These reparative institutional solutions require new democratic framing, emphasizing substantial socioeconomic *redistributions*—for example, major medical debt abolition for Black and other modest-income Americans. This health care social justice also requires dramatically new antiracist medical diagnostics and clinical implementations. These major clinical reforms can help dismantle the negatively racialized health hierarchies, such as those demanded in Black antiracist movements for decades. For example, in 1969, the new Black Panther Party's members set up a medical clinic in Oakland, California, offering free sickle-cell testing alongside their free breakfast educational programs. While white FBI memos labeled that organization as politically "subversive," *it was not*, and by 1972, the Black Panthers had tested over 500,000 Black Americans—thereby forcing the US government to launch its own medical screening program. The Black Panther Party's thirteen free clinics (1969–1975) provided much holistic health care while periodically documenting white iatrogenic medical harm using public health databases. Indeed, the Party provided an important social justice model for future community-based participatory health care and research (Nelson, 2011).

In addition, contemporary mutual aid networks, like the National Black Food and Justice Alliance, have integrated collective Black food sovereignty efforts with better trauma-informed health care, in this process rejecting white health care's commonplace hyper-individualism. Moreover, the "Black Food Geographies" perspective of the Black scholar Ashanté Reese similarly highlights healthy food sovereignty as part of substantial Black counterframing on individual and community health matters (Reese, 2019). In addition, empowering Black communities through a variety of important health screening programs, such as cancer screenings, has proven effective in improving Black health care access (Emory Health care, 2022). In fact, numerous Black community organizations have worked hard to increase the representation of Black health care workers and perspectives across numerous areas of US health care, an essential change necessary in providing truly equitable US health care delivery.

Culturally sensitive Afrocentric care

Afrocentric, community-based health care solutions currently offer an important counter to the problematic, white-run US health care system. For example, African-origin Sankofa programs operating in several communities demonstrate significant use of a Ghanaian-derived epistemology, emphasizing Black African health care knowledge and collective healing as responses to the health damage often caused by mainstream Western medical practices. This African-origin knowledge and practice challenge Western medical perspectives that have periodically pathologized Black

American bodies. Rooted in African medical traditions, this Sankofa health care approach offers human healing that can confront the negative legacies of the Western epistemic violence often intertwined with Western health care (see Osafo et al., 2015; Youngman and Patin, 2024; Patin et al., 2024). This innovative Sankofa approach can even be applied in US medical practice by situating the Sankofa health knowledge and care within certain broader health care reform societal movements, such as the one called "Medicare for All." The individuals and groups associated with this movement advocate for major health care reforms, such as a national US single-payer health insurance program publicly funded for Americans of all ages. Many Black community leaders explicitly recognize that this novel Medicare movement can bring better health care for Black communities, in part as a form of reparative justice for the white medical damage done to Black America (Black Lives Matter Global Network Foundation, 2024).

The Sankofa approaches emphasize holistic and community-centered health and health care. For instance, Maryland's Sankofa Health Center implements this framework in their efforts by integrating conventional medical treatments with holistic practices such as meditation, yoga, massage, and acupuncture therapeutics—that is, a truly multidimensional wellness approach addressing the whole person's body and mind. These health care efforts also go beyond standard clinical practices to include broader community health education and even group therapy. All this is done from a strong Black counter-framed lens (Sankofa Health Center, 2025). Georgia's Sankofa Medical Charities also implements this type of expanded and innovative health care access for their African American population, one underserved by white-run health facilities (Sankofa Medical Charities, 2025). Chicago's Sankofa Wellness Center similarly emphasizes community control of its local health care, coupling disease treatment with wellness programming (Rush University System for Health, 2024). Collectively, the numerous Sankofa centers represent an essential Black health care intervention set up to remedy persisting racial inequalities in US health care. Drawing on significant African knowledge and healing traditions, they offer a critical African American counter-framed addition or alternative to white medical and health care dominance. In all Western societies, "race" only exists in the minds of the dominant Western oppressing groups that define themselves as "white" and is socially imposed on subordinated others as "Black" or "nonwhite". These social conceptions are then dominant across a society because the dominant group makes them so.

Transforming US Medical education: Building a more reparative curriculum

At a minimum, white health care leaders need to incorporate into their dominant US health care framework a broad, racially reparative knowledge. This includes greatly expanded medical education that recognizes the centuries-old African and African American wellness traditions, implements comprehensive antiracist training for all health care workers, and significantly reduces racial inequality in current health care delivery. For example, the pathbreaking "Structural Racism and Health Equity"

(SRHE) course at the University of California, Los Angeles' (UCLA) School of Medicine represents a major curricular shift addressing systemic racial inequalities. This innovative SRHE curriculum has made important use of critical social science perspectives—antiracist theory, queer theory, and the diversity, equity, and inclusion (DEI) perspective—to analyze how deeply systemic racial, class, and gender oppressions shape US health care education and health care practices. Their educational sessions have examined topics such as environmental racism, Western medical colonialism, and queer health justice (UCLA David Geffen School of Medicine, 2025; Baker et al., 2025). This SRHE course draws, in part and conceptually, on the systemic racism theory and research tradition. This innovative educational approach utilizes this theory and research to provide a detailed analysis of how US racial hierarchies, the overarching white racial framing, and continuing institutionalized discrimination still generate major racial inequalities in the predominantly white-run—especially at its top—US health care system.

Significantly, the courageous UCLA Medical School's SRHE curriculum has had success in reducing the implicit and hidden racial biases of white medical students, thereby highlighting the significant potential of an authentic antiracist health care curriculum. However, this curricular innovation has become a lightning rod for white supremacist attacks. Arch-conservative media outlets and Republican politicians have sharply criticized the SRHE educational approach, accusing these important educational innovations of politicizing medicine and even violating civil rights laws. In fact, the right-wing *Wall Street Journal* and other right-wing media have criticized the SRHE course, and some far-right groups have filed federal complaints and lawsuits against such antiracism training in medical schools (Mazer, 2024). In response to these white supremacist pressures, UCLA has scaled back the SRHE course to some degree, even as their faculty and students have reported white harassment and administrative suppression of their racial justice advocacy for US health care (Daily Bruin Staff, 2025).

In addition, far-right Republicans and their corporate allies have led a major coordinated legislative and legal campaign to dismantle the modest DEI programs and the more aggressive antiracist programs across higher education, including medical education. In 2025, the US Department of Health and Human Services, then led by white right-wing politicians, issued a warning to medical schools not to engage in so-called "race-conscious" admissions and educational policies, explicitly targeting curricula addressing "systemic racism" (Nixon Peabody, 2025). Undoubtedly, these far-right attacks are already undermining decades of US progress in addressing the systemic roots of racial inequality in the country's health care system (see Streed et al., 2025).

To effectively challenge systemic racial discrimination in US health care education and practice, we must counter not only the implicit white racial biases but also the overt white racial biases shaping how health care is taught and delivered. One recent study showed that explicit antiracism training significantly reduces US health care providers' negative racial framing and leads to better patient outcomes (UCLA David Geffen School of Medicine, 2025). Yet, despite this strong empirical evidence for their success, such antiracist curricula and associated health care programs are rare. Most US medical schools actually do not offer any substantial education on the extensive systemic racism in US health care, or in US society more broadly. In my view, greatly expanding these antiracist curricular efforts is essential for developing a multiracial democratic lens in this country's medical education. There are many explicit examples of why this is also important for improving the everyday practice of US health care workers. For example, incorporating Black feminist analyses, such as their critiques of biased white male health politics, can reveal how Black women's pain is too often dismissed not just because of white racist attitudes but also due to entrenched male gender stereotypes as well. This dismissal of Black women's pain is very hard to counter in a mostly white-controlled medical system like the one in the US which is currently moving politically in a more authoritarian direction. If history is one's guide, only large-scale antiracist civil rights movements can successfully counter such a hostile antiblack political climate.

Conclusion: Reparative, epistemic, and institutional change

In his provocative book *Necropolitics*, the scholar Achille Mbembe powerfully theorizes the global political-economic landscape that, for centuries, has been shaped by escalating white Western dominance, societal conflicts, and substantial inequality, including the periodic emergence of white racist, fascist, and nationalist forces creating much *human oppression* and *death*. He argues critically that contemporary Western democracy has revealed its evil "nocturnal body"—its dark underside rooted in the centuries-long white capitalistic and colonizing legacies of violence and domination. This underside regularly weakens its official democratic ideals (Mbembe, 2019). In Mbembe's terms, as noted earlier, this white Western power organizes who gets to live and who must die—a necropolitical logic thoroughly embedded in US health care's enduring racial hierarchy. From US slavery to the present day, white-controlled capitalistic institutions have privileged white health while exposing Black and other nonwhite Americans to disproportionate suffering and premature death.

A recent US Urban Institute report, "Envisioning a New Health System Rooted in Equity," dramatically reimagines the US health care system by prioritizing racial and economic equality over its current emphasis on expanding capitalistic profits:

> *"Envisioning a health system that places health, wellness, and equity at its core is a radical but necessary exercise to help America achieve its democratic ideals. This exercise requires imagining a society that has dismantled and repaired a legacy of racism and segregation and that prioritizes the health and social needs of its people over the profit-seeking incentives of major health care industry groups—particularly the needs of people who have been made vulnerable by generations of racism and other forms of injustice. In short, it requires envisioning the pathways and political processes that can realize racial and economic justice and equity ... Reenvisioning the US health system around a more equitable,*

> *salutogenic approach poses a values question: What and whom do we prioritize—a corporate model of health care that benefits mainly those who hold the most power, or a community-based model in which health care is a public good and a human right?*
>
> **(Iton and Francis, 2023, pp. 1, 29)."**

Clearly, this current corporate model of health care cannot be fully dismantled without confronting both its material and ideological dimensions. This dual focus on unjust white material enrichments and on the continuing ideological rationalization of the white racial frame is essential to any successful project working for major US health and health care justice. Only by also dismantling the many systemic, intergenerational, and epistemic dimensions of contemporary white health care racism, as theorized in the systemic racism framework, can we hope to move beyond superficial societal reforms and toward real societal health care transformation.

Unmistakably, those seeking more real democracy in the US and other Western countries must work to implement major systemic changes in this racially oppressive reality, including significant wealth redistribution and genuinely democratic community health and health care benefits. Only by aggressively confronting the major racial injustices and embedding real antiracist principles throughout US health care policy and practice can we begin to repair the large-scale individual and collective harms inflicted by centuries of systemic health-related racism. Thus, the path forward must be one led by a systemic reparative epistemology and implemented through a democratic, community-driven health care transformation.

References

Adler, R., (2023). 'Pasteur Begins Developing Germ Theory and Microbiology'. *EBSCO Research Starters: History*. Available at: EBSCOhost.

Aetna Insurance Company, 1853. Slave Insurance Policy for 21-year-old Enslaved "House Servant" Named Lucinda, University Archives, Hartford, CT.

Agency for Healthcare Research and Quality, 2023. 2023 National Healthcare Quality and Disparities Report. Rockville, MD: AHRQ. Available at: https://www.ahrq.gov/research/findings/nhqrdr/nhqdr23/index.html.

Alang, S.M., Carter, C.R., Blackstock, O., 2023. Past is prologue: dismantling colonial legacies to advance black health equity in the United States. Health Equity 7 (1), 831–834. doi:10.1089/heq.2023.0226.

American Cancer Society, 2025. Cancer facts and figures for African American/Black people 2025, American Cancer Society, Atlanta, GA.

American Public Media Research Lab, 2023. 'The Color of Coronavirus: COVID-19 Deaths by Race and Ethnicity in the U.S.' Available at: https://www.apmresearchlab.org/covid/deaths-by-race.

Anderson, K.O., Green, C.R., Payne, R., 2009. Racial and ethnic disparities in pain: causes and consequences of unequal care. J. Pain 10 (12), 1187–1204.

Armstrong, K., Ravenell, K.L., McMurphy, S., Putt, M., 2007. Racial/ethnic differences in physician distrust in the United States. Am. J. Public Health 97 (7), 1283–1289. doi:10.2105/AJPH.2005.080762.

Baker, C., Bueno, S., Carrillo, L., Flores, E., Hing, M., Kaur, G., Kennedy, D., LeTran, K., Rodriguez, D., Samra, S., Sheikh, H., Wells, L., 2025. Medical education as liberatory praxis: experiences from the UCLA "structural racism and health equity" curriculum. Soc. Sci. Med. 376, 118094. doi:10.1016/j.socscimed.2025.118094.

Benjamin, R., 2019. Race after Technology: Abolitionist Tools for the New Jim Code, Polity Press, Cambridge.

Black Lives Matter Global Network Foundation, (2024). 'BLM Foundation Endorses Medicare For All Legislation'. *Black Lives Matter*. Available at: https://blacklivesmatter.com/blm-global-network-foundation-endorses-medicare-for-all-legislation/. (Accessed: 20 June 2025).

Bowleg, L., 2023. The white racial frame of public health discourses about racialized health differences and "disparities": what it reveals about power and how it thwarts health equity. Front. Public Health 11, 1187307. doi:10.3389/fpubh.2023.1187307.

Braveman, P.A., Arkin, E., Proctor, D., Kauh, T., Holm, N., 2022. Systemic and structural racism: definitions, examples, health damages, and approaches to dismantling. Health Aff. 41 (2), 171–178. doi:10.1377/hlthaff.2021.01394.

Brown, C.E., Marshall, A.R., Snyder, C.R., et al, 2023. Perspectives about racism and patient-clinician communication among Black adults with serious illness. JAMA Netw. Open 6 (7), e2321746. doi:10.1001/jamanetworkopen.2023.21746.

Brown, E.R., 1979. Rockefeller Medicine Men: Medicine and Capitalism in America, University of California Press, Berkeley.

Burgess, D.J., van Ryn, M., Dovidio, J., Saha, S., 2007. Reducing racial bias among health care providers: lessons from social-cognitive psychology. J. Gen. Intern. Med. 22 (6), 882–887. doi:10.1007/s11606-007-0160-1.

Caldwell, J.T., Ford, C.L., Wallace, S.P., Wang, M.C., Takahashi, L.M. (2017) 'Racial and ethnic residential segregation and access to health care in rural areas', CDC Stacks. Available at: https://stacks.cdc.gov/view/cdc/44048. (Accessed 28 June 2025).

Centers for Disease Control and Prevention, 2017. Vital signs: racial disparities in age-specific mortality among Blacks or African Americans—United States, 1999–2015. MMWR 66 (17), 444–456.

Centers for Disease Control and Prevention, 2025. 'Tuskegee Study—timeline'. U.S. Department of Health and Human Services. Available at: https://www.cdc.gov/tuskegee/timeline.htm. (Accessed: June 27, 2025)

Centers for Medicare and Medicaid Services, 2024. National health expenditure projections 2024–2033. Health Aff. 43 (6), 1–15. Available at: https://www.healthsystemtracker.org/brief/what-drives-health-spending-in-the-u-s-compared-to-other-countries/. (Accessed: 29 June 2025).

Chartis, 2025 'Black History Month 2025: the healthcare workforce and diversity initiatives'. Available at: https://www.chartis.com/insights/black-history-month-2025-healthcare-workforce-and-diversity-initiatives. (Accessed: 28 June 2025).

Cooper Owens, D., 2017. Medical Bondage: Race, Gender, and the Origins of American Gynecology, University of Georgia Press, Athens.

Cox, D.G., 2021. Race, Reconstruction, and the invention of "Negro superstition," 1862–1877. J. Am. Stud. 55 (5), 1125–1152.

Craemer, T., 2015. Estimating slavery reparations: present value comparisons of historical multigenerational reparations policies. Soc. Sci. Q. 96 (2), 639–655.

Crouch, M. (2024) 'N.C.'s new medical debt relief plan: 11 must-dos for hospitals'. *North Carolina Health News*. Available at: https://www.northcarolinahealthnews.org/2024/08/13/nc-medical-debt-relief-plan/.

Curtin, P.D., 2010. Germs and Jim Crow: the impact of microbiology on public health in the American South. J. Hist. Med. Allied Sci. 65 (3), 305–332.

Daily Bruin Staff (2025) 'Medical community claims unilateral UCLA leadership support in Israel-Hamas war', *Daily Bruin*. Available at: https://dailybruin.com/2025/06/03/medical-community-claims-unilateral-ucla-leadership-support-in-israel-hamas-war (Accessed: 29 June 2025).

Daniels, J., Schulz, A.J., 2008. Constructing whiteness in health disparities research. In: Zuberi, T., Bonilla-Silva, E., Eds., White Logic, White Methods: Racism and Methodology. Rowman and Littlefield, Lanham, MD, pp. 89–107.

Dow, W. (2025) 'What do the looming cuts to medicaid really mean?' *UC Berkeley School of Public Health*. Available at: https://publichealth.berkeley.edu/articles/news/commentary/what-do-cuts-to-medicaid-really-mean (Accessed 9 September 2025).

Du Bois, W.E.B., 1906. The Health and Physique of the Negro American: report of a Social Study Made Under the Direction of Atlanta University; Together with the Proceedings of the Eleventh Conference for the Study of the Negro Problems, Held at Atlanta University, on May the 29th, 1906. Atlanta, GA: atlanta University Press.

Edwards-Grossi, E., Willoughby, C.D.E., 2024. Slavery and its afterlives in US psychiatry. Am. J. Public Health 114, S250–S257. doi:10.2105/AJPH.2023.307554.

Emory Healthcare (2022) 2022 Community Health Needs Assessment Implementation Plan. Atlanta, GA: emory Healthcare. Available at: https://www.emoryhealthcare.org/-/media/Project/EH/Emory/ui/pdfs/community/2022-chna-ip-12-22.pdf. (Accessed: 10 June 2025).

Ezenwa, M.O., et al, 2013. Racial disparities in pain management in primary care. Pain Manag. Nurs. 14 (2), 97–104.

Feagin, J.R., 2006. Systemic Racism: A Theory of Oppression, Routledge, New York.

Feagin, J.R., 2020, The White Racial Frame, 3rd edn. Routledge, New York.

Feagin, J.R., 2025, Racist America: Roots, Current Realities, and Future Reparations, 5th edn. Routledge, New York.

Feagin, J.R., Bennefield, Z., 2014. Systemic racism and U.S. Health Care. Soc. Sci. Med. 103, 7–14. doi:10.1016/j.socscimed.2013.09.006.

Figueroa, J.F., Frakt, A.B., Jha, A.K., et al, 2025. Structural Discrimination in Nonprofit Hospital Community Benefit Spending in the US. JAMA Health Forum 6 (2), e245679.

FitzGerald, C., Hurst, S., 2017. Implicit bias in health care professionals: a systematic review. BMC Med. Ethics 18 (1), 19. doi:10.1186/s12910-017-0179-8.

Flexner, A., 1910. Medical Education in the United States and Canada: a Report to the Carnegie Foundation for the Advancement of Teaching. Bulletin Number Four. New York: carnegie Foundation for the Advancement of Teaching. Available at: https://archive.carnegiefoundation.org/publications/pdfs/elibrary/Carnegie_Flexner_Report.pdf.

Gender Equity Policy Institute, (2025). *Maternal Health: reproductive Justice and Health Equity*. Institute for Women's Policy Research Federal Policy Agenda 2025. Available at: https://iwpr.org/wp-content/uploads/2025/02/Maternal-Health_IWPR-Federal-Policy-Agenda-2025-1.pdf (Accessed: 29 May 2025).

Graetz, N., Esposito, M., Irons, D., McKetta, S., 2023. Historic inequities in neighborhood redlining and contemporary declines in urban life expectancy: a longitudinal ecological study. Soc. Sci. Med. Available at: 320, 115765. doi:10.1016/j.socscimed.2023.115765.

Grogan, C.M., Park, S.E., 2017. The racial divide in state medicaid expansions. J. Health Polit., Policy Law 42 (3), 539–572.

Hagiwara, N., Kron, F.W., Scerbo, M.W., Watson, G.S., 2023. A call for systemic change: implicit racial bias in health care. Health Equity 7 (1), 1–8. doi:10.1089/heq.2022.0148.

Hanna-Attisha, M., LaChance, J., Sadler, R.C., Champney Schnepp, A., 2016. Elevated blood lead levels in children associated with the flint drinking water crisis: a spatial analysis of Risk and public health response. Am. J. Public Health 106 (2), 283–290. doi:10.2105/AJPH.2015.303003.

Hartman, M., Martin, A.B., Benson, J., Catlin, A., the National Health Expenditure Accounts Team, 2025. National health care spending in 2024: growth accelerates as prices and enrollment rise. Health Aff. 44 (6), 1021–1032. doi:10.1377/hlthaff.2025.00001.

Hing, A.K., 2019. The right to vote, the right to health: voter suppression as a determinant of racial health disparities. J. Health Disparities Res. Pract. 12 (6), 50–61. Available at: https://digitalscholarship.unlv.edu/jhdrp/vol12/iss6/5/. (Accessed 29 June 2025).

Hixenbaugh, M. (2024) 'Hundreds of Unclaimed Bodies Were Sent to a Texas Medical School. Families Say They Weren't Told', *NBC News*, Available at: https://www.nbcnews.com/news/us-news/texas-medical-school-unclaimed-bodies-rcna123456.

Hoffman, K.M., Trawalter, S., Axt, J.R., Oliver, M.N., 2016. Racial bias in pain assessment and treatment recommendations, and false beliefs about biological differences between Blacks and Whites. Proc. Natl. Acad. Sci. U. S. A. 113 (16), 4296–4301. doi:10.1073/pnas.1516047113.

Iton, A.B., Francis, D. (2023) 'Envisioning a New Health System Rooted in Equity'. Washington, DC: urban Institute. Available at: https://www.urban.org/research/publication/envisioning-new-health-system-rooted-equity

Kaiser Family Foundation (2024) 'Survey on Racism, Discrimination, and Health', Available at: https://www.kff.org/racism-survey/ (Accessed: 9 May 2025).

Kyere, E. (2020) 'Enslaved people's health was ignored from the country's beginning, laying the groundwork for today's health disparities', *The Conversation*, Available at: https://theconversation.com/enslaved-peoples-health-was-ignored-from-the-countrys-beginning-laying-the-groundwork-for-todays-health-disparities-140153.

LaVeist, T.A., Isaac, L.A., Williams, K.J., 2013. The state of research on racial/ethnic discrimination in the receipt of health care. Med. Care Res. Rev. 70 (5 suppl), 55S–75S. doi:10.1177/1077558713499808.

Marks, J., 2017. Is Science Racist?, Polity, Medford, MA.

Mazer, B. (2024) 'Does Med School Have a DEI Problem?', *The Atlantic*, Available at: https://www.theatlantic.com/health/archive/2024/06/ucla-woke-medical-school-dei/678606/ (Accessed: 29 June 2025).

Mbembe, A., 2019. Necropolitics (S. Corcoran, trans.). Duke University Press, Durham.

Merck and Co., Inc, (2025) 'History of the Program'. Mectizan Donation Program. Available at: https://mectizan.org/history-of-the-program/. (Accessed: 4 June 2025).

Michigan Civil Rights Commission, 2017. The Flint Water Crisis: systemic Racism Through the Lens of Flint. Lansing, MI: michigan Department of Civil Rights. Available at: https://www.michigan.gov/-/media/Project/Websites/mdcr/mcrc/reports/2017/flint-crisis-report-edited.pdf. (Accessed 28 June 2025).

Minna Stern, A. (2020) 'Forced Sterilization Policies in the US Targeted Minorities and Those with Disabilities – and Lasted into the 21st Century', *The Conversation*. Available at: https://theconversation.com/forced-sterilization-policies-in-the-us-targeted-minorities-and-those-with-disabilities-and-lasted-into-the-21st-century-141293.

Mustakeem, S.M., 2016. Slavery at Sea: Terror, Sex, and Sickness in the Middle Passage, University of Illinois Press, Urbana.

NBC News, (2024a). 'Dealing the Dead: University of North Texas Health Science Center Used Unclaimed Bodies—Disproportionately Black and Poor—for Dissection, Research, and Commercial Leasing Without Consent', Available at: https://www.nbcnews.com/news/us-news/university-north-texas-unclaimed-bodies-medical-research-investigation-rcna171080. (Accessed: 20 June 2025).

NBC News (2024b) 'Hundreds of unclaimed bodies were sent to a Texas medical school. Now, NBC News is publishing their names', Available at: https://www.nbcnews.com/news/us-news/unclaimed-bodies-texas-medical-school-names-released-rcna173310. (Accessed 27 June 2025).

NCpedia, n.d. 'Jim Crow.' Available at: https://www.ncpedia.org/anchor/jim-crow. (Accessed 27 June 2025).

Nelson, A., 2011. Body and Soul: The Black Panther Party and the Fight against Medical Discrimination, University of Minnesota Press, Minneapolis.

Newkirk II, V.R. (2016) 'America's Health Segregation Problem', *The Atlantic*. Available at: https://www.theatlantic.com/politics/archive/2016/05/americas-health-segregation-problem/483219/

Nixon Peabody, L.L.P. (2025) 'HHS issues guidance on nondiscrimination to medical schools receiving federal funds', *Nixon Peabody Insights*. Available at: https://www.nixonpeabody.com/insights/alerts/2025/05/15/hhs-issues-guidance-on-nondiscrimination-to-medical-schools-receiving-federal-funds (Accessed: 22 March 2025).

Nong, P., Creary, M., Platt, J., Kardia, S., 2023. A critical analysis of white racial framing and comfort with medical research. AJOB Empir. Bioeth. 14 (2), 65–73. doi:10.1080/23294515.2022.2160506.

North Carolina Civil War and Reconstruction History Center, (n.d.) 'Civil War Hospitals and Freedmen's Bureau Hospitals in North Carolina'. (Accessed 26 April 2025).

Northington Gamble, V., 1997. Under the Shadow of Tuskegee: african Americans and Health Care. Am. J. Public Health 87 (11), 1773–1778. doi:10.2105/AJPH.87.11.1773.

Obermeyer, Z., Powers, B., Vogeli, C., Mullainathan, S., 2019. Dissecting racial bias in an algorithm used to manage the health of populations. Science 366 (6464), 447–453. doi:10.1126/science.aax2342.

Opara, I.N., Riddle-Jones, L., Allen, N., 2022. Modern day drapetomania: calling out scientific racism. J. Gen. Intern. Med. 37 (1), 225–226. doi:10.1007/s11606-021-07181-8.

Ortega, A.N., Roby, D.H., 2021. Ending structural racism in the US health care system to eliminate health care inequities. JAMA 326 (7), 613–615.

Osafo, J., Akotia, C.S., Andoh-Arthur, J., Quarshie, E.N.-B., 2015. Sankofa: a multisite collaboration on paediatric HIV disclosure in Ghana. AIDS 29 (Suppl 1), S35–S46. https://www.ncbi.nlm.nih.gov/pmc/articles/PMC4880360/.

PatientEngagementHIT (2022) 'Black, Hispanic patients see lower paxlovid, COVID-19 treatment access'. Available at: https://www.techtarget.com/patientengagement/news/366584608/Black-Hispanic-Patients-See-Lower-Paxlovid-COVID-19-Treatment-Access (Accessed: 20 May 2025).

Patin, B., Youngman, T., Hauser, E., 2024. Toward an etiology of harm for knowledge organization: onto-epistemic injustice in classificatory systems of record. Knowl. Organ. 51 (7), 495–513. doi:10.5771/0943-7444-2024-7-495.

Patton, S. (2024) 'Once upon a time, enslaved Africans were AN insurance commodity', *Black Enterprise*. Available at: https://www.blackenterprise.com/enslaved-africans-were-insurance-commodities-juneteenth/ (Accessed: 4 June 2025).

Randall, V.R., 1993. Racist Health care: reforming an unjust health care system to meet the needs of African-Americans. Health Matrix J. Law Med. 3 (1), 127–184.

Reese, A., 2019. Black Food Geographies: Race, Self-Reliance, and Food Access in Washington, D.C, University of North Carolina Press, Chapel Hill.

Roberts, C.E., 2017. To Heal and to Harm: medicine, Knowledge, and Power in the Atlantic Slave Trade (PhD thesis). Harvard University. See abstract at https://dash.harvard.edu/entities/publication/7974a138-1b84-49bf-a227-204611e7ad2c.

Roberts, D., 2011. Fatal Invention: how science, politics, and big business re-create race in the 21st century, The New Press, New York.

Robinson, P., (2023). 'Trillions owed in reparations for slavery says report.', *NY Carib News*. Available at: https://reparationscomm.org/reparations-news/trillions-owed-in-reparations-for-slavery-says-report/. (Accessed 3 June 2025).

Rush University System for Health (2024) 'Sankofa Wellness Center Breaks Ground'. Available at: https://www.rush.edu/news/sankofa-wellness-center-breaks-ground (Accessed: 20 June 2025).

Rushovich, T., Nethery, R.C., Krieger, N., 2024. 1965 US voting rights act impact on Black and Black versus White infant death rates in Jim Crow states, 1959–1980 and 2017–2021. Am. J. Public Health 114 (3), 300–308. doi:10.2105/AJPH.2023.307518.

Sankofa Health Center, (2025). 'About Sankofa Health'. Available at: https://www.sankofahealthcenter.com/about (Accessed 19 June 2025).

Sankofa Medical Charities (2025) 'Sankofa Medical Charities'. Available at: https://atlantakidneyspa.com/about/sankofa-medical-charities/ (Accessed: 8 May 2025).

Shead, S., 2022. ' "Granny" Midwife to Nurse-Midwife: the Decline of Southern Black Midwifery in the 20th Century". Hist. Perspect. Santa Clara Univ. Undergrad. J. Hist. Series II, 27, Article 9. Available at: https://scholarcommons.scu.edu/historical-perspectives/vol27/iss1/9 (Accessed: 1 July 2025).

Sidel, V.W., Sidel, R., 2006, Social Problems, 5th edn. W.W. Norton, New York.

Simbo.ai, (2024). 'Examining the Root Causes of Black Medical Professionals Leaving Academic Medicine and the Impact on Healthcare Diversity'. Available at: https://www.simbo.ai/blog/black-medical-professionals-leaving-academic-medicine-and-the-impact-on-healthcare-diversity-1116694/. (Accessed 28 June 2025).

Smith, D.B., 2016. The Power to Heal: Civil Rights, Medicare, and the Struggle to Transform America's Health Care System, Vanderbilt University Press, Nashville.

Smith, L.T., 2012, Decolonizing Methodologies: Research and Indigenous Peoples, 2nd edn. Zed Books, London.

Spivak, G.C., 1988. Can the Subaltern Speak?. In: Nelson, C., Grossberg, L., Eds., Marxism and the Interpretation of Culture. University of Illinois Press, Urbana, pp. 271–313.

Stern, A.M., (2020) 'Forced Sterilization Policies in the US Targeted Minorities and Those with Disabilities – and Lasted into the 21st Century', *The Conversation*, Available at: https://theconversation.com/forced-sterilization-policies-in-the-us-targeted-minorities-and-those-with-disabilities-and-lasted-into-the-21st-century-141293.

Streed, C.G., Guerra, P., De Maio, F., 2025. Facing political attacks on medical education—the future of diversity, equity, and inclusion in medicine. N. Engl. J. Med. 392 (10), 941–944. doi:10.1056/NEJMp2500489.

Summerside, M., 2017. From henrietta to HeLa: a history of racism and a reality of distrust in the American Medical system. Univ. Colo. Honors J. Gender and Ethnic Studies section 37–51. Available at: https://www.colorado.edu/honorsjournal/archives/2017/gender-and-ethnic-studies-2017. (Accessed: 29 June 2025).

Swanson, J.M., Arnold, L.E., Molina, B.S.G., et al, 2020. Diagnosis and treatment of ADHD in the United States: update by sex and race. J. Atten. Disord. 24 (1), 9–19. doi:10.1177/1087054718816163.

Texas Woman's University, n.d. 'Autonomy Revoked: the Forced Sterilization of Women of Color in 20th Century America'. Available at: https://twu.edu/media/documents/history-government/Autonomy-Revoked–The-Forced-Sterilization-of-Women-of-Color-in-20th-Century-America.pdf. (Accessed 26 April 2025).

The Greenlining Institute, 2021. Algorithmic Bias Explained: how Automated Decision-Making Becomes Automated Discrimination. Oakland, CA: the Greenlining Institute. Available at: https://greenlining.org/wp-content/uploads/2021/04/Greenlining-Institute-Algorithmic-Bias-Explained-Report-Feb-2021.pdf. (Accessed: 28 June 2025).

Travers, J.L., et al., (2023). 'Nursing homes serving Black residents have greater hospitalizations, emergency department visits', *NYU Rory Meyers College of Nursing News*. Available at: https://nursing.nyu.edu/news/nursing-homes-serving-black-residents-have-greater-hospitalizations-emergency-department-visits. (Accessed 28 March 2025).

Tuana, N., 2007. The Epistemology of Ignorance. In: Sullivan, S., Tuana, N., Eds., Race and Epistemologies of Ignorance. SUNY Press, Albany, pp. 1–20.

Twinamatsiko, A.J., 2024. 'Legitimizing Health Injustice Through Colorblindness'. O'Neill Institute for National and Global Health Law, Georgetown University Law Center. Available at: https://oneill.law.georgetown.edu/wp-content/uploads/2024/07/240701-HPLI-Affirmative-Action-Report_8.5x11-FINAL.pdf. (Accessed: 28 June 2025).

UCLA David Geffen School of Medicine, (2025). 'HEALS Curriculum: structural Racism and Health Equity.' Available at: https://medschool.ucla.edu/education/heals. (Accessed 10 June 2025).

Vann, R.N., 2022. Mortality disparities between Black and White Americans mediated by income and health behaviors. Prev. Med. Rep. 25, 101688. doi:10.1016/j.pmedr.2021.101688.

Vela, M.B., Erondu, A.I., Smith, N.A., Peek, M.E., Woodruff, J.N., Chin, M.H., 2022. Eliminating explicit and implicit biases in health care: evidence and research needs. Annu. Rev. Public Health 43, 477–493.

Williams, D.R., Mohammed, S.A., 2009. Discrimination and racial disparities in health: evidence and needed research. J. Behav. Med. 32 (1), 20–47. doi:10.1007/s10865-008-9185-0.

Williams, D.R., Mohammed, S.A., 2013. Racism and health I: pathways and scientific evidence. Am. Behav. Sci. 57 (8), 1152–1173. doi:10.1177/0002764213487340.

Wrigley-Field, E., 2024. Stolen lives: redress for Slavery's and Jim Crow's ongoing theft of lifespan. RSF Russell Sage Found. J. Soc. Sci. 10 (2), 88–116. doi:10.7758/RSF.2024.10.2.04.

Youngman, T., Patin, B., 2024. Spreading Sankofa: addressing digital epistemicide through LIS education. J. Digit. Soc. Res. 6 (3), 10–25. doi:10.33621/jdsr.v6i3.33238.

Zephyrin, L., et al. (2024) 'Advancing Racial Equity in U.S. Health Care', The Commonwealth Fund, Available at: https://www.commonwealthfund.org/publications/fund-reports/2024/apr/advancing-racial-equity-us-health-care. (Accessed 28 June 2025).

Zhilkova, A., Alsabahi, L., Olson, D., Maru, D., Tsao, T.-Y., Morse, M.E., 2023. 'Hospital segregation, critical care strain, and inpatient mortality during the COVID-19. medRxiv. doi:10.1101/2023.12.04.23299393.

Zippia, 2025. 'Doctor Demographics and Statistics in the U.S.' Available at: https://www.zippia.com/doctor-jobs/demographics/ (Accessed 26 June 2025).

CHAPTER

2 Intersections between colonialism, medical racism, and race-based medical practice

Faadiel Essop[a] and Saleha Suleman[b]

[a]*Centre for Cardio-Metabolic Research in Africa (CARMA), Division of Medical Physiology, Faculty of Medicine and Health Sciences, Stellenbosch University, Cape Town, South Africa*

[b]*Department of Public Health, University of Cape Town, Cape Town, South Africa*

Focusing on scientific racism: An introduction

Around the world, the pages of history are stained with instances of scientific racism, with scientists using pseudoscientific theories and methods to perpetuate racial inequalities, justify racial hierarchies, and promote inherent beliefs of racial superiority. How do we begin to grasp the idea of "scientific racism"? Garrod (2006) provides some useful insights by stating: "While science has the reputation of objectively testing theories using the scientific method, scientific racism is the exact opposite. It seeks to create definitions of race and culture based on opinion and extremely questionable evidence … ." He further notes (Garrod, 2006) that, "The important distinction between science and scientific racism is that the creation of separate racial categories was not solely for the sake of a biological system of classification, but rather for political means …"

Thus "theories" of scientific racism helped to justify, validate, and preserve the prevailing social order (Calliste et al., 2000). Moreover, although the notions of scientific racism existed for centuries (albeit referred to by different terms such as Aryanism, Teutonism, or Nordicism), the underlying basis of racial supremacy has remained the same (Jackson and Weidman, 2005). As global awareness becomes increasingly cognizant of the damaging effects of colonialism and the demand for equality grows, scientific racism must be put under the spotlight and critically examined. Here, its historical roots need to be scrutinized to ascertain its harmful effects on colonized peoples and also to assist in identifying remaining structural legacies. Scientific racism's origins stretch back to the 17th century and helped create an impetus in Western Europe and North America to categorize human populations into different "races." Such attempts triggered extensive research endeavors aiming to provide scientific proof for racial differences, including the measurements of physical characteristics like skull shapes, skin color, and other facial features such as nose lengths (Scientific Racism, n.d.). Moreover, such beliefs were supported by

Decolonizing Medical Education, Research and Practice. https://doi.org/10.1016/B978-0-443-44998-7.00011-9

key philosophers of the Enlightenment Age, together with the spawning of Social Darwinist and eugenics movements during the 19th and 20th centuries. Hence, this influence extended to the social sphere with (false) fears of interracial marriages, based on the views of 19th-century physicians who claimed that those with "mixed blood" were likely to exhibit shorter lifespans and also be more susceptible to disease (Tucker, 2007). These ideas persisted into the 20th century, where geneticists warned that interracial marriages would cause "genetic disharmonies" (Tucker, 2007) and, more recently, when a prominent geneticist explained that the relatively high infant mortality rate in African Americans occurred as a result of the "hybrid incompatibilities caused by white genes that were undetected" (Tucker, 2007).

Scientific racism grew exponentially during the Age of Enlightenment, when modern Western science became a major rationale to replace faith, superstitions, and indigenous knowledge. During this period, racial categorizations became hierarchical, with beliefs that Europeans were the most evolved and intelligent (Skibba, 2019), while other "racial" groups were less endowed in terms of vital behavioral and/or cognitive characteristics (Tucker, 2007). This was further developed into a "scientific" notion that evolution was a struggle between races (Jackson and Weidman, 2005), and that some would do better because of their innate racial characteristics, which provided them with increased immunity from environmental influences. In line with such arguments, some physicians during the 19th and 20th centuries concluded that the freeing of slaves would lead to mental, moral, and physical deterioration, substantiated by a comparison of the anatomical differences in Black and White populations (Brandt, 1978).

Such biases extended more widely to eventually influence social attitudes, unequal power symmetries between the Global North and South, public policy, biomedical education and research, and clinical management practices. For example, a recent publication highlights how inequities in research authorship result in the dominance of peer-reviewed publications in English, and how it is predominantly researchers from the Global North who garner scientific awards (Pai and Abimbola, 2024). Considering this, the current chapter will focus on the notion of scientific racism to stimulate meaningful dialog regarding its profound impact and misuse. Moreover, we hope this chapter will also aid scientists, researchers, educators, medical professionals, and policy makers recognizing how science and medicine have been employed as political tools.

The roots of scientific racism and its links to colonialism

The determination of the precise origins of scientific racism is complex, as various overlapping influences collectively contributed to its establishment. A crucial question to consider is how perceived physical differences between people (e.g., skin color, hair texture) morphed into an essentialist, racialized paradigm. The Scientific Revolution and Enlightenment periods in Western Europe (spanning the 16th to 18th centuries) are closely linked to the colonization of non-European territories and the subjugation of indigenous peoples. Here, the ideological justification included the notion of civilized, full-fledged, and rational Europeans being given *carte blanche* to overwhelm the so-called backward, inferior, and ignorant indigenous peoples (Omi and Winant, 2014).

The initial impetus for colonial expansion during the 15th and 16th centuries was rooted in economic, religious, and political motivations, with science playing a more significant role during later periods (Omi and Winant, 2014). For example, Omi and Winant (2014, p. 113) noted that, "In practice, of course, the seizure of territories and goods, the introduction of slavery through the encornienda [Spanish system of forced labor] and other forms of coerced native labor, and then through the organization of the African slave trade—not to mention the practice of outright extermination—all presupposed a worldview which distinguished Europeans, as children of God and fully-fledged human beings, from 'others.'"

However, as rationalism and science eventually triumphed over religion in Western Europe, this enabled a renewed justification to continue and expand the subjugation and exploitation of colonized populations. The well-traveled French physician François Bernier (1620–1688) likely made one of the earliest efforts to group humans based on physical attributes. He classified human beings into four races based on their distinct physical traits and his own personal observations (Bernier, 2001). Although this represented a shift away from the classic Biblical explanations for human differences, he did not articulate a clear racial hierarchy (Harpham, n.d.a). The Swedish taxonomist Carl Linnaeus (1707–1778), in his *Systema Naturae,* was the first to conceive a human taxonomy by categorizing four main "varieties" based on geography, skin color, and stereotypical traits (Linnaeus, 1758). This included *Homo sapiens europaeus* (Europe, "white" skin color), *Homo sapiens americanus* (Americas, "red" skin color), *Homo sapiens asiaticus* (Asia, "brown" skin color), and *Homo sapiens africanus* (Africa, "black" skin color). Of note, when his work was translated into English in 1792, the term "varieties" was rendered as "sub-species," thereby suggesting a deeper and more fixed difference between such categories (The Linnean Society of London, n.d.). The French naturalist Georges-Louis Leclerc, Comte de Buffon (1707–1788), differed with Linnaeus and described six races, where he postulated that physical variations (e.g., skin color and skull sizes) occurred because of environmental influences (Buffon, 1954). However, he proposed that Europeans constitute the ideal form and that differences displayed by non-Europeans occurred because of degeneration (due to environmental factors) from this idyllic appearance (Classification, n.d.).

Johann Friedrich Blumenbach (1752–1840) was a German physician and naturalist who categorized humans into five races consistent with geographic regions, i.e., Caucasian, Mongolian, Malayan, Ethiopian, and American (Bendyshe, 1865b, 1865a, no pagination). He investigated differences in skulls and was the first to employ the term "Caucasian" (still commonly used as a synonym for "White"), as he found what he described as the most beautiful and symmetrical skull in this region (modern-day Georgia). Notably, although he did not arrange the different races in a hierarchical manner and merely in a geographic way, he held the view that

Caucasians represent the ancestral human configuration from which the others degenerated (or deviated). For example, he stated (Bendyshe, 1865b, p. 2): "For in the first place, that [Caucasian] stock displays, [...] the most beautiful form of the skull from which, as from a mean and primeval type, the others diverge by most easy gradations on both sides to the two ultimate extremes (that is, on the one side the Mongolian, on the other the Ethiopian). Besides, it is white in colour, which we may fairly assume to have been the primitive colour of mankind, since, as we have shown above, it is very easy for that to degenerate into brown, but very much more difficult for dark to become white."

George Cuvier (1769–1832) was a French zoologist who worked at the prestigious Museum of Natural History in Paris and is revered as a pioneer in paleontology. He developed the use of comparative anatomy in animal taxonomical studies and identified three human races. These were Caucasian ("white"), Mongolian ("yellow"), and Ethiopian ("black"), using craniometric measurements (facial angles of skulls). He supported the idea of a racial hierarchy, with Caucasians being the most civilized, while noting that "negroes" were more ape-like and barbaric in nature (Harpham, n.d.b). Notably, Cuvier was also responsible for the dissection and display of the bodily remains of Saartjie Baartman, a Khoisan woman from South Africa, who was put on public display and humiliated in Europe for her unique body features (Ras, 2017). Here, Ras (2017) writes: "Baartman's existence and identity was purely based on the European view of the grotesque body—wholly other to European women. To Europeans Saartjie was primitive, bestial and crudely sexual."

Slightly distinct from George Cuvier, Dr. Samuel Morton (1799–1851) focused specifically on craniology (instead of skin color) to reinforce the idea of white supremacy and black inferiority. He believed that the different races were created separately as distinct species (polygenism) and that the physical and mental attributes of each race were fixed. Morton performed numerous measurements on skull sizes as a marker for intellectual and moral capacities. He published his findings in two major works: *Crania Americana* in 1839 and *Crania Aegyptiaca* in 1844 (Renschler and Monge, 2008). He claimed that such measurements showed that Blacks possessed smaller brains and hence were less intelligent than whites. However, his work was subsequently criticized by Dr. Stephen Jay Gould (1941–2002), who reported several methodological errors that skewed the data in favor of Caucasians (Gould, 1996). In contrast, others defended Morton's measurements and questioned Gould's findings in this regard (Renschler and Monge, 2008; Mitchell, 2018).

Joseph Arthur de Gobineau (1853–1915) expanded Blumenbach's views on different races beyond physical characteristics to include unequal moral and intellectual attributes, with Whites considered a superior race and civilization (Nale, 2014). He wrote a four-volume essay on the inequality of races and White racial superiority and expressed concerns regarding the weakening of Whites by mixing their blood with other races. Here, "blood" meant more than a physical substance but constituted a far more encompassing nature, as Nale (2014, p. 119) explained: "The blood is in this important way always more than blood. It is not merely a physical substance in the veins but also as the values distinct to a people. This blood, so long as it is pure, is manifested in the social bond or spirit of the people. More than just a bodily fluid, it is the primary force behind Gobineau's historical project, for the blood and values of a people lie at the heart of degeneration … What dies in blood mixing is the spiritual texture of a collective group, the values residing in the 'veins' of the body politic."

This entrenchment of scientific racism not only impacted research practices but also began to influence the clinical management of patients. For example, Dr. Samuel Cartwright (1793–1863)—an internationally recognized Southern physician—ascribed so-called mental illnesses to Black slaves (Willoughby, 2018). He believed Blacks suffered from neurological illnesses such as "laziness" and "mischief" and developed the theory of *drapetomania* (literally translated as "runaway madness" from Greek), which he believed "afflicted" slaves. Alarmingly, he proposed that this mental condition could be cured by suitable punishments of the slaves. Of note, Cartwright was not an outlier, and his ideas and practices simply reflected the dominant views of his era. He also conducted spirometry experiments and concluded that the lung function of Black slaves was inferior when compared to whites. Here, Braun (2014, p. 99) noted that, "Perhaps the most significant experiments for the future of spirometry were those of plantation physician and slaveholder Samuel Cartwright in the US South. Drawing explicitly on Jefferson's interpretive framework, Cartwright built his own spirometer to study differences in lung capacity in slaves and Whites, and to quantify it precisely. According to Cartwright, 'the deficiency in the negro' was '20%'. Defining difference as 'deficiency' … ."

In parallel, such ideas were supported by prominent philosophers of the Enlightenment Age who made foundational contributions to contemporary Western philosophy and thought. For example, while the French philosopher and historian Voltaire (1694–1778; born François-Marie Arouet) championed freedom, civil liberties, and tolerance, he still perpetuated some racist views. He writes that (Harpham, n.d.c), "The negro race is a species of men as different from ours as the breed of spaniels is from that of greyhounds … Their eyes are not formed like ours. The black wool on their heads and other parts has no resemblance to our hair; and it may be said that if their understanding is not of a different nature from ours, it is at least greatly inferior. They are not capable of any great application or association of ideas, and seem formed neither for the advantages nor abuses of our philosophy. They are a race peculiar to that part of Africa, the same as elephants and monkeys."

Likewise, the Scottish philosopher and historian David Hume (1711–1776), whose empirical approach laid the foundations for the modern scientific method, stated in 1753 (Hume, 1985) that, "I am apt to suspect the negroes, and in general all the other species of men (for there are four or five different kinds) to be naturally inferior to the whites. There never was a civilized nation of any other complexion than white, nor even any individual eminent either in action or speculation. No ingenious manufactures amongst them, no arts, no sciences. On the other hand, the most rude and barbarous of the whites, such as the ancient Germans, the present Tartars, have still something eminent about them, in their valour, form of government, or some other particular. Such a uniform and constant difference could not happen, in so many countries and ages, if nature had not made an original distinction betwixt [between] these breeds of men. Not to mention our colonies, there are negroe slaves dispersed all over Europe, of which none ever discovered any symptoms of ingenuity."

Immanuel Kant (1724–1804), one of the major intellectuals during the Enlightenment and a foundational thinker in modern Western philosophy, regarded whites as near perfect while considering non-Whites and "negros" (Black people) as inferior (Allais, 2016). He also wrote that even if Blacks were set free (after enslavement), they never accomplished anything worthy in disciplines like science and the arts. Kant postulated that different races displayed fixed hereditary attributes (e.g., intellectual ability) and described Blacks as being "stupid" (Allais, 2016). Thus the coexistence of Enlightenment ideals advocating equality and human freedom, simultaneously with racial theories and colonialism, accentuates a major contradiction. While Western Europe championed liberty, equality, and freedom, it rationalized the subjugation of non-Europeans through scientific racism and philosophical views, supporting racial superiority. In agreement, Allais (2016, p. 2) noted that, "Kant's racism and Kant's theory of race raise questions both for the history of philosophy and for philosophy. It highlights a centrally troubling feature of the enlightenment: the period in which European thinkers developed ideas about human equality and autonomy while (some) European states were involved in colonizing and enslaving large parts of the world. This is something to which we should pay attention in trying to understand both the enlightenment and the development of the modern idea of race."

Although Charles Darwin's (1809–1882) focus was on animals and biological evolution, there is some uncertainty about whether he espoused racist views or if his work was misused by others who supported scientific racism (Schwartz, 2009). For example, Darwin was strongly opposed to slavery, expressed support for abolitionists, and was not in favor of racial hierarchies. However, at times, he also used pejorative language. For example, he referred to non-Europeans as "savages," although some may argue that he was simply employing prevailing terminologies in line with societal norms. It is therefore opportune to consider some statements made by Darwin in this regard. For example, in his *Descent of Man* (Darwin, 1882) he states, "At some future period, not very distant as measured by centuries, the civilised races of man will almost certainly exterminate, and replace, the savage races throughout the world."

Moreover, Darwin (1871, pp. 133–134) displays some Eurocentrism when he notes that, "With savages, the weak in body or mind are soon eliminated; and those that survive commonly exhibit a vigorous state of health. We civilised men, on the other hand, do our utmost to check the process of elimination; we build asylums for the imbecile, the maimed, and the sick; we institute poor laws; and our medical men exert their utmost skill to save the life of everyone to the last moment."

During the 19th and early 20th centuries, two overlapping ideological streams further entrenched scientific racism. At this time, Social Darwinism extended Darwin's ideas by applying them to the human context, with the British anthropologist Herbert Spencer (1820–1903) first coining the term "survival of the fittest" (Spencer, 1884). Thus they employed pseudoscientific findings (as discussed in this chapter), philosophical arguments, social class, and wealth to identify the "fittest" (i.e., Whites). This outlook meant that human society was now competitive by nature, while also justifying inequality, racial hierarchies, *laissez-faire* capitalism, and colonialist practices.

The eugenics movement took this a step further by promoting active interventions to ensure the welfare of the "fittest" stock. Here, Francis Galton (1822–1911), Darwin's first cousin, was the first to coin the term "eugenics" (Paul and Spencer, 1995). He is regarded as the pioneer of this field and promoted the idea of childbearing among the "fittest" to improve the gene pool, while simultaneously rejecting the "unfit" stock of the lower classes and people of color (Paul and Spencer, 1995). This is referred to as "positive eugenics," while the opposite holds for "negative eugenics," i.e., limiting reproduction among the "unfit." Galton argued this as he believed that traits such as intelligence, health, and morality were hereditary in nature, and that interbreeding with the "unfit" stock would therefore result in the deterioration of the White race (Galton, 1869). Such ideas were also supported by eminent scientists such as Karl Pearson (1857–1936), who was a British mathematician and biostatistician with a profound influence on contemporary statistical methods such as correlation and Chi-squared tests. For example, Pearson (1905, p. 21): "What I have said about bad stock seems to me to hold for the lower races of man. How many centuries, how many thousand of years, have the Kaffir or the negro held large districts in Africa undisturbed by the white man? Yet their intertribal struggles have not yet produced a civilization in the least comparable with the Aryan. Educate and nurture them as you will, I do not believe that you will succeed in modifying the stock."

Furthermore, regarding White and Black intermixing, he states (Pearson, 1905, p. 22), "If you bring the white man into contact with the black, you too often suspend the very process of natural selection on which the evolution of a higher type depends. You get superior and inferior races living on the same soil, and that coexistence is demoralizing for both. They naturally sink into the position of master and servant…"

Of note, eugenics was relatively widespread in Europe and North America during this period, for example, with the arranging of international eugenics congresses, the active role of prestigious peer-reviewed publications in this regard, and with the support of private donors such as the Carnegie Foundation and Dr. John Harvey Kellogg (1852–1943) (Durst, 2015; Schambra, 2021; Lombardo, 2024). For example, Dr. Kellogg (brother of William Keith Kellogg, developer of the well-known cereal brand) established the Race Betterment Foundation in Battle Creek, Michigan, arranged several eugenics conferences, and wrote numerous articles on eugenics and racial superiority (Markel, 2017). Moreover, prestigious medical journals also played a part in establishing eugenics. For example, in a recent article in the New England Journal of Medicine (Lombardo, 2024, p. 872), it is stated: "It is worth remembering that, as physician involvement in eugenics was being chronicled in the Journal, advocacy for eugenics fed popular contempt for people with disabilities, people from marginalized racial or ethnic groups, immigrants, and poor people. It fueled the frenzy for laws that permitted the sterilization of thousands of people and provided a template for the murderous ambitions of Hitler. Attitudes that were similar, and in some cases identical, to those embraced by the eugenists are still very much a part of life in the United States today. The job of confronting them is a challenge that will require the commitment of physicians and their professional sounding boards, such as the New England Journal of Medicine. Perhaps this article, recognizing this journal's role enabling and endorsing eugenic thought in the United States, can provide a starting point for that effort."

It is, therefore, clear that the harmful idea of scientific racism played a key role in colonialism and also laid the foundation for unethical actions and legal practices such as forced sterilization, laws preventing mixed marriages, racial purity laws, the Holocaust in Nazi Germany, and the current genocide being committed by the state of Israel in Gaza (Palestine).

Some clinical and scientific applications of scientific racism

Scientific racism was not (and is still not) limited to justifying hierarchies in social and political spheres. There are many well-known instances of clinical and scientific applications of scientific racism, from the time of the Enlightenment to current discriminatory practices in research, public policy, and health care. One notorious example is the subconscious belief that Black patients possess a biologically higher pain tolerance than their white counterparts. Here, some describe in detail how such false beliefs by both laypersons and medical professionals continue to exacerbate disparities regarding the undertreatment of Black patients for pain management (Hoffman et al., 2016).

The Tuskegee Syphilis Study (1932–1972) is another dreadful example of the application of scientific racism in the research domain. For this study, 600 Black men from Macon County (Alabama, United States [US]) were misled into believing that they were being treated for syphilis and informed that they suffered from "bad blood." However, in truth, treatment was actively withheld to study the natural progression of syphilis without any therapeutic interventions (Brandt, 1978). Alarmingly, the study continued despite penicillin treatment becoming available during the mid-1940s and carried on for several decades until it was stopped due to the actions of a whistleblower in 1972. This unethical study was based purely on racist assumptions regarding the value of Black lives, and it tragically resulted in more than one hundred preventable deaths, as well as the long-term suffering of those who survived (Brandt, 1978). Thus it is not surprising that Black Americans display relatively high levels of mistrust in health care systems in the US.

Eugenic theories also enabled the mass forced sterilization of thousands of women considered "unfit" to reproduce, many of whom were Black, Indigenous, or immigrants. This occurred in various countries, including the US, the United Kingdom, Canada, and Germany (Dyck, 2014; Reilly, 2015). Moreover, these ideas were legally embraced in numerous American states and were justified by propagating the false idea of an inherent genetic inferiority of certain races (Reilly, 2015). Proponents of eugenics described such human rights violations as "necessary evils," likening them to gardens that need weeding or trees that require pruning (Dyck, 2014). The repercussions of such targeted sterilizations were profound, as they stripped affected individuals of their basic reproductive rights and bodily autonomy, thereby violating fundamental human rights.

The structural legacy of scientific racism continues to cast its dark shadow on Black people and marginalized countries (collectively referred to as the Global South) even in contemporary times. Examples include the lack of patent sharing for COVID-19 vaccines during the recent pandemic and vaccine hoarding by affluent (and White) countries in the Global North (Stancil, 2023). Moreover, the production of a 100% effective pre-exposure prophylaxis for HIV, costing $42,000 a year, is completely out of reach for most people residing in the Global South (Stancil, 2023; Pai and Abimbola, 2024). Another example includes mpox, and as it currently impacts African populations, the most popular and scaremongering images being circulated are of people of color infected with the virus and covering their faces. A recent opinion piece explains how such "mugshots" reinforce the stigma that diseases originate in Africa. Here, Downs (2024) writes about the belief that yellow fever was originally brought to Europe due to contact with Africans: "When yellow fever threatened to invade London in 1845 from British ships arriving from Africa, medical and government authorities did not know that mosquitos were the culprit that caused this vector borne disease." Thus it was theorized that the disease was transmitted because of human contact with these ships. He further notes (Downs, 2024) that, "Scottish doctor James Ormiston McWilliam … took copious notes of the African people's racial features, reinforcing the notion that that disease carriers needed to be visualized to be understood." Together, this raises the rhetorical question of whether people of color are only good enough to participate in clinical trials, but then not worthy of receiving access to such life-saving treatments once approved (Pai and Abimbola, 2024).

The persistence of race-based medical practice cannot be explained solely by outdated science or individual prejudice. As Feagin and Bennefield (2014) argue, these practices are sustained by systemic racism within US medicine, and it is seen as a deeply rooted institutional order that privileges White norms and interests while marginalizing racialized groups (Feagin and Bennefield, 2014). Feagin's framework highlights how racial hierarchies are maintained through what they term *white racial framing*: the routine centering of whiteness as the default standard in research, policy, and clinical care (Feagin, 2006). Within this system, inequities are reproduced through institutional inertia (the slow or absent correction of racist assumptions) and the minimization of suffering among communities of color. Understanding race-based medical practice through this lens situates it not as an anomaly within an otherwise neutral science, but as a predictable outcome of a health care system that is historically structured by racial domination.

Feagin's broader theory of systemic racism situates these patterns within an overarching racialized social order that extends far beyond individual institutions. In his book *The White Racial Frame* (2023), he argues that racism is not merely an attitudinal problem but rather a foundational organizing principle of Western societies. It is embedded in their economic structures, political systems, and cultural narratives. He further argues that the white racial frame operates as a shared interpretive schema that normalizes white experiences, perspectives, and interests as universal while rendering others as deviant or deficient (Feagin, 2023). Within medicine, this frame shapes research priorities, clinical norms, and definitions of "normal" physiology, often without explicit intent. It also reproduces structural inequities through institutional design and policy inertia. Understanding medicine as part of this broader racialized order allows for a shift from focusing on individual prejudice to analyzing how racial hierarchies are systematically maintained through knowledge production, funding, and institutional governance.

No biological basis for race

Considering this historical background, the question arises whether the notion of scientific racism is valid or not. Here, genetic data and sociological analyses conclusively show that there is no scientific basis for racial categorizations and that it is a social and political construct. In support, the sociologist Dr. Wendy Roth (2016) highlights the complexity of racial classifications and argues that they consist of multiple dimensions. She identified several factors that eventually influence the nature of such racial classifications. These include racial identity (how an individual self-identifies), racial self-classification (the "race" selected when completing a survey or questionnaire), observed race (how others classify an individual), reflected race (an individual's belief about how she is classified by others), phenotype (physical traits relevant to race such as skin color, hair texture, nose and lip shape, eye color and its various gradations), and racial ancestry (reference to an individual's ancestors in terms of race) (Roth, 2016). Roth (2016, p. 1331) alerts readers to the complexity of racial groupings by stating, "Yet in all societies, understanding the diverse dimensions of race shows how racial inequalities are constructed in multiple ways—from within and without, by aspects of our experiences, our appearance, our interactions, and our family history, and by no single one of these on its own."

Such complexity was demonstrated during the apartheid years in South Africa, where individuals could be reclassified into different racial groups through various "tests" and bureaucratic decisions (Posel, 2001). Moreover, during the early 20th century, state legislation enforced the "one-drop rule" for racial classification in the US, meaning that a single drop of "Black blood" rendered an individual "Black" (Lujan and DiCarlo, 2021). Thus any individual with a single ancestor of African descent was classified as "Black" (or "negro") to thereby eliminate the possibility of a fair-skinned "Black" person passing as a White person.

Genetic data analyses provide compelling evidence against scientific racism. A major conclusion of the Human Genome Project was that the sequencing of the entire human genome provided no scientific basis for the concept of distinct "races"

(Collins and Mansoura, 2001). Here, the DNA sequence between any two individuals is 99.9% similar, while the remaining 0.1% difference does not vary in line with popularly employed socio-politically constructed racial classifications (Kwok et al., 1996; Jorde et al., 2000; Collins and Mansoura, 2001). Furthermore, the renowned evolutionary biologist Dr. Richard Lewontin (1929–2021) performed a genetic analysis on variation in different "races" and found that there is far greater variation *within* such groupings than between them (Lewontin, 1972). He writes that, "It is clear that our perception of relatively large differences between human races and subgroups, as compared to the variation within these groups, is indeed a biased perception and that, based on randomly chosen genetic differences, human races and populations are remarkably similar to each other, with the largest part by far of human variation being accounted for by the differences between individuals. Human racial classification is of no social value and is positively destructive of social and human relations. Since such racial classification is now seen to be of virtually no genetic or taxonomic significance either, no justification can be offered for its continuance."

Similarly, other genetic research work that analyzed human molecular diversity, employing numerous microsatellite loci and polymorphic restriction sites in sixteen populations around the world, came to similar conclusions (Barbujani et al., 1997). Here, they found only minor changes in genetic variation between populations on the same continent and between populations residing on different continents. The authors (Barbujani et al., 1997, p. 4518) concluded their study by stating that, "The differences among human groups, even very distant ones and no matter whether the groups are defined on a racial or on a geographical basis, represent only a small fraction of the global genetic diversity of our species."

Recent advances in epigenetics further illuminate how the biological consequences attributed to "race" are, in fact, the embodied outcomes of racism and structural inequality. Kuzawa and Sweet (2009) argue that racial disparities in cardiovascular and other chronic diseases arise not from innate genetic differences but from the cumulative effects of social adversity, psychosocial stress, and structural discrimination that have become biologically embedded across generations (Kuzawa and Sweet, 2009). Through processes such as DNA methylation and histone modification, chronic exposure to racism can influence gene expression and physiological function, shaping disease risk independently of genetic ancestry (Kuzawa and Sweet, 2009). In later work, Kuzawa and Thayer (2011) extended this analysis to propose that epigenetic mechanisms represent a key pathway for understanding how social and environmental conditions during critical developmental periods can produce intergenerational patterns of health inequality (Kuzawa and Thayer, 2011). Together, these findings challenge reductionist genetic explanations for racial health disparities and underscore that it is racism, not race, that becomes written into the body.

Relatedly, the Swedish Nobel Laureate Svante Pääbo (1955–) found that when individuals are evenly sampled from across the globe for genetic studies, a pattern emerged where gradients of allele frequencies span the entire world compared to one with discrete clusters (as "races") (Serre and Pääbo, 2004). They concluded their study (2004, p. 1679) by stating, "Our results show that when individuals are sampled homogeneously from around the globe, the pattern seen is one of gradients of allele frequencies that extend over the entire world, rather than discrete clusters. Therefore, there is no reason to assume that major genetic discontinuities exist between different continents or 'races.'"

Of note, a question may arise regarding some scientific studies that found differences between Black and White patients in terms of the distinct nature of disease onset and clinical management. However, a critical examination of such studies quickly reveals that this premise is unfounded. This will be further elucidated by discussing several relevant examples. Focusing on the kidney, its function can be assessed by measuring circulating creatinine levels, which can then be used in a specific mathematical equation to determine the degree of functionality [termed the "estimated glomerular filtration rate" (eGFR)] (Khazanchi et al., 2023). However, in line with the notion of scientific racism, an additional factor was introduced into this equation for Black persons. This was based on a belief that Black individuals possess increased muscle mass and, hence, a higher circulating creatinine level compared to Whites. In practice, this would mean that for two individuals with the same creatinine levels, the equation would categorize Black patients at a lower stage for chronic kidney disease when compared to Whites, significantly influencing their treatment modalities. Considering there is no clear biological basis for such a race-based equation and related clinical guidelines, a task force under the supervision of the National Kidney Foundation and the American Society of Nephrology recently recommended removing the race factor from such calculations (Delgado et al., 2021). Here, they (Delgado et al., 2021, p.3011) recommended that, "The CKD-EPIcr_R equation that was developed without the use of the race variable be implemented immediately, including in all laboratories." Moreover, the report (Delgado et al., 2021, p.2994) also stated that, "This unified approach, without specification of race, should be adopted across the United States. High-priority and multistakeholder efforts should implement this solution."

Although this race factor has now been removed from clinical assessments, some researchers calculated that if the race factor did not form part of this equation, then one-third of the African American patients would have been assigned a more severe stage of chronic kidney disease (Ahmed et al., 2021). In agreement, Khazanchi et al. (2023, p. 10) noted that, "Removing race adjustment from the MDRD eGFR equation could result in an additional 3.3 million Black Americans receiving a stage 3 CKD [chronic kidney disease] diagnosis, 300,000 more qualifying for a nephrologist referral, and 31,000 becoming eligible for transplant evaluation and inclusion on a waitlist."

A similar scenario exists for cancer. For example, American and European guidelines for prostate cancer include screening criteria that are differentiated by race, despite the same weaknesses in these measurements as highlighted earlier, i.e., issues of racial self-identification and the unfounded assertion of scientific racism. Thus some have recently called for the adoption of race-neutral prostate cancer screening algorithms, while emphasizing that race and ethnicity do not constitute determinants of such risk (Arenas-Gallo et al., 2024). The authors also argue that greater emphasis should instead be placed on social determinants of health and their effects vis-à-vis the risk of prostate cancer onset and development.

Such clinical management practices also extend to the functional evaluation of other organs. For example, race-based adjustments for lung functional assessments of Black patients (supposedly possessing lower lung capacities) meant that they were undertreated and displayed worse clinical outcomes compared to their White counterparts (Schluger et al., 2022). Thus some argued for the removal of race-based spirometry assessments and suggested that greater emphasis should be placed on the role of environmental and nutritional factors in this instance. The American Thoracic Society subsequently released a formal statement asserting that race or ethnicity should no longer be used as adjustment factors when performing pulmonary functional tests (Bhakta et al., 2023). Likewise, the American Heart Association removed race as a risk predictor when they recently released a new calculator (Predicting Risk of Cardiovascular Disease Events [PREVENT]) for determining the 10- and 30-year risk of heart failure, acute myocardial infarction, and stroke (Khan et al., 2023).

Some may also argue that variations in specific genes (polymorphisms) may explain the onset of unique pathological complications occurring in Black and White patients. For example, the incidence of kidney failure is relatively high among Blacks in the US compared to Whites, and such racial disparity has been linked to polymorphic variations in the apolipoprotein L1 (*APOL1*) gene (Cerdeña et al., 2021). However, this is not a robust link, as the frequency of *APOL1* alleles differs markedly across the African continent, with significant variations in allele frequency between Nigerian and Cameroonian populations, even though descendants of both would be considered to be "African American." Moreover, such alleles are not only found in Western African populations but also in others from Europe, the US, Pakistan, and Latin America, indicating significant admixing (Cerdeña et al., 2021). In support, some stated (Cerdeña et al., 2021, p. 859) that, "Although some Black Americans carry high levels of West African ancestry owing to the shameful history of chattel slavery, wide variation in APOL1 "risk" allele frequency is expected among all individuals racialized as "Black." It is therefore inappropriate to conclude that APOL1 mutations increase risk for kidney disease among all Black Americans or to limit research or testing of APOL1 genotypes to Black individuals: Such determinations may constrain clinical decision making and misdirect resources to the detriment of patients." The difference in kidney failure incidence, therefore, likely occurs due to a multitude of reasons, such as structural racism, wealth disparities, employment levels, socio-economic status, access to health care, toxic environmental factors, and the nature of neighborhoods, which can be incorrectly simplified into racial differences.

Sickle cell disease is another genetically linked condition that is associated with Black people. It occurs due to a mutation that alters the normal adult hemoglobin's structure, leading to crescent-shaped red blood cells. Notably, those diagnosed with sickle cell disease have historically suffered from structural discrimination and the delayed development of novel therapeutic interventions due to limited research funding (Kidane Gebremeskel et al., 2025). However, sickle cell disease does not only affect Black people or those originating from Africa; it is also found in the Mediterranean, the Middle East, and India. This mutation is most prevalent in geographic regions where malaria was endemic, as sickle cell anemia offers some protection and a survival advantage against malaria for heterozygous carriers.

In contrast, cystic fibrosis (CF) is often regarded as a disease that largely afflicts White individuals (Wu et al., 2024). Here, mutations in the CF transmembrane conductance regulator (*CFTR*) gene cause CF if inherited from both parents. Since Blacks are less likely to display known gene variants, this can often lead to a delayed diagnosis and the undertreatment of such patients. Moreover, Blacks are generally understudied and underrepresented in clinical trials. Hence, *CFTR* variants from Black patients are missing from mutation panels used to clinically screen for this complication (Wu et al., 2024). Thus a shift from systemic racism to an inclusive approach for the correct diagnosis and clinical management of CF is required (Wu et al., 2024): "As illustrated by current findings, overcoming racial/ethnic inequities intrinsic to CF screening, diagnosis, and treatment remains a significant unmet medical need. It must be acknowledged that these processes are affected by systemic racism, which is embedded in nearly every thread of U.S. society."

However, racial management practices are still applied to some diseases. For example, there is a belief that Blacks display relatively high salt sensitivity (being more prone to retain sodium) and relatively lower plasma renin levels. This is known as the Slavery Hypertension Hypothesis, which postulates that this increased retention of sodium evolved as a survival mechanism in West Africans during the slave trade. However, this hypothesis has been criticized by others, as the predicted patterns of increased sodium retention are not reflected in West African populations, and there are also variations in plasma renin levels within Black populations (Savage and Panofsky, 2023). This has resulted in the development of racialized hypertension treatment guidelines, with angiotensin-converting enzyme (ACE) inhibitors not prescribed as a first-line treatment for Black patients (James et al., 2014). Here, some have criticized the racialized findings of the Antihypertensive and Lipid-Lowering Treatment to Prevent Heart Attack Trial (ALLHAT), which informed the development of such guidelines. Savage and Panofsky, 2023, p. 796) state, "We argue that the differential racial recommendations crafted by JNC 8 and informed by ALLHAT are shaped and discredited by five interlocking issues in ALLHAT's design and analysis and the incorporation of outcomes into JNC 8; namely (1) poorly defined categories/measures; (2) post hoc analysis and reformulation of the study question; (3) artificial study design; (4) overstatement of drug differences; and (5) selective uptake of evidence." The authors, therefore, concluded that ALLHAT was rooted in the refuted Slavery Hypertension Hypothesis and that it suffers from presuppositions regarding racial differences that informed the study design and data interpretations. Hence, strong advocacy for the abolition of race adjustments is required, and a call should be made for researchers to instead focus on the genuine determinants of racialized disparities (e.g., stress, living conditions, and racism) that can be observed with hypertension (Savage and Panofsky, 2023; Rabay et al., 2024).

Recent debates on digital and algorithmic racism reveal how scientific racism adapts to new technological contexts. The critique by Vyas et al. (2020) marked a turning point in challenging "race correction" across clinical algorithms, showing that such adjustments, once justified as objective, have reproduced structural inequities in diagnosis and treatment (Vyas et al., 2020). Yet, as medicine increasingly turns to artificial intelligence (AI) and big data analytics, these racial biases risk becoming further embedded in automated systems. Benjamin (2022) argues that algorithmic decision-making often inherits the assumptions and exclusions of the societies that design it, thereby reproducing racial hierarchies under the guise of neutrality (Benjamin, 2023). Within AI-driven health care, such biases shape everything from diagnostic imaging and triage systems to predictive models for disease risk. This evolution signals a digital continuation of colonial logic: a data infrastructure that encodes inequality while presenting itself as impartial science.

Recommendations: Way ahead in terms of decolonization of science and medicine

In 2025, Jiménez-Trejo and colleagues extended this conversation by declaring, "scientific racism as a public health emergency of global concern." This piece reframed the persistence of racial inequities in medicine as a continuing structural crisis rather than a historical residue. The authors argued that scientific racism spans every layer of medical knowledge production, from the composition of editorial boards and authorship imbalances between low- and high-income countries to inequities in research funding, bias in health care systems, and resistance to anti-racist reform (Jiménez-Trejo et al., 2025). Crucially, they positioned epistemic exclusion itself—the question of who produces medical knowledge and whose bodies are measured—as a determinant of global health. In naming scientific racism as a contemporary public health emergency, they underscored that dismantling these systems is not simply an ethical imperative but a scientific necessity for achieving equity in health research and practice.

What, then, is the solution to eradicating scientific racism? Although this is a complex question with multifaceted answers, we propose that the decolonization of medical education, research, and clinical practice represents a key factor in this process. As discussed earlier in this chapter, it is clear that although colonial occupation has ended, there is an embedded structural dimension that persists in terms of medical education, research, and clinical practices. This phenomenon is referred to as "coloniality," which is distinct from the term "colonialism" (Quijano, 2007; González and Sánchez-Lasaballett, 2023). Hence, there is a robust impetus for formerly colonized territories to "decolonize" their thinking, training, organizational structures, and practices in this regard.

We therefore propose five initial recommendations: curriculum reform, acknowledging historical wrongs, promoting diversity in research and practice, decolonizing scientific research, and institutional changes. These recommendations will be briefly discussed below. We are aware that this essential decolonization process will likely occur relatively slowly, as it involves challenging pre-existing and well-embedded notions of "doing things," empowering students and researchers to think critically and more holistically, and also attempting to counteract the historically ingrained Eurocentric and racially biased foundations of such disciplines and practices.

In terms of curriculum reform, this could, for example, focus on making students aware of the history of scientific racism and then critically think about or question subconscious biases and privileges. Another way is to include more African-centered health perspectives in remodeled curricula, including, for example: (1) teaching students about a balance between indigenous knowledge/medicine and Western knowledge/medicine, (2) providing historical examples regarding the significant contributions of non-Western scientists and clinicians (not usually acknowledged), and (3) formulating assignments that require students to critically assess such topics.

Furthermore, since traditional medical textbooks often prioritize Western medical knowledge, it would be useful to revise academic materials to include diverse cultural perspectives and global health challenges. There should also be a focus on the role of historical contexts in health care and on including the voices of marginalized communities to ensure a more global understanding of medicine and health care. For example, some work done by the current authors of this chapter demonstrates that incorporating conversations on medical racism and coloniality into biomedical sciences curricula can promote greater inclusivity by fostering an environment of open discussion and learning in the classroom (unpublished data). Moreover, we found that such initiatives better equipped students to challenge and mitigate medical racism and coloniality in their future careers and to become advocates for social justice.

Acknowledging historical injustices is a crucial aspect of the decolonization of medicine. For example, in 2008, the American Medical Association (AMA) issued a formal apology for its role in promoting the notion of scientific racism, including its exclusion of Black physicians (Aluko, 2008). Since then, the AMA has launched various initiatives to address health disparities and promote diversity within the medical profession. The American Psychological Association also apologized to people of color for its role in this context (American Psychological Association, 2021). Several prestigious medical journals have taken a similar approach. For example, recently, The Lancet published a paper that illuminated the journal's role in perpetuating colonialist attitudes in research and global health (Khan, 2024). These are important steps in recognizing the responsibility of professional associations and journals in perpetuating inequalities, as well as emphasizing their role in promoting more equitable research practices through inclusivity and representation.

As earlier discussed, clinical research has historically underrepresented or altogether excluded minority groups and, hence, contributed to medical practices that may not be fully applicable to all patients (New Scientist, 2018; Bibbins-Domingo and Helman, 2022). For instance, although Americans of African descent are more likely to suffer from respiratory ailments, only 1.9% of all respiratory disease-related studies included such participants (Scientific American, 2018). The bias of language when stigmatizing "others" who are different from what would typically be expected by clinicians is another aspect to be addressed in this regard (Ofri, 2018). Thus scientists need to make a more concerted effort to promote diversity in research by including more representative participant cohorts in clinical trials. This should include not only participants from so-called different racial and ethnic groups but also from a range of socio-economic backgrounds. It is vital that the number of participants in clinical trials is truly representative of the diversity of the populations being studied. Furthermore, funding agencies should empower Indigenous-led research by providing grants for initiatives that prioritize investigations focusing on the health needs of previously disadvantaged groups. For example, the First Nations Health Authority in Canada has introduced a research fund for such purposes (Government of Canada, n.d.).

We also recommend the decolonization of medical research endeavors, with one idea being a shift of the power dynamics back into the hands of researchers. This would mean that, for internationally funded research projects, funders should rely on the expertise of research teams who take into consideration local contextual factors where the study is being pursued. Moreover, instead of dictating specific outcomes and/or research inputs, funders should enable researchers in the Global South to take the lead in setting agendas, rather than utilizing a "copy-and-paste" mechanism from other studies that can result in inaccurate and less impactful results.

We also propose that another way to decolonize medical research is to rethink traditional research metrics. For example, recent evidence has indicated that impact factors and citation counts often prioritize Western journals or those published in English (Collyer, 2018; MacGregor-Fors et al., 2020; Gomez et al., 2022). This is, of course, related to researchers in the Global North having access to far greater resources, leading to increased academic productivity and research outputs, and garnering more citations, therefore resulting in a greater impact (MacGregor-Fors et al., 2020). This has been coined the nationality citation bias, or "bibliometric coloniality," with a related phenomenon where authors in the Global North seldom reference sources from the Global South (Mzileni and Heleta, 2023). The Declaration on Research Assessment is therefore advocating for fairer and more inclusive evaluation criteria and considering a wider range of contributions (Cagan, 2013). More journals should therefore consider moving toward greater inclusive metrics in this regard.

A related frontier for decolonization lies in the governance of digital health technologies and AI. As medicine increasingly adopts algorithmic systems for diagnosis, triage, and treatment prediction, the risk of embedding racial and colonial biases within these tools grows. Algorithmic design often reflects the data it is trained on, and currently, these datasets are overwhelmingly drawn from Western populations and filtered through White-dominant institutional lenses. This reproduces exclusion in new forms—marginalized groups become invisible in data, while existing inequities are automated through "objective" technologies. Therefore decolonizing medical research today also means developing frameworks for algorithmic accountability and inclusive data governance. Training health professionals to recognize algorithmic bias, expanding datasets to represent diverse populations, and ensuring meaningful participation from researchers and institutions in the Global South are crucial to preventing the digital replication of scientific racism.

Our last recommendation focuses on institutional change. We suggest that this can begin by reassessing the value of research metrics in terms of performance evaluations and for new appointments. Institutions should also promote increased awareness regarding the roots and development of the concept of scientific racism, the colonial legacy of structural racism, and its devastating impact on medical education, research, and clinical practices.

Conclusions and future perspectives

The intersections of colonialism, scientific racism, and race-based medical practice reveal how medicine has long been entangled with systems of power and oppression. Here, medical science was historically mobilized to justify slavery, colonial expansion, and the exclusion of marginalized communities from health care. These legacies persist in contemporary practices such as race-based clinical care, biased research methodologies, and inequitable curricula that reinforce outdated notions of biological difference.

Yet, this history also provides an opportunity. Acknowledging medicine's complicity is the first step toward transformation. Ongoing institutional reckonings, ranging from apologies by professional bodies to curriculum reform in medical schools, indicate a growing willingness to confront these harms. However, meaningful change demands more than symbolic gestures. It requires structural shifts in how knowledge is produced, taught, and applied—from dismantling race-based corrections in clinical tools, to embedding critical historical and social perspectives into training, to valuing Indigenous and local knowledge systems alongside biomedicine.

Looking forward, we must reimagine medicine as a field truly grounded in equity, justice, and inclusivity. Practically, this means prioritizing patient care that recognizes race as a social determinant rather than a biological reality and cultivating health professionals who can address structural inequities. The future of medicine depends on building curricula, research practices, and clinical systems that are conscious of historical inequities and globally representative. Only then can medicine begin to undo the harms of its past and contribute to a fairer, more humane health care system for all.

References

Ahmed, S., et al., 2021. Examining the potential impact of race multiplier utilization in estimated glomerular filtration rate calculation on African-American care outcomes. J. Gen. Intern. Med. 36 (2), 464–471. doi:10.1007/s11606-020-06280-5.

Allais, L., 2016. Kant's Racism. Philos. Pap. 45 (1–2), 1–36. doi:10.1080/05568641.2016.1199170.

Aluko, Y., 2008. American Medical Association apologizes for racism in medicine. J. Natl. Med. Assoc. 100 (10), 1246–1247. doi:10.1016/S0027-9684(15)31496-6.

American Psychological Association, 2021. *Apology to People of Color for APA's Role in Promoting, Perpetuating, and Failing to Challenge Racism, Racial Discrimination, and Human Hierarchy in U.S.* Available at: https://www.apa.org/about/policy/racism-apology. (Accessed 15 August 2024).

Arenas-Gallo, C., et al., 2024. Race-based screening under the public health ethics microscope—the case of prostate cancer. N. Engl. J. Med. 391 (5), 468–474. doi:10.1056/NEJMms2402322.

Barbujani, G., et al., 1997. An apportionment of human DNA diversity. Proc. Natl. Acad. Sci. U. S. A. 94 (9), 4516–4519. doi:10.1073/pnas.94.9.4516.

Bendyshe, T. (1865a). The Anthropological Treatises of Blumenbach and Hunter. Anthropological Society of London, London.

Bendyshe, T. (1865b). The History of Anthropology. Memoirs of the Anthropological Society of London. Anthropological Society of London, London.

Benjamin, R., 2022. Viral justice: How we grow the world we want. Princeton University Press.

Benjamin, R. (2023). Viral Justice: How We Grow the World We Want. Princeton University Press, Princeton.

Bernier, F., 2001. A new division of the Earth. Hist. Workshop J. 51 (1), 247–250. doi:10.1093/hwj/2001.51.247.

Bhakta, N.R., et al., 2023. European Respiratory Society/American Thoracic Society technical statement: standardisation of the measurement of lung volumes, 2023 update. Eur. Respir. J. 62 (4), 2201519. doi:10.1183/13993003.01519-2022.

Bibbins-Domingo, K., Helman, A. (Eds.), 2022. 'Why diverse representation in clinical research matters and the current state of representation within the clinical research ecosystem'. Improving Representation in Clinical Trials and Research: Building Research Equity for Women and Underrepresented Groups. National Academies Press, Washington, DC. pp. 23–76. doi:10.17226/26479.

Brandt, A.M., 1978. 'Racism and research: the case of the Tuskegee Syphilis Study', Hastings Cent Rep. 8 (6). pp. 21–29. PMID: 721302.

Braun, L., 2014. Breathing Race into the Machine: The Surprising Career of the Spirometer from Plantation to Genetics. University of Minnesota Press, Minneapolis, MN.

Buffon, C.D., 1954. Oeuvres Philosophiques de Buffon (Edited by J. Piveteau), Presses Universitaires de France.

Cagan, R., 2013. San Francisco declaration on research assessment. Dis. Model. Mech. doi:10.1242/dmm.012955.

Calliste, A.M., Dei, G.J.S, Aguiar, M., (Eds.), 2000. Anti-racist feminism: Critical race and gender studies. Fernwood Publishing, Halifax, NS.

Cerdeña, J.P., Tsai, J., Grubbs, V., 2021. APOL1, Black race, and kidney disease: turning attention to structural racism. Am. J. Kidney Dis. 77 (6), 857–860. doi:10.1053/j.ajkd.2020.11.029.

Classification, n.d. Race deconstructed: science and the making of difference. Available at: https://exhibits.lib.unc.edu/exhibits/show/race-deconstructed/classification. (Accessed 28 August 2025).

Collins, F.S., Mansoura, M.K., 2001. The human genome project. Revealing the shared inheritance of all humankind. Cancer 91 (1 Suppl), 221–225. doi:10.1002/1097-0142(20010101)91:1+<221::aid-cncr8>3.3.co;2-0.

Collyer, F.M., 2018. Global patterns in the publishing of academic knowledge: Global North, Global South. Curr. Sociol. 66 (1), 56–73. doi:10.1177/0011392116680020.

Darwin, C., 1882. The Descent of Man.

Darwin, C.R., 1871. The Descent of Man, and Selection in Relation to Sex. second ed. John Murray, London.

Delgado, C., et al., 2021. A unifying approach for GFR estimation: recommendations of the NKF-ASN task force on reassessing the inclusion of race in diagnosing kidney disease. J. Am. Soc. Nephrol. 32 (12), 2994–3015. doi:10.1681/ASN.2021070988.

Downs, J., 2024. Graphic mpox images to educate the public are deeply problematic. STAT. Available at: https://www.statnews.com/2024/08/19/mpox-photos-mugshots-africa-stigma/. (Accessed 5 June 2026).

Durst, D.L., 2015. *John Harvey Kellogg (1852-1943) and the Magazine 'Good Health': medical and Religious Rhetoric in the Popular Promotion of Eugenics, Nomocracy in Politics*. Available at: https://thenomocracyproject.wordpress.com/2015/02/12/john-harvey-kellogg-1852-1943-and-the-magazine-good-health-medical-and-religious-rhetoric-in-the-popular-promotion-of-eugenics-by-dennis-l-durst/. (Accessed 28 August 2025).

Dyck, E., 2014. History of eugenics revisited. Can. Bull. Med. Hist. 31 (1), 7–16.

Feagin, J., 2006. Systemic Racism. Routledge, New York. doi:10.4324/9781315880938.

Feagin, J., Bennefield, Z., 2014. Systemic racism and U.S. health care. Soc. Sci. Med. 103, 7–14. doi:10.1016/j.socscimed.2013.09.006.

Feagin, J.R., 2023. The White Racial Frame. third ed. Routledge, New York. doi:10.4324/9780429353246.

Galton, F., 1869. Hereditary Genius: An Inquiry into Its Laws and Consequences. Macmillan and Co, London. doi:10.1037/13474-000.

Garrod, J.Z., 2006. A brave old world: an analysis of scientific racism and BiDil. McGill J. Med. 9 (1), 54–60.

Gomez, C.J., Herman, A.C., Parigi, P., 2022. Leading countries in global science increasingly receive more citations than other countries doing similar research. Nat. Hum. Behav. 6 (7), 919–929. doi:10.1038/s41562-022-01351-5.

González, Y., Sánchez-Lasaballett, E., 2023. Storytelling, precarity and decolonizing practices. Globalizations 1–15. doi:10.1080/14747731.2023.2248867.

Gould, P., 1996. Space, time and the human being. Int. Soc. Sci. J. 48 (150), 449–460. doi:10.1111/j.1468-2451.1996.tb00099.x.

Government of Canada, n.d. Indigenous Health Equity Fund. Available at: https://www.sac-isc.gc.ca/eng/1721758041536/1721758068269. (Accessed: 28 August 2025).

Harpham, G.G., n.d.a. François Bernier, Theories of Race. Available at: https://www.theoriesofrace.com/1. (Accessed 28 August 2025).

Harpham, G.G., n.d.b. Georges Cuvier, Theories of Race. Available at: https://www.theoriesofrace.com/24. Accessed: 28 August 2025.

Harpham, G.G., n.d.c. Voltaire (François-Marie Arouet), Theories of Race. Available at: https://www.theoriesofrace.com/2/. (Accessed 28 August 2025).

Hoffman, K.M., et al., 2016. Racial bias in pain assessment and treatment recommendations, and false beliefs about biological differences between blacks and whites. Proc. Natl. Acad. Sci. U. S. A. 113 (16), 4296–4301.

Hume, D., 1985. Essays, Moral, Political, and Literary. In: Miller, E. (Ed.), Liberty Fund Inc. https://oll-resources.s3.us-east-2.amazonaws.com/oll3/store/titles/704/0059_Bk.pdf. (Accessed 5 June 2026).

Jackson Jr, J.P., Weidman, N.M., 2005. The origins of scientific racism. J. Blacks High. Educ. 50, 66–79.

James, P.A., et al., 2014. 2014 evidence-based guideline for the management of high blood pressure in adults. JAMA 311 (5), 507. doi:10.1001/jama.2013.284427.

Jiménez-Trejo, F., et al., 2025. Scientific racism as a public health emergency of global concern. Lancet Reg. Health Am. 44, 101049. doi:10.1016/j.lana.2025.101049.

Jorde, L.B., et al., 2000. The distribution of human genetic diversity: a comparison of mitochondrial, autosomal, and Y-chromosome data. Am. J. Hum. Genet. 66 (3), 979–988. doi:10.1086/302825.

Khan, M.S., et al., 2024. The Lancet and colonialism: past, present, and future. Lancet 403 (10433), 1304–1308. doi:10.1016/S0140-6736(24)00102-8.

Khan, S.S., et al., 2023. Novel prediction equations for absolute risk assessment of total cardiovascular Disease incorporating cardiovascular-kidney-metabolic health: a scientific statement from the American Heart Association. Circulation 148 (24), 1982–2004. doi:10.1161/CIR.0000000000001191.

Khazanchi, R., Soled, D.R., Yearby, R., 2023. Racism-conscious praxis: a framework to materialize anti-oppression in medicine, public health, and health policy. Am J Bioeth. 23 (4), 31–34. doi:10.1080/15265161.2023.2186521.

Khazanchi, R., Soled, D.R., Yearby, R., 2023. Racism-conscious praxis: a framework to materialize anti-oppression in medicine, public health, and health policy. Am. J. Bioeth. 23 (4), 31–34. doi:10.1080/15265161.2023.2186521.

Kidane Gebremeskel, A.S., et al., 2025. The use of race and ethnicity in sickle cell disease research. BMC Med. Res. Methodol. 25 (1), 63. doi:10.1186/s12874-025-02513-5.

Kuzawa, C.W., Sweet, E., 2009. Epigenetics and the embodiment of race: developmental origins of US racial disparities in cardiovascular health. Am. J. Hum. Biol. 21 (1), 2–15. doi:10.1002/ajhb.20822.

Kuzawa, C.W., Thayer, Z.M., 2011. Timescales of human adaptation: the role of epigenetic processes. Epigenomics 3 (2), 221–234. doi:10.2217/epi.11.11.

Kwok, P.-Y., et al., 1996. Increasing the information content of STS-based genome maps: identifying polymorphisms in mapped STSs. Genomics 31 (1), 123–126. doi:10.1006/geno.1996.0019.

Lewontin, R.C., 1972. The Apportionment of Human Diversity. In: Dobzhansky, T., Hecht, M.K., Steere, W.C. (Eds), Evolutionary Biology. Springer, New York, NY. doi:10.1007/978-1-4684-9063-3_14.

Linnaeus, C., 1758. Systema Naturae per Regna Tria Naturae, secundum Classes, Ordines, Genera, Species, cum Characteribus, Differentiis, Synonymis, Locis. Systema Naturae, tenth ed. Laurentii Salvii, Holmiae.

Lombardo, P.A., 2024. "Ridding the race of his defective blood"—eugenics in the *Journal*, 1906–1948. N. Engl. J. Med. 390 (10), 869–873. doi:10.1056/NEJMp2307346.

Lujan, H.L., DiCarlo, S.E., 2021. The racist "one drop rule" influencing science: it is time to stop teaching "race corrections" in medicine. Adv. Physiol. Educ. 45 (3), 644–650. doi:10.1152/advan.00063.2021.

MacGregor-Fors, I., et al., 2020. Urban bird ecologists cite more publications from the global North; why?. J. Urban Ecol. 6 (1). doi:10.1093/jue/juaa027.

Markel, H. 2017. The Kelloggs: The Battling Brothers of Battle Creek, second ed. Michigan Historical Review.

Mitchell, P.W., 2018. A new take on the 19th-century skull collection of Samuel Morton, Penn Today. 4 October. Available at: https://penntoday.upenn.edu/news/new-take-on-infamous-Morton-skulls. (Accessed: 5 June 2026).

Mzileni, P., Heleta, S., 2023. Tackling the Global North's bibliometric coloniality, University World News. Available at: https://www.universityworldnews.com/post.php?story=2023030708041595. (Accessed 22 August 2024).

Nale, J., 2014. Arthur de Gobineau on blood and race. Crit. Philos. Race 2 (1), 106–124. doi:10.5325/critphilrace.2.1.0106.

New Scientist, 2018. We must try harder to avoid biological bias in clinical trials. Available at: https://www.newscientist.com/article/mg23831843-900-we-must-try-harder-to-avoid-biological-bias-in-clinical-trials/. (Accessed 22 August 2024).

Ofri, D., 2018. Your medical chart might be biased: here's what doctors should do about it, Slate. Available at: https://slate.com/technology/2018/05/medical-charts-are-biased-but-we-can-fix-that.html. (Accessed 22 August 2024).

Omi, M., Winant, H., 2014. Racial Formation in the United States. Routledge. doi:10.4324/9780203076804.

Pai, M., Abimbola, S., 2024. Science should save all, not just some. Science 385 (6709), 581–581. doi:10.1126/science.ads2151.

Paul, D.B., Spencer, H.G., 1995. The hidden science of eugenics. Nature 374 (6520), 302–304. doi:10.1038/374302a0.

Pearson, K., 1905. National Life from the Standpoint of Science. A. and C. Black. Available at: Biodiversity Heritage Library. (Accessed 5 June2026).

Posel, D., 2001. Race as common sense: racial classification in 20th-century South Africa. Afr. Stud. Rev. 44 (2), 87. doi:10.2307/525576.

Quijano, A., 2007. Coloniality and modernity/rationality. Cult. Stud. 21 (2–3), 168–178. doi:10.1080/09502380601164353.

Rabay, C.J., et al. 2024. Clinicians' perspectives on race-specific guidelines for hypertensive treatment. Soc. Sci. Med. 351 116938. doi:10.1016/j.socscimed.2024.116938.

Ras, I.F., 2017. Broken bodies and present ghosts: Ubuntu and African women's theology. HTS Teol. Stud. 73 (3), 1–7. doi:10.4102/hts.v73i3.4651.

Reilly, P.R., 2015. Eugenics and involuntary sterilization: 1907–2015. Annu. Rev. Genomics Hum. Genet. 16 (1), 351–368.

Renschler, E.S., Monge, J., 2008. The Samuel George Morton cranial collection—historical significance and new research. Expedition 50, 30–38.

Roth, W.D., 2016. The multiple dimensions of race. Ethn. Racial Stud. 39 (8), 1310–1338. doi:10.1080/01419870.2016.1140793.

Savage, L.C., Panofsky, A., 2023. The self-fulfilling process of clinical race correction: the case of Eighth Joint National Committee Recommendations. Health Equity 7 (1), 793–802. doi:10.1089/heq.2023.0064.

Schambra, W.A., 2021. Carnegie's Midnight Confession, Hudson. Available at: https://www.hudson.org/human-rights/carnegie-s-midnight-confession. (Accessed 28 August 2025).

Schluger, N.W., Dozor, A.J., Jung, Y.E.G., 2022. Rethinking the race adjustment in pulmonary function testing. Ann. Am. Thorac. Soc. 19 (3), 353–356. doi:10.1513/AnnalsATS.202107-890PS.

Schwartz, J.S., 2009. How Charles Darwin's early years led him to revolutionize biological thought. Evol. Educ. Outreach 2 (3), 569–572. doi:10.1007/s12052-009-0150-3.

Scientific American, 2018. Clinical trials have far too little racial and ethnic diversity. Available at: https://www.scientificamerican.com/article/clinical-trials-have-far-too-little-racial-and-ethnic-diversity/. (Accessed 22 August 2024).

Scientific Racism, n.d. Harvard Library. Available at: https://library.harvard.edu/confronting-anti-black-racism/scientific-racism. (Accessed 8 August 2024).

Serre, D., Pääbo, S., 2004. Evidence for gradients of human genetic diversity within and among continents. Genome Res. 14 (9), 1679–1685. doi:10.1101/gr.2529604.

Skibba, R., 2019. The Disturbing Resilience of Scientific Racism,Smithsonian Magazine. Available at: https://www.smithsonianmag.com/science-nature/disturbing-resilience-scientific-racism-180972243/. (Accessed 6 August 2024).

Spencer, H., 1884. The Man versus the State: Containing “The New Toryism,” “The Coming Slavery,” “The Sins of Legislators,” and “The Great Political Superstition.” Reprinted from The Contemporary Review. With a Postscript. William and Norgate, London.

Stancil, K., 2023. Hoarding lifesaving knowledge while millions die: the political economy of global COVID-19 vaccine apartheid. Master of Arts (MA) thesis. University of Kentucky. Available at: https://uknowledge.uky.edu/geography_etds/97/. (Accessed: 5 June 2026).

The Linnean Society of London, n.d. *Linnaeus and race*. Available at: https://www.linnean.org/learning/who-was-linnaeus/linnaeus-and-race. (Accessed: 5 June 2026).

Tucker, W.H., 2007. The ideology of racism: misusing science to justify racial discrimination. UN Chronicle 44 (3), 18–19. Available at: https://www.un.org/en/chronicle/article/ideology-racism-misusing-science-justify-racial-discrimination. (Accessed: 5 June 2026).

Vyas, D.A., Eisenstein, L.G., Jones, D.S., 2020. Hidden in plain sight—reconsidering the use of race correction in clinical algorithms. N. Engl. J. Med. 383 (9), 874–882. doi:10.1056/NEJMms2004740.

Willoughby, C.D.E., 2018. Running away from drapetomania: Samuel A. Cartwright, medicine, and race in the Antebellum South. J. South. Hist. 84 (3), 579–614.

Wu, M., et al., 2024. Racial inequities and rare CFTR variants: impact on cystic fibrosis diagnosis and treatment. J. Clin. Transl. Endocrinol. 36, 100344. doi:10.1016/j.jcte.2024.100344.

CHAPTER

Race, racism, and the US production and exportation of medical science

3

Jaya Aysola and Michelle Munyikwa
Perelman School of Medicine, University of Pennsylvania, Philadelphia, PA, United States

Introduction

Despite long-standing evidence and repeated calls to action, we still have not achieved racial health equity in the United States (US). This lack of progress is, in part, due to the inability to fully embrace and address structural and institutional racism as the root cause of such inequities. Even though foundational evidence on the role of racism in health care and health was introduced nearly 100 years ago by pioneers like Dr. James McCune Smith and subsequently many others (Jones, 2000; Fullilove, 1998; La Veist, 1996; Williams, 1999; Jackson et al., 1996; Mckenzie, 2003), interventions addressing racism remain elusive. Thus we have yet to produce upstream interventions at the optimal scale to mitigate health and health care disparities.

Advancing health equity requires critical investments to evaluate and reengineer our systems to dismantle structural racism and discrimination. We posit that the science we produce, impart, and apply must be reimagined with a multicultural lens that accurately attributes the root causes of race-based inequities in health care and health. In this chapter, we will make this argument with illustrative examples, followed by strategic recommendations to guide actions to achieve health equity. Our discussion will center on the US, given that it has been the largest funder and exporter of medical science (Boroush, 2020). However, the role of structural racism in the production, dissemination, and application of medical science is not isolated to the US. Medical science served colonialism and European empires' brutal oppression of Indigenous populations and extraction of their resources by constructing the concept of race, arbitrary hierarchies, and "white" superiority (Amster, 2022; Anderson et al., 2023). Whether in the US, the Global North, or the Global South, health inequities stem from the legacy of settler colonialism and its role in the invention of race and the notion of "white" superiority (Ditomaso, 2024).

Decolonizing Medical Education, Research and Practice. https://doi.org/10.1016/B978-0-443-44998-7.00014-4

Defining the challenge

The literature and discourse on health equity often conflate or misrepresent common terms and concepts; however, a shared language and understanding of the challenge is essential to developing solutions. We therefore begin with some definitions. Health care disparities are differences in care after accounting for equal access, patient preferences, and clinical need. Such disparities result from provider biases (unconscious and conscious interpersonal racism and/or discrimination) coupled with biases intrinsic to health care institutions (institutional racism) and the medical field at large (structural racism) (Institute of Medicine Committee et al., 2003; Payne and Hannay, 2021; Maina et al., 2018; Fitzgerald and Hurst, 2017).

Health disparities are differences in health outcomes and the burden of disease that stem from the inequitable distribution of "social determinants of health" (SDoH) or upstream social and structural factors that influence health (Gee, 2008; Mendez et al., 2011; Williams and Collins, 2001; Alexis et al., 2020). Mounting evidence reveals that racism at multiple levels (interpersonal, institutional, and structural) acts in concert across multiple social systems to create inequities in health (Williams and Jackson, 2005; Williams et al., 2019a, 2019b; Williams and Mohammed, 2013a, 2013b; Bailey et al., 2017; Gee and Ford, 2011). The structural racism intrinsic to US laws, policies, and structures embedded within our social systems, coupled with and justified by interpersonal racism and discrimination, together serve as the antecedent to differences in SDoH (Yearby, 2020; Zambrana and Williams, 2022; Yearby et al., 2022). The role of the US in upholding social structures forged through imperialism has ensured that these SDoH are a global phenomenon, albeit with different features in local contexts. Hence, the legacies of colonialism are visible throughout interconnected countries in both the Global North and South (Paradies, 2016; Wispelwey et al., 2023; Czyzewski, 2011). These legacies particularly influence not only the clinical capabilities available in differently resourced countries but also the mechanisms for knowledge production and how interventions, especially in the Global South, are structured along lines of difference (Petryna, 2005, 2007).

We will here do an in-depth focus on one social system governing equitable health care access and quality, and detail the role of structural racism and discrimination (SRD) in terms of the production, sharing, and application of medical knowledge that governs contemporary health care systems.

SRD and the knowledge we produce

Faulty underpinnings of race in medical science versus the realities of human variation

The manufactured division of humans into racial categories is consistently used in studies that inform medical knowledge without adequate scrutiny. The discrete and arbitrary racial classifications poorly reflect human variation, as humans share 99.9% of their DNA, with more genetic and phenotypic variation within the same racial and ethnic populations than between them. Biologic variations stem from genome and environment interactions, local and regional biologic changes over time, and genetic exchanges among populations (Fuentes et al., 2019). Both genomic and phenotypic variation occur in clines or gradations of change across geographic regions. As Frank Livingstone stated, "there are no races, only clines" (Livingstone and Dobzhansky, 1962). Growing genetic evidence continues to underscore clinal variation in change and refute the notions of distinct human races (Fujimura et al., 2014).

Lack of representation among scientists and study populations

The health services and medical research thus established are informed by the cultural perspectives of the scientists and the populations they study. The dominant historical perspective of scientists has been, and continues to be, represented by a non-Hispanic white perspective. This singular perspective impacts the research questions and interventions posed, the hypotheses generated, and the interpretation of findings. This biased perspective is often defined as neutral and objective, even while it represents a culturally and socially specific point of view. It also shapes what is considered evidence and whose experiences matter to the broader project of scientific knowledge production. The result is both biased and myopic science, which is only a component of the problem. Medical knowledge produced in the US with this narrow perspective has historically served to pathologize, stigmatize, and/or intentionally injure historically marginalized populations (Washington, 2006; Byrd et al., 2002; Davis and Reid, 1999; Lawrence, 2000; Gutiérrez and Fuentes, 2009). Moreover, colonialism and the creation of white hegemony have historically promulgated science that serves to dehumanize and injure oppressed populations (Amster, 2022; Tilley, 2016).

A distinct challenge still looms: the lack of diverse representation in study populations is far-reaching, spanning observational cohort studies, clinical trials, and research in health services. One illustrative example is the genesis of co-pays (out-of-pocket co-payment made by patients for medical services). From 1971 to 1986, a federally funded experiment evaluated the impact of cost on health care utilization and subsequent health outcomes. It found that individuals required to pay for services, as compared to those whose costs were fully covered, spent 30% less on health care with no significant differences in health outcomes. However, this study population lacked diverse representation. Subsequent studies with more diverse populations revealed that co-pays among BIPOC (Black, Indigenous, and People of Color) populations curtailed essential health care use and impacted health outcomes (Choudhry et al., 2014; Remler and Greene, 2009b). Of note, we employ racial terms here not to support the notion of distinct physiological differences or the existence of different "races," but rather to better understand and expose the ongoing effects of structural racism. These differential effects of co-pays by race and ethnicity persisted even after accounting for income but may be explained by the persistent racial wealth gap of at least 30% between BIPOC and non-Hispanic white individuals in the US (Chandra and Skinner, 2003; Darity et al., 2018). Co-pays, therefore, represent an example of structural racism that creates differential uptake of needed services and adherence to prescription medications by race and ethnicity (Franklin, 2014; Remler and Greene, 2009a).

Studies that lack representativeness continue to introduce structural biases and shape the way we deliver health care (Chowdhury et al., 2019). This also includes genomic research, where socially constructed categories of race and ethnicity are poor proxies for ancestry. The biological differences between racial groups are due to racism, not differences in genetic predisposition (Fuentes et al., 2019). However, representation of diverse populations of varied ancestries remains important, especially to combat diagnostic biases stemming from stereotypical patterns of disease attributed to race, such as cystic fibrosis or sickle cell disease (Amutah et al., 2021). Research on genomic databases has consistently shown minimal representation of BIPOC populations and the ancestries they may represent across all disease areas. In contrast, European populations represent over 75% of all genomic research while comprising merely 15% of the global population (Landry et al., 2018). This is even more concerning considering that the greatest genetic diversity manifests in African and African diasporic populations, and that the genetic variants found outside Africa are largely a subset of those found within it (Fuentes et al., 2019; Sirugo et al., 2019). In other words, genetic diversity decreases in populations located further from Africa, suggesting that medical science would be far better served if genomic studies centered on studying populations in Africa. Despite this, the opposite holds true, where genomic research—akin to medical research at large—is centered on European populations.

This may occur due to various reasons, including the lack of intention—deliberate or unaware—among genomic researchers. Researchers are not immune to the ingrained and permeating nature of white centricity or the centering of the white race throughout all societal systems, including knowledge production by the medical sciences. This can also, in part, be attributed to the lack of trust that minoritized communities display regarding the intent of genomic and other scientific research, ownership of their data, and the lack of benefits accruing as a result of such scientific endeavors (Benjamin, 2016, 2014; Reardon and Tallbear, 2012). This represents some of the outcomes of structural and institutional racism.

Disparities in funding diverse scientists, community-based study questions, and global health

SRD exerts influence over the processes and incentives intrinsic to the institutions that fund the questions posed, as well as the scientists provided with an opportunity to research those questions. Despite evidence that diversity promotes innovation and enhances scientific inquiry (Page, 2019; Freeman and Huang, 2014; Alshebli et al., 2018; Adams, 2013; Powell, 2018), there remains a lack of racial and ethnic parity in National Institutes of Health (NIH) funding for research, with evidence that BIPOC researchers often propose community and population-level studies that are not prioritized by the NIH (Ginther et al., 2011; Corbyn, 2011; Hoppe et al., 2019).

To address structural racism as the root of this disparity, the NIH launched the UNITE initiative with a commitment to changing the extramural ecosystem such that diversity, equity, and inclusion would be reflected in funded research and the biomedical workforce (Collins et al., 2021). UNITE introduced dedicated funding for new scientific questions focused on racism. Often, funding for new areas of research goes to career researchers who shift from topic to topic in response to funding interests, and such "health equity tourism" can have a detrimental impact on the science we produce to result in meaningful changes (Lett et al., 2022). Historically, NIH funding also exhibited preferences for either niche, nonpragmatic randomized controlled trials or data analyses that served to describe health care and health disparities nationally without meaningful hypotheses of the underlying root causes (De Meulemeester et al., 2018; Fox et al., 2022). In addition, other sources of funding have competing interests. For example, pharmaceutical-funded trials with misaligned incentives have prompted the examination of race-based differences in outcomes without a sufficient hypothesis-driven rationale. An example of this is the introduction of BiDil as a heart failure medication exclusively for Black patients (Brody and Hunt, 2006; Pollock, 2012).

Challenges with research priorities and "Colonial Science"

While the US remains a dominant force in medical science production, its global leadership role has declined and has taken further hits due to recent changes in federal funding for research (Tollefson and Van Noorden, 2024). This provides an opportunity to reimagine how to conduct medical research, what areas to focus on, and whom to include as equal partners in the process of producing medical knowledge. For example, the 90–10 rule remains a challenge (Research, 2001; Kilama, 2009) where only 10% of global health research is devoted to conditions accounting for 90% of the global disease burden, while 90% of research funding is focused on the 10% of diseases centered in the US and other countries in the Global North (Research, 2001). In addition to this extreme misalignment between research priorities and the global burden of disease, countries in the Global North often engage in "helicopter research," both within their countries and in other countries in the Global South (Haelewaters et al., 2021; Lambert et al., 2024; Adame, 2021). Colonial science can also be referred to as "helicopter research," "parachute research," or "parasite research," and describes a practice where researchers from wealthier countries conduct studies in other countries (largely the Global South) with little involvement of local researchers and/or community members, and then simply leave once they have extracted the necessary information they targeted (Odeny and Bosurgi, 2022; Schutte and Abdool Karim, 2023; Mcintosh et al., 2023). This practice also occurs domestically in the US, where well-resourced elite institutions conduct studies in certain populations or less-resourced institutions without engaging in nonhierarchical partnerships with such communities and institutions (Lambert et al., 2024). The result of this lack of representation and equal partnership is poor science, as such research may perpetuate disparities and cause potential harm to the very populations that the researchers are aiming to help. For example, helicopter research may result in irrelevant research questions and poor research design, while extracting resources and stifling the capacity of local institutions and populations (Else, 2025). In response, some global health journals have instituted policies to uphold ethical standards when conducting international research, calling for ethical partnerships and equal authorships (Mcintosh et al., 2023; Rees et al., 2022).

Poor conceptualization of race and racism

The conceptualization of race and racism matters to the science we generate and how it is interpreted within the health care setting. However, the knowledge created within medicine, health services research, and public health has not routinely applied the Public Health Critical Race Praxis (PHCRP) (Ford and Airhihenbuwa, 2018). This framework details four key principles that should be considered in our conceptualizations of race and racism: (1) structural racism evolves over time and context; (2) a field's accumulative understandings, including its prior and current racial biases, are contained within its evidence base and impact knowledge production; (3) race is a social construct, and socio-political factors give meaning to differences in phenotypic characteristics often used to assign race; and (4) research must be assessed for unintended racial bias in its methods and interpretation of results through community engagement and involvement of diverse perspectives (Ford and Airhihenbuwa, 2018).

The conflation of race with socioeconomic or place-based indicators, or the consideration of race as a biomedical rather than a social construct, results in studies that misinterpret race-based differences in outcomes. It also perpetuates stereotypes that health disparities are due to genetics, socioeconomic status, and/or patient behaviors, rather than structural racism. Disinvested neighborhoods, voter disenfranchisement, wealth gaps, and health care disparities can be considered distinct downstream consequences of structural racism (Gee and Ford, 2011; Yearby, 2020). Considering this common upstream cause, significant associations between race and place-based conditions may exist, but it should not be inferred that disinvested neighborhoods or socioeconomic status alone cause racial and ethnic disparities in health. It is critical to unpack why place-based differences by race exist in the US, as an illustrative example of how historical and sociopolitical forces that enact racism can shape population health, rather than differences innate to place or race. In the US, deliberate policies that created racial residential segregation are a fundamental cause of racial disparities in health (Williams and Collins, 2001; Rothstein, 2017). Historical policies such as redlining (practice of denying financial support, for e.g. loans, to individuals residing in neighborhoods characterized by specific racial categories) and contemporary discriminatory practices, such as mortgage application denials, continue to segregate neighborhoods by race as a starting point (Gee, 2008; Mendez et al., 2011; Beyer et al., 2016) (Cherian, 2014; Wheeler and Olson, 2015). Thereafter, racial segregation by neighborhoods serves as a foundation for other policies that fundamentally influence the daily lives of populations, with significant material consequences in terms of their health. For example, anchored to where one lives, there are clear differences in educational and economic

FIG. 3.1

Examples of how levels of racism across multiple social systems collectively impact health in the US.

opportunities (Wang et al., 2022; Orfield and Lee, 2005; Vivian, 2017; Massey et al., 1987; Galster and Keeney, 1988); differential exposures to environmental contaminants and toxins (Roesler, 2021; Shkembi et al., 2024); and differences in policing (Kurwa, 2022; Siegel et al., 2019; Bass, 2001). Fig. 3.1 provides an illustrative example of how this occurs in the US, with levels of racism intersecting across multiple social systems, beginning with where one lives. Similar forces play out on the world stage where discussions of significant inequities in health among certain nations—despite ample evidence—are rarely attributed to colonialism and imperialism.

SRD and the knowledge we impart

Imparting biased knowledge in our learning environments is a natural consequence of structural racism embedded in the knowledge we produce. We risk perpetuating the cycle of structural racism in knowledge production if we impart a flawed understanding of race and racism to future researchers and medical practitioners.

Prior work highlights the potential role of undergraduate medical education in perpetuating physician biases through misrepresenting race/ethnicity (Nieblas-Bedolla et al., 2020; Tsai et al., 2016; Amutah et al., 2021; Aysola et al., 2021; Park et al., 2022). Findings from a study we conducted analyzing preclinical curricula in US medical schools revealed five key misrepresentations of race when imparting medical knowledge. The first is the use of imprecise terms, such as "African American" to describe anyone with African ancestry, disregarding differences between first- and second-generation immigrants from the Caribbean or Africa and those whose ancestors were enslaved in the US, and/or antiquated terms such as "Caucasian."

This imprecision in defining racial categories is especially problematic when carried forward to describe differences in disease prevalence and is the second most common way race is misrepresented when imparting medical knowledge. Race-based differences in disease burden or outcomes are often presented without any context and hence prime learners to attribute such differences to genetic predisposition and/

or consequences of personal behaviors, rather than racism. An illustrative example of this is how medical classrooms in the US commonly discuss the disproportionate burden of Type 2 diabetes among the US Akimel O'odham (also known as Pima) people, without sufficient explanations of historical and social causes. Despite high degrees of genetic similarity, the Akimel O'odham living in Mexico exhibit significantly lower rates of Type 2 diabetes and obesity than those living in the US (Schulz et al., 2006). Here, a historical insult, rather than genetic predisposition explains this pattern (Schulz et al., 2006; Baier and Hanson, 2004). Historically, many members of the US Akimel O'odham communities were master water engineers, and the tribe lived off the Gila River with only one documented case of diabetes (Knowler et al., 1990; Schulz and Chaudhari, 2015). However, due to the expansion of Euro-American settlements their livelihood was threatened by the diversion of the Gila and Salt Rivers and the construction of the Gila and Roosevelt Dams (Lewis and Hestand, 2006). The US military subsequently supplied them with calorie-dense, nutrient-poor surplus foods such as white flour, cheese, refined sugar, lard, and canned food. This program did not offer fresh produce until 1996 (Smith-Morris, 2004). The Akimel O'odham have since sought increased access to and protection of their water sources. Their efforts led to the Arizona Water Settlement Act of 2004 and ongoing local actions for water rights (Lewis and Hestand, 2006). As they continue to advocate for better living conditions, the impact of these historical events on the Pima populations in the US is undeniable. It is therefore essential for learners to understand such context when discussing disease prevalence. However, when imparting medical knowledge, disease prevalence discussions rarely include sociopolitical insults, colonialism, and migration patterns.

Race-based epidemiologic data presented without explanation perpetuates race-based diagnostic bias or the belief that race itself puts patients at risk for diseases. This is the third pattern of misrepresenting race that we found when examining medical curricula. Rather than presenting race as being correlated with social factors that shape disease or acknowledging race as an imperfect proxy for ancestry or family history that may predispose one to disease, the educators we observed portrayed race itself as an essential—biologic—causal mechanism. Lecturers in the US frequently connected diseases to specific racial groups. For example, medical students are primed to view sickle cell disease as affecting only Black people, rather than as common in populations at risk for malaria across the globe (Amutah et al., 2021; Aysola et al., 2021; Kato et al., 2018). Another troubling example is when physicians are primed to perceive cystic fibrosis as a disease of white people, which leads to overlooking this diagnosis in a Black patient (Garcia, 2004; Rubin, 2021). Delayed diagnoses of Black and other persons of color with cystic fibrosis continue to occur, despite newborn screening tests, because minoritized geographic ancestries often carry cystic fibrosis allele variants excluded from newborn screening panels employed in the US (Wu et al., 2024). This exclusion stems, once again, from the lack of representation in genomic research and the focus on capturing genetic diversity among those of European descent, which is merely a subset of the genetic diversity among those who live in, or descend from, Africa, which holds the greatest genetic diversity (Sirugo et al., 2019).

Another consequence borne out of discussing disease prevalence without sufficient context is pathologizing race. Race, from its conception, has served to pathologize certain groups and perpetuate the myth of white superiority. Pathologizing race is the fourth pattern of misrepresentation that continues in medical school classrooms today. Pathologizing race by highlighting increased disease burdens exclusively in historically marginalized groups without context, precision, or accuracy is a common challenge. We see this with presentations of differences between races in life expectancy in the US and differences between nations in mortality rates across the globe (Amutah et al., 2021; Aysola et al., 2021). Illustratively, the differential mortality in the early COVID-19 pandemic among BIPOC communities in the US was initially attributed to biological differences, obscuring the structural reasons—such as differential participation in "essential worker" industries—that drove the pandemic, in addition to unequal access to health care (Mcclure et al., 2020).

The fifth pattern our study highlighted was teaching race-based clinical guidelines without information about their origin and evidentiary basis. Race-based clinical guidelines are a natural consequence of the misuse of race in producing and imparting medical science. These guidelines use race as a biologic marker for disease or a proxy for genetic predisposition and perpetuate the notion that race is a biologic category. Research conducted with a flawed understanding of race informs flawed guidelines. Pervasive in medicine, such guidelines have endorsed the use of racial categories in the diagnosis and treatment of common conditions such as renal disease and hypertension, despite their grounding in misguided scientific inquiry and interpretation of data (Vyas et al., 2020; Douglas, 2003; Bloche, 2004). A historical but critical example that we highlight from examining the preclinical curricula is the upward adjustment for persons designated as Black or African American in estimating the glomerular filtration rate (GFR), which raises the threshold for concern for Black patients only (Williams et al., 2021; Hsu et al., 2021; Inker et al., 2021). For example, a patient with one Black parent and one white parent whose creatinine level is 2.8 mg per deciliter would have an estimated GFR (eGFR) of 18 mL per minute per 1.73 m^2 if identified as white and 21 mL per minute per 1.73 m^2 if identified as Black. As a white patient, she would qualify to be added to the waiting list for a kidney transplant, but as a Black patient, she would not—a distinction that magnifies well-established racial and ethnic disparities in renal transplant referrals (Epstein et al., 2000). If the patient identified as both races or mixed race, the clinician would be left to make a binary choice (Amutah et al., 2021; Aysola et al., 2021). The US exported race correction of eGFR, with many countries conducting studies to evaluate and ultimately recommend against its use (Rocha et al., 2020; Umeukeje et al., 2022). Pediatrics never adopted this race correction of eGFR and further challenged the science supporting its use in adulthood. Beginning in 2020, a growing movement emerged to eliminate race-based guidelines, with medical societies interrogating such guidelines within their respective fields, together with federally funded reviews of the evidence (Siddique et al., 2024; Tipton et al., 2023). Despite this, progress has been slow to de-implement the race-based tools and algorithms and unlearn the faulty underpinnings of race in medicine. After interviewing several clinicians across a myriad of specialties, we found that several factors contribute to the slow pace of de-implementing race-based guidelines. Clinicians still hold a misunderstanding of what race is and what it is not. In addition, there is a need for medical societies to endorse the removal of race corrections and provide specific details regarding how such modifications should be made (Aysola et al., 2024).

With the US being the largest exporter of medical science, a crucial challenge is the absence of plurality globally in terms of medical knowledge produced, imparted, and applied. A subsequent study of medical curricula related to Asian American populations builds upon this challenge alongside similar themes of misrepresentation of race more generally. We found the omission of information about Asians altogether within medical curricula, presenting prevalence data about Asians as a singular, aggregated group. This is done despite substantial heterogeneity, a lack of consistency, and sometimes even inaccuracy in the language used to describe Asian American populations and subgroups (Park et al., 2022).

A common argument for the imprecise use of race in the medical curriculum is that board examinations test students on race-based guidelines and racial heuristics (Ripp and Braun, 2017). Such exams should be reformed in due course. However, in the interim, educators should comprehensively discuss the antecedents of disease to allow students to strategically recognize race-based patterns in such exams, without perceiving them as absolute or as stemming from biological differences. There may be relevant physiological differences among humans that correlate with ancestral background; however, these differences do not correlate well enough with the social categories of race and ethnicity to justify their teaching and use in medicine. This lack of correlation, however, does not imply that race should not be used in medicine or medical education. Unequal treatment in health care due to structural and unconscious racism can be measured and eliminated only if we continue to discuss race (Amutah et al., 2021; Aysola et al., 2021).

SRD and the knowledge we apply

The knowledge we produce and impart naturally informs our policies and practices. Many aspects of the health care delivery system stem from studies centered around cultural norms that exclude many populations. The pandemic revealed this with the reimagination of care delivery models (e.g., virtual visits, eliminating cost-sharing for COVID-19 care), resulting in a significant reduction in no-show rates and increased vaccinations (Bressman et al., 2022; Muppavarapu et al., 2022; Adepoju et al., 2022; Lau and Knudsen, 2022; Ritchie and Leff, 2022).

Inclusion and representation

The lack of representation in the health care workforce, researchers, and leadership stems from barriers to inclusivity. Cultural environments that lack inclusivity prevent critical representation at all levels of the organization by hindering equal recruitment, advancement, and retention of diverse trainees and employees (Aysola et al., 2023; Cohen et al., 2002; Lett et al., 2019; Goode and Landefeld, 2018; Aysola et al., 2018; Guevara et al., 2021; Alexis et al., 2020). For example, it is insufficient to diversify the scientific community without addressing the lack of equitable support for the scientific achievements that such diversity produces. An analysis of all US PhDs spanning three decades found that BIPOC scientists innovate at higher rates than their white counterparts, but their novel contributions do not equally translate into higher rates of academic positions and/or promotions and academic career advancement (Hofstra et al., 2020). This diversity paradox, where the novel contributions of BIPOC scientists do not earn them academic positions and/or promotions, is one example of the lack of parity in sponsorship and promotion of BIPOC faculty with novel and high-impact research (Aysola et al., 2018; Hofstra et al., 2020). Without such sponsorship, BIPOC faculty often do not advance to leadership positions that allow for increased influence over processes that impact how medical knowledge is produced, disseminated, and applied.

Legacy of scientific racism

Applying race as a biological construct in clinical practice dates back to the precolonial era, where it was used to justify the enslavement and the oppressively brutal treatment of Black populations, as well as the maltreatment of other populations of color (Washington, 2006; Byrd et al., 2002; Davis and Reid, 1999; Lawrence, 2000; Gutiérrez and Fuentes, 2009). The legacy of this deeply troubling period permeates the field and informs race-based clinical guidelines, which often exacerbate preexisting health care disparities (Vyas et al., 2020; Wright et al., 2022). The legacy of scientific racism includes eugenics and its influence on mainstream genetics, which has only recently been acknowledged (Jackson et al., 2023). Even without using race, algorithms may introduce bias as they reflect the racism intrinsic to our systems. For example, an algorithm developed to enroll patients into a chronic disease management program used health care costs as a proxy for health care needs and disproportionately enrolled more white patients (who were less ill) into the program over Black patients who were more severely ill. This occurs because clinicians typically spend more time on white patients than Black patients with equivalent needs. The authors found that using a measure that better approximates health care needs eliminated the disparity (Obermeyer et al., 2019).

Recommendations

Dismantling structural racism embedded in the medical science we produce, impart, and apply requires several simultaneous efforts to correct its course. Foundational knowledge and a greater understanding of the politics of generating medical science in an inequitable world are key to the training and skill set of those whom we entrust as medical researchers. To produce such a scientific workforce, we need system-level interventions that transform the foundations of the field that novice researchers enter. These skills are also essential for physicians who consume and apply medical knowledge. To impart such skills, we must transform the way we educate and evaluate future physicians. Fig. 3.2 outlines key actions, while Table 3.1 provides supporting recommendations and key resources. Corrective actions fall into three critical overarching principles that will now be discussed.

The first step is ensuring a multicultural, rather than singular cultural, perspective among medical researchers and educators, research funders, and the health system's clinical and administrative workforce at all levels of the organization, including executive leadership positions. This includes diversifying who and what we fund, as well as diversifying executive leadership and the boards that govern health care delivery systems. The pervasive racism within the academic research enterprise cannot be undone by the singular white perspective that introduced it. Key to achieving a diverse leadership body is translating existing evidence and guidelines into strategic actions to track and address inequities in recruitment, retention, compensation, evaluation, and advancement within medical organizations. Investments in the recruitment of a diverse workforce alone are insufficient, as such recruitment, without investments in creating inclusive cultures, will not yield the benefits of diversity that are aimed for.

The second is the appropriate use of race and ethnicity in identifying racism as *the* main underlying cause of health care and health inequities. Medical curricular improvements, changes to standardized exams, improved training for researchers, and optimized health system data collection should be accompanied by addressing barriers due to racism in high-quality, accessible care. In addition, physician trainees,

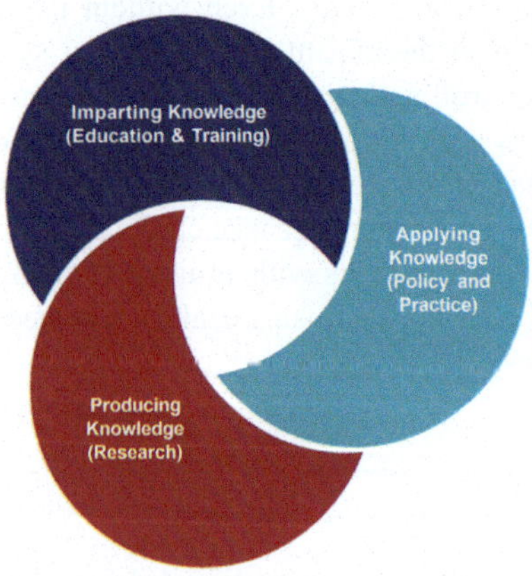

Imparting Knowledge (Education & Training)
- Identify and correct the misuse of race and ethnicity in medical school (UME) and physician training (GME)
- Revisit the use and representation of race/ethnicity in medical board exams
- Recruit, retain, and advance diverse medical educators and trainees

Applying Knowledge (Policy and Practice)
- Implement evidence-based REAL data collection
- Remove race-based correction factors
- Revisit policies that impact equitable health care access
- Revisit workforce policies to ensure inclusivity
- Stratify safety and professionalism reports by employee demographic characteristics
- Diversify healthcare workforce at all levels, up through executive leadership
- Expand and reimagine community investments

Producing Knowledge (Research)
- Hire, retain, promote, and fund diverse research scientists
- Diversify study populations and ensure representativeness
- Engage in non-hierarchical partnerships with communities within and across nations to co-produce science
- Advance racial and ethnic parity in publications, grants, and promotions
- Expand NIH funding for science impacting health equity
- Expand CITI or any form of human subject research training to include:
  - Appropriate use and interpretation of race and ethnicity
  - Role of racism in health
- Incorporate equity into research training infrastructure (e.g. CTSA)

Abbreviations: UME: undergraduate medical education; GME: graduate medical education; REAL or REaL: Race/Ethnicity, Ancestry, and Language, or Race (including Hispanic), Ethnicity (granular), and Language; CITI: Collaborative Institutional Training Initiative; CTSAs: Clinical and Translational Science Awards

FIG. 3.2

Actions to mitigate racism in medical research, education, and policy and practice.

Table 3.1 Recommendations to address structural racism and discrimination (SRD) in medicine.

Recommendation	Suggested corrective actions	Resources/References
Recruit, hire, retain, and advance a diverse, representative workforce within all areas and at all levels of the medical-industrial complex, academic research, and health care enterprise	*Enact antiracist recruitment and hiring practices, such as:* • Diverse representation on recruitment and selection committees. • Holistic review and other inclusive hiring practices • Priming techniques and other antibias interview evaluations, such as scorecards • Monitoring and evaluation of the research training pipeline, with targeted hiring and retention interventions at key transitions	Racial and Ethnic Diversity at Medical Schools—Why Aren't We There Yet? New England Journal of Medicine. 2021 Nov 4;385(19):1732-4. Holistic review – shaping the medical profession one applicant at a time. The New England Journal of Medicine. 2013 Apr 25;368(17):1565. Bridging the gap: holistic review to increase diversity in graduate medical education. Academic Medicine. 2019 Aug 1;94(8):1137-41.
	Achieve pay parity • Conduct compensation audits and remedy any identified pay gaps • Do not ask about the candidate's current or prior salary history (21 states and 21 localities have salary history bans) • Institute transparent compensation frameworks • Institute policies for equitable compensation for global research collaborators and assistants	How to identify–and fix–pay inequality at your company. Harvard Business Review. 2020. Promising Practices for Understanding and Addressing Salary Equity at U.S. Medical Schools. AAMC (Association of American Medical Colleges) Report. 2019. Stronger together: advancing equity in global health research partnerships. BMJ Global Health 2025;10:e019601.

Continued

Table 3.1 Recommendations to address structural racism and discrimination (SRD) in medicine—cont'd

	Revisit workforce policies and create systems of accountability to ensure inclusivity • Create equity in access to opportunities for advancement, sponsorship, and promotion, including for those born outside of the Global North • Combat the myth of meritocracy, which attributes success or failure to individual abilities or merits and ignores disparities in opportunities • Revisit institutional values to ensure alignment with health equity and Public Health Critical Race Praxis (PHCRP) principles • Encourage funders to enact accountability measures for training and capacity-building in research projects • Elevate and support training for research mentoring • Stratify safety and professionalism reports by employee demographic characteristics to identify disparities in reporting • Recognize and stop academic bullying and the weaponization of professionalism standards	Perceptions of factors associated with inclusive work and learning environments in health care organizations: a qualitative narrative analysis. JAMA Network Open. 2018 Aug 3;1(4). Beyond the cultural myth of medical meritocracy. Medical education vol. 54,1 (2020): 46-53. Assessment of perceptions of professionalism among faculty, trainees, staff, and students in a large university-based health system. JAMA Network Open. 2020 Nov 2;3(11). Bullying in Academic Medicine: Experiences of Women Physician Leaders. Academic Medicine, 98(2):255-263, February 2023.

	Advance racial and ethnic parity in publications and grants • Identify and address bias in the review process • Diversify editorial boards and review committees • Reframe how impact is allocated to publications in journals in the US and Western Europe • Adopt policies that support ethnic partnerships and promote authorship between nations	Role of funders in addressing the continued lack of diversity in science and medicine. Nat Med 27, 1859–1861 (2021). Non-White scientists appear on fewer editorial boards, spend more time under review, and receive fewer citations. Proceedings of the National Academy of Sciences. 2023 Mar 28;120(13). Countering helicopter research with equitable partnerships. The Lancet Global Health, Volume 11, Issue 7, e1007–e1008. Challenges and Potential Solutions to Ensure Meaningful Partnerships. Hypertension. 2023 Sept. 2023.
Diversify study populations, ensure representativeness, and involve community partners in all study phases	*Diversify study populations and ensure representativeness* • Efforts for broader representation must be made without conflating race and ethnicity with ancestry, or place-based deprivation, or socioeconomic characteristics *Engage community members as partners in all study phases in nontokenizing ways* • Meaningful engagement involves the community in defining the problem, the hypotheses, potential interventions, and the interpretation of outcomes	Misrepresenting race—the role of medical schools in propagating physician bias. New England Journal of Medicine. 2021 Mar 4; 384(9):872-8. It should not require a pandemic to make community engagement in research leadership essential, not optional. Journal of Clinical and Translational Science 5: e95, 1–7. Why Increased Diversity in Dermatologic Clinical Trials Is Not Inherently Ethical—An Evidence-Based, Community-Engaged Approach to Diverse Trial Recruitment. JAMA Dermatol. 2022 Oct 1;158(10):1219. doi: 10.1001/jamadermatol.2022.3109. PMID: 35976641.

Continued

Table 3.1 Recommendations to address structural racism and discrimination (SRD) in medicine—cont'd

Impart and assess evidence-based knowledge about race and racism	*Impart evidence-based medical knowledge when it comes to race and racism*
	• Use evidence-based standards for discussing race, Hispanic ethnicity, and ancestry (granular ethnicity)
	• Provide context about the potential social and historical causes of differences in health by race and ethnicity, given that both race and ethnicity are poor proxies for ancestry and genetic predisposition
	• Avoid stigmatizing or pathologizing language when discussing health inequities, such as "vulnerable," or language that centers norms on Whiteness
	• Revisit the use and representation of race and ethnicity in medical board exams
	• Incorporate equity training and consultation services into the Clinical and Translational Science Awards (CTSA) Program
	• Expand the Collaborative Institutional Training Initiative (CITI) and human subject research training to include:
	  • Appropriate use and interpretation of race and ethnicity
	  • Role of racism in health
		Race, Ethnicity, and Language Data: Standardization for Health Care Quality Improvement. Institute of Medicine, 2009. Washington, DC: The National Academies Press.
		Misrepresenting race—the role of medical schools in propagating physician bias. New England Journal of Medicine. 2021 Mar 4;384(9):872-8.
		Representation of Asian American Populations in Medical School Curricula. JAMA Network Open. 2022 Sep 1;5(9).
		Advancing Health Equity: Guide to Language, Narrative, and Concepts. AAMC (Association of American Medical Colleges) and AMA (American Medical Association) Report, 2021.

Address root causes of health disparities with significant upstream community investments	*Expand and reimagine community investments*
	• Investments that address the systems maintaining inequities and provide economic opportunities to those historically marginalized due to structural racism
	  ◦ Reparations for Black Americans who are descendants of enslaved persons and/or significant investments
	  ◦ Address insufficient amounts in reparations for Native Americans.
	  ◦ Health care credits for insurance, prescriptions, and mental health services
	  ◦ Housing reparations or credits for home ownership
	  ◦ Credits for higher education
		The Case for Health Reparations. Frontiers in Public Health. 2021 2021-July-08;9.
		Reparations as a public health priority—a strategy for ending black–white health disparities. New England Journal of Medicine. 2020 Nov 26;383(22):2101-3.
		Examining the Role of Family History of US Enslavement in Health Care System Distrust Today. Ethnic Dis. 2021 Jul 15;31(3):417-424.
		Housing discrimination as a basis for Black reparations. Public Affairs Quarterly. 2007;21(3):255-273.
		Higher Education Redress Statutes: A Critical Analysis of States' Reparations in Higher Education. Wash. & Lee L. Rev. 2022; 79:1829.

Eliminate barriers to high-quality, accessible health care	*Revisit policies that impact equitable health care access*
	• Remove race-based correction factors
	• Eliminate co-pays and cost-sharing for all health care services and prescription medications
	• Addressing health care policies that have created a racially segregated, two-tier care system
	  ◦ Greater accountability and oversight of disproportionate share hospital (DSH) payments and Medicaid provider reimbursements
	  ◦ Accounting for social barriers to care and optimal health within pay-for-performance models
	• Close racial disparities in geographic access to primary care.
	• Scrutinize innovations aiming to improve care (e.g., precision medicine) for differences in utility, access, and representation
		Hidden in plain sight—reconsidering the use of race correction in clinical algorithms. New England Journal of Medicine. 2020 Aug 27;383(9):874-82.
		Eliminating race-based medicine. Pediatrics. 2022 Jul 1;150(1).
		Structural Racism in Historical and Modern US Health Care Policy: Study examines structural racism in historical and modern US health care policy. Health Affairs. 2022 Feb 1;41(2):187-94.
		Precision medicine in 2030—seven ways to transform healthcare. Cell. 2021 Mar 18;184(6):1415-9.
		Precision medicine and the problem of structural injustice. Med Health Care Philos. 2023 Sep;26(3):433-450.

public health practitioners, performance improvement teams, and other learners and health care professionals should be equipped with the necessary skills to integrate an equity lens into traditional quality improvement methods and develop countermeasures and interventions that apply a systems-deficit rather than population-deficit lens (Aysola and Myers, 2018). A systems-deficit lens calls for a shift in focus from our current state, where our interventions often center on deficits of disinvested communities or marginalized populations. Instead, interventions to improve health care systems and overall health should focus on how to fix biases within societal systems and/or the historical and contemporary structures, policies, and practices that continue to disinvest in communities and marginalize populations.

Third is strategic community engagement and investments. We must involve the community in the science we produce not only as participants but as consultants and critical partners in every phase of scientific inquiry (Asabor et al., 2022; Seward et al., 2021). This includes ending colonial science and replacing "helicopter research" with true nonhierarchical partnerships by researchers in the Global North with those in the Global South, allowing local communities and populations to define their research priorities (Benjamin, 2009). Community investments must redress historical marginalization with upstream corrective actions such as reparations (Williams and Collins, 2004). Reparations need not be an elusive strategy and should not be limited to historically enslaved persons and Indigenous populations within the US. Countries in the Global North must critically evaluate their historical legacies and contemporary actions in creating and perpetuating health inequities across the globe. However, such evaluation is merely the starting point and must be followed by corrective actions and reparations (Shelton et al., 2021). Rather than lamenting immigration that stems from countries previously colonized, the Global North should consider how best to restore economic and political stability in those countries as part of corrective actions and view migration and access to public goods within colonial powers as a form of reparation (Goldstone, 2024).

References

Adame, F., 2021. Meaningful collaborations can end 'helicopter research'. Nature 10. doi:10.1038/d41586-021-01795-1.

Adams, J., 2013. The fourth age of research. Nature 497, 557–560.

Adepoju, O.E., Chae, M., Liaw, W., Angelocci, T., Millard, P., Matuk-Villazon, O., 2022. Transition to telemedicine and its impact on missed appointments in community-based clinics. Ann. Med. 54, 98–107.

Alexis, D.A., Kearney, M.D., Williams, J.C., Xu, C., Higginbotham, E.J., Aysola, J., 2020. Assessment of perceptions of professionalism among faculty, trainees, staff, and students in a large University-based health system. JAMA Netw. Open 3, e2021452.

Alshebli, B.K., Rahwan, T., Woon, W.L., 2018. The preeminence of ethnic diversity in scientific collaboration. Nat. Commun. 9, 5163.

Amster, E.J., 2022. The past, present and future of race and colonialism in medicine. CMAJ 194, E708–e710.

Amutah, C., Greenidge, K., Mante, A., Munyikwa, M., Surya, S.L., Higginbotham, E., Jones, D.S., Lavizzo-Mourey, R., Roberts, D., TSAI, J., 2021. Misrepresenting Race—the Role of Medical Schools in Propagating Physician Bias, Mass Medical Soc.

Anderson, N., Nguyen, M., Marcotte, K., Ramos, M., Gruppen, L.D., Boatright, D., 2023. The long shadow: a historical perspective on racism in medical education. Acad. Med. 98, S28–s36.

Asabor, E.N., Cohen, J.M., Aysola, J., 2022. Why increased diversity in dermatologic clinical trials is not inherently ethical—an evidence-based, community-engaged approach to diverse trial recruitment. JAMA Dermatol. 158, 1219.

Aysola, J., Barg, F.K., Martinez, A.B., Kearney, M., Agesa, K., Carmona, C., Higginbotham, E., 2018. Perceptions of factors associated with inclusive work and learning environments in Health Care Organizations: a qualitative narrative analysis. JAMA Netw. Open 1, e181003.

Aysola, J., Getachew, Y., Chand, M., Siddique, S., Keddem, S., 2024. Implementing race-conscious medicine within healthcare systems: a qualitative analysis of race removal from clinical algorithms and decision tools. In: APHA 2024 Annual Meeting and Expo, APHA.

Aysola, J., Murdock, H.M., Lett, E., Williams, C., Wade Jr., R., Higginbotham, E.J., 2023. Operationalizing inclusion: moving from an elusive goal to strategic action. Epidemiol. Rev. 45, 140–145.

Aysola, J., Myers, J.S., 2018. Integrating training in quality improvement and health equity in Graduate Medical Education: two curricula for the price of one. Acad. Med. 93, 31–34.

Aysola, J., Surya, S.L., Jones, D.S., 2021. The role of medical schools in propagating physician bias. Reply. N. Engl. J. Med. 384, e97.

Baier, L.J., Hanson, R.L., 2004. Genetic studies of the etiology of type 2 diabetes in Pima Indians: hunting for pieces to a complicated puzzle. Diabetes 53, 1181.

Bailey, Z.D., Krieger, N., AGÉNOR, M., Graves, J., Linos, N., Bassett, M.T., 2017. Structural racism and health inequities in the USA: evidence and interventions. Lancet 389, 1453–1463.

Bass, S., 2001. Policing space, policing race: social control imperatives and police discretionary decisions. Soc. Justice 28, 156–176.

Benjamin, R., 2009. A lab of their own: genomic sovereignty as postcolonial science policy. Policy Soc. 28, 341–355.

Benjamin, R., 2014. Race for cures: rethinking the racial logics of 'Trust' in biomedicine. Sociol. Compass 8, 755–769.

Benjamin, R., 2016. Informed refusal:toward a justice-based bioethics. Sci., Technol. Hum. Values 41, 967–990.

Beyer, K.M., Zhou, Y., Matthews, K., Bemanian, A., Laud, P.W., Nattinger, A.B., 2016. New spatially continuous indices of redlining and racial bias in mortgage lending: links to survival after breast cancer diagnosis and implications for health disparities research. Health Place 40, 34–43.

Bloche, M.G., 2004. Race-based therapeutics. N. Engl. J. Med. 351, 2035–2037.

Boroush, M., 2020. Research and Development: US Trends and International Comparisons. Science and Engineering Indicators 2020. NSB-2020-3. *National Science Foundation.*

Bressman, E., Werner, R.M., Childs, C., Albrecht, A., Myers, J.S., Adusumalli, S., 2022. Association of telemedicine with primary care appointment access after hospital discharge. J. Gen. Intern. Med. 37, 2879–2881.

Brody, H., Hunt, L.M., 2006. BiDil: assessing a race-based pharmaceutical. Ann. Fam. Med. 4, 556–560.

Byrd, W.M., Clayton, L.A., Schissel, C., 2002. An American Health Dilemma: A Medical History of African Americans and the Problem of Race-beginnings to 1900, Springer.

Chandra, A., Skinner, J., 2003. Geography and racial health disparities. Public Goods Regul. Health Econ. eJournal. NBER Working Paper No. w9513. Available from: https://ssrn.com/abstract=382444.

Cherian, M., 2014. Race in the mortgage market: an empirical investigation using HMDA data. Race, Gender & Class 21 (1-2), 48–63. http://www.jstor.org/stable/43496959. Accessed 13 June 2026.

Choudhry, N.K., Bykov, K., Shrank, W.H., Toscano, M., Rawlins, W.S., Reisman, L., Brennan, T.A., Franklin, J.M., 2014. Eliminating medication copayments reduces disparities in cardiovascular care. Health Aff. 33, 863–870.

Chowdhury, S., Laux, T., Morse, M., JENKS, A., Stonington, S., Jain, Y., 2019. Democratizing evidence production—a 51-year-old man with sudden onset of dense hemiparesis. N. Engl. J. Med. 381, 1501–1505.

Cohen, J.J., Gabriel, B.A., Terrell, C., 2002. The case for diversity in the health care workforce. Health Aff. 21, 90–102.

Collins, F.S., Adams, A.B., Aklin, C., Archer, T.K., Bernard, M.A., Boone, E., Burklow, J., Evans, M.K., Jackson, S., Johnson, A.C., Lorsch, J., Lowden, M.R., Nápoles, A.M., Ordóñez, A.E., Rivers, R., Rucker, V., Schwetz, T., Segre, J.A., Tabak, L.A., Hooper, M.W., Wolinetz, C., 2021. Affirming NIH's commitment to addressing structural racism in the biomedical research enterprise. Cell 184, 3075–3079.

Corbyn, Z., 2011. Black Applicants Less Likely to Win NIH Grants, Nature Publishing Group.

Czyzewski, K., 2011. Colonialism as a broader social determinant of health. Int. Indig. Policy J. 2, 1–14.

Darity Jr, W., Hamilton, D., Paul, M., Aja, A., Price, A., Moore, A., Chiopris, C., 2018. What we get wrong about closing the racial wealth gap. *Samuel DuBois Cook Center on Social Equity and Insight Center for Community Economic Development*, 1, 1-67.

Davis, S.M., Reid, R., 1999. Practicing participatory research in American Indian communities. Am. J. Clin. Nutr. 69, 755S–759S.

De Meulemeester, J., Fedyk, M., Jurkovic, L., Reaume, M., Dowlatshahi, D., Stotts, G., Shamy, M., 2018. Many randomized clinical trials may not be justified: a cross-sectional analysis of the ethics and science of randomized clinical trials. J. Clin. Epidemiol. 97, 20–25.

Ditomaso, N., 2024. The invention of race and the persistence of racial hierarchy: white privilege, White supremacy, and White colorblindness. Soc. Personal. Psychol. Compass 18, e12953.

Douglas, J.G., 2003. Questioning race-based hypertension management—reply. Arch. Intern. Med. 163, 1744–1745.

Else, H., 2025. Top universities warned against unfair research partnerships on their doorstep. Nature. doi:10.1038/d41586-025-00159-3.

Epstein, A.M., Ayanian, J.Z., Keogh, J.H., Noonan, S.J., Armistead, N., Cleary, P.D., Weissman, J.S., David-Kasdan, J.A., Carlson, D., Fuller, J., 2000. Racial disparities in access to renal transplantation—clinically appropriate or due to underuse or overuse?. N. Engl. J. Med. 343, 1537–1544.

Fitzgerald, C., Hurst, S., 2017. Implicit bias in healthcare professionals: a systematic review. BMC Med. Ethics 18, 1–18.

Ford, C.L., Airhihenbuwa, C.O., 2018. Commentary: just what is critical race theory and what's it doing in a progressive field like public health?. Ethn. Dis. 28, 223–230.

Fox, M.P., Murray, E.J., Lesko, C.R., Sealy-Jefferson, S., 2022. On the need to revitalize descriptive epidemiology. Am. J. Epidemiol. 191, 1174–1179.

Franklin, N.K.C.K.B.W.H.S.M.T.W.S.R.L.R.T.A.B.A.J.M., 2014. Eliminating medication copayments reduces disparities in cardiovascular care. Health Aff. 33, 863–870.

Freeman, R.B., Huang, W., 2014. Collaboration: strength in diversity. Nature 513, 305.

Fuentes, A., Ackermann, R.R., Athreya, S., Bolnick, D., Lasisi, T., Lee, S.H., Mclean, S.A., Nelson, R., 2019. AAPA statement on race and racism. Am. J. Phys. Anthropol. 169, 400–402.

Fujimura, J.H., Bolnick, D.A., Rajagopalan, R., Kaufman, J.S., Lewontin, R.C., Duster, T., Ossorio, P., Marks, J., 2014. Clines without classes: how to make sense of human variation. Sociol. Theory 32, 208–227.

Fullilove, M.T., 1998. Comment: abandoning “race” as a variable in public health research–an idea whose time has come. Am. J. Public Health 88, 1297–1298.

Galster, G.C., Keeney, W.M., 1988. Race, residence, discrimination, and economic opportunity: modeling the nexus of urban racial phenomena. Urban Aff. Q. 24, 87–117.

Garcia, R.S., 2004. The misuse of race in medical diagnosis. Pediatrics 113, 1394–1395.

Gee, G.C., 2008. A multilevel analysis of the relationship between institutional and individual racial discrimination and health status. Am. J. Public Health 98, S48–S56.

Gee, G.C., Ford, C.L., 2011. Structural racism and health inequities: old issues, new directions. Du Bois Rev. 8, 115–132.

Ginther, D.K., Schaffer, W.T., Schnell, J., Masimore, B., Liu, F., Haak, L.L., Kington, R., 2011. Race, ethnicity, and NIH research awards. Science 333, 1015–1019.

Goldstone, Z., 2024. Migration as reparation for colonialism. Res Publica 30, 763–781.

Goode, C.A., Landefeld, T., 2018. The lack of diversity in healthcare. J. Best Pract. Health Prof. Divers. 11, 73–95.

Guevara, J.P., Wade, R., Aysola, J., 2021. Racial and ethnic diversity at medical schools—Why aren't we there yet?. N. Engl. J. Med. 385, 1732–1734.

Gutiérrez, E.R., Fuentes, L., 2009. Population control by sterilization: the cases of Puerto Rican and Mexican-Origin women in the United States. Latino (a) Res. Rev. 7, 85–100.

Haelewaters, D., Hofmann, T.A., Romero-Olivares, A.L., 2021. Ten simple rules for Global North researchers to stop perpetuating helicopter research in the Global South. PLoS Comput. Biol. 17, e1009277.

Hofstra, B., Kulkarni, V.V., Munoz-Najar Galvez, S., He, B., Jurafsky, D., Mcfarland, D.A., 2020. The diversity-innovation paradox in science. Proc. Natl. Acad. Sci. U. S. A. 117, 9284–9291.

Hoppe, T.A., Litovitz, A., Willis, K.A., Meseroll, R.A., Perkins, M.J., Hutchins, B.I., Davis, A.F., Lauer, M.S., Valantine, H.A., Anderson, J.M., 2019. Topic choice contributes to the lower rate of NIH awards to African-American/black scientists. Sci. Adv. 5, eaaw7238.

Hsu, C.-Y., Yang, W., Parikh, R.V., Anderson, A.H., Chen, T.K., Cohen, D.L., He, J., Mohanty, M.J., Lash, J.P., Mills, K.T., Muiru, A.N., Parsa, A., Saunders, M.R., Shafi, T., Townsend, R.R., Waikar, S.S., Wang, J., Wolf, M., Tan, T.C., Feldman, H.I., GO, A.S., 2021. Race, genetic ancestry, and estimating kidney function in CKD. N. Engl. J. Med. 385, 1750–1760.

Inker, L.A., Eneanya, N.D., Coresh, J., Tighiouart, H., wang, D., Sang, Y., Crews, D.C., Doria, A., Estrella, M.M., Froissart, M., Grams, M.E., Greene, T., Grubb, A., Gudnason, V., Gutiérrez, O.M., Kalil, R., Karger, A.B., Mauer, M., Navis, G., Nelson, R.G., Poggio, E.D., Rodby, R., Rossing, P., Rule, A.D., Selvin, E., Seegmiller, J.C., Shlipak, M.G., Torres, V.E., Yang, W., Ballew, S.H., Couture, S.J., Powe, N.R., Levey, A.S., 2021. New creatinine- and cystatin C–based equations to estimate GFR without race. N. Engl. J. Med. 385, 1737–1749.

Institute of Medicine Committee On, U., Eliminating, R. & Ethnic Disparities In HEALTH, C. 2003. *In*: SMEDLEY, B. D., STITH, A. Y. & Nelson, A. R. (eds.) *Unequal Treatment: confronting Racial and Ethnic Disparities in Health Care.* Washington (DC): national Academies Press (US). Copyright 2002 by the National Academy of Sciences. All rights reserved.

Jackson, C.S., Turner, D., June, M., Miller, M.V., 2023. Facing our history—building an equitable future. Am. J. Hum. Genet. 110, 377–395.

Jackson, J.S., Brown, T.N., Williams, D.R., Torres, M., Sellers, S.L., Brown, K., 1996. Racism and the physical and mental health status of African Americans: a thirteen year national panel study. Ethn. Dis. 6, 132–147.

Jones, C.P., 2000. Levels of racism: a theoretic framework and a gardener's tale. Am. J. Public Health 90, 1212–1215.

Kato, G.J., Piel, F.B., Reid, C.D., Gaston, M.H., Ohene-Frempong, K., Krishnamurti, L., Smith, W.R., Panepinto, J.A., Weatherall, D.J., Costa, F.F., Vichinsky, E.P., 2018. Sickle cell disease. Nat. Rev. Dis. Primers 4, 18010.

Kilama, W.L., 2009. The 10/90 gap in sub-Saharan Africa: resolving inequities in health research. Acta Trop. 112 Suppl 1, S8–s15.

Knowler, W.C., Pettitt, D.J., Saad, M.F., Bennett, P.H. 1990. Diabetes mellitus in the Pima Indians: incidence, risk factors and pathogenesis.

Kurwa, R., 2022. Policing, property, and the production of racial segregation. In: McCabe, B.J., Rosen, E. (Eds.), The Sociology of Housing: How Homes Shape Our Social Lives. University of Chicago Press, pp. 291–304.

La Veist, T.A., 1996. Why we should continue to study race … but do a better job: an essay on race, racism and health. Ethn. Dis. 6, 21–29.

Lambert, W.M., Camacho-Rivera, M., Boutin-Foster, C., Salifu, M., Riley, W.J., 2024. Ending "domestic helicopter research". Cell 187, 1823–1827.

Landry, L.G., Ali, N., Williams, D.R., Rehm, H.L., Bonham, V.L., 2018. Lack of diversity in genomic databases is a barrier to translating precision medicine research into practice. Health Aff. 37, 780–785.

Lau, J., Knudsen, J., 2022. Reducing disparities in telemedicine: an equity-focused, public health approach. Health Aff. 41, 647–650.

Lawrence, J., 2000. The Indian health service and the sterilization of Native American women. Am. Indian Q. 24, 400–419.

Lett, E., Adekunle, D., Mcmurray, P., Asabor, E.N., Irie, W., Simon, M.A., Hardeman, R., Mclemore, M.R., 2022. Health equity tourism: ravaging the justice landscape. J. Med. Syst. 46, 17.

Lett, E., Murdock, H.M., Orji, W.U., Aysola, J., Sebro, R., 2019. Trends in racial/ethnic representation among US medical students. JAMA Netw. Open 2, e1910490.

Lewis, R.B., Hestand, J.T., 2006. Federal reserved water rights: gila River Indian community settlement. J. Contemp. Water Res. Educ. 133, 34–42.

Livingstone, F.B., Dobzhansky, T., 1962. On the non-existence of human races. Curr. Anthropol. 3, 279–281.

Maina, I.W., Belton, T.D., Ginzberg, S., Singh, A., Johnson, T.J., 2018. A decade of studying implicit racial/ethnic bias in healthcare providers using the implicit association test. Soc. Sci. Med. 199, 219–229.

Massey, D.S., Condran, G.A., Denton, N.A., 1987. The effect of residential segregation on black social and economic well-being. Soc. Forces 66, 29–56.

Mcclure, E.S., Vasudevan, P., Bailey, Z., Patel, S., Robinson, W.R., 2020. Racial capitalism within public health—how occupational settings drive COVID-19 disparities. Am. J. Epidemiol. 189, 1244–1253.

Mcintosh, K., Messin, L., Jin, P., Mullan, Z., 2023. Countering helicopter research with equitable partnerships. Lancet Glob. Health 11, e1007–e1008.

Mckenzie, K., 2003. Racism and Health: Antiracism Is an Important Health Issue, British Medical Journal Publishing Group.

Mendez, D.D., Hogan, V.K., Culhane, J., 2011. Institutional racism and pregnancy health: using home mortgage disclosure act data to develop an index for mortgage discrimination at the community level. Public Health Rep. (Washington, D.C.: 1974) 126 (Suppl 3), 102–114.

Muppavarapu, K., Saeed, S.A., Jones, K., Hurd, O., Haley, V., 2022. Study of impact of telehealth use on clinic "No Show" rates at an academic practice. Psychiatr. Q. 93, 689–699.

Nieblas-Bedolla, E., Christophers, B., Nkinsi, N.T., Schumann, P.D., Stein, E., 2020. Changing how race is portrayed in medical education: recommendations from medical students. Acad. Med. 95, 1802–1806.

Obermeyer, Z., Powers, B., Vogeli, C., mullainathan, S., 2019. Dissecting racial bias in an algorithm used to manage the health of populations. Science 366, 447–453.

Odeny, B., Bosurgi, R. 2022. Time to End Parachute Science, Public Library of Science, San Francisco, CA USA.

Orfield, G., LEE, C. 2005. Why segregation matters: poverty and educational inequality.

Page, S.E., 2019. The Diversity Bonus: How Great Teams Pay off in the Knowledge Economy, Princeton University Press.

Paradies, Y., 2016. Colonisation, racism and indigenous health. J. Popul. Res. 33, 83–96.

Park, P.S.U., Algur, E., Narayan, S., Song, W.B., Kearney, M.D., Aysola, J., 2022. Representation of Asian American populations in medical school curricula. JAMA Netw. Open 5, e2233080.

Payne, B.K., Hannay, J.W., 2021. Implicit bias reflects systemic racism. Trends Cogn. Sci. 25, 927–936.

Petryna, A., 2005. Ethical variability: drug development and globalizing clinical trials. Am. Ethnol. 32, 183–197.

Petryna, A., 2007. Clinical trials offshored: on private sector science and public health. BioSocieties 2, 21–40.

Pollock, A., 2012. Medicating Race: Heart Disease and Durable Preoccupations with Difference, Duke University Press.

Powell, K., 2018. These labs are remarkably diverse - here's why they're winning at science. Nature 558, 19–22.

Reardon, J., Tallbear, K., 2012. Genomics, anthropology, and the construction of whiteness as PROPERTYYOUR DNA is our history. Curr. Anthropol. 53, S233–S245.

Rees, C.A., Sirna, S.J., Manji, H.K., Kisenge, R., Manji, K.P., 2022. Authorship equity guidelines in global health journals. BMJ Glob. Health 7 (10).

Remler, D.K., Greene, J., 2009a. Cost-sharing: a blunt instrument. Annu. Rev. Public Health 30, 293–311.

Remler, D.K., Greene, J., 2009b. Cost-sharing: a blunt instrument. Annu. Rev. Public Health 30, 293–311.

Research, G.F.F.H., 2001. The 10/90 report on health research. Global forum for health research.
Ripp, K., Braun, L., 2017. Race/ethnicity in medical education: an analysis of a question bank for step 1 of the United States medical licensing examination. Teach. Learn. Med. 29, 115–122.
Ritchie, C., Leff, B., 2022. Home-based care reimagined: a full-fledged health care delivery ecosystem without walls. Health Aff. 41, 689–695.
Rocha, A.D., Garcia, S., Santos, A.B., Eduardo, J.C.C., Mesquita, C.T., Lugon, J.R., Strogoff-De-Matos, J.P., 2020. No race-ethnicity adjustment in CKD-EPI EQUATIONS IS REQUIRED FOR ESTIMATING glomerular filtration rate in the brazilian population. Int. J. Nephrol. 2020, 2141038.
Roesler, S., 2021. Racial segregation and environmental injustice. Envtl. L. Rep. 51, 10773.
Rothstein, R., 2017. The Color of Law: A Forgotten History of How Our Government Segregated America, Liveright Publishing.
Rubin, R., 2021. Tackling the misconception that cystic fibrosis is a "White People's Disease". JAMA 325, 2330–2332.
Schulz, L.O., Bennett, P.H., Ravussin, E., Kidd, J.R., Kidd, K.K., Esparza, J., Valencia, M.E., 2006. Effects of traditional and western environments on prevalence of type 2 diabetes in Pima Indians in Mexico and the US. Diabetes Care 29, 1866–1871.
Schulz, L.O., Chaudhari, L.S., 2015. High-risk populations: the Pimas of Arizona and Mexico. Curr. Obes. Rep. 4, 92–98.
Schutte, A.E., Abdool Karim, Q., 2023. Reimagining global hypertension research: from helicopter science to meaningful partnerships. Hypertension 80, 2239–2242.
Seward, N., Hanlon, C., Hinrichs-Kraples, S., Lund, C., Murdoch, J., Salisbury, T.T., Verhey, R., Shidhaye, R., Thornicroft, G., Araya, R., 2021. A guide to systems-level, participatory, theory-informed implementation research in global health. BMJ Glob. Health, 6 (12), 1–16.
Shelton, R.C., Adsul, P., Oh, A., Moise, N., Griffith, D.M., 2021. Application of an antiracism lens in the field of implementation science (IS): recommendations for reframing implementation research with a focus on justice and racial equity. Implement. Res. Pract. 2, 26334895211049482.
Shkembi, A., Smith, L.M., Neitzel, R.L., 2024. Linking environmental injustices in Detroit, MI to institutional racial segregation through historical federal redlining. J. Expo. Sci. Environ. Epidemiol. 34, 389–398.
Siddique, S.M., Tipton, K., Leas, B., Jepson, C., Aysola, J., Cohen, J.B., Flores, E., Harhay, M.O., Schmidt, H., Weissman, G.E., 2024. The impact of health care algorithms on racial and ethnic disparities: a systematic review. Ann. Intern. Med. 177, 484–496.
Siegel, M., Sherman, R., Li, C., Knopov, A., 2019. The relationship between racial residential segregation and black-white disparities in fatal police shootings at the city level, 2013–2017. J. Natl. Med. Assoc. 111, 580–587.
Sirugo, G., Williams, S.M., Tishkoff, S.A., 2019. The Missing Diversity in Human Genetic Studies. Cell 177, 26–31.
Smith-Morris, C.M., 2004. Reducing diabetes in Indian country: lessons from the three domains influencing Pima diabetes. Hum. Organ. 63, 34–46.
Tilley, H., 2016. Medicine, empires, and ethics in Colonial Africa. AMA J. Ethics 18 (7), 743–753.

Tipton, K., Leas, B.F., Flores, E., Jepson, C., Aysola, J., Cohen, J., Harhay, M., Schmidt, H., Weissman, G., Treadwell, J. 2023. Impact of healthcare algorithms on racial and ethnic disparities in health and healthcare.

Tollefson, J., Van Noorden, R., 2024. The US is the world's science superpower - but for how long?. Nature 634, 770–774.

Tsai, J., Ucik, L., Baldwin, N., Hasslinger, C., George, P., 2016. Race matters? Examining and rethinking race portrayal in preclinical medical education. Acad. Med. 91, 916–920.

Umeukeje, E.M., Koonce, T.Y., Kusnoor, S.V., Ulasi, I.I., Kostelanetz, S., Williams, A.M., Blasingame, M.N., Epelbaum, M.I., Giuse, D.A., Apple, A.N., Kaur, K., González Peña, T., Barry, D., Eisenstein, L.G., Nutt, C.T., Giuse, N.B., 2022. Systematic review of international studies evaluating MDRD and CKD-EPI estimated glomerular filtration rate (eGFR) equations in Black adults. PLoS One 17, e0276252.

Vivian, L., 2017. The influence of school segregation on students' educational achievement. Perspectives 9, 11.

Vyas, D.A., Eisenstein, L.G., Jones, D.S., 2020. Hidden in Plain Sight—reconsidering the Use of Race Correction in Clinical Algorithms, Mass Medical Soc.

Wang, G., Schwartz, G.L., Kim, M.H., White, J.S., Glymour, M.M., Reardon, S., Kershaw, K.N., Gomez, S.L., Inamdar, P.P., Hamad, R., 2022. School racial segregation and the health of Black children. Pediatrics 149 (5), 24–33.

Washington, H.A., 2006. Medical Apartheid: The Dark History of Medical Experimentation on Black Americans from Colonial Times to the Present, Doubleday Books.

Wheeler, C.H., Olson, L.M., 2015. Racial differences in mortgage denials over the housing cycle: evidence from US metropolitan areas. J. Hous. Econ. 30, 33–49.

Williams, D.R., 1999. Race, socioeconomic status, and health the added effects of racism and discrimination. Ann. N. Y. Acad. Sci. 896, 173–188.

Williams, D.R., Collins, C., 2001. Racial residential segregation: a fundamental cause of racial disparities in health. Public Health Rep. 116, 404–416.

Williams, D.R., Collins, C., 2004. Reparations: a viable strategy to address the enigma of African-American Health. Am. Behav. Sci. 47, 977–1000.

Williams, D.R., Jackson, P.B., 2005. Social sources of racial disparities in health. Health Aff. (Millwood) 24, 325–334.

Williams, D.R., Lawrence, J.A., Davis, B.A., 2019a. Racism and health: evidence and needed research. Annu. Rev. Public Health 40, 105–125.

Williams, D.R., Lawrence, J.A., Davis, B.A., Vu, C., 2019b. Understanding how discrimination can affect health. Health Serv. Res. 54 (Suppl 2), 1374–1388.

Williams, D.R., Mohammed, S.A., 2013a. Racism and health I: pathways and scientific evidence. Am. Behav. Sci. 57, 1152–1173.

Williams, D.R., Mohammed, S.A., 2013b. Racism and health II: a needed research agenda for effective interventions. Am. Behav. Sci. 57, 1200–1226.

Williams, W.W., Hogan, J.W., Ingelfinger, J.R., 2021. Time to eliminate health care disparities in the estimation of kidney function. N. Engl. J. Med. 385, 1804–1806.

Wispelwey, B., Osuagwu, C., Mills, D., Goronga, T., Morse, M., 2023. Towards a bidirectional decoloniality in academic global health: insights from settler colonialism and racial capitalism. Lancet Glob. Health 11, e1469–e1474.

Wright, J.L., Davis, W.S., Joseph, M.M., Ellison, A.M., Heard-Garris, N.J., Johnson, T.L., Equity, T.A.B.C.O., 2022. Eliminating race-based medicine. Pediatrics, 150 (1), 1–10.

Wu, M., Davis, J.D., Zhao, C., Daley, T., Oliver, K.E., 2024. Racial inequities and rare CFTR variants: impact on cystic fibrosis diagnosis and treatment. J. Clin. Transl. Endocrinol. 36, 100344.

Yearby, R., 2020. Structural racism and health disparities: reconfiguring the social determinants of health framework to include the root cause. J. Law Med. Ethics 48, 518–526.

Yearby, R., Clark, B., Figueroa, J.F., 2022. Structural racism in historical and modern US health care policy. Health Aff. 41, 187–194.

Zambrana, R.E., Williams, D.R., 2022. The intellectual roots of current knowledge On racism and health: relevance to policy and the National equity discourse. Health Aff. 41, 163–170.

CHAPTER

4 Colonial laboratories as imperial sites of knowledge production: The case of the Medical Research Institute (MRI) in Nigeria

Adedamola Adetiba

Centre for History of Science, Technology and Medicine, Faculty of Biology, Medicine, and Health, University of Manchester, Manchester, England, United Kingdom

Introduction

It is not only humans that die in Africa; institutions and systems do too. The question that will, therefore, be interrogated in this chapter is: "Why do medical research institutions exist only temporarily in sub-Saharan Africa?" To pose it differently: what forms of structural violence are perpetuated within global health that render progress in research institutions short-lived? Existing histories and anthropologies of medical research often tie the temporalities of medicine and science in Africa to economic shifts triggered by the structural adjustment programs of the 1980s. What happened in medical research was an extension of the consequences of African governments' health liberalization policies which absolved states of social responsibilities (Loewenson, 1993). Health care thus became a commodity—a burden for individuals to bear. This resulted in decreasing investments in health care sectors, where infrastructures that once stood as epicenters of medical research became shadows of themselves. This shift resulted in such institutions instead specializing in clerical and administrative tasks rather than the actual research work of studying the etiology, underlying causes, and risk factors driving disease onset and progression. These were institutions rooted in the colonial period—specifically those created by colonial states as part of scientific colonialism and which played prominent roles in vector-borne disease control programs (Geissler and Molyneux, 2011). Until the 1980s, such institutions showed signs of progress and promise, facilitating the emergence of enduring scientific communities, composed mainly of African epidemiologists, entomologists, bacteriologists, and laboratory assistants. With the new economic model, institutions once bustling with surveillance and entomological work suddenly fell silent or, in some cases, ceased to exist. Foreign institutions have since become the

Decolonizing Medical Education, Research and Practice. https://doi.org/10.1016/B978-0-443-44998-7.10004-3

dominant players in medical research, funding and leading work that was previously the domain of government research institutions.

By examining the history of one of Nigeria's pioneering government-owned and funded medical research institutions as a case study, this chapter argues that the temporality of such institutions was a product of the imperial and global health systems' inherent structural violence. In other words, these institutions ceased to exist not because they failed, but because they were not originally designed to exist permanently, or (more precisely) beyond the expectations set by the custodians of imperial state power. Even if such institutions survived the colonial period, they failed to overcome a system that kept them as perpetual appendages within a network tethered to metropolitan Europe.

The focus of this narrative is the Medical Research Institute (MRI) located in Lagos, Nigeria, and established in 1910. It was conceived due to the imperial state's interest in enhancing the capacities of its West African dependencies to contribute meaningfully to tropical medicine research. The laboratory was also part of a grander agenda to serve as a hub, akin to a helipad or conduit, supporting tropical medicine in metropolitan Britain. Thus it functioned as a helipad, an extended faculty or laboratory for experts affiliated with institutions like the London and Liverpool Schools of Tropical Medicine, granting them access to viable research sites to explore and understand tropical diseases in regions marked by endemic and epidemic cases. As a conduit, the imperial state consolidated a network for the transfer of research materials—an infrastructure it had worked assiduously to establish since the late 19th century.

Histories of tropical medicine recount the expeditionary activities of British institutions conducting entomological research in the natural habitats of disease vectors during the 1890s and early 1900s (Dumett, 1968). Scientists affiliated with these institutions collected mosquitoes and mosquito larvae, as well as blood and urine smears, for further investigation within laboratory settings in Britain. With laboratories like the MRI, the imperial state and its scientific clients were able to mitigate the risk of specimen contamination by studying materials close to the environments from which they were sourced.

What is discernible from colonial reports on the MRI is the temporality of scientific work conducted by both visiting scientists and those employed either temporarily or permanently. The recurrent movement toward the MRI from British metropolitan institutions determined the length of time devoted to researching specific diseases. The MRI was a master of many trades, home to many visitors, and host to many projects. It exhibited the capacity to host research but lacked a framework or process to sustain it. This makes it difficult to label the laboratory as an authority in any specific domain of expertise, not even as an expert institution for health issues peculiar to the local population. This lack of focus, due to operational fluctuations and its susceptibility to external influences, meant that outside expediencies became the driving force for developments within the MRI. Between 1910 and 1930, the MRI was the epicenter of vector-borne disease research; a position it relinquished in the 1940s when imperial and external interests shifted elsewhere. The temporality and fragility of this laboratory is what I imagine as structural violence: the elusive agendas that deterred it from fulfilling its full potential. The term *structural violence* was coined by the peace scholar Johan Galtung and defined as "when violence is embedded within systems or structures and becomes evident in unequal power relations, assigning different chances to different people based on class and status" (Galtung, 1969).

It is important to establish enduring research institutions in Africa. This becomes especially urgent when one considers the significance of African-based research in the global effort to drastically decrease the disease burden on the continent. For example, there is a pressing need to accurately capture the malaria burden of disease and mosquito distributions across the tropical regions of Africa. This is not work that should be left exclusively to a few centers of excellence outside the African continent. Recent studies have identified the gap between malaria research investment and the national malaria burden as a barrier to achieving the World Health Organization's (WHO) 2030 global target on malaria reduction (Head et al., 2017). These studies underscore the imperative for stakeholders, both national and global, to increase the funding pool for malaria research, which would support more entomological studies on mosquito distribution within the heartland of the epidemic. Such efforts should, in turn, lead to the development of new tools and techniques to decrease mosquito populations and intensify the clinical discovery of vaccines and drugs that could revolutionize malaria therapies (malERA Consultative Group on Vaccines, 2011). In sum, this strand of scholarship is a call for more targeted funding from existing donors and a challenge for more role players to become involved in financing research. Currently, sixty per cent of malaria research funding comes from the United States' (US) National Institutes of Health (NIH) and the Gates Foundation (Head et al., 2017).

Another dimension to this reflection on funding is the allocation of resources to a specific category of experts and institutions, specifically those with global reputations in malaria research. Such individuals and groups have long been at the forefront of malaria studies. However, in recent years, there has been growing advocacy among scientists in (and from) the Global South for the decolonization of malaria research to decentralize the networks that produce most of the knowledge, data, and technologies used in the global campaign against malaria. These are calls for inclusion and equality that are gaining momentum following a new US-led Malaria Initiative. Here, the US government's President's Malaria Initiative (PMI) awarded a US$30 million grant in 2021 to fund a data optimization project for malaria control and elimination. Most of the consortium institutions that received the grant were based in the US, the United Kingdom (UK), and Australia, with not a single African institution participating in this program. In an open letter, African scientists called out the marginalization of African institutions and launched a campaign to make the global health landscape more inclusive (Erondu et al., 2021). These scientists argued that the current structure of global health research funding often favors partnerships with Western institutions regarding health challenges that are characteristically African, while dismissing the expertise and agency of institutions and experts in Africa.

The inequality in global malaria research can be traced back to the early histories of malaria in tropical Africa. In other words, recent developments that characterize global health research are rooted in structures shaped by imperial expediencies. From the inception of tropical medicine as a medical specialty to the point when malaria became a global health agenda under the WHO, research conducted in malaria-endemic regions was dominated by experts affiliated with European institutions. A close reading of colonial archives reveals that imperial and global health stakeholders consistently favored foreign experts to coordinate and facilitate malaria research, often at the expense of local expertise. This tendency is particularly evident in British Nigeria, where the imperial government privileged established experts from metropolitan institutions over those in the colonial medical service when seeking to expand its knowledge of malaria transmission and control strategies.

The decolonial turn and global health research

The issues discussed in this chapter strongly resonate with key themes regarding the scholarship of decolonizing global health research. This body of work critiques the current research landscape, advocating for the need to eliminate some of the pervasive wrongs that facilitate epistemic injustice. These challenges can be understood as forms of structural violence within global health research. These are barriers that impede African research institutions from taking local ownership of research initiatives. The most noticeable wrongs identified in decolonial studies are systems and requirements that discriminate against who should (or should not) participate in global health research. Scholars argue that academic global health is configured in ways that marginalize researchers who should be the primary producers of knowledge, given their location and affiliation within environments where such global health challenges are endemic (Bhakuni and Abimbola, 2021, p. 1466). These researchers are excluded when the gatekeepers of the field deem their research focus and epistemic lens inappropriate or unnecessary. Hence, to free global health research of such wrongs, the frameworks through which researchers and their work are recruited and evaluated, and the lens through which their positionality is perceived, must change if we are to achieve a just and equitable research landscape (Besson, 2022).

While the exclusion of African scientists and institutions from global health research is an apparent reality, there is a tendency among major stakeholders who finance and coordinate research programs to dismiss such criticisms as excessively harsh. This dismissal often arises when isolated cases of support for African institutions are brought to the fore. However, the research funding landscape is indeed changing, with funding institutions becoming more reflective and self-critical about their approaches and strategies. Hence, there are pockets of funding aimed at strengthening research capacities across the continent. Some of these funds are provided by high-income countries (HICs) and support research, capacity-building, and technology-transfer programs (Head et al., 2017). One cannot deny that these efforts have had a significant impact on Africa's research landscape. Here, a key highlight is the rise of research centers of excellence that continue to lead clinical and surveillance investigations of infectious diseases across the continent. As a result, African institutions are increasingly contributing to global health knowledge while offering regionally relevant solutions.

Despite this progress, the call for inclusion and equality extends beyond current efforts. It is a call to decentralize global health research to the extent that it dismantles the lens through which scientists and institutions on the continent are perceived when urgent problems need to be solved. Providing human capacity and technological support is important, but more could be achieved if such institutions are regarded as equal partners in addressing global health emergencies. For example, a considerable amount of research within the continent is facilitated through transnational research partnerships, which often create uneven power relations between colleagues from HICs and low- and middle-income countries (LMICs) (Lawrence and Hirsh, 2020, p. 520).

Since most grants are initiated externally, scientists in HICs are more likely to act as principal investigators and ultimately determine the research focus and composition of partnerships. There also remains a tendency to collaborate with only a select few on the continent—typically those whose institutions have developed a historical rapport with foreign experts. This imbalance is further entrenched when some funders require local applicants to have mentors or institutional support from the Global North as a criterion for accessing funding. Others may not state this explicitly but still prioritize experts affiliated with institutions that have longstanding ties to their usual funding networks. There is a rationale behind this: funders want assurance that recipients have the collegial networks and institutional environment necessary to execute projects successfully. However, the current system elevates certain experts and institutions on the continent as gatekeepers while systematically excluding others. These local partnerships could be leveraged to develop a robust global malaria surveillance infrastructure.

There is also the issue of sustainability. Many thriving research institutions are closely tethered to institutions and experts in HICs. This raises a serious concern about whether they would be capable of existing independently of the support they receive and the collaboration they are part of. Moreover, what would happen to such institutions if there were a policy shift in their supporting institutions that then required them to secure partnerships elsewhere? In agreement, in recent years, there has been a notable decline in funding for global health research, which continues to affect the capacity of HIC institutions and experts to sustain longstanding alliances and partnerships with scientists in LMICs (Yach et al., 2025).

The decolonization of global health research, therefore, requires serious efforts to address, remove, and/or dismantle structures that reify power inherited from colonial episodes (Chaudhuri et al., 2021). In practical terms, this would mean a shift of power toward local ownership, with leaders in LMICs at the helm of global health research and their colleagues in HICs providing ancillary support (Costello and Zumla, 2000).

Imperial anxieties and early interests in malaria research

The history of malaria research provides valuable insights for enhancing our understanding of the persistent issues surrounding inclusion and equality in global health research. Some of these issues date back to the late 19th century, when tropical medicine emerged as a distinct medical specialty. The historiography of this period reveals a compelling tale of how and why this line of inquiry evolved. Here, it is known that the major works that defined malaria control during the next century emerged during this period as part of efforts by trading companies, the military, and imperial governments to remove obstacles to their respective interests in the tropical colonies (Harrison, 1992).

Early histories detail that malaria represented a major health problem and one where key stakeholders needed solutions at all costs. New networks, structures, and institutions were therefore formed mainly due to such imperial concerns and expediencies (Dumett, 1968; Worboys, 1976, 1979, 1988). In the British Empire, such developments were facilitated by the horrifying experiences of soldiers, traders, and missionaries who sojourned and settled in the tropics during the 19th century. Labels such as the "dark and dank continent" and "Whiteman's Grave" were some of the ways in which the tropics were perceived and imagined by the British public (Curtin, 1961, 1997). Early colonial reports detail the relatively high rates of mortality and their impact on the British imperial project (Dumett, 1968). For example, during the 19th century, malaria hindered colonial officials from residing in the tropics for two consecutive years, and they were required to embark on leave every six months. They subsequently returned to the tropics after being away for such a period.

Existing histories have examined how these concerns and anxieties impacted tropical medicine. Such developments shaped tropical medicine by facilitating the emergence of medical specialties. The first institutions involved in investigations into tropical diseases were hospitals affiliated with trading companies and the military. As both had a heavy presence in the colonies and were severely impacted by tropical diseases, they invested considerable efforts in treating and researching these illnesses. Here, 19th-century history informs us that the West African Squadron of the British government (stationed in this part of the continent) continued to enforce the abolition of the slave trade and expanded British influence in the region (Dumett, 1968). It goes further to emphasize the considerable health problems that military officers had to endure, including tropical diseases like malaria. For example, between 1837 and 1943, the invaliding rate (i.e., discharging someone from duty due to illness) was as high as 33% (Dumett, 1968). This influenced the early responsibilities of the military medical corps, specifically their interests in the study of tropical diseases. They aimed to attenuate the incidence of deaths among British troops through ideas and knowledge produced from research conducted in this context (Curtin, 1997). The historian Philip Curtin noted that such problems encountered by British troops could lead to ways of improving the health conditions of British officials in other locations, such as Sierra Leone, and constituted a "change of order" (Curtin, 1997). Thus military medicine played a crucial role in ensuring the success of British troops and significantly contributed to British gains during the partition of Africa in the post-1885 period.

Likewise, Myles Osborne and Susan Kent demonstrate in *Africans and Britons in the Age of Empire* that the triumph of military medicine is evident in the British penetration, occupation, and administration of Africa. They highlight specific details of how advancements in disease treatment assisted the British army's campaign against the Asante people (Osborne and Kent, 2015). Other works also show the contributions of military medicine to the development of tropical medicine in places such as India. For example, Ronald Ross (1857–1932), who was renowned for his work on malaria transmission was in fact an army major in India (Harrison, 1992).

Doctors affiliated with trading companies were also involved in tropical medicine research during the 19th century. For example, Mark Harrison examined the works of doctors affiliated with the East India Company on tropical diseases (a field of study he termed "tropical hygiene") during a precursory period to the formalization of tropical medicine (Harrison, 1992). These doctors, like their colleagues in the military and the Indian Medical Service, were on a mission to address one of the most significant challenges facing their enterprise in the tropical world.

What is notable about this period is that the institutions at the heart of early research into tropical diseases were often located in tropical colonies rather than in metropolitan Britain. The key distinction is that they were imperial institutions that exhibited technocratic traits while researching tropical diseases. Researchers were not primarily from such areas but instead relocated to the tropics after years of studying or working in the metropole as medical students or officers. The tropics thus became the theater of knowledge production, the point through which knowledge was assembled. Thus it became the space where Western knowledge and other knowledges about tropical diseases converged, with such institutions facilitating this assemblage.

Postcolonial scholarship on medicine helps to reinforce this point. For instance, in *The Calcutta Chromosome*, Amitav Ghosh crafted a fictional representation of how Western theories of malaria drew upon Indian folk medicine (Ghosh, 1996). Ghosh wrote about Ross's work as being influenced by indigenous knowledge in India while he was researching malaria transmission. When we connect Ghosh's work with that of Michael Worboys—who explained the historical processes that ushered in the tropical medicine specialty in the 19th century—knowledge emerging from this network offered utility to medical officers working in British metropolitan institutions. For example, this is evident when considering the relationship between Ross and Sir Patrick Manson (1844–1922) (renowned as the father of tropical medicine) (Worboys, 2022). Here, Manson's early misconceptions of tropical medicine were corrected through his proximity to experts affiliated with the Indian Medical Service and others in China.

Developments in the 1890s brought about significant changes, particularly in the establishment of new institutions in metropolitan Britain, as well as in its tropical colonies. Michael Worboys wrote extensively on these early years, specifically on how developments in tropical medicine extended Britain's constructive imperialism policy (Worboys, 1979). As part of this policy, the colonial government facilitated the establishment of two key institutions in tropical medicine, namely the London School of Tropical Medicine and the Liverpool School of Tropical Medicine. The important point to consider in this development is how resources from tropical colonies were mobilized to facilitate the establishment of such institutions and then sustain such research activities. It is important to note that resources mobilized from the colonies supported not only these two institutions but also served the needs of major scientists and research groups investigating tropical diseases.

British metropolitan scientists were favored for several reasons, one of which is historically connected to the ways the Colonial Office (CO) perceived the delivery of health care in colonial dependencies during the 19th century. The hierarchy in tropical medicine resulted from the CO's varied agendas for doctors in colonial and metropolitan institutions (Haynes, 1999). Colonial doctors were recruited purposefully for administrative convenience—to employ a corps of medical practitioners who were unemployed in Britain. The government saw this as a viable way to provide jobs for unemployed medical practitioners and to fill vacancies in its ambitious empire. Here, the primary role of medical officers was to provide primary health care to the colonial population, rather than to advance research into tropical diseases. The job of laboratory research fell within the jurisdiction of British metropolitan institutions and a few technocrats who occasionally visited colonial laboratories to survey tropical diseases.

During the colonial period, the delineation of responsibilities in imperial medicine played out in various ways in malaria research. First, it is imperative to provide insight regarding the structures that enabled this. The establishment of tropical schools and efforts to strengthen and coordinate their operations was a consequence of the constructive imperialism ideology of Joseph Chamberlain (1836–1914), one of the chief architects of British imperialism. In 1903, the CO established a research coordinating institution with the primary responsibility of mobilizing funds to advance research in tropical medicine. These funds were drawn from colonial territories and were chiefly allocated to work conducted at the University of Cambridge, the Liverpool and London Schools of Tropical Medicine, and the University of London (Haynes, 1999, p. 206). One reason such institutions were able to attract funds from the colonies for tropical medicine research was the enormous power they wielded and their proximity to the seat of imperial authority. Here, the committee consisted of respected persons from both tropical medicine and the broader British health care system and were individuals with whom metropolitan scientists and institutions were familiar and could easily trust, as such projects were regarded as crucial for the success of the imperial state. For example, one of the committee members was Sir Patrick Manson, who had a collegial relationship with Ronald Ross and his institution, the Liverpool School of Tropical Medicine. He also worked closely with the London School of Tropical Medicine—which he helped establish in 1899 (Haynes, 2001). This suggests that metropolitan institutions ascended to hegemonic heights because of the social power they had accumulated by the time tropical medicine emerged as a distinct medical specialty.

The invention of hierarchies in malaria research

One of the common trends in the history of malaria research is that both stakeholders in the CO and colonial territories agreed on the need for hierarchies and a division of responsibilities in combating and eradicating malaria. The institutions actually empowered to conduct research in tropical medicine were based in Britain (Wilkinson and Power, 1998). These institutions were entrusted with the responsibility of periodically visiting tropical colonies to survey the environment (specifically puddles, ponds, and swamps) to ascertain the distribution of the Culex and Anopheles mosquitoes across the region. In support, the Royal Society, in 1898, commissioned a study on malaria transmission in Freetown (Sierra Leone) to determine the best way to ensure the survival of European residents (Dumett, 1968, p. 165). During 1899, the CO influenced the establishment of the London and Liverpool tropical schools and supported malaria expeditions that followed shortly thereafter (Worboys, 2022). The imperial government was desperate to find a lasting solution to malaria (for economic purposes) and was able to mobilize viable institutions to do the job.

By the 20th century, hierarchies began to emerge around the concept of research laboratories. This does not mean that laboratories did not exist prior, but rather that their importance became a subject of policy discussion at the level of the imperial government. Two years after Joseph Chamberlain left office as Colonial Secretary, his successor, Victor Bruce (Earl of Elgin) (1849–1917), issued a circular in 1906 to colonial administrators recommending two policies to improve the fight against tropical diseases (, 1906). Firstly, he advocated for extending the training period for medical officers from three to 6 months to enhance research capabilities within the colonial medical service. His call in the circular was for colonial administrators to encourage medical officers to spend a significant amount of time in the tropical schools before embarking on work in the colonies. Secondly, he recommended the establishment of laboratories in tropical colonies where research on tropical diseases could be conducted.

The original intention of the empire at this time was to decentralize laboratory research to the extent that medical officers would be able to conduct independent investigations. The events that followed this circular were so consequential that they shaped the empire's approach to tropical medicine for the next five decades. Colonial administrators, while acknowledging the importance of independent research, communicated clearly to the committee and the CO why this was not a pragmatic path forward. They argued that it was not feasible logistically, as they were already understaffed and hence could not afford to delegate a medical officer to spend considerable time in training and then devote themselves entirely to research. The Governor of Northern Nigeria stated categorically that the protectorate lacked the staff and resources to support such a scheme (, 1906, p. 11). The Principal Medical Officer of Lagos, Dr. Henry Strachan (1859–1921), was particularly concerned about the impracticality of training medical officers to become specialists in tropical medicine. He noted that he could not identify anyone suitable for such a role and that short-term training would not be sufficient (, 1906, p. 12).

On this note, he recommended the establishment of specialized research institutions within colonies where research could take place. It is, therefore clear that hierarchies were formed as a matter of logistical expediency; the training afforded to medical officers did not necessarily guarantee them a seat at the table. The events that followed marked the beginning of the exclusion of local experts from malaria research. Strachan recommended that an expert should manage the research laboratory—someone not necessarily a medical officer but with the requisite skills and experience in animal and plant biology, and capable of surveying insect populations and assessing disease endemicity (, 1906, p. 4). As mentioned earlier, he believed it would not be feasible for a medical officer to possess both the time and the skills required to conduct meaningful research. For instance, he claimed to have tested the value of a particular medical officer's skills after completing the compulsory course at one of the tropical schools. He was both astonished and disappointed that little or nothing emerged from the research conducted by such trained experts. He, therefore preferred someone who had not necessarily undergone the training program (introduced by the CO) to lead the research.

This event is significant because it laid the foundation for both local experts and external stakeholders to rely on technocratic expertise in coordinating research and interventions aimed at malaria eradication. Although not directly part of this study, this gesture and posture of empire lingered throughout the colonial period and became the basis for the formation of the Nigerian Malaria Service in the 1950s. We observe that local laboratory institutions became the conduits through which metropolitan institutions accessed tropical colonies and conducted research on local populations. These institutions were developed as pathways for collecting research samples (bodily fluids such as urine and blood, alongside insects) and for substantiating popular theories on malaria. This demonstrates that institutions perceived as the forefront of medical research, perhaps due to their location on the continent, cannot be accurately characterized as "local". Therefore when discussing issues such as epistemic equality, there is a risk of mislabeling institutions on the continent that are empowered by philanthropic or global health initiatives. Some of these institutions may have imperial origins, having been established within the continent for logistical convenience and with a clear agenda of data collection and surveillance—not necessarily to improve public health, but as part of broader surveillance projects orchestrated by HICs.

In its 1908 report, the advisory committee appeared to agree with Strachan's recommendation. Lord Elgin stipulated that the director of the research laboratory should not be a medical officer but a technocrat, i.e., someone seasoned in medical research and capable of expanding the field of tropical medicine (, 1906, p. 4). He further emphasized the need for the director to align closely with the authorities in the territory where the laboratory was situated, to understand the prevailing diseases ravaging the area. The key point here is that the director and the laboratory were expected to conduct research tailored to the needs of the local population. While the intention is clear and valid (the need to link medical research to public health), one could posit that this marked the beginning of a persistent reliance on external institutions for solutions to local problems. It could be argued that the director was "local" in the sense that recruitment for the position was financed through local funds and that successful candidates were expected to work closely with local stakeholders. However, this was far from the case, as many of these positions ultimately became tied to British metropolitan institutions, which ultimately defined the agendas and leadership of the institutes.

The activities of the MRI reveal the power imbalances in international malaria research during the first few decades of the 20th century. The institute was part of Strachan's proposal to the committee, which he envisioned as operating outside the purview of the colonial medical service. In fact, this was the very justification for its establishment in the first instance. Strachan imagined that a seasoned expert could run the institute in tropical medicine, i.e., someone who could compensate for the limitations of the colonial medical service in conducting basic research in the field. This vision aligns closely with Haynes's argument that medical officers were primarily focused on public health, leaving laboratory research to metropolitan institutions (Haynes, 1999). Research work, therefore fell squarely within the exclusive remit of the newly established laboratories in Britain.

Studying the etiology of malaria in the tropics

In this section of the chapter, the early history of the MRI will be explored, along with how its vision and mission (which were shaped by imperial and colonial authorities) influenced the nature and direction of its research activities. The grand scheme behind the establishment of this local laboratory was to develop a robust research architecture across the empire, comprising laboratories in the colonies that could inform the ambitious research studies being carried out in British metropolitan institutions. These laboratories were envisioned as pathways through which ideas and research materials could be accessed and transmitted to experts and institutions in Britain. They were intended to serve as viable platforms for teaching and researching tropical medicine. The question is: to what extent did the MRI meet this expectation? The answer, although seemingly straightforward, helps us appreciate the ways local laboratories were imagined and utilized by metropolitan scientists and the imperial government, and the extent to which this vision detached them from surrounding local realities.

The agency of the MRI, whether as local collaborators for work done by metropolitan colleagues, as partners with the colonial medical service, or both, was primarily determined by the personality of its director. The director was expected to be a seasoned researcher and someone who had made significant contributions to the field of tropical medicine. Moreover, they were also expected to be someone who managed to establish robust partnerships with metropolitan institutions and colleagues (Simpson to Egerton, 10th July 1908). This vision unfolded in various ways depending on who was in control of the institute. Dr. William Medlycott Graham was appointed as the first director of the institute in 1910, with his remarkable experience with insects playing a key role in securing the position. At the time, he was affiliated with the African Entomological Committee and the Natural History Museum (Entomological Research Committee, Tropical Africa, 1911). Though Graham was based in the Gold Coast, he conducted entomological studies across West Africa. He particularly focused on places such as Lagos, where he identified various mosquito species. Some of his findings were published in the *Annual Magazine of Natural History,* where he classified and cataloged the diverse mosquito species found in Lagos (Graham, 1910, p. 264). His appointment was based on the strength of such earlier works. In support, Sir William Simpson (1855–1931), one of the founding members of the London School of Tropical Medicine, recommended Graham for the position because he was a well-respected entomologist who had conducted extensive work on insects across West Africa. Simpson wrote in his reference: "Graham's collection of biting flies, beetles, and mosquitoes for the Gold Coast is unique, and I know of no one outside the British Museum who has more intimate knowledge of the insects which play such an important part in the causation of disease in man, animal, and plants" (Simpson to Egerton, 10th July 1908).

With Graham's appointment, the first box was ticked; i.e., the institute would be headed by a figure well-known and respected within tropical medicine networks. The next challenge was preparing the facility to be ready and conducive for research work. Around the time that Graham was appointed, a decision was made to establish the institute in Yaba (Lagos, Nigeria). Yaba would go on to play a central role in the history of medical research during the remainder of colonial rule, hosting some of the most important research conducted in British West Africa. In fact, institutions such as the Rockefeller Yellow Fever Commission Laboratory were also located here, together with the MRI (Vaughan, 2018). Yaba currently hosts the Nigerian Institute of Medical Research, which evolved from the West African Council of Medical Research, an affiliate of the Medical Research Council in the UK. It was also central to colonial urban renewal programs; e.g., during 1917 and 1952, it served as the site for rehabilitation schemes for displaced people. These schemes were introduced against the backdrop of "slum dwelling", "overcrowding", and "insanitary conditions" on Lagos Island (Davies, 2022). While there is no definitive evidence explaining why Yaba was chosen as the site for these research institutions, it is known that Lagos was selected due to its viable transportation infrastructure. Like other aspects of the colonial extractive enterprise, transportation—specifically the railway system in Lagos—could easily facilitate the movement of research materials from across the country. The idea behind establishing the MRI in Lagos was to enable the efficient collection of samples and specimens from across British West Africa.

Graham was able to develop the preliminary structure of the MRI, and the buildings within the compound were designed to reflect his interests in entomology. One of the first functional facilities was the museum, a spacious area situated within the director's bungalow. The primary focus of the museum was to collect fauna and pathological specimens transported from across West Africa (Nigeria. Medical Research Institute, 1910, p. 2). The aim was to gather materials, preserve them, and subsequently make them available to visitors for identification purposes. Graham's interests prevailed in the facility, and within its first year of operation the museum had hosted collections of mosquitoes from across West Africa, as well as mosquito larvae from Lagos. This further entrenched the entomological study of mosquitoes and complemented the work being done by the Entomological Research Committee (Tropical Africa) in West Africa.

With this system in place, a division of labor was set in motion. While the MRI connected with various extractive agents within the imperial network tasked with collecting and transferring pathological materials, metropolitan scientists were able to focus on teaching and interpreting the information derived from these new materials to students in Britain. The institute's museum collected and labeled such materials, carefully stored them in museum containers, and prepared them for those requiring access (Nigeria. Medical Research Institute, 1912, p. 2).

Graham also facilitated the establishment of a bacteriological laboratory, which housed sterilizing and incubation rooms, an operating theater, a reaction stall, and a stall for large animals (Nigeria. Medical Research Institute, 1912, p. 2). These facilities were utilized in diverse ways, and their functionality was shaped by the research interests of the director, visiting researchers, and the relatively modest demands of colleagues affiliated with the West African Medical Staff. In the second annual report, Graham noted how the institute's attention was split across these lines. Thus, while he continued to work on mosquitoes, he was also called upon by some of his colleagues to investigate pathological materials—particularly internal organs, blood smears, and urine—to understand the physiological changes associated with blackwater fever infection (Nigeria. Medical Research Institute, 1912, p. 2).

A significant shift in the MRI's activities occurred shortly after Graham left office. Gradually, the institute's focus on malaria became interwoven with requests to investigate other diseases. Some of these were prompted by the yellow fever epidemic, which had begun to spread in places like Sekondi in the Gold Coast. After Graham stepped down as director and returned to London, he was replaced by Dr. Albert Ezra Neale (c. 1880–c. 1950), a junior medical officer of the West African Medical Staff who was seconded to the institute to continue its work during Graham's absence (Nigeria. Medical Research Institute, 1912, p. 2). This was a contingency arrangement and not intended to be a long-term one. Reading through Neale's report, it seems that he was still operating under Graham's shadow—mainly because he continued the investigations left unfinished by his predecessor. For example, he worked closely with medical officers to collect additional urine samples to complete the study of blackwater fever that was initiated by Graham. He also continued to examine water samples for mosquito larvae and other waterborne parasites, enriching the institute's mosquito collections and providing vector control recommendations to the Medical Officer of Health in Lagos.

Graham's influence at the MRI was undeniable, and though short-lived, his work on mosquitoes shaped the institute's early direction. However, the momentum he had built was lost after his departure, and malaria research (central under his tenure) gradually lost its primacy and made way for new focus areas. Notably, his departure coincided with the onset of the yellow fever epidemic, where new researchers arrived, new networks converged, and fresh agendas emerged. This shift was no accident; it was embedded in the outward-facing philosophy that defined the MRI's institutional character. From the outset, the empire had envisioned the MRI as operating beyond the narrow scope of the colonial medical service, driven instead by ideas and priorities shaped in distant metropolitan institutions. Given the diversity of these external influences, the institute could pivot easily depending on which expertise was most prominent at any given moment—just as it had aligned with Graham's interests in malaria during his leadership.

What this reveals is the underlying temporality introduced by the imperial system: a mode of operation in which the institute's focus was never fixed but continually reoriented by shifting external forces and personalities. One way to appreciate such temporality is by comparing developments in colonial laboratories to those in metropolitan schools during the same period. For example, malaria research was not deprioritized in metropolitan institutions such as the London and Liverpool schools, even as research agendas were changing at the MRI. Following the outbreak of the First World War, a building within the London School was dedicated to the study of malaria chemotherapy and dysentery (Wilkinson and Power, 1998). These institutions demonstrated significant flexibility in scope and focus, accommodating new research agendas alongside those initially developed during their formative years. Such flexibility could be interpreted as a sign of progress—at least for the tropical schools—but not for the MRI. The tropical schools were able to develop new research streams, financed through resources from either the imperial government or colonial territories.

In contrast, the MRI underwent drastic shifts in agenda in response to significant developments occurring outside its boundaries. As the tropical schools were expanding and creating new research paths (Wilkinson and Power, 1998, p. 281), this led to frequent changes at the MRI, i.e., both in terms of its focus and agenda, which were not optimal to ensure sustainability. This occurred as colonial laboratories lacked the capacity to pursue multiple research directions simultaneously and thus had to adopt a scale of preference, i.e., prioritizing and deprioritizing research based on agendas set by external institutions and stakeholders. Such fluctuations would leave an indelible impact on the MRI and on the tropical colonies' ability to advance malaria research. This development would eventually shift the locus of malaria research to the tropical schools in Britain and limit the colonies' capacity to engage in substantive work until the 1940s. One could also argue that this shift would contribute to the eventual demise of the colonial laboratory—particularly as external stakeholders began to redirect their attention elsewhere.

The MRI as a site for "everything"

The phase following Dr. William Graham's departure marked a pivotal transition, once again underscoring the fragility of institutions built upon external agendas. The historical episodes that shaped the MRI after Graham left office will next be considered, together with shifts that fundamentally reframed its research priorities. Although malaria was pivotal to the focus of the MRI, official conversations (particularly at the level of the imperial government) were increasingly concerned with yellow fever and the limited knowledge surrounding it. Therefore several surveillance efforts were undertaken within the colonial service in the Gold Coast to help colonial authorities and the imperial government understand the spread and etiology of the disease. Graham's first report shows that this development had little impact on the work being conducted at the MRI (Nigeria. Medical Research Institute, 1912). However, after his departure the new leadership steered the MRI in other directions, with this period characterized by the cultivation of new ties with scientists in the tropical schools and those affiliated with the newly constituted Yellow Fever Commission.

One of the more striking features of the MRI's early history is the temporality of its leadership, staffing, and research agendas. The directors who succeeded Graham (particularly during the first 3 years) had notably short tenures. Graham himself served as director for less than a year, and the same was true for his successor. During 1913, the MRI had two directors in a single calendar year. Dr. Arthur Hutton (c. 1880–c. 1950) assumed the role of Acting Director at the beginning of the year, followed by Dr. John Macfie (09-1879–10-1948) who succeeded him (Nigeria Medical Research Institute, 1913). Both were medical officers affiliated with the West African Medical Service and were recruited based on their expertise in yellow fever and trypanosomiasis, respectively. They were also seconded to the Yellow Fever Commission to investigate the disease's etiology. In its first report to the Colonial Secretary, the Commission noted that "staffs of the Medical Research Institute at Lagos, with investigators of the commission, have observed many yellow fever cases with great care," and that "scientific observations of the highest importance were made in Lagos" (, 1908, p. 4).

What influenced such temporalities experienced at the MRI? Here, it is important to revisit the point made earlier, namely that the imperial government, and not colonial authorities, defined what was considered important and worthy of study. As such, agendas were set externally, and the MRI remained in a continuous state of transition, responding to priorities determined on a larger stage. The contradiction here is that, although colonial governments funded the MRI, key decisions about its operations and research priorities were made by institutions outside its immediate context. Chief among these was the Advisory Committee, whose objective was to advance research in tropical medicine—much of which had begun before colonial laboratories were established. This clarification is crucial for understanding the undulations within the MRI, i.e., the frequent changes in leadership and the recurring disruptions in research focus.

As noted earlier, a key event that impacted the MRI's work following Dr. Graham's departure was the yellow fever epidemic. In response, the Yellow Fever Commission was established in 1913 "to study the nature and relative frequency of the fevers occurring among Europeans, natives, and others in West Africa, especially about Yellow Fever and its minor manifestations" (Yellow Fever Commission (West

Africa), 1913). The Commission designated Lagos as "the local center of investigation" (Yellow Fever Commission (West Africa), 1913)—not necessarily because of institutional robustness, but rather due to the MRI's availability and the abundance of cases that could be studied without hindrances. One way to think of the MRI is as a pathway to access the world around it. From time to time, it housed various institutions and their agendas, acting as a host for their work—a reservoir and laboratory for collecting and interpreting research materials (blood smears, urine, insects, etc.).

Some historical context will now be provided regarding this claim. The first recorded case of yellow fever was reported in Lagos in 1912. By 1913, hospital records showed that 34 cases were treated during the year, ten of which were Europeans, resulting in four deaths; three were Syrians, with two deaths; and 21 were natives, with zero mortality. This triggered a wave of panic. Everyone within the imperial government was attempting to rationalize the nature of such deaths and, by extension, the etiology of the disease. Yellow fever suddenly became the new malaria—perceived as having the potential to obstruct imperial business. The colonial annual report for 1913 documented casualty cases among government officials, distinguishing between the death tolls of "natives" and European personnel (Southern Nigeria Medical Department, 1913, p. 14). Beyond this, the outbreak reactivated the African immunity theory, which had previously shaped malaria control strategies earlier in the century. The Principal Medical Officer of Health reported that symptoms in native cases were variable, far from typical, and generally not severe. He observed: "There were no deaths. In several instances, the disease was characterized by fever, jaundice, albuminuria, etc. A concurrent malarial infection complicated some cases … One of the most severe cases was detected during the routine examination of passengers traveling on the railway. This patient suffered from a moderate fever and was apparently feeling so well that he was able to undertake a long journey" (Nigeria Medical and Sanitary Department, 1913, p. 23). He further noted that "the mildness of the disease in natives may not unreasonably be attributed to the possible fact that practically all are attacked in childhood."

The fixation on yellow fever served as a convenient distraction from a more pressing health crisis. During 1913, relatively high infant mortality rates were recorded among the African population, unrelated to yellow fever. Thus one could argue that yellow fever was not the greatest threat to Africans at the time. At the height of the yellow fever epidemic, 49% of deaths recorded in Southern Nigeria were of children under 5 years old and 1,100 children died in Lagos that year. Most of these children died of malaria, as they were highly susceptible to the disease, and the environments in which they lived were natural breeding grounds for mosquitoes. According to the 1913 annual medical report, there was not a single death reported from yellow fever among the African population, and most of the complications of this disease occurred because some patients displayed a concurrent malarial infection (Nigeria Medical and Sanitary Department, 1913, p. 16). The government's main observation (and perhaps its chief complaint) was that residents kept water-holding containers around their homes, which served as ideal breeding sites for mosquitoes.

Hence, for that year, approximately 20,000 notices in this regard were issued to residents in Lagos (Nigeria Medical and Sanitary Department, 1913, p. 23).

Of note, such African health issues were never intended to be the focus of the MRI. From 1913 onwards, it became the frontline site of the British campaign against yellow fever, hosting investigators from the Yellow Fever Commission and serving as a recipient of blood smears and urine samples from patients suspected of infection. One of the scientists hosted at the MRI was Dr. Harald Seidelin (1878–1932), a staff member of the Liverpool School and former director of the Medical Institute in Yucatan, Mexico. Seidelin had conducted a series of medical expeditions on yellow fever in the Gold Coast. Prior to his work in Accra, he had been involved in complex research on yellow fever at the institute in Yucatan (Sowell, 2015, p. 96). As Sowell argues, it was in Yucatan that Seidelin discovered a protozoan-like body, which he named *Paraplasma flavigenum*, as a possible vector of yellow fever (Yellow Fever Commission (West Africa), 1913, p. 4). The expedition in Accra merely served to authenticate his theory. The Commission appointed Seidelin to join other scientists at the MRI to replicate the work he had conducted in Accra. Alongside the MRI director, he participated in a series of investigations, including research aimed at authenticating his discovery of *Paraplasma flavigenum*. The team pursued microscopic examinations of bodily materials to identify the presence of "Seidelin bodies" and their possible connection to the yellow fever virus (Yellow Fever Commission (West Africa), 1913, p. 4). The findings from these experiments were shared with members of the Royal Society of Medicine in 1913 and published in the seventh edition of the Society's proceedings the following year (Nigeria. Medical Research Institute, 1914).

The MRI and the mobility of colonial expertise

The MRI also served as an incubatory space for graduates from Britain to develop competence in tropical medicine research before eventual remobilization across the length and breadth of the empire. This has been a defining characteristic of the tropical medicine specialty since its inception. Tropical colonies were sites of discovery—not just of microbes, but of experts. The history of tropical medicine reveals remarkable breakthroughs in disease transmission and the emergence of specialists in places like India during the late 19th and early 20th centuries (Harrison, 1992). The establishment of laboratories such as the MRI in Lagos operationalized this nurturing process. The MRI became a breeding ground for colonial doctors to consolidate their research careers. This dynamic further contributed to the laboratory's inability to retain competent personnel who could guarantee its permanence. The MRI became a site of discovery—or perhaps a transitional phase—for such doctors, i.e., a space where experts were identified and then remobilized to replicate such work elsewhere.

One of the colonial doctors involved in the early activities of the MRI was John Macfie, who took over as director after Dr. Graham left Lagos. He was a medical officer affiliated with the West African Medical Staff, having completed his medical training at the University of Edinburgh shortly thereafter. Before embarking on his trip to Lagos, he undertook a diploma course at the Liverpool School of Tropical Medicine. Regular tours between 1910 and 1922 characterized his career in West Africa. Prior to assuming the role of director, he had been actively involved in epidemiological fieldwork, studying malaria, yellow fever, trypanosomiasis, and other diseases. During 1914, Dr. Macfie was commissioned alongside two colleagues from the West African Medical Staff—Dr. E.J. Wyler (c. 1880–1950) and Dr. Andrew Connal (1883–1949)—to work with Seidelin on his theory of yellow fever (Yellow Fever Commission (West Africa), 1913, p. 4). Macfie led the team in a series of investigations that included research aimed at mapping Seidelin's *Paraplasma flavigenum* in blood smears sent to the institute by colonial doctors and in mosquitoes caught by sanitary inspectors (Yellow Fever Commission (West Africa), 1913, p. 4). These early studies were instrumental in discrediting *Paraplasma flavigenum* as the cause of yellow fever. The microorganism was not found in either the blood samples or the mosquitoes examined.

The careers of medical officers offer insight into what the tropical environment—alongside its constituent elements of laboratories, microbes, and diseases—meant to both the officers and the imperial project. As medical students, the tropics appealed as a study site—an opportunity to engage with tropical medicine in its natural setting. This explains why many medical officers specializing in tropical medicine were highly mobile, exhibiting the ability to traverse the empire in search of answers. For example, Dr. Macfie left his position as director after a few months to take up a post as a pathologist in Accra (Nigeria. Medical Research Institute, 1914, p. 19). Likewise, Dr. H. Sinclair Coghill (1880–1919) worked briefly as assistant director of the MRI after working at the London School of Tropical Medicine, where he had first been a student and later a demonstrator (Nigeria. Medical Research Institute, 1914, p. 19).

As much as they contributed to the expanding field of tropical medicine, the activities of medical officers reveal that the research of tropical diseases was often prioritized over actually lowering their disease burden. Hence, very little was done in terms of actual disease control during the times when intensive research efforts were pursued. Tropical medicine, therefore thrived as a research enterprise within tropical colonies and at the MRI, but it was rarely translated into practice (Johnson, 2010). Beyond the structural barriers that rendered control efforts ineffective, the imperial state's priority was to provide an enabling research environment for its medical officers.

Conclusion

The early history of tropical medicine—specifically of an institution like the MRI—has serious implications regarding how we imagine inequality in global health research. Inequality is not limited to the sidelining of researchers from underresourced settings in authorship and funding. Therefore decolonization should not be reduced to diversifying authorship or restructuring research finance processes and schemes. The history of the MRI reveals foundational problems within the research field, which rendered institutions in LMICs perpetually functioning as appendages or satellites to metropolitan institutions in the Global North. Invariably, LMICs and their research institutions continue to serve as viable tools for the legitimization of knowledge rather than for actual public health work.

This history reinforces the notion that research institutions in underresourced settings were never really neutral spaces. As much as they were shaped by the personalities that led them and by certain local exigencies, they were often guided or influenced by British metropolitan institutions to which they were historically tied. These structures played a role in setting the agendas of colonial research institutions and in tilting them to mirror external interests. Hence, the MRI was created to facilitate research agendas on behalf of metropolitan institutions—a responsibility it effectively fulfilled during its early history.

The MRI's establishment was born out of imperial expediency, and its continued relevance was shaped by the interests pursued by both the imperial government and metropolitan institutions. By the 1940s, the MRI had become less functional within the colonial medical services as the landscape of medical research had radically changed. The work once carried out by the MRI on yellow fever was now being conducted by the Rockefeller Yellow Fever Research Laboratory. During 1949, a new institution, headed by Professor Leonardo Bruce-Chwatt, was established in Lagos to facilitate research work on malaria. However, like the MRI this facility was coordinated by a technocrat who was recruited from outside Nigeria and who was in frequent partnership with experts in metropolitan institutions and the WHO. It served as the major role player in terms of malaria surveillance and a series of pilot studies on malaria eradication programs in Nigeria. However, the institution did not last long, i.e., not beyond the tenure of its director who left to take up a leadership position in the WHO during 1958.

References

Report of the Advisory Committee for the Tropical Diseases Research Fund, 1906. Darling and Son, England.

Report of the Advisory Committee for the Tropical Diseases Research Fund, 1908. Darling and Son, England.

Besson, E.S.K., 2022. How to identify epistemic injustice in global health research funding practices: a decolonial guide. BMJ Glob. Health 7 (4), e008549.

Bhakuni, H., Abimbola, S., 2021. Epistemic injustice in academic global health. Lancet Glob. Health 9 (10), e1465–e1470.

British Online Archives, 1908. *Medical Research Institute, 1907–1929*. Simpson to Egerton.

Chaudhuri, M.M., Mkumba, L., Raveendran, Y., Smith, R.D., 2021. Decolonising global health: beyond 'reformative' roadmaps and towards decolonial thought. BMJ Glob. Health 6 (7), e006371.

Costello, A., Zumla, A., 2000. Moving to research partnerships in developing countries. BMJ 321 (7264), 827–829.
Curtin, P.D., 1961. "The White Man's Grave": image and reality, 1780–1850. J. Br. Stud. 1 (1), 109–110.
Curtin, P.D., 1997. Disease and Empire: The Health of European Troops in the Conquest of Africa. Cambridge University Press, Cambridge.
Davies, L., 2022. Yaba housing scheme and the colonial 're-planning' of Lagos, 1917–1952. Plan. Perspect. 37 (2), 267–292.
Dumett, R.E., 1968. The campaign against malaria and the expansion of scientific medical and sanitary services in British West Africa, 1898–1910. Afr. Hist. Stud. 1 (2), 153–197.
Entomological Research Committee (Tropical Africa), 1911. Instructions to Collectors. H.M. Stationery Office, London.
Erondu, N.A., Aniebo, I., Kyobutungi, C., et al., 2021. Open letter to international funders of science and development in Africa. Nat. Med. 27, 742–744.
Galtung, J., 1969. Violence, peace, and peace research. J. Peace Res. 6 (3), 167–191. doi:10.1177/002234336900600301.
Geissler, W., Molyneux, C., 2011. Parasite lost: remembering modern times with Kenyan government medical scientists. In: Geissler, W., Molyneux, C. (Eds.), Evidence, Ethos, and Experiment. Berghahn Books, Oxford.
Ghosh, A., 1996. The Calcutta Chromosome: A Novel of Fevers, Delirium and Discovery. Picador, London.
Graham, W.M., 1910. On new species of West-African Cullicidae. Ann. Mag. Nat. Hist. Series 8. 5 (25–30), 264.
Harrison, M., 1992. Tropical medicine in 19th-century India. Br. J. Hist. Sci. 25 (3), 299–318.
Haynes, D.M., 1999. The social production of metropolitan expertise in tropical diseases: the imperial state, colonial service and the tropical diseases research fund. Sci. Technol. Soc. 4 (2), 205–238.
Haynes, D.M., 2001. Imperial Medicine: Patrick Manson and the Conquest of Tropical Disease. University of Pennsylvania Press, Philadelphia.
Head, M.G., Goss, S., Gelister, Y., Alegana, V., Brown, R.J., Clarke, S.C., Fitchett, J.R., Atun, R., Scott, J.A.G., Newell, M.L., Padmadas, S.S., 2017. Global funding trends for malaria research in sub-Saharan Africa: a systematic analysis. Lancet Glob. Health 5 (8), e772–e781.
Johnson, R., 2010. The West African Medical Staff and the administration of imperial tropical medicine, 1902–14. J. Imp. Commonw. Hist. 38 (3), 419–439.
Lawrence, D.S., Hirsch, L.A., 2020. Decolonising global health: transnational research partnerships under the spotlight. Int. Health 12 (6), 518–523.
Loewenson, R., 1993. Structural adjustment and health policy in Africa. Int. J. Health Serv. 23 (4), 717–730.
malERA Consultative Group on Vaccines, 2011. A research agenda for malaria eradication: vaccines. PLoS Med. 8 (1), e1000398.
Nigeria Medical and Sanitary Department, 1913. Annual Medical and Sanitary Report, 1913. Wellcome Collection. Government Printer, Lagos.
Nigeria Medical Research Institute, 1913. Annual Report on the Medical Research Institute, 1909. Wellcome Collection. Government Printer, Lagos.
Nigeria. Medical Research Institute, 1910. Annual Report on the Medical Research Institute, 1909. Government Printer, Lagos.

Nigeria. Medical Research Institute, 1912. Annual Report on the Medical Research Institute. Government Printer, Lagos.

Nigeria. Medical Research Institute, 1914. Annual Report on the Medical Research Institute. Government Printer, Lagos.

Osborne, M., Kent, S.K., 2015. Africans and Britons in the Age of Empires, 1660–1980. Routledge, Abingdon.

Southern Nigeria Medical Department, 1913. Annual Medical Report, 1913. Wellcome Collection. Waterlow, London.

Sowell, D., 2015. Medicine in the Periphery: Public Health in Yucatan, Mexico, 1870–1960. Lexington Books, Maryland.

Vaughan, M., 2018. A research enclave in 1940s Nigeria: the Rockefeller Foundation yellow fever research institute at Yaba, Lagos, 1943–49. Bull. Hist. Med. 92 (1), 172–205.

Wilkinson, L., Power, H., 1998. The London and Liverpool schools of tropical medicine 1898–1998. Br. Med. Bull. 54 (2), 281–292. doi:10.1093/oxfordjournals.bmb.a011688.

Worboys, M., 1979. *Science and British colonial imperialism, 1895–1940* (PhD thesis). University of Sussex.

Worboys, M., 1976. The emergence of tropical medicine: a study in the establishment of a scientific speciality. In: Lemaine, G., et al., (Eds.), Perspectives on the Emergence of Scientific Disciplines. Mouton, The Hague and Paris, pp. 76–98.

Worboys, M., 2022. Manson, Ross and colonial medical policy: tropical medicine in London and Liverpool, 1899–1914. In: Lewis, M., MacLeod, R. (Eds.), Disease, Medicine and Empire. Routledge, London, pp. 21–37.

Worboys, M., 1988. Manson, Ross, and colonial medical policy: tropical medicine in London and Liverpool, 1899–1914. In: MacLeod, R., Lewis, M., (Eds.), Disease, Medicine and Empire: Perspectives on Western Medicine and the Experience of European Expansion. Routledge, London, pp. 21–37.

Yach, D., Ron, A., Nitzan, D., 2025. The golden age of global health is over. What follows? J. Epidemiol. Glob. Health 15 (1), 98.

Yellow Fever Commission (West Africa), 1913. First Report. HMSO, London.

CHAPTER 5

When evidence misrepresents: Epistemic failures of evidence-based medicine and the BiDil controversy

Wandile Ganya[a] and Wanele Ganya[b]

[a]*Division of Cardiology, Department of Medicine, Faculty of Medicine and Health Sciences, Stellenbosch University, Cape Town, South Africa*

[b]*Division of Nephrology, Department of Medicine, Faculty of Medicine and Health Sciences, Stellenbosch University, Cape Town, South Africa*

Highlights

- EBM's use of extensional population statistics cannot license direct, individual-level inference because such generalizations lack the counterfactual invariance required for modal justification.
- This failure reflects EBM's implicit Humean metaphysics, which treats co-variation as causal relevance and thereby generates systematic category mistakes, exemplified by race-based inference in cases such as BiDil.
- A metaphysically adequate alternative requires that clinical generalizations be grounded in invariant causal structure—captured through structural realism, dispositional essentialism, and interventionist criteria for modal stability.
- High-impact clinical and regulatory claims must, therefore, be conditioned on mechanistic coherence and demonstrated invariance across contexts, eliminating extensional classifiers (such as race) that lack causal and modal grounding.

Introduction

Evidence-based medicine (EBM) has established itself as the dominant methodological framework in contemporary clinical practice. It seeks to regulate therapeutic and diagnostic decisions through recourse to aggregated data, primarily derived from randomized controlled trials (RCTs), metaanalyses, and large-scale observational

Decolonizing Medical Education, Research and Practice. https://doi.org/10.1016/B978-0-443-44998-7.00015-6

studies. EBM positions itself as a corrective to the perceived arbitrariness of clinical judgment, offering instead a model of inference grounded in bare statistical generalizations and hierarchies of evidence. The promise is both epistemic and practical: that the systematic deployment of population-level evidence will enhance clinical reliability, minimize bias, and ensure objectivity in clinical decision-making (Tonelli, 2006; Sackett et al., 1996).

Yet this promise rests on philosophical assumptions that are rarely made explicit and, as we shall argue, are deeply problematic. While existing critiques of EBM have drawn attention to its overreliance on frequentist inference, its neglect of mechanistic reasoning, and its limited capacity to accommodate individual variation, these objections typically remain at the methodological or epistemological level. They address how EBM reasons, but not what it presupposes about the nature of properties, the status of bare statistical generalizations, and the conditions under which they can legitimately support direct inference in clinical contexts.

This chapter argues that EBM implicitly relies on an extensionalist metaphysics, wherein individuals are grouped into reference classes based on surface-level resemblances such as age, sex, or race, and inferences across these classes are taken to be justified if they exhibit statistical robustness (used interchangeably with statistical regularity hereinafter). On this view, bare statistical generalizations are allowed by virtue of their empirical frequency alone, irrespective of whether the properties involved are causally operative or explanatorily relevant.

Such metaphysical commitments are inadequate to the epistemic demands of clinical judgment. Clinical reasoning, we argue, is of necessity modal: it requires the capacity to evaluate what would happen to a particular patient under hypothetical intervention. This, in turn, depends on generalizations that are *modally robust*, that is, generalizations that remain stable under plausible, counterfactual variation in background conditions. This is not a demand for invariance under arbitrary variation, but under variation that is both plausible and epistemically salient. A correlation that holds in one population may fail in another if the underlying causal structure differs. It follows that statistical regularity, however well established, is insufficient for inference unless grounded in an account of the mechanisms and dispositional properties that sustain the regularity across contexts (Cartwright, 2007; Woodward, 2003).

EBM systematically fails to respect these modal and metaphysical constraints. Its standard inferential practices rely on what we term *bare statistical generalizations*: frequency claims that are not anchored in explanatory or causal structure. Such generalizations are *modally fragile*; they break down under contextual variation and cannot support the counterfactual reasoning that clinical decision-making demands (Cartwright, 2007; Woodward, 2003; Ganya and Ganya, 2025). While this fragility is especially visible in ethically sensitive domains such as race-based diagnostic thresholds, it is symptomatic of EBM's broader inferential architecture.

In diagnosing this failure, this chapter advances a metaphysical alternative grounded in epistemic structural realism and dispositional essentialism. Structural realism, in its epistemic form, holds that what science tracks (when successful) is not merely the empirical surface of phenomena but the underlying causal structures that generate observable patterns. Dispositional essentialism, meanwhile, maintains that many clinically salient properties are irreducibly dispositional: their identity and explanatory role are defined by what they tend to do under specific conditions. Regarding this view, the clinical relevance of a property cannot be inferred from co-variation alone; it must be established by showing that the property forms part of a stable causal nexus that supports modal inference.

These metaphysical commitments converge in an interventionist account of causation, particularly as developed by Woodward (2003). Interventionism provides a formal characterization of causal relevance in terms of stability under counterfactual manipulation. It thereby offers a criterion for distinguishing epistemically warranted generalizations from those that merely report coincidental or context-sensitive patterns. In this framework, the legitimacy of a clinical inference depends not on statistical regularity per se but on whether the generalization remains invariant under the kinds of hypothetical interventions that clinical reasoning routinely invokes.

This chapter extends prior work (Ganya, 2024; Ganya and Ganya, 2025) and proceeds as follows. We will initially introduce the problem of modal fragility and diagnose the epistemic shortcomings of bare statistical generalizations. Thereafter, we situate this fragility within a broader extensionalist ontology and trace its roots to a Humean conception of causation. We subsequently articulate the metaphysical costs of this framework and argue for its replacement with a modal account grounded in structural realism and dispositional essentialism. We next develop this alternative framework and demonstrate how it underwrites counterfactual inference in clinical contexts. The following section then applies the framework to concrete cases of misclassification, focusing on diagnostic thresholds, algorithmic prediction, and race-based medicine. We thereafter anticipate and address key objections and conclude by restating the necessity of metaphysical reform for epistemically and ethically adequate clinical reasoning.

Why bare statistical generalizations fail: A modal critique

The problem of modal fragility arises from a failure to distinguish between merely descriptive regularities and those that possess explanatory or causal significance. In its standard evidential practices, EBM makes frequent appeal to generalizations derived from population-level studies. These generalizations are typically assumed to warrant direct inference to individual cases by virtue of their statistical robustness (Sackett et al., 1996; Howick, 2011). However, such inferences are vulnerable to epistemic failure when they are not grounded in a structure that ensures their modal resilience, that is, their stability under plausible variations in background conditions (Ganya and Ganya, 2025).

Consider the clinician who reads that "patients aged 40-65 with elevated C-reactive protein (CRP) levels are 40% more likely to experience adverse cardiac events within 6 months." On the face of it, this looks like a statistically robust generalization and is relevant to cardiovascular care. But the clinician's epistemic task, we argue, is not merely to classify a patient as falling within a reference class and then apply a frequency, but rather to determine whether the statistical tendency observed in the population would apply to this individual under possible intervention. Would this patient suffer a cardiac event if left untreated? Would they avoid one if treated with medication *x*? These are modal questions (i.e., questions focusing on what is possible, necessary, or impossible), and they are not answered by frequencies alone unless those frequencies are grounded in a mechanism or causal structure that remains stable across the relevant counterfactual worlds (Pearl, 2009; Deaton and Cartwright, 2018).

This issue is not merely theoretical; it reflects the structure of direct inference, which is central to clinical reasoning. Clinical decisions must be justified not only by what has occurred in aggregate populations but also by what would occur in a specific individual under hypothetical interventions. Inference of this sort requires generalizations that are projectable: they must apply across a range of cases that share the relevant underlying structure, not merely surface-level attributes. When this condition fails, inference becomes speculative rather than evidentially warranted (Goodman, 1955; Cartwright, 2007).

As intimated in the foregoing section, to say that a generalization is modally fragile is to claim that it supports inference in some contexts but fails to do so in others, without any principled account of where or why the breakdown occurs. Its application relies on background features that are neither transparent nor tracked by the generalization itself. Consequently, the generalization licenses projection without warrant, and inference becomes indistinguishable from assumption (Tonelli, 2006; Stegenga, 2018).

The source of this failure lies in the ontological assumptions implicit in extensional grouping. When generalizations are formed over extensional predicates, such as age, sex, race, or biomarker values, there is a risk of conflating descriptive similarity with explanatory sameness. Yet, these are not equivalent. Descriptive similarity is epistemically relevant only when it tracks properties that are causally or dispositionally efficacious. Hence, we argue that if the shared features of the reference class are not responsible for the outcome, then the generalization is explanatorily inert, and its use in inference is a category mistake—this relates to the problem of projectability (Goodman, 1955).

Not all predicates that describe empirical regularities support inductive inference. As Goodman argues, "green" is projectable, whereas "grue" is not, despite their identical empirical profiles up to a point (Goodman, 1955). The distinction is not merely semantic but epistemological: it concerns whether the predicate is entrenched in practices of explanation and inference that reflect real structure. In clinical reasoning, the analog to projectable predicates is able to track stable causal relations. As Cartwright notes, capacities are the explanatory units of scientific inference (Cartwright, 2007). Pearl likewise provides a formal model of causation that distinguishes correlation from causal relevance using directed acyclic graphs and interventions (Pearl, 2009). Generalizations that rely on superficial predicates fail to meet this standard. They are not merely inductively weak; they are epistemically defective insofar as they lack modal significance.

The challenge is compounded by the fact that background conditions in clinical settings are almost always epistemically opaque (Ganya, 2024). Biological systems are complex, historically contingent, and multiply determined. Patients are embedded in contexts defined not only by physiology but also by environmental, social, and economic variables that vary widely and are rarely captured by trial designs. Without an account of how and why a generalization holds—an account that explains its dependence on background conditions—inference becomes an exercise in projection without justification (Vandenbroucke, 2008; Ioannidis, 2005).

The problematic use of race-based adjustments in clinical algorithms illustrates this issue vividly. Race, operationalized as a variable in clinical inference, functions as an extensional predicate: individuals are grouped based on self-identified categories, and generalizations are formed over that grouping (Haslanger, 2012; Strand, 2021). However, race is not a causally efficacious property per se. It functions as a proxy for diverse variables such as genetic ancestry, social determinants, environmental exposures, and access to care (Duster, 2007). The generalizations drawn from race-based classifications are doubly fragile: race is not itself a causal variable, and the features it proxies for differ widely across individuals.

Extensional predicates are acceptable only insofar as they track intensional kinds. When inference is based on co-variation across groupings that lack causal unity, error is not only possible but likely. This is not a mere failure of empirical accuracy; it is a metaphysical mistake—treating empirical frequencies as if they had the modal force of causal laws (Anjum et al., 2020).

The consequence is a clinical epistemology that systematically overstates the inferential reach of statistical generalizations. It applies population-level correlations to individuals without ensuring that the generalizations are grounded in causal mechanisms that remain stable across contexts. EBM, by failing to distinguish between extensional correlation and intensional relevance, constructs a model of inference that is deeply vulnerable to modal fragility (Bluhm, 2005; Howick, 2011).

In the next section, we examine how this epistemic fragility stems from EBM's underlying metaphysical commitments, particularly its adherence to an extensionalist ontology.

Extensionalist ontology in EBM

The critique advanced thus far locates the epistemic limitations of EBM not merely in the limitations of its statistical methods but in a set of metaphysical commitments. These commitments pertain to the nature of properties, the status of regularities, and the grounds of inferential legitimacy. In particular, EBM appears committed to an extensionalist ontology—one in which surface-level similarities among individuals justify generalization and in which regularities are treated as sufficient for inference without the need for explanatory or modal grounding (Tonelli, 2006; Stegenga, 2018; Howick, 2011).

In formal semantics, an extensional context is one in which the substitution of co-referential expressions preserves truth value. Extensional classification, by analogy, refers to the grouping of entities based on observable or measurable features, such as demographic markers, biomarker levels, or diagnostic labels. These classifications are operationalized in reference classes used to generate population-level statistical generalizations. The assumption is that shared extensional properties are sufficient for predictive inference: if two individuals are similar in the relevant observable respects, then outcomes observed in one may be expected in the other (Sackett et al., 1996; Bluhm, 2005).

However, the problem arises when extensional similarity is treated as sufficient for inferential projection. That is, when membership in a reference class, defined by purely extensional criteria, is taken to warrant inferences about unobserved outcomes, future events, or the effects of interventions. This move commits a well-known error: it conflates statistical co-variation with explanatory or causal relevance. In Goodman's (1955) terms, it fails to distinguish between projectable and nonprojectable predicates. A predicate is projectable only when it corresponds to a property that participates in a causal or explanatory structure. Mere surface-level resemblance does not guarantee that an outcome will follow unless the similarity tracks a modally robust property—one that supports the inference under plausible background variation (Goodman, 1955; Cartwright, 2007).

This distinction has profound implications for clinical reasoning. Diagnostic categories such as "hypertension" or "diabetes" are often defined by thresholds drawn from population data. However, such thresholds, unless tied to underlying causal mechanisms, may reflect artifacts of data modeling rather than genuine boundaries in nature (Deaton and Cartwright, 2018). The risk is that patients are grouped based on features that correlate with outcomes but do not causally explain them, thereby rendering the generalization modally unstable and inferentially unreliable (Vandenbroucke, 2008; Ioannidis, 2005).

The inferential architecture of EBM, we argue, systematically collapses the distinction between statistical correlation and causal relevance. This collapse is not a matter of poor implementation but rather reflects a set of implicit metaphysical commitments. EBM treats regularities over extensional predicates as sufficient for clinical inference, provided they are statistically significant. In so doing, it adopts a conception of properties as categorical and context-insensitive, defined solely by their empirical co-occurrence with outcomes. This conception is broadly Humean in its ontology: causation is reduced to regularity, and properties are treated as passive rather than as sources of causal efficacy—see section below (Mumford and Anjum, 2011; Woodward, 2003).

Such a view excludes intentional features such as modal profile, causal role, or dispositional potential. As a result, clinical inference becomes a matter of pattern extrapolation rather than explanation. The generalization that "patients with trait *y* are more likely to benefit from treatment *z*" is treated as self-justifying, without regard to whether trait y plays any explanatory role in the causal pathway that links treatment z to the outcome.

This metaphysical commitment is often implicit. It manifests in the way EBM ranks evidence, elevating randomized trials and metaanalyses over mechanistic reasoning or individualized judgment, and in its reliance on statistical thresholds and predictive algorithms that abstract away from the underlying causal structure (Bluhm, 2005; Greenhalgh et al. 2014; Howick, 2011).

The result is a conception of clinical evidence that privileges extensional regularities while neglecting the modal and explanatory requirements of sound inference. The next section will elaborate on this critique by showing how the Humean metaphysics underlying EBM fails to support the counterfactual reasoning essential to clinical judgment.

The metaphysical minimalism of EBM

The metaphysical commitments underlying EBM extend beyond its reliance on extensional classification. At a foundational level, EBM is structured by a broadly Humean metaphysics, in which causation is conceived not as a real relation in the world but as a pattern of regular co-occurrence among events. This conception carries substantial implications for the kinds of generalizations EBM treats as evidentially legitimate truth-claims and for the epistemic status accorded to observed statistical associations (Tonelli, 2006; Stegenga, 2018).

According to the Humean view, the world consists of local matters of fact, with regularities emerging from the arrangements of particular events. Causation, in this view, is not a fundamental feature of reality but a descriptive shorthand for stable succession. Laws of nature are understood not as necessary relations but as the most economical summaries of the patterns we observe (Cartwright, 2007; Bird, 2007). This metaphysical minimalism renders explanation dispensable and treats statistical regularities as the primary currency of scientific knowledge.

Such a metaphysics, we argue, is well aligned with the core practices of EBM. Clinical trials aim to detect statistically significant differences in outcomes between treatment and control groups. Metaanalyses aggregate effect sizes across studies in the hope of producing more precise estimates. Clinical guidelines are then constructed by identifying statistically robust associations that generalize over large populations (Sackett et al., 1996; Howick, 2011). The implicit assumption is that once a regularity has been confirmed through appropriate statistical controls, no further explanation is needed. The regularity itself becomes the warrant for intervention in clinical decision-making.

However, this model inherits all the limitations of the Humean framework from which it derives. Most notably, it cannot account for the normativity of clinical inference. Causal claims in EBM are not merely descriptive; they are action-guiding. To assert that a drug causes a reduction in blood pressure is to license an intervention; it is to assert that administering the drug will, under relevantly similar conditions, bring about a particular outcome. This is a modal claim: it involves a commitment to what would happen under hypothetical interventions (Woodward, 2003).

The Humean framework, by denying that causation involves modal necessity, is structurally incapable of underwriting this kind of reasoning. In terms of the Humean view, the fact that a treatment has produced a given outcome in the past does not, by itself, license the inference that it would produce the same outcome under different background conditions. There is no ontological basis for projecting the regularity into nearby possible worlds (Goodman, 1955; Worral, 1989; Ganya and Ganya, 2025). This is precisely the kind of inferential move that clinical judgment requires, and it is one that EBM, when grounded in Humean metaphysics, is unable to justify.

The consequences are especially acute when applied to individual patients. A statistically robust generalization may be true in the aggregate, yet fail when applied to a particular case where the background conditions differ in ways not captured by the reference class. Clinical inference, if it is to be epistemically legitimate, must account for this possibility. It must evaluate whether the generalization remains stable under plausible counterfactual variation, a task that Humeanism is structurally unequipped to perform (Anjum et al., 2020; Ganya and Ganya, 2025).

Moreover, EBM's frequent reliance on statistical models without mechanistic elaboration leads to a systematic conflation of correlation with causation. Inferences are drawn from observed co-variation without specifying the underlying dispositional properties or causal structures that would explain why the regularity holds (Cartwright, 2007; Pearle, 2009). This practice results in a kind of epistemic opacity, where predictive accuracy masks a lack of explanatory depth (Ganya, 2024).

The result is a conception of medical inference that is precise in its quantification, yet fragile in its application. It offers statistical data without the modal structure necessary to guide clinical judgment. The reliance on Humean regularity renders EBM incapable of distinguishing between modally robust generalizations—those that remain valid across relevant contexts—and accidental correlations that are context-sensitive and unprojectable (Deaton and Cartwright, 2018; Ioannidis, 2005).

The metaphysical framework of EBM reduces causation to observable patterns and properties to categorical observables. It strips away the modal and dispositional content that clinical reasoning presupposes. As we shall argue in the following section, overcoming these limitations requires a metaphysical reorientation: one grounded in structural realism and dispositional essentialism, where generalizations are permitted only when they reflect the explanatory architecture of the domain.

A modal metaphysics for clinical inference

If the epistemic limitations of EBM arise from its extensionalist classification and Humean metaphysics, then an adequate corrective must be sought in a metaphysical framework that restores modal structure and explanatory depth to clinical inference. Structural realism and dispositional essentialism offer two such frameworks. Each affirms the reality of causal structure and the modal richness of properties, and together they provide a foundation for understanding when generalizations are epistemically warranted and projectable to individual cases.

Structural realism, in its epistemic variant, holds that what persists across theoretical change is not the intrinsic nature of entities but the relational structures instantiated among them (Worrall, 1989). Here, the aim of scientific inquiry is not to capture the hidden essences of phenomena but to identify the invariant networks of relations that underpin causal and explanatory success. In the context of EBM, this yields an important constraint: generalizations must not merely describe statistical co-variation but must reflect the causal architecture of biological systems in a manner that supports reliable inference under hypothetical variation.

This structuralist orientation is continuous with Woodward's interventionist theory of causation, according to which a causal claim is warranted only if it supports counterfactuals under suitable interventions (Woodward, 2003). That is, A causes B if, under a specific set of background conditions, intervening on A would bring about a change in B. But for such counterfactuals to have inferential force, they must be grounded in real and stable mechanisms. The role of structural realism is to justify the metaphysical integrity of these mechanisms: they are not mere summaries of empirical regularities but real features of the world that confer modal resilience to our generalizations (Woodward, 2003; Ganya and Ganya, 2025).

This explanatory demand is made more precise by Russo and Williamson's epistemic theory of causality (Russo and Williamson, 2007). Their account rejects both monistic reduction, where causes are identified solely with probabilistic dependencies or mechanistic chains, and ontological pluralism, which posits distinct types of causes tracked by distinct forms of evidence. Instead, they propose an integrated epistemology, wherein causal claims are warranted only when supported by both probabilistic and mechanistic considerations. The former establishes that a putative cause makes a difference; the latter explains why such a difference is to be expected under variation. Importantly, these two evidentiary strands converge on a single causal judgment, which is robust not merely in virtue of observed frequency but because it is modally stable, considering background theory and explanatory architecture (Russo and Williamson, 2007).

EBM, in its dominant formulations, remains largely insensitive to these constraints. The legitimacy of generalizations is often taken to reside in their statistical robustness (Sackett et al., 1996). Yet, this conception treats evidential support as extensional, ignoring the modal and explanatory conditions that would allow such generalizations to project to new cases (Deaton and Cartwright, 2018; Ioannidis, 2005). A treatment effect may hold in a trial population yet fail to generalize to a new patient if the underlying mechanism is absent or differently instantiated. Without identifying the structural conditions that mediate the effect, there is no principled way to assess its stability under intervention. In such cases, statistical robustness masquerades as explanatory adequacy.

Structural realism clarifies the source of this fragility. The failure of generalization is not merely a matter of empirical error but of inferential misalignment: the absence of a stable relational structure that licenses counterfactual projection. The epistemic theory of causality provides the normative framework for this diagnosis. It holds that statistical and mechanistic evidence are not alternative paths to causal knowledge but jointly necessary components of a unified evidential base. Only where both converge do we have grounds for asserting a causal relation that is not merely predictive but explanatory (Russo and Williamson, 2007).

This integration of structural realism with an epistemic theory of causality thus imposes a principled standard on clinical inference. A generalization is evidentially permitted only if it is grounded in structures that remain invariant across the counterfactual space relevant to the clinical judgment. Absent such grounding, the inference is modally fragile and normatively defective. On this view, what EBM requires is not a statistical recalibration but a metaphysical reform: a shift from extensional description to modal explanation, from observed regularities to causally structured generalizations.

This account is grounded in *dispositional essentialism*, the metaphysical thesis that properties are individuated not solely by their categorical features but by their modal profiles, that is, by the dispositions they confer upon their bearers across a range of counterfactual circumstances (Bird, 2007; Mumford and Anjum, 2011). This view stands in marked contrast to the Humean conception, according to which properties are inert and context-independent, and causal relations are merely contingent regularities among such properties. For the dispositional essentialist, by contrast, a property is the power to behave in certain ways under certain conditions; its causal tendencies are metaphysically internal to it.

This ontological shift has direct implications for medical concepts. Consider "hypertension": it is not merely the possession of a particular blood pressure reading but a tendency, under suitable physiological constraints, to contribute causally to outcomes such as endothelial damage, left ventricular hypertrophy, or stroke. Its identity as a pathological condition is constituted by its dispositional role within a broader causal nexus, not merely by a numerical threshold.

The same holds for pharmacological properties. A drug's efficacy cannot be regarded as a monolithic, context-insensitive effect, inferable purely from aggregate data. Rather, its therapeutic power is a manifestation of a dispositional profile—its capacity to produce specific physiological changes in organisms that instantiate the requisite receptive dispositions. For example, a beta-blocker exerts its antihypertensive effect via antagonism at β-adrenergic receptors, thereby modulating sympathetic nervous system activity. However, this causal tendency is not unconditional: it depends upon the presence, density, and functional integrity of the target receptors, as well as downstream signaling pathways. Should these mediating conditions be absent or altered, as in cases of receptor polymorphism or comorbid pathophysiology, the drug's dispositional power will fail to be actualized. Its efficacy, then, is not a simple empirical generalization but a context-dependent expression of latent causal structure.

Dispositional essentialism thereby furnishes a principled explanation for why clinical generalizations must be assessed for modal robustness. To assert that a treatment will succeed in a new patient is to claim, implicitly, that the dispositional configuration responsible for the observed effect in prior cases is replicated in the new case. It is to claim more than mere similarity in observable traits; it is to claim that the underlying powers, structures, and conditions that mediate the causal effect are present once again.

This requirement is systematically omitted by conventional formulations of EBM, which treat evidential warrant as a function of population-level co-variation. In doing so, they obscure the modal content of inference, reducing prediction to extrapolation from statistical patterns over extensional groupings. The epistemic result is a form of projection error: generalizations that are not anchored in dispositional similarity may appear robust in aggregate but fail in application (Tonelli, 2006; Stegenga, 2018).

Indeed, as Anjum et al. (2020) have emphasized, clinical judgment often hinges on a fine-grained understanding of interacting causal powers specific to the individual. This is particularly salient in contexts of multimorbidity, polypharmacy, or atypical physiological profiles—precisely those cases in which statistical generalizations are least reliable. From a dispositionalist standpoint, such variation is not epistemic noise to be smoothed out by averaging but a signal of the causal complexity that responsible inference must accommodate. Clinical judgment, on this view, is a form of modal discernment: a capacity to recognize, in the individual case, the presence or absence of the dispositional alignments necessary for a given intervention to succeed.

Although structural realism and dispositional essentialism emerge from distinct philosophical lineages—one broadly epistemological, the other ontological—they converge on a foundational metaphysical insight: the projectability of generalizations in the sciences depends not on surface-level regularities but on the modal stability of the properties and relations involved. Structural realism elucidates the epistemic reliability of certain generalizations by locating them within stable relational structures. Dispositional essentialism, in turn, accounts for the explanatory force of those structures by grounding them in the modal natures of the entities that instantiate them.

Applied to clinical inference, this convergence yields a principled alternative to the extensionalist and Humean presuppositions that tacitly undergird EBM. Here, the evidential force of a generalization derives not from statistical co-variation alone but from its embedding within a causally robust architecture. Such architecture is constituted by structural relations among entities whose dispositional properties determine their potential to bring about effects under specified conditions. It is this joint structure—modal, not merely empirical—that licenses counterfactual reasoning of the sort routinely employed in clinical decision-making.

Crucially, this framework also provides a rationale for differentiating forms of evidence in terms of their epistemic depth. Mechanistic evidence, for instance, does not merely supplement statistical findings with additional detail; rather, it discloses the dispositional basis of observed regularities. Its epistemic significance lies in its capacity to identify the causal scaffolding that determines whether and how an effect generalizes across contexts. Likewise, individual case studies or qualitative investigations can be epistemically illuminating when they reveal the specific configuration of causal powers operative in atypical or multimorbid patients. What they uncover is not anecdotal residue but context-specific instantiations of a more general dispositional ontology.

The implication is clear. If clinical generalizations are to be both epistemically justified and ethically defensible, they must be grounded in metaphysical assumptions that acknowledge the modal character of the clinical domain. This does not entail the rejection of statistical methods, but it does demand their subordination to a framework in which inferential legitimacy is determined by explanatory relevance and causal invariance, rather than by frequency alone. Absent such a framework, inference risks becoming unmoored from the very structures that confer stability and predictive power.

Having established this integrated metaphysical foundation, we can now examine how its absence manifests in practical contexts, specifically how the failure to track dispositional and structural features results in inferential missteps and, in some cases, ethical harm.

Race, thresholds, and the ethics of metaphysical error

The preceding sections have developed a metaphysical framework capable of distinguishing modally permitted generalizations from those that are merely extensional or statistically robust. We now turn to assess how the absence of such a framework leads to concrete failures in clinical inference—failures that are not only epistemic but also ethically consequential.

Clinical inference is shaped not only by empirical data but also by the metaphysical commitments embedded in its categories, thresholds, and algorithms. Once situated within a broadly Humean and extensionalist framework—one that treats properties as inert, kinds as extensional aggregates, and generalizations as justified by statistical co-variation—the epistemic and ethical consequences of such assumptions become clear. Misclassification is not merely a practical error; it is a metaphysical one, stemming from a failure to align clinical categories with the dispositional and structural features that actually govern health outcomes.

Diagnostic thresholds exemplify this problem. Numerical cut-offs, such as blood pressure above 140 mmHg or fasting glucose beyond 7.0 mmol/L, are typically justified by population-level risk curves derived from large-scale studies (Vandenbroucke, 2008). From an extensional perspective, these thresholds are unproblematic: they serve as proxies for disease risk. Yet, from a dispositionalist standpoint, this practice reifies statistical artifacts into putative natural kinds. Two patients may occupy the same side of a threshold but differ radically in their causal dispositions, such as hormonal regulation or inflammatory profiles, none of which is captured by the threshold alone. Structural realism compounds this concern: unless a threshold corresponds to a point at which the underlying causal architecture shifts, it lacks modal grounding and cannot support reliable generalization. Most thresholds are determined not by mechanistic discontinuities but by pragmatic concerns, such as sensitivity and specificity. The result is an epistemic illusion of precision: thresholds appear objective but mask dispositional heterogeneity, thereby licensing systematically unstable inferences (Tonelli, 2006; Cartwright, 2007; Bluhm, 2005; Howick, 2011).

A similar metaphysical error underpins the use of algorithmic models in clinical settings. These models, often constructed via machine learning, identify patterns of association in training data and extrapolate them to new cases. Yet, in doing so, they replicate and magnify the extensionalist presuppositions of standard EBM (Stegenga, 2018; Greenhalgh et al., 2014). Predictive success is grounded in statistical fit rather than causal understanding. Variables such as CRP levels or smoking status are treated as independent predictors, without modeling the relational or dispositional context in which they exert their effects. When employing this approach, prediction becomes detached from explanation. Dispositional essentialism challenges this by insisting that reliable inference depends on tracking the causal powers instantiated by individual patients. Structural realism adds that these powers must be embedded in stable relational frameworks if predictions are to be projectable. A model trained on one population may fail in another not due to bias alone, but because it never represented the dispositional architecture in the first place. Formal precision in the absence of modal depth becomes epistemically vacuous (Deaton and Cartwright, 2018; Ionnidis, 2005).

Nowhere is the ethical and metaphysical cost of such misclassification clearer than in the case of race-based medical inference (Strand, 2021; Duster, 2007). The United States Food and Drug Administration's 2005 race-specific approval of BiDil, a combination therapy for heart failure, was justified based on subgroup analyses showing efficacy in self-identified Black patients (Sankar and Kahn, 2005; Kahn, 2004; Temple and Stockbridge, 2007). The implicit assumption was that race could serve as a proxy for the dispositional properties relevant to treatment response. Yet, this assumption is indefensible on both structural realist and dispositionalist grounds. From a structural perspective, the observed correlation lacks explanatory anchoring: no mechanism was identified that would make race, as socially defined, causally relevant to the drug's action. The biological heterogeneity of the racial category and the absence of mechanistic specificity render the generalization modally fragile. From a dispositional standpoint, the category of race fails to pick out a stable cluster of causal powers. It is not an essence-bearing kind but a socially constructed label applied to a demographically diverse population. Treating it as if it were a dispositional essence constitutes a category mistake in the strongest metaphysical sense.

Subsequent analyses suggested that the observed benefits may have been due to variables such as socioeconomic status, comorbid conditions, or underuse of angiotensin-converting enzyme inhibitors—factors with genuine causal salience. However, under the extensional logic of EBM, these were conflated with race. The result was a generalization that lacked counterfactual robustness: it did not hold across nearby possible worlds in which patients were racially categorized differently but shared the same relevant dispositions. Interventionist theories of causation underscore the problem: race is not a manipulable variable and cannot serve as the target of an ideal intervention. It cannot, therefore, license counterfactual claims of the form, "Had this patient not been Black, their response to BiDil would have differed" (Sankar and Kahn, 2005; Kahn, 2004; Temple and Stockbridge, 2007).

What BiDil illustrates is not simply a failure of classification but a metaphysical error endemic to a certain model of evidence. The race-based generalization satisfied the evidential standards of EBM—statistical significance in a subgroup—but failed to meet the modal criteria required for explanatory adequacy and ethical legitimacy. It projected a regularity ungrounded in structure, mistook co-variation for causation, and embedded that error in institutional practice (Strand, 2021; Duster, 2007). In doing so, it exemplifies a broader problem: a clinical epistemology that, by ignoring intensional content, licenses inferences from predicates that do not carve nature at its joints. The remedy, as argued throughout, is not the abandonment of empirical methods but the revision of the metaphysical framework within which evidence is interpreted. Only by grounding inference in modal and dispositional structure can medicine avoid the epistemic and ethical distortions that arise from reifying statistical aggregates into causal kinds.

Objections

The argument developed in this chapter challenges the metaphysical infrastructure of EBM, proposing a modal and dispositional alternative that reorients the conditions under which statistical generalizations may support clinical inference. While much of the preceding analysis already anticipates likely lines of resistance, three central objections warrant more direct engagement. These objections concern the normative standard proposed, the relevance of metaphysical critique, and the sufficiency of methodological reform. By clarifying and responding to each, we can further elucidate the scope and necessity of the critique.

A first objection contends that the demand for modal robustness, structural invariance, and dispositional individuation sets an unreasonably high epistemic bar for clinical inference. Medicine, it is argued, frequently operates under conditions of epistemic constraint, limited mechanistic understanding, urgent decision-making, and the necessity of action in the face of uncertainty. From this perspective, extensional generalizations and statistical associations are often the best available guides to practice. To insist that such generalizations be grounded in a fully articulated modal or causal architecture risks paralyzing decision-making or delegitimizing widely used heuristics. Yet this objection misconstrues the normative function of the proposed framework. As earlier clarified, the demand for modal robustness is not an absolutist standard requiring complete mechanistic transparency. Rather, it functions as a regulative ideal—a criterion by which to distinguish between generalizations that are likely to succeed in counterfactual projection and those that are context-bound or fragile. Modal grounding, in this sense, is not a veto on action under uncertainty but a tool for calibrating the evidential weight of statistical regularities. It enables clinicians to assess whether a given generalization retains its epistemic warrant across relevant contexts and to act with appropriate modal humility when such stability is uncertain. Far from paralyzing judgment, the framework refines it.

A second objection targets the metaphysical register of the critique itself. Here, EBM is a methodological program concerned with evidence hierarchies, trial design, and risk stratification. It does not assert, nor does it require, any particular ontological commitments. As such, appeals to structural realism, dispositional essentialism, or interventionist causation are philosophically interesting but epistemically orthogonal. However, we previously argued that this characterization underestimates the extent to which methodological norms are underwritten by metaphysical assumptions. To treat population-level statistical generalizations as sufficient grounds for clinical inference is, implicitly, to adopt a Humean conception of causation (as constant conjunction), an extensional account of properties (as context-insensitive), and a regularity theory of laws (as descriptive generalizations rather than modal constraints). These are not neutral choices. They shape what counts as legitimate evidence, what justifies inference, and how causal claims are interpreted in practice. The metaphysical critique advanced here does not impose a theoretical overlay onto EBM; it renders explicit the latent ontology already operative within its evidentiary practices. Making these commitments transparent is a prerequisite for epistemic accountability.

A final objection grants that EBM's inferential weaknesses may be real but contends that they are better addressed through empirical refinement, improved statistical methods, better stratification, and more representative datasets, rather than through metaphysical reform. On this view, the central problem lies not in EBM's ontology but in its implementation. Yet, while methodological improvements are valuable, they are insufficient. Without a principled distinction between projectable and non-projectable predicates, or between modally robust and context-sensitive generalizations, even the most sophisticated statistical models remain vulnerable to inferential failure. The BiDil case exemplifies this point: race-based efficacy was inferred not because of statistical error but because of metaphysical opacity. The relevant dispositional properties were neither identified nor tracked, and race functioned as an extensional placeholder rather than a causally operative kind. No amount of statistical refinement could correct the underlying ontological mistake. The error was not empirical; it was metaphysical.

These objections serve a clarificatory function. They illuminate the conceptual stakes of the critique and underscore its relevance to both theory and practice. The central claim remains that EBM, as currently constituted, relies on an impoverished metaphysical framework that confuses correlation with causation, statistical robustness with modal stability, and extensional similarity with dispositional identity. Rectifying this confusion requires more than methodological vigilance; it requires metaphysical reorientation. Only by grounding inference in modal, dispositional, and structural principles can clinical reasoning fulfill its epistemic and ethical aims: to justify action under uncertainty, to respect the individuality of patients, and to ensure that the generalizations we act upon are not only statistically informed but metaphysically legitimate.

Recommendations

Regulatory agencies, together with guideline-producing bodies and health-technology assessment organizations, can translate the framework developed in this chapter into implementable standards by adopting a risk-based approach to evidential and classificatory requirements. Rather than treating all statistical findings as equally actionable, regulators should distinguish between low- and high-risk uses of population-level evidence and scale expectations for causal and mechanistic justification accordingly.

First, evidential standards should be revised so that population-level statistical findings are not, by themselves, treated as sufficient grounds for high-impact clinical claims. Substantial evidence of effectiveness will continue to rest on adequate and well-controlled investigations, but where sponsors or guideline developers advance claims about subgroup-specific benefit, causal interpretation, or broad transportability, they should be required to provide a coherent biological and clinical rationale for those claims. Regulatory guidance should specify that, for such claims, sponsors must: (1) describe the hypothesized mechanisms or pathways linking intervention, target, and outcome; (2) explain why these mechanisms are expected to operate in the target population or subgroup; and (3) identify background factors that may limit generalizability. Review templates and advisory committee procedures should be updated to include explicit questions about biological plausibility, causal coherence, and stability of effect across clinically relevant contexts. This would make it procedurally explicit that statistically significant associations, without a minimally articulated rationale for why they should persist under relevant clinical variation, are not sufficient on their own to support strong claims about individual-level benefit or harm.

Second, regulators should introduce specific safeguards for high-risk forms of classification and stratification, including the use of race and other demographic variables, coarse thresholds, and algorithmic risk scores in regulated products, labeling, and decision-support tools. For any such variable or reference class used to define an indication, dosing recommendation, subgroup effect, or automated triage rule, sponsors should be required to: (1) justify the clinical and biological relevance of the class; (2) demonstrate, where feasible, that the grouping is associated with differences in underlying mechanisms or clearly specified risk pathways, rather than serving solely as a proxy for uncharacterized social or environmental factors; and (3) provide sensitivity analyses or additional data exploring whether alternative, more proximal variables could better capture the relevant causal structure. Particular scrutiny should be applied to race, broad age bands, and single-value biomarker thresholds. Regulatory guidance should state that race, as a social category, will not ordinarily be accepted as a stand-alone biological classifier, and that its use must be justified in terms of clearly articulated mechanisms, plausible links to pharmacokinetics or pharmacodynamics, or well-characterized patterns of risk that cannot be represented by more specific variables.

To operationalize these safeguards, agencies should require that clinical prediction models, algorithmic decision-support systems, and other software as medical devices submitted for review include documentation describing: (1) the clinical rationale for each input variable; (2) the expected mechanisms or pathways through which these variables relate to the predicted outcome; and (3) evidence that the model's performance is stable across key subgroups and care settings. Independent validation studies, conducted in populations and settings distinct from the training data, should be encouraged or required for high-risk tools, with a defined mandate to assess whether model variables and groupings behave in ways consistent with their proposed causal or clinical interpretation. Where a variable or classification appears to function only as a coarse proxy for unanalyzed complexity, or where model performance is highly unstable across contexts, regulators should be prepared to restrict the scope of labeling, require additional post-marketing studies, or withhold approval until further justification is provided.

Although regulators do not directly control clinical governance, documentation standards, or professional curricula, they can influence these domains by shaping expectations regarding how evidence is used in practice. Guidance documents and labeling standards should encourage the design of decision-support tools, risk calculators, and clinical algorithms in ways that facilitate context-sensitive, mechanistically informed judgment rather than automatic rule-following. For example, regulators can recommend that tools include clear statements about the populations and settings in which their performance has been demonstrated, the clinical assumptions under which their use is appropriate, and the kinds of patient characteristics that may limit their validity. In parallel, agencies can collaborate with professional societies, accreditation bodies, and funding agencies to promote educational initiatives that strengthen clinicians' and researchers' ability to interpret statistical findings while considering causal reasoning, biological mechanisms, and social determinants of health.

Finally, research governance and regulatory science programs should be aligned with this framework. Calls for proposals, guidance on trial design, and reporting standards can be revised to encourage the collection and analysis of mechanistic and structural data alongside outcome statistics, particularly in areas where subgroup effects, precision medicine claims, or complex algorithms are central. Sponsors making claims about generalizability or transportability across populations should be expected to present some form of structured causal assessment, such as explicit discussion of effect modifiers, biologically relevant differences between study and target populations, and the conditions under which the observed effects are likely to fail. Regulatory science initiatives can prioritize methods for integrating mechanistic, epidemiological, and algorithmic evidence into coherent causal judgments, and convene cross-sector fora—bringing together clinicians, methodologists, philosophers of science, patients, and regulators—to periodically review and refine these standards as scientific and methodological tools develop.

Through this combination of risk-based evidential requirements, strengthened safeguards for high-risk classifications, explicit expectations for the justification of variables and groupings in models, and coordinated research-policy efforts, regulatory agencies can begin to incorporate the modal and causal concerns developed in this chapter into their everyday practices. The result is not an impracticable demand for complete mechanistic transparency but a structured requirement that the strongest and most consequential clinical claims rest on evidence that is not only statistically robust but also biologically and clinically interpretable and sufficiently stable across the contexts in which it is intended to guide action.

Conclusion

EBM, as currently constituted, rests on a metaphysical framework that treats statistical regularities as sufficient for clinical inference while neglecting the modal and causal structures required for explanatory adequacy. This chapter has argued that such an extensionalist and Humean conception renders EBM epistemically illegitimate: it cannot support the counterfactual reasoning that clinical judgment requires, nor can it discriminate between stable, projectable generalizations and context-bound correlations. In contrast, a metaphysically informed approach, grounded in structural realism and dispositional essentialism, treats clinical properties as individuated by their modal profiles and allows generalizations only when they reflect invariant causal architecture. Reconstructing EBM along these lines demands a reorientation of its inferential norms: from co-variation to causal relevance, from statistical robustness to modal stability, and from extensional reference classes to dispositional identity. This is not merely a philosophical correction but an epistemic and ethical necessity. Inference that fails to respect the ontology of clinical properties obscures the individuality of patients and perpetuates epistemic injustice. A reformed EBM must, therefore, aim not merely at prediction but at understanding—an understanding that is explanatory, counterfactual, and metaphysically grounded (refer to summarizing Fig. 5.1).

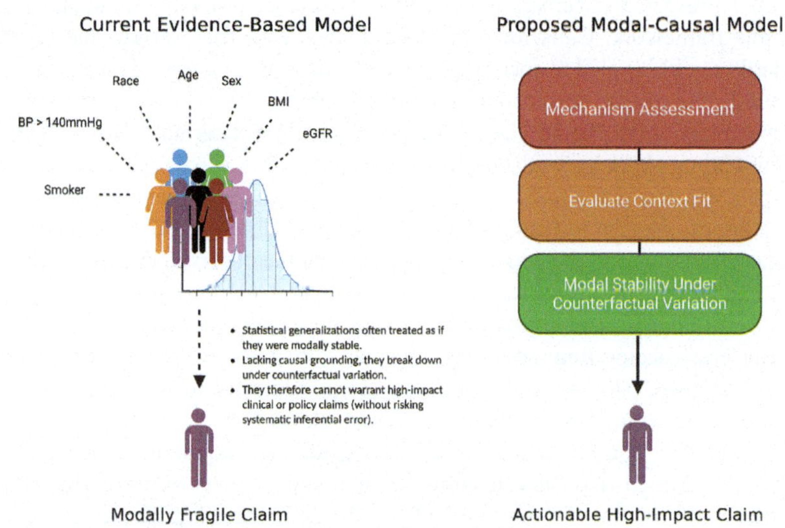

FIG. 5.1 Current EBM model versus newly proposed model.

Glossary

Bare statistical generalization A statistical generalization treated as evidentially sufficient just because of observed frequency, without support from mechanisms, causal structure, or modal facts.

Category mistake An error in which something is referred to as if it belonged to a fundamentally different kind or category (e.g., "the number two is blue"), yielding a distinctive type of inappropriateness.

Causal mechanism An organized system of entities and activities that produces a phenomenon in a regular way; mechanisms underwrite causal explanations in much of the current philosophy of science.

Causal relevance The property of a factor that genuinely makes a difference to whether or how an effect occurs within a given causal structure, not merely correlating with it.

Counterfactual reasoning Reasoning about what would happen under hypothetical circumstances or interventions ("If X were done, Y would occur").

Clinical inference Reasoning in medicine from evidence (trials, mechanisms, cases) to judgments about diagnosis, prognosis, or treatment for particular patients.

Disposition/dispositional property A property characterized by what its bearer is inclined or able to do under certain conditions (e.g., solubility, fragility), often contrasted with purely categorical properties.

Dispositional essentialism The view that at least some properties essentially confer certain dispositions or causal powers, and that the laws of nature are grounded in these dispositional essences.

Epistemic injustice A distinctive kind of injustice in which someone is wronged as a knower (e.g., when prejudice leads others to give a speaker less credibility than they deserve, or when collective interpretive resources fail them).

Epistemic opacity A situation in which the processes or assumptions behind an inference, model, or algorithm are not transparent or accessible to an agent who relies on them.

Epistemic structural realism A version of structural realism which holds that what we can know (and what is preserved across theory change) is the structure of relations described by science, not the intrinsic natures of underlying entities.

Evidence-based medicine (EBM) A clinical movement and methodology that aims to base medical decisions on systematically gathered evidence (randomized controlled trials, metaanalyses), often organized in evidence hierarchies.

Extensionalist ontology An ontological outlook that treats classification and properties in terms of observable or "extensional" features alone (who falls under which predicate), without incorporating modal or causal content.

Extensional predicate A predicate that groups things purely by surface or observational features, regardless of whether those groupings correspond to underlying causal kinds.

Humean metaphysics A broadly Hume-inspired view where the world is a mosaic of local matters of fact, laws are just summaries of regularities, and there are no necessary connections or irreducible causal powers.

Instantiate To show a specific and concrete example of an abstract idea; for e.g., a clinical trial can instantiate a theory by demonstrating that it works in practice.

Interventionism A theory that analyzes causation in terms of what would happen under ideal interventions. A causes B if appropriately manipulating A (while holding other things fixed) would change B in a stable way.

Intensional kind A kind individuated by its underlying nature or role, often causal or dispositional, rather than by mere extensional overlap in observable features.

Mechanistic evidence Evidence that identifies or supports the existence of mechanisms, entities, and activities organized to produce a phenomenon, thus explaining how an effect is brought about.

Modal fragility The property of a generalization that fails under small or plausible changes in background conditions; such a generalization does not support stable counterfactuals.

Modal grounding The underlying features (such as laws, structures, or dispositions) that explain why certain possibilities and necessities obtain, and what makes modal claims true.

Modal profile The pattern of what is possible, necessary, or impossible for a thing or property across nearby possible situations (how it would behave under various conditions).

Modal robustness The property of generalization that remains true across a range of relevant counterfactual scenarios, making it reliable for prediction and intervention.

Monistic reduction Reducing complex phenomena to one type of substance or reality.

Projectable Predicate A predicate that can legitimately figure in inductive inferences because of its entrenched, successful use in past inductions, as opposed to gerrymandered ones.

Reference class A group of individuals defined by certain properties (age range, biomarker, diagnosis), relative to which statistical claims are made and from which inferences about individuals are drawn.

Scientific realism The view that mature, well-confirmed scientific theories are true and that their central unobservable posits really exist.

Statistical robustness The refers to the stability of a statistical relationship under various tests, such as different samples, model specifications, or reasonable perturbations of assumptions.

Structural invariance The property of a causal or relational structure remaining the same under certain interventions or changes in background conditions, which allows it to support counterfactual predictions.

Structural realism A view in the philosophy of science that we should be realistic about the structure described by our best theories (the relational patterns among entities), rather than about the intrinsic nature of the entities themselves.

Threshold effect/diagnostic threshold A cut-off point in a continuous measure (blood pressure, glucose levels) used to classify cases ("hypertensive" vs. "nonhypertensive"), typically derived from population risk curves rather than genuine mechanistic discontinuities.

References

Anjum, R.L., Rocca, E., Andersen, M.B., 2020. Rethinking Causality, Complexity and Evidence for the Unique Patient: A Cause-health Resource for Healthcare Professionals and Researchers. Springer, Cham.

Bird, A., 2007. Nature's Metaphysics: Laws and Properties. Oxford University Press, Oxford.

Bluhm, R., 2005. From hierarchy to network: a richer view of evidence for evidence-based medicine. Perspect. Biol. Med. 48 (4), 535–547.

Cartwright, N., 2007. Causal Powers: What are They?, Why Do We Need Them? What Can Be Done with Them and What Cannot? Contingency and Dissent in Science Project. London School of Economics and Political Science, London.

Deaton, A., Cartwright, N., 2018. Understanding and misunderstanding randomized controlled trials. Soc. Sci. Med. 210, 2–21.

Duster, T., 2007. Medicalisation of race. Lancet 369 (9562), 702–704.
Ganya, W., 2024. The credence of statistical generalisations should not categorically license outright rational belief. S. Afr. J. PhiloS. 43 (3), 283–288.
Ganya, W., Ganya, W., 2025. The modal fragility of bare statistical generalisations in medicine. Curr. Allergy Clin. Immunol. 38 (4), 203–209.
Goodman, N., 1955. Fact, Fiction, and Forecast. Harvard University Press, Cambridge.
Greenhalgh, T., Howick, J., Maskrey, N., 2014. Evidence based medicine: a movement in crisis? BMJ 348, g3725.
Haslanger, S., 2012. Resisting Reality: Social Construction and Social Critique. Oxford University Press, Oxford.
Howick, J., 2011. The Philosophy of Evidence-Based Medicine. Wiley-Blackwell, Oxford.
Ioannidis, J.P.A., 2005. Why most published research findings are false. PLoS Med. 2 (8), e124.
Kahn, J., 2004. How a drug becomes ethnic: law, commerce, and the production of racial categories in medicine. Yale J. Health Poly L. Ethics 4, 1.
Mumford, S., Anjum, R.L., 2011. Getting Causes from Powers. Oxford University Press, Oxford.
Pearl, J., 2009. Causality: Models, Reasoning, and Inference, 2nd ed. Cambridge University Press, Cambridge.
Russo, F., Williamson, J., 2007. Interpreting causality in the health sciences. Int. Stud. Philos. Sci. 21 (2), 157–170.
Sackett, D.L., Rosenberg, W.M.C., Gray, J.A.M., Haynes, R.B., Richardson, W.S., 1996. Evidence-based medicine: what it is and what it isn't. BMJ 312, 71.
Sankar, P., Kahn, J., 2005. BiDil: race medicine or race marketing? Health Aff. 24 (Suppl 1), W5–455.
Stegenga, J., 2018. Medical Nihilism. Oxford University Press, Oxford.
Strand, N.K., 2021. Racial myths and regulatory responsibility. J. Law Med. Ethics 49 (2), 231–240.
Temple, R., Stockbridge, N.L., 2007. BiDil for heart failure in black patients: the US food and drug administration perspective. Ann. Intern. Med. 146 (1), 57–62.
Tonelli, M.R., 2006. The philosophical limits of evidence-based medicine. Acad. Med. 81 (6), 1234–1240.
Vandenbroucke, J.P., 2008. Observational research, randomised trials, and two views of medical science. PLoS Med. 5 (3), e67.
Woodward, J., 2003. Making Things Happen: A Theory of Causal Explanation. Oxford University Press, Oxford.
Worrall, J., 1989. Structural realism: the best of both worlds? Dialectica 43 (1–2), 99–124.

PART

II

Solutions

CHAPTER

Knowing the dead: Pedagogies of forensic humanitarian action in postcolonial contexts

6

Keegan O. Meiring[a,b], Kathryn Smith[c], and Amanda Alblas[a,b]

[a]*Biological Anthropology Research Unit (BARU), Division of Clinical Anatomy, Department of Biomedical Sciences, Faculty of Medicine and Health Sciences, Stellenbosch University, Cape Town, South Africa*

[b]*Victim Identification Service Unit (ViSUN), Division of Clinical Anatomy, Department of Biomedical Sciences, Faculty of Medicine and Health Sciences, Stellenbosch University, Cape Town, South Africa*

[c]*VIZ.Lab, Department of Visual Arts, Faculty of Arts and Social Sciences, Stellenbosch University, Stellenbosch, South Africa*

We need to have been brought up in noble habits if we are to be adequate students of noble and just things.

Aristotle, Nicomachean Ethics, I.4, 1095b4–6, trans. Joe Sachs.

Introduction

Across the world, the management of the dead has become one of the defining humanitarian challenges of our time. Be it in times of disaster, conflict, or other quotidian events, the sobering reality is that countless individuals die without known identities, all while their families are left searching. Therefore, the obligation to restore identity, uphold dignity, and provide answers for the living is not only a scientific question but also a profoundly moral one.

In times of conflict, the Geneva Conventions of 1949 and their Additional Protocols provide a framework regarding how the dead must be searched for, collected, identified where possible, and treated with respect (Clegg, 2021; ICRC, 2020a). The influence of these provisions extends beyond the battlefield by informing a universal ethic of dignity and remembrance that underpins the contemporary field of Forensic Humanitarian Action (FHA). Originally developed by the Argentine Forensic Anthropology Team (Spanish: Equipo Argentino de Antropología Forense, EAAF) and later formalized through the work of the International Committee of the Red Cross (ICRC) (Tidball Binz and Cordner, 2022), FHA reframes forensic expertise as a humanitarian duty grounded in both international human rights and humanitarian law (Parra et al., 2020; Ubelaker et al., 2019). The global application

Decolonizing Medical Education, Research and Practice. https://doi.org/10.1016/B978-0-443-44998-7.00008-9

of these rationales extends from the identification of tsunami victims in Southeast Asia (Sumathipala et al., 2006) to the recovery of missing persons in Latin America (Spradley, 2014) and the resolution of migrant deaths in the Mediterranean (M'charek and Casartelli, 2019; Missing Migrants Project, 2025). In situations that rarely attract international attention, such as the mortuaries of South Africa, FHA is a useful tool in conceptualizing the thousands of unclaimed and unidentified bodies that remain in storage each year (Gibbon et al., 2023, Gibbon et al. (2025)).

The Republic of South Africa is underwritten by a democratic constitution that enshrines dignity as a foundational human right. Unfortunately, the medico-legal system often struggles to extend this right to all decedents. Underresourced mortuaries, outdated legislation, and the enduring legacies of colonial anatomical governance mean that many of those marginalized in life are again marginalized in death. These limitations are indicative of the challenges faced by the Global South. However, when juxtaposed with the role of higher education institutions (HEIs), it necessitates a reframing of forensic and medical science education. This is because these studies cannot be easily disentangled from the histories of colonial medicine and the unequal geographies of knowledge production. For example, human remains acquired in South Africa during apartheid for medical education and research were frequently those of vulnerable individuals that were acquired without direct consent, and as such, are not neutral teaching tools (Ajam, 2019; Fuhr and de Klerk, 2024; Ramgoolam, 2017). These *anatomized specimens* are historical artifacts representative of past inequalities and the disproportionate extraction from the powerless in a racialized system (Gibbon et al., 2025). It is in this context that the need for FHA becomes evident, as without a rights-based and humanitarian approach, the treatment of the dead risks reproducing historical injustices rather than repairing them. FHA therefore offers more than a technical framework by providing a normative bridge between the Geneva Conventions' obligations of dignity (Gaggioli, 2018; ICRC, 2020b) and the decolonial imperative to confront the legacies of racialized science (Chaussee et al., 2022; Go, 2025). This means shifting pedagogy away from silent, depersonalized teaching traditions and toward models that emphasize consent, cultural sensitivity, and structural justice.

Against this backdrop, the Victim Identification Service Unit (ViSUN) at the Faculty of Medicine and Health Sciences at Stellenbosch University (Cape Town, South Africa) emerges as both a local innovation and a case study of global relevance. ViSUN operates at the interface of forensic practice and pedagogy by assisting the South African Police Service (SAPS) and Forensic Pathology Services (FPS) via the analysis of unidentified skeletal remains when conventional methods fail. To prepare students for this responsibility, the chapter will explore how curricular reforms can cultivate not only technical competence but also ethical reflexivity and humanitarian sensitivity. Moreover, ViSUN's pedagogical model raises critical questions such as: Can forensic education be decolonized while it continues to rely on structurally marginalized bodies? How can students be trained not only in technical skills but also in ethical reasoning and historical awareness? And what would it mean to frame forensic pedagogy itself as a form of FHA? These questions also relate to a broader institutional perspective on how universities (for example, through frameworks such as the United Nations Sustainable Development Goals (SDGs) 2030 (https://sdgs.un.org/goals) and Stellenbosch University's Vision 2040) can position themselves as humanitarian role players accountable not only to the state but also to communities and families.

To engage meaningfully with these questions, the chapter proceeds in two ways. It begins by tracing the pedagogical, institutional, and humanitarian logics that structure forensic education, attending to the histories and power relations that stabilize such arrangements. The chapter then turns to decolonization as a mode of re-thinking, i.e., examining how forensic pedagogy can reckon with its material and ethical dependencies while being reimagined as FHA.

Forensic humanitarian action: From global frameworks to Southern African practice

Across continents, FHA has emerged as one of the most profound expressions of global solidarity. This interdisciplinary commitment to restoring dignity to the dead and providing answers to the living is defined by the ICRC as humanitarian work grounded in neutrality, impartiality, and independence. The integration of the principles of international humanitarian law and international human rights law into the empirical and ethical practice of forensic science represents a humanitarian transformation of scientific authority (ICRC, 2024).

The contemporary scope of FHA spans governmental, intergovernmental, and civil society initiatives, reflecting a near-universal moral consensus that every person deserves to be identified, named, and remembered. Among its most visible initiatives is the International Organization for Migration's Missing Migrants Project (2014–present), which has documented over 78,961 migrant deaths worldwide since 2014 (https://missingmigrants.iom.int/). Data from medico-legal authorities, humanitarian agencies, and media sources expose how irregular migration routes (e.g., from the Sahel to the Aegean) produce unacknowledged humanitarian crises. Through this visibility, FHA transforms what might otherwise be perceived as isolated tragedies into a systemic moral problem, compelling both public and policy attention (Bird and Shangraw, 2024).

The European Cooperation in Science and Technology (COST) Action on Migrant Disaster Victim Identification (MDVI) (COST, 2023) constitutes a research and coordination framework aimed at strengthening transnational responses to migrant deaths along the Mediterranean and Atlantic routes. Each year, thousands perish attempting such crossings, yet only a fraction of these individuals are ever identified. Unfortunately, this is due to fragmented forensic infrastructure, inconsistent standards, and limited inter-country collaboration. Additionally, post-conflict identification missions under EULEX in Kosovo (2008–2014) applied FHA principles to locate and identify those missing from the Yugoslav wars, balancing judicial mandates with family-centered humanitarian obligations. COST MDVI mobilizes an interdisciplinary network of forensic scientists, anthropologists, human rights legislators, and data ethicists across Europe and neighboring regions to address this gap (COST, 2023). The efforts can now be expanded upon with technological advances in artificial intelligence, drone mapping, craniofacial analysis, and digital databasing; therefore, promoting a more ethical, sustainable, and rights-based identification system (Bélisle-Pipon and Ravitsky, 2025; Victor et al., 2024). FHA also fosters knowledge exchange and capacity building among member countries, highlighting how it can evolve into a continental platform for policy harmonization and humanitarian accountability.

The Americas offer another vital geographical location for FHA's development. Here, the Argentine Forensic Anthropology Team (EAAF), founded in 1984, is widely regarded as the field's progenitor (García-Mancuso et al., 2020; Sanford, 2008). Emerging from Argentina's "Dirty War" and the search for the disappeared, the EAAF pioneered a model of forensics in the service of human rights. Over the past 40 years, this model has informed investigations in more than 50 countries, from the Balkans to South Sudan (INRAP, 2025; Kim and Rosenblatt, 2023; Ubelaker et al., 2019). Its Latin American counterparts, such as the Guatemalan Forensic Anthropology Foundation, extended this vision by excavating mass graves from civil conflict and enabling the voices of survivors and families to reshape transitional justice processes. Organizations such as these have also assisted the South African Missing Persons Task Team, operating under the National Prosecuting Authority. This provides a long-running example of FHA-like practice applied to the exigencies of past political violence and transnational repatriation. Through the Exile Repatriation Project, led by the Department of Sport, Arts and Culture in cooperation with the Missing Persons Task Team and the Truth and Reconciliation Committee unit, the purpose of this collaborative effort is to actively identify, exhume, and return the remains of those who died in exile under South Africa's apartheid regime (Nelson Mandela Foundation, 2025; South African History Online, 2020). The program involves archival research, cemetery inspection, grave mapping, forensic investigation, and transnational coordination linked to previous and ongoing missions to Angola, Zambia, and Zimbabwe. These missions underscore how FHA was never only a technical discipline but also a political and ethical movement committed to naming victims of structural violence and state repression.

In North America, Operation Identification at Texas State University (2012–present) exemplifies the localization of FHA principles within a university context (NBC News, 2024). Based in South Texas, Operation Identification combines community partnerships, student-led exhumations, and anthropological expertise to recover and identify migrants who die crossing the United States–Mexico border (Spradley, 2014). Through careful fieldwork, archival matching, and DNA collaboration with Mexican authorities, Operation Identification has returned hundreds of individuals to their families, transforming forensic education into humanitarian praxis (Siegert et al., 2024). This model demonstrates how FHA can thrive even in politically charged environments by reframing the management of migrant deaths not as a border-control issue but as an ethical imperative.

Beyond these flagship projects, regional networks continue to expand FHA's reach across all inhabited continents. In Asia and the Pacific, the International Centre for Humanitarian Forensics (ICHF) was established in 2018. It serves as the world's first dedicated training hub for humanitarian forensics, advancing best practices in mass fatality management, disaster victim identification, and human rights investigations (ICHF, 2022). The collaboration between the ICRC and the National Forensic Sciences University of India aims to develop innovative capacity-building efforts to mitigate the effects of disasters, such as the 2004 Indian Ocean tsunami. This disaster necessitated the creation of a coordinated multinational humanitarian forensic team, ranging in expertise and geographical origin. Working across Indonesia, Sri Lanka, and Thailand the transformative power of coordinated humanitarian forensics enabled the successful identification of over 3,000 victims through combined DNA, dental, and fingerprint analysis (Adamovic et al., 2023; Kumar, 2022; Singh, 2022). Over time, this collaboration led to the eventual institutionalization of the ICHF (ICHF, 2022). In Oceania, disaster victim identification teams in Australia and the Pacific incorporated FHA methods into national emergency frameworks, particularly following events such as the 2009 Samoa tsunami and the 2018 Papua New Guinea earthquake, where the return of remains was managed with cultural and community engagement at its core. Together, these initiatives reveal that FHA has evolved from a specialized, reactive practice into a globally coordinated humanitarian framework anchored in moral purpose and technical excellence.

Across the African continent, FHA is gaining traction as both a moral framework and a practical response to the enduring legacies of conflict, migration, and state fragility. In East Africa, the ICRC's Nairobi Regional Delegation (ICRC, 2025) and the Kenya National Commission on Human Rights (Kenya National Commission on Human Rights, n.d.) have pioneered initiatives on the dignified management of the dead, developing national protocols for body recovery and identification in the wake of terror attacks and natural disasters. In West Africa, post-conflict forensic efforts in Nigeria and Sierra Leone have integrated FHA principles with community reconciliation and culturally sensitive burial practices. This uptake of FHA principles was in response to insurgencies, pandemics, and mass-fatality events that have plagued this region of the continent, extending back to its colonial history (Moon, 2014). In North Africa, the search for missing persons following the Libyan civil wars and Mediterranean migration crises brought together the ICRC, United Nations agencies, and local forensic teams. Their purpose is to locate and identify the victims, while Egyptian authorities established morgue-based DNA programs for migrants lost at sea (International Organization for Migration (IOM), 2025). Meanwhile, in Central Africa, the Democratic Republic of Congo and Chad hosted training missions linking forensic recovery to transitional justice, transforming exhumation from a technical act into a process of social healing (Office of the High Commissioner for Human Rights, 2022; Southern African Development Community (SADC), 2024). Collectively, these initiatives mark the rise of an African humanitarian forensic imagination by adapting global standards to diverse cultural, religious, and political landscapes.

Therefore, whether enacted in the deserts of Texas, the mass graves of Central America, the floodplains of Southeast Asia, or the archives of the ICRC in Geneva, each instance of FHA enacts a similar ethos. Quintessentially, the management of the dead is inseparable from the pursuit of human dignity, accountability, and education. While these international efforts demonstrate the universal reach of FHA, their moral and methodological imperatives acquire urgency in postcolonial contexts. For example, in South Africa, the crisis of unidentified and unclaimed bodies mirrors, in quieter but equally devastating form, the humanitarian emergencies witnessed at global borders.

Southern African commitment

The recent conflict in northern Mozambique's Cabo Delgado province, which intensified in the early 2020s, revealed the region's fragile capacity to manage the dead amid overlapping crises of violence, displacement, and poverty. Though South Africa was not a direct participant in the war, its humanitarian and forensic systems were inevitably affected (Adebayo, 2023). The side effects of this conflict resulted in the influx of displaced persons, unrecorded deaths along migration routes, and the silent accumulation of unidentified bodies in local mortuaries (African Center for Strategic Studies, 2025; United Nations High Commissioner for Refugees, n.d.). These developments exposed how the boundaries between war and peace are porous in southern Africa and how the consequences of conflict reverberate far beyond its immediate geography. The Mozambican experience thus offers a cautionary lens through which South Africa should reassess its approach to humanitarian forensic action (Sangweni, 2023). This would be not only in moments of acute crisis but also in the slow emergencies produced by poverty, inequality, and infrastructural neglect (Zavattaro et al., 2021). It demonstrates that even where death is not directly caused by conflict, failures in death management can entrench structural violence and undermine public trust (Bhutto, 2024; JMU, 2023). From a policy perspective, FHA therefore functions as a framework for regional moral accountability, requiring advance preparedness, cross-border coordination, and a human-rights–based approach to the dignified management of the dead (JMU, 2023) across southern Africa.

The International Committee of the Red Cross (ICRC)

Within this evolving regional landscape, the ICRC was instrumental in advancing FHA across southern Africa. Their involvement in the region has transformed global humanitarian principles into context-specific practices that bridge borders, disciplines, and state capacities (Ozcelik, 2026; Phetha and Ndlovu, 2026). Recognizing that deaths related to migration, displacement, and violence seldom respect political boundaries, the ICRC promoted a route-based, multidisciplinary approach that links forensic experts, border authorities, and community actors across the region. Its engagement has extended beyond conflict zones to include Mozambique, Botswana,

Namibia, Eswatini, and Lesotho, where the organization supported training programs, cross-border coordination, and the development of national frameworks for the dignified management of the dead (ICRC, n.d.). For example, the ICRC's humanitarian forensic teams worked alongside local authorities during the Cabo Delgado conflict in Mozambique and subsequent displacement crises, assisting in documenting missing persons, managing temporary burial sites, and promoting family tracing mechanisms (Demeyere et al., 2025; Serradinho, 2024). These experiences informed subsequent capacity-building in neighboring countries, where migration, disaster, and health emergencies have similarly strained forensic infrastructure.

Between 2016 and 2018, the ICRC (2018) implemented a pilot project in South Africa and Zimbabwe to operationalize such lessons by focusing on three interdependent objectives: (1) restoring identity and dignity to deceased migrants; (2) improving information-sharing among families, state institutions, and forensic practitioners; and (3) supporting families with psychosocial accompaniment and the repatriation of remains where feasible. Rather than supplanting national structures, the project complemented and strengthened existing medico-legal systems, offering technical training and fostering cross-border dialog. Collectively, these efforts reveal how humanitarian forensics functions as both a diplomatic and ethical enterprise. It situates the treatment of the dead within a broader regional ecology of responsibility. However, despite such advances, the humanitarian crisis of unidentified and unclaimed bodies remains most acute in South Africa, where overstretched mortuaries retain thousands of untraced remains each year (Latchanna, 2025; Manyane, 2024). Johannesburg is South Africa's commercial and migratory hub and stands at the intersection of poverty, mobility, and mortality. However, recent waves of semigration (a trend of people moving to other regions within their own country) that commenced during the COVID-19 pandemic (Jagganath, 2025) have shifted this problem to other parts of the country. One notable example includes Cape Town and the larger Western Cape province of South Africa, which already experiences some of the highest murder rates across the globe (Western Cape Government, 2025) and remains susceptible to historical plights associated with (and symptomatic of) South Africa's colonial past.

A "pracademic" continuum

Following the regional coordination led by the ICRC, the University of the Witwatersrand's Identification Program (Wits ID) and the Western Cape Cold Case Consortium (W4C) represent two complementary expressions of FHA within South Africa's complex medico-legal landscape. Both initiatives confront the same humanitarian dilemma of an accumulation of thousands of unclaimed and unidentified bodies in mortuaries each year, despite contrasting social, geographic, and institutional contexts.

The Wits' Identification Program exemplifies the formal integration of humanitarian forensics into state infrastructure and higher education. Based within the Forensic Pathology Services Medico-Legal Mortuary and operating through the Human

Decedent Identification Unit (HDIU) and Human Variation and Identification Research Unit (HVIRU) (University of the Witwatersrand, n.d.), the program occupies the administrative and logistical heart of South Africa's identification efforts. Its position in the northern region of South Africa, with the highest concentration of unclaimed bodies, makes it both the epicenter of the nation's forensic crisis and a vital node in the broader southern African humanitarian network. The Wits ID initiative works closely with government departments, law enforcement agencies, and the ICRC. Such positioning, therefore, aligns its technical practices with global standards for identification, data ethics, and family liaison. Between 2018 and 2020, the HDIU processed 385 unidentified decedents, achieving 87 identifications primarily through fingerprint and DNA analysis (Keyes et al., 2016, 2022). Beyond its statistical achievements, the program functions as a training hub for postgraduate students and practitioners, embedding FHA principles into curricula and professional routines. It thus bridges the humanitarian ethos of the ICRC with the operational realities of a state-run forensic system, translating international humanitarian standards into locally sustainable practice.

The W4C in Cape Town provides a contrasting model. A collaboration between academics variously affiliated with Stellenbosch University and the University of Cape Town and the Salt River medico-legal facility (now closed and replaced by the Observatory Forensic Pathology Institute, or OFPI), W4C reflects the Western Cape's unique intersection of cosmopolitan diversity and entrenched inequality. The province's coastal geography and economic disparities produce a demographic pattern of unidentified deaths that differs from South Africa's inland urban sprawl. Here, many cases involve migrants, the unhoused, and individuals who die anonymously on the city's periphery. The W4C, therefore, developed a collaborative, interdisciplinary model that draws on forensic anthropology, genetics, isotopes, and facial reconstruction to reconstitute identities that elude conventional procedures. As reported by Gibbon et al. (2025), a pilot study investigating five unresolved, complex cases demonstrated the viability of collaborative, multimodal analysis and new approaches to presenting data in an environment of limited state capacity. The consortium's participants embody a networked humanitarianism that mirrors Cape Town's social fabric.

While both the Wits ID and W4C operate within a local university ecosystem and collaborate closely with the Forensic Pathology Services (FPS), their emphases differ in scope and orientation. Wits ID embodies the formal institutional model that is anchored in accredited academic training, postgraduate research, and transparent partnerships with investigative and journalistic stakeholders. In contrast, the W4C emerged as a transdisciplinary outreach initiative aimed at capacity building among local practitioners (e.g., undertakers and forensic officers), by improving postmortem data collection and community-based documentation. However, rather than constituting a distinct paradigm, W4C can be viewed as an extension of Wits ID's applied humanitarian ethos. The more recent Science for Justice (S4J) (Stellenbosch University, 2024, 2025b; Thompson, 2026) initiative arguably represents a deeper interdisciplinary evolution that bridges forensic science, data governance, and social justice. All of this is done to develop locally responsive datasets and to apply forensic expertise beyond conventional medico-legal contexts. Together, these initiatives reveal a continuum of South African forensic-humanitarian engagement that integrates pedagogy, practice, and policy innovation.

Viewed through this lens, ViSUN extends this continuum of forensic humanitarian practice in the South African context. The W4C initially drew inspiration from the Sutherland restitution process (Gibbon et al., 2023), which foregrounded community-engaged science and the ethics of restorative collaboration. Since then, it has evolved into a model for applying those principles within the complex, paper-based environment of South Africa's FPS. In contrast, the Western Cape's more digitized infrastructure (supported by the LiveLink system that connects FPS facilities and enables inter-institutional collaboration) creates conditions for more integrated, cross-sectoral work. Together, these contexts highlight how South African FHA practice adapts to differing infrastructural and institutional realities. Namely, the W4C's engagement arises from necessity and improvization, while ViSUN's model leverages digital coordination to link education, ethics, and operational practice. However, both can transform identification work into an educational encounter to train students not only in technical proficiency but also in reflexivity and moral reasoning.

The case study

Operating from the Division of Clinical Anatomy within the Faculty of Medicine and Health Sciences and housed in the new Biomedical Research Institute (BMRI) at Stellenbosch University (Thompson, 2023), ViSUN arose from increasing requests by the metropolitan and provincial sectors of the SAPS and the Western Cape FPS for assistance with unidentified, decomposed, or skeletonized human remains. Such cases are often beyond the analytical capacity of conventional mortuary procedures and require anthropological and anatomical expertise to interpret skeletal morphology, taphonomic alterations, and indicators of trauma. Working in formal collaboration with the SAPS Forensic Services in Plattekloof in Cape Town, the unit follows a clearly defined workflow. Referrals from FPS trigger processes of photographic documentation, external examination, maceration, and detailed anthropological profiling to generate biological reports that can guide identification and narrow searches in missing-persons databases. Through this, ViSUN provides a vital public service that restores not only names but also dignity to those who have remained voiceless within the system.

Participation in the unit's operations forms a component of Stellenbosch University's BSc Honors program in Anatomy, where students engage in mock scene recovery, skeletal analysis, and trauma interpretation under close supervision. This experiential learning model introduces students to the ethical and procedural dimensions of forensic work, including chain-of-custody management, legal documentation, and the moral implications of handling unclaimed remains. Through these encounters, students learn that every forensic case is also a social narrative and a reflection of how inequality, migration, and history shape who becomes separated from a legal identity.

The operations of ViSUN are governed by dual logic. Firstly, as a utilitarian imperative to support state forensic systems and secondly, as a deontological obligation (an ethical concept using rules to distinguish right from wrong) to uphold the dignity of the dead. From a utilitarian perspective, ViSUN plays an indispensable role in resource optimization. By narrowing the scope of unidentified cases and providing actionable data, it reduces state expenditure on prolonged storage, redundant testing, and protracted investigations (Maromo, 2024). Its partnership with SAPS and FPS makes it an extension of the state's forensic infrastructure, filling gaps created by institutional inefficiency and infrastructural strain. Here, the dead become a dataset: unidentified human remains are a logistical problem to be solved as efficiently as possible. This logic is further embedded in teaching, as human remains are regarded as a pedagogical asset in developing students' technical competencies and professionalism (Billings et al., 2024; Hutchinson et al., 2020; Kramer et al., 2019). This utilitarian framing obscures deeper ethical questions, especially when graduate unemployment remains rampant (Mahlungu, 2025) and justice is delayed (Khan, 2025; Mabuza, 2025). Most notably, it assumes that the use of human remains in education and investigation is unproblematic if outcomes are socially beneficial (Morris, 2014). It presumes that the dead are educational tools, available for forensic processing by virtue of their administrative status as "unclaimed." Herein lies the deontological tension. While ViSUN performs a humanitarian function sanctioned under the South African Criminal Law (Forensic Procedures) Amendment Act 37 of 2013, its practices are ethically ambivalent. It reflects a broader challenge in a post-apartheid South Africa about how to transform institutions tasked with delivering public value but without replicating colonial epistemologies or necropolitical practices (Elgerud et al., 2025). By embedding humanitarian principles within higher education, ViSUN transforms the forensic laboratory into a space for epistemic renewal. It merges the institutional rigor of Johannesburg's forensic networks with Cape Town's culture of collaborative innovation, creating a model of decolonial scientific practice in which identification is both a technical and ethical act.

The stakes of the unclaimed

Each year, hundreds of bodies languish in state mortuaries without names, documents, or familial recognition. In the Western Cape region of South Africa, it was reported that over 300 bodies lie across its state facilities (Solomons, 2021), compared to the 3,000 reported nationally (Parliament of the Republic of South Africa, 2024). These individuals are often isolated during life, with many being victims of violence, homelessness, or mental illness (City of Cape Town, 2022). The difficulty of determining their identity is compounded as personal identification documentation is usually absent or non-existent. When conventional forensic methods of identification (i.e., fingerprinting or DNA) yield no results, and the next of kin cannot be traced, the deceased are frequently subjected to pauper burials (Serra, 2024). However, this crisis is not simply about mismanaged mortuaries or understaffed pathology units. It is symptomatic of deeper structural failures that span forensic science, health policy, legal frameworks, and broader state capacity. At the intersection of poverty, violence, and institutional inertia lies a medico-legal system struggling to fulfill its most fundamental obligation: to serve the dead and return them to the living through identification and justice.

The inability to identify and humanely process the deceased cannot be divorced from South Africa's broader political and socio-economic landscape. Three interconnected forces underpin the present crisis. First, there is the systemic under-resourcing of the medico-legal infrastructure (BusinessLive, 2024). State mortuaries are often ill-equipped, with outdated facilities, limited cold storage capacity, and staff working under extreme caseload pressures. The FPS still faces significant challenges in post-apartheid South Africa (Parliamentary Monitoring Group, 2019), such as poor interdepartmental coordination, data sharing, and case throughput. Technological innovations such as real-time biometric access, isotope analysis, or automated missing persons integration are unevenly distributed or absent altogether (Smith and Horne, 2024). Second, the scale of violent death and disappearance in South Africa remains among the highest globally. For example, for the 2022–2023 period, the murder rate was 56 per 100,000 (ISS, 2023). High levels of gang violence in historically disenfranchised communities like the Cape Flats of Cape Town have turned certain urban zones into spaces where the premature death of vulnerable persons is not only expected but institutionally normalized (Bowers Du Toit, 2014; Meyer, 2019). Third, these failures are rooted in South Africa's colonial and apartheid histories of racial governance, spatial displacement, and social marginalization (Breckenridge, 2014; Daily Maverick, 2025). The bureaucratic and geographic systems created by the apartheid regime aimed to isolate racialized communities and devalue their lives. While dismantling the legislative apparatus of racial segregation, post-apartheid South Africa has yet to fully confront how these histories continue to structure life and death. The very fact that certain individuals remain unclaimed is evidence of contemporary inequality and a derivative of generational ruptures, community disintegration, and state withdrawal.

The following sections will, therefore, interrogate the ethics, pedagogy, and epistemology of forensic identification in the South African context. This will be achieved by posing a central question: can forensic education be decolonized if it continues to work with structurally marginalized bodies within bureaucracies shaped by racial and colonial legacies? Rather than dismissing this pedagogical practice as inherently exploitative, the remainder of this chapter argues that, in contexts of state failure, such educational engagements can become acts of restorative justice. However, this requires critical reflexivity, as the handling of human remains in classroom settings demands not only procedural rigor but also ethical transparency. As such, the development of forensic skills should also be accompanied by relevant education in history, law, and moral reasoning.

Historical and legislative entanglements

South Africa's early anatomical legislation is grounded in colonial imperatives of control, scientific rationality, and racial exploitation. The Anatomy Act 3 of 1911 (introduced under the Union of South Africa, as cited in Pillay et al. (2017)) was designed to regulate the dissection and use of human bodies for educational purposes (Masters and Shapiro, 1951). It formalized what had been an ad hoc and often coercive practice, where hospitals, prisons, and poorhouses became supply lines for medical schools (Hoole, 2018). The 1911 Act provided the legal basis for the state to take possession of unclaimed bodies, allowing them to be transferred to anatomical institutions for dissection without consent from kin or community. This model was not unique to South Africa, as it followed precedents set in Great Britain's Anatomy Act of 1832, which authorized the use of the bodies of the poor for medical science. Sprouting from the Age of Enlightenment came the establishment of a socio-legal distinction between the bodily remains of the affluent (worthy of burial) and those of the impoverished (available for study) (McManners, 1981). However, in colonial South Africa, the racialized nature of society meant that this distinction became explicitly racial. *Some* bodies (mostly "Black") were disproportionately represented in anatomical collections and postmortem education, further entrenching the epistemic subjugation of the colonized as raw material for colonial science (Agarwal, 2025; Mbembe, 2020; Morris, 2022; Walters and Jansen, 2022; Alblas and Gibbon, 2026). Of note, we here employ these historical racial terms to argue our point, whilst being mindful that there is no biological basis for such classifications and that it is nothing but a socio-political construct used to oppress people. The Anatomy Act was revised in 1959 to reflect the more systematic racial governance introduced under apartheid. Although it retained the 1911 framework's emphasis on state possession of unclaimed bodies, the 1959 version was introduced in tandem with other legislation (i.e., the Population Registration Act of 1950, the Group Areas Act [1950], and the Separate Amenities Act [1953]) that sought to segregate all aspects of life (and death, by extension) along racial lines. The consequence was a fully racialized anatomical economy where mostly Black bodies were not only politically marginalized but also institutionally repurposed to serve the pedagogical needs of medical schools (Finn et al., 2022; Labuschagne and Mathey, 2000; Tobias, 1980). The Population Registration Act (No. 30 of 1950) formalized the system of racial categorization that underpinned the apartheid regime. These categories also governed how bodies were processed after death. The racial classification assigned on one's identity document not only determined one's burial location but could also influence decisions around autopsy, anatomical dissection, and morgue prioritization (Bowker and Star, 2000; Saayman, 2020). Such ideas and praxis subsequently informed the statistical assumptions used by forensic pathologists and anthropologists when estimating biological profiles. These ideas are underpinned by a pseudo-scientific belief that race could be reliably determined through skeletal morphology. Thus race was both an administrative category and a scientific artifact. It operated as a mechanism of social control and as a heuristic within medico-legal discourse, one that continues to persist within forensic anthropology until today (Gibbon et al., 2023).

Despite the end of apartheid and the adoption of the 1994 Constitution, many of the statutory instruments governing medico-legal death remain unchanged or only partially reformed. For example, the National Health Act (No. 61 of 2003) attempts to regulate health services within a democratic framework. However, there are still many critical gray areas regarding unclaimed bodies, informed consent, and body donation (Meiring et al., 2024). Unclaimed bodies are often treated as state custodial property and subjected to disposal via pauper burial or, in the case of anatomical institutions, transfer to medical faculties for educational use. The net result is a continuity of colonial logic, where bodies that are unclaimed by family (often due to poverty, displacement, or bureaucratic failure) are absorbed into a state management system that prioritizes logistical expediency. In practice, this means that disenfranchised groups continue to serve as the pedagogical substratum of health sciences (Walters and Jansen, 2022), while their communities remain peripheral to decision-making processes regarding bodily handling and restitution (Rassool and Gibbon, 2024).

The current legal frameworks are, therefore, inadequate to ensure the dignity and traceability of unclaimed and unknown persons. The state retains no obligation to conduct outreach efforts in search of potential family members; however, attempts by staff at the Tygerberg FPS have resulted in some successes in decreasing the number of deceased individuals in this Western Cape mortuary (Cape Times, 2025; Stellenbosch University, 2025a). The process of declaring a body unclaimed or unidentified is technically straightforward but ethically opaque. Once a body is entered into the state system, a fixed holding period (minimum of 30 days) is observed (NHA, Regulation 180, Section 10). If identification efforts fail (i.e., DNA, fingerprinting, or biological profiling), the body is assigned a state case number and scheduled for pauper burial. Families often only become aware of the death weeks or months after burial, when legal complications or external investigations arise. This process evokes concerns of human rights violations and is misaligned with international norms of posthumous dignity, such as those outlined by the ICRC (ICRC, 2020) and the United Nations Guidelines on the Management of the Dead (2024).

Dignity: "a concept which talks about a person's intrinsic worth - a value of all people, which they are born with as human beings", is enshrined in the South African Constitution (Chapter 2, Section 10). Whether this governance extends to the unclaimed and unknown deceased in this respect is unknown. This legislative *ignorantia* places researchers and students in positions where they are compelled to operate in a legal vacuum. While institutions such as Stellenbosch University emphasize transformation and human dignity in policy documents (Stellenbosch University, 2019), their forensic units remain tethered to archaic legislative frameworks that prevent a full transition from state-based utility to human-centered justice.

Pedagogy, practice, and ethical reform

The training of and transfer of skills to forensic practitioners is not a neutral exercise but a deeply political and ethical endeavor. The ways students are taught to engage with human remains profoundly shape the contours of forensic practice. In the South African context, HEIs carry a particular burden of ensuring that the violence of the past is not replicated in their pedagogical formats. Arguably, current curricular practices in forensic anthropology and clinical anatomy have the potential for greater engagement with the ethical, historical, and political dimensions of bodily science. Drawing on the example of ViSUN and broader national challenges in medico-legal education, this section advocates for further transformation of the forensic curriculum.

The current curriculum of the Forensic Anthropology module in the BSc Honors (Anatomy) program at Stellenbosch University is robust, with exposure to both the didactic (e.g., anatomical dissection, osteology, trauma analysis, taphonomic processes, and postmortem examination) and administrative/workflow (e.g., chain of custody, case intake, maceration, skeletal profiling, and report drafting) components of the victim identification process. Although building such competencies is essential, equal consideration of the ethical and philosophical concerns of this profession (i.e., moral reasoning, historical consciousness, and epistemic reflexivity) is indispensable.

Comparable curricular tensions exist, for example, the Forensic Anthropology Research Centre at the University of Pretoria and the Wits BHSc (Hons) Forensic Sciences programs both include ethical reasoning modules, but still operate largely within procedural frameworks, like many Global North institutions. Internationally, programs such as the University College London Institute of Archeology's MSc in Bioarchaeological and Forensic Anthropology have experimented with decolonial and historically situated pedagogy. This was achieved by embedding modules that interrogate the colonial roots of anatomical science and the ethics of human remains curation. These examples reveal a growing recognition that technical competence must be accompanied by moral and historical literacy. In South Africa, HEIs such as Stellenbosch University, the University of Pretoria, the University of Cape Town, and the University of the Witwatersrand are gradually experimenting with more integrative four-field approaches (Kehoe, 2013). This approach aims to connect biological, cultural, archeological, and linguistic perspectives (Hicks, 2013). While these remain uneven, emerging initiatives like S4J suggest how the ethos of Lorde's warning (Bowleg, 2021; Ferri, 2022) of "nothing about us without us" may be countered. However, achieving this requires several key areas to be addressed:

Ethical reasoning beyond procedural compliance

Students are commonly taught to respect and obey chain-of-custody protocols, avoid contamination, and maintain confidentiality. Only a few programs, however, compel students to engage in moral reasoning about the social lives of the dead. For example, the University of Western Australia's Master of Forensic Anthropology includes a dedicated unit on Ethics and Research Methods in Forensic Science, prompting students to reflect on dignity, consent, and population-specific standards. Similarly, Texas State University's Forensic Anthropology Center integrates reflective discussions on the human rights dimensions of decomposition research. Embedding such approaches within ViSUN or other similar programs would help address questions such as: "What are my views about the analysis process?", "Who benefits from this identification?", and "How can this scientific procedure be socially reflective?"

Historical contextualization of anatomical and forensic practice

Although students learn the mechanics of maceration and skeletal reconstruction, they lack the historical context of such operations. We argue that the use of unclaimed bodies in HEIs provides an opportunity to recreate the existing and past narrations of exploitation and inequality. International analogs offer valuable pedagogical cues. The EAAF in Buenos Aires, foregrounds the social and historical responsibilities of future practitioners by situating all its training within the history of political disappearance and transitional justice. Likewise, in India, the All-India Institute of Medical Sciences integrates historical awareness and medical jurisprudence into its forensic medicine curriculum, linking ethical and historical instruction. The inclusion of similar contextual discussions within ViSUN's curriculum would enable alignment with global movements that seek to repair the historical disconnection between anatomy, ethics, and justice.

Forensic history as ethical pedagogy

The University of the Western Cape's Forensic History Project offers a clear South African example of this pedagogical shift in practice. Together with the associated Kronos special issue and related postgraduate teaching, it shows how the forensic encounter can be approached as more than technical recovery, identification and evidentiary interpretation (Rousseau et al., 2018; University of the Western Cape, n.d.). The project and associated courses (in which Kathryn Smith was a guest) examined efforts to locate, identify, and return the remains of people who had disappeared or been killed in contexts of political conflict. It also asked how post-1994 South Africa has understood the politically missing person, missingness, human remains, and memorialization. Its value lies in the exposure of HEIs pedagogical obligation. Future practitioners must therefore learn to recognize that human remains may carry unresolved histories, especially where political violence, racial science, repatriation, family claims and institutional responsibility intersect. Scientific competence and procedural compliance remain necessary, but alone, they are insufficient. They must be joined to historical accountability, epistemic humility, and ethical responsibility. In this sense, the project reframes the university as more than just a site of technical and theoretical training.

Critical race theory and epistemic justice

Given that forensic anthropology still utilizes race (expressed as "ancestry" or "population affinity") as a category in biological profiling (albeit operationally rather than ontologically), it is remarkable how few curricula engage students in the social construction of race, the history of scientific racism, and/or the ethics of classification. The University College London's Institute of Archaeology, through its Decolonizing Archaeology initiative (University College London, 2022), and the University of Bradford, through reflective coursework on the history of anthropometry (University of Bradford, n.d.), demonstrate how critical race theory and epistemic justice can be integrated within technical modules. These curricular experiments resonate with South Africa's ongoing need to train practitioners who can critically evaluate the inherited racial taxonomies embedded in forensic methods.

It is our view that such omissions have real-world consequences regarding how the dead are understood, classified, and narrated within forensic practice. To address such deficits, a decolonized curriculum in forensic anthropology must be more than a diversification of content and instead, constitute a paradigm shift. The Features of Science module within Stellenbosch University's Faculty of Medicine and Health Sciences provides a local model for reflective pedagogy (Essop, 2024, 2025). Similar reflexive learning environments exist at the University College London and the University of Western Australia, where students engage through journaling, group dialog, and participatory learning. Incorporating such methods at ViSUN (and similar programs) would better align with transformative learning (Mezirow, 1997) and situated learning (Lave and Wenger, 1991). Both of these learning models emphasize that learners' social contexts and frames of reference shape their ethical action. The following four intersecting frameworks, each exemplified by global analogs, could guide this transformation:

1. **African bioethics:** Drawing on Ubuntu philosophies, this approach emphasizes relationality and communal responsibility (refer to chapter 13 of this book volume). The ICRC Pretoria Delegation and African Centre for Medicolegal Systems (ACMS) provide models of humanitarian forensics, where ethics is inseparable from dignity and social repair. Reflective journals and discussion-based modules could prompt ViSUN students to evaluate their ethical stances and engage in cultural and spiritual considerations.
2. **Restorative justice:** The EAAF operationalizes restorative practices through family engagement and truth-seeking, while the University of Tennessee's Forensic Anthropology Center collaborates with human rights missions (USA FBI, 2019). These examples demonstrate how technical procedures can become reparative acts. ViSUN could adopt similar group tasks and interdisciplinary dialogs to examine how historical violence shapes the ongoing presence of unclaimed remains.
3. **Genealogies of scientific racism:** Advanced primarily through the humanities, such as history, philosophy of science, and museum studies (Msimang, 2021; Rassool, 2015; Walters, 2023), this area of research has yet to be explored

extensively in the context of FHA by South African HEIs. Here, lessons can be taken from international HEIs; for example, the University College London's Decolonizing Science initiative and Bradford's History of Science units foreground critical examination of craniometry and ancestry estimation. ViSUN could integrate comparable case-based seminars that confront the racialized histories of local forensic practice.

4. **Community engagement and participatory learning:** Programs such as Texas State University's Forensic Anthropology Center (Adams et al., 2024), Australia's Facility for Taphonomic Experimental Research (AFTER, n.d.), and Argentina's EAAF (Albrecht, 2024) demonstrate participatory models linking laboratories, families, and communities. Entities similar to ViSUN could develop comparable engagement frameworks through public dialogs and collaborative workshops, extending situated learning beyond the laboratory.

Structural embedding is required to operationalize such frameworks. Establishing core modules that include ethical reasoning, genealogies of science, and community engagement, coupled with transdisciplinary co-teaching, would help institutionalize transformation. The current pedagogical and institutional logic of forensic education in South Africa remains largely utilitarian, focused on efficient case resolution, resource management, and workforce readiness. This logic mirrors trends observed at the National Forensic Sciences University (India) and other state-linked forensic institutions, where procedural training can eclipse moral inquiry. Yet the fundamental question persists: how should we relate to the dead, particularly when they cannot consent or resist? Here, deontological ethics offers a critical counterbalance. It invites educators and students to ask not only "what works?" but "what is just?" and "what is owed to those whose lives were already shaped by marginalization?"

The following curricular tools are recommended to facilitate such a shift, with many provided by international exemplars:

- **Case-based ethics modules:** Modeled after the University of Western Australia's Forensic Ethics and Professional Practice course, these modules immerse students in real dilemmas involving race, consent, and structural violence.
- **Critical reflective journals:** As used at the University College London and within Stellenbosch University's Features of Science course, journals cultivate self-awareness and moral reasoning.
- **Interdisciplinary co-teaching:** Following the example of EAAF and ACMS, diverse instructors could jointly facilitate reflection on practice, law, and community engagement.
- **Community feedback loops:** Structured dialogs with affected families, echoing EAAF's participatory methods, create authentic, ethical encounters.
- **Ritual and commemorative practices:** Inspired by humanitarian forensics globally and remembrance ceremonies in Latin American training programs, moments of silence or memorialization could be integrated to honor the dead and cultivate moral imagination.

Together, curricula are not merely vessels for knowledge transmission but ways that contribute to worldview formation. The responsibility of forensic educators, particularly in the post-colonial world and in post-apartheid South Africa, is immense. The human body has long been a site of racial classification, state violence, and scientific extraction. By learning from global institutions such as EAAF (Argentina), University College London, FACTS (USA), the Facility for Taphonomic Experimental Research (Australia), the All-India Institute of Medical Sciences (India), and ACMS/ICRC (South Africa); ViSUN can model a curriculum that is not only scientifically rigorous but also ethically restorative and historically conscious. These tools should not only enrich the intellectual life of the curriculum but also cultivate the moral imagination necessary for working with the dead in relational, dignified, and transformative ways.

Reframing the role of the university

Universities are not passive repositories of knowledge but rather historical role players, ideological spaces, and laboratories of possibility. In societies emerging from colonial and racialized rule, universities are charged with the dual responsibility of redressing historical injustices and cultivating new forms of knowing and being. As one of the country's oldest and most historically Afrikaans institutions, Stellenbosch University carries both the burden of its past complicity in apartheid and the promise of its present transformation mandate. The university's Vision 2040 and Strategic Framework, and Institutional Plan (Fataar, 2022) mark a decisive institutional turn toward social repair by articulating commitments to human dignity, justice, and equity, and the transformation of knowledge systems in support of a just society.

While these aspirations are compelling, they demand material and pedagogical anchoring, especially for those disciplines historically entangled with biopolitical governance. Here, disciplines such as forensic anthropology occupy a liminal moral space where education, research, and service intersect with histories of violence and exclusion. In this context, university-based forensic units like the ViSUN can serve as laboratories for the realization of Vision 2040, particularly when their operations are oriented toward epistemic justice, curricular decolonization, and community legitimacy.

South Africa is not unique in this discussion, as at a global level, universities are also grappling with similar challenges. Here, the University College London's Institute of Archeology, the University of Bradford, and the University of Western Australia have begun to reframe forensic anthropology teaching to include critical engagement with colonial legacies and the ethics of race and representation. Likewise, the EAAF exemplifies how a research institution can pair technical excellence with a human rights mission centered on dignity, commemoration, and restorative justice. Within Africa, the ACMS (working in collaboration with the ICRC Pretoria Delegation) embodies a humanitarian model of forensic science that integrates professional training with social accountability.

ViSUN occupies a complex institutional position at Stellenbosch University as it provides a critical service to the SAPS and FPS by assisting in the identification of human remains when conventional methods fail. It also trains students in anthropological methods, instills procedural rigor, and strengthens the country's forensic capacity. These outputs alone are valuable but may not fulfill the university's transformation mandate entirely. For example, ViSUN's existing operational model, aligned with state priorities of case resolution and efficiency, reflects what Foucault (2007) terms *governmentality*. This Foucauldian concept frames the rationalities and practices through which institutions manage life and death in the name of order and knowledge. Thus ViSUN's work, while ostensibly technical, enacts a subtle form of governance over the dead by inversely determining whose remains are investigated, how they are classified, and how they re-enter the social order.

In terms of fulfilling its public mandate, ViSUN is a founding member of the S4J working group, within the inaugural cohort of Stellenbosch University's "Public Squares" initiaitive (Stellenbosch University, 2024). S4J is a transdisciplinary platform that aligns scientific inquiry with questions of justice, legitimacy, and social responsiveness. This affiliation offers a structural mechanism for transitioning toward participatory models of forensic practice, where community stakeholders, ethicists, and legal experts co-shape decision-making processes. The inclusion of non-academic voices parallels the participatory ethos of institutions such as EAAF and ACMS, where community collaboration is central to the legitimacy of forensic work.

From a Foucauldian perspective, ViSUN's parastatal role situates it within broader biopolitical regimes where universities function as governmental actors, mediating between the state's regulatory frameworks and society's moral expectations. Such arrangements, though necessary for capacity-building, raise critical questions of accountability: "To whom is the university answerable?" "Whose interests does the university serve?" Drawing on Policy Network Theory (Rhodes, 1997), ViSUN's embeddedness within a network of state and institutional actors can be viewed as both enabling and constraining. These policy communities often privilege stability, technical expertise, and managerial efficiency, which are essential for forensic accuracy but potentially antithetical to transformative change.

To resist this narrowing of purpose, ViSUN could cultivate what Foucault (as cited in Demetriou, 2016) describes as counter-conduct, i.e., the capacity for reflexive resistance within institutions to challenge dominant rationalities. Counter-conduct, in this context, means embedding critique and care into the everyday routines of forensic practice. Systematically, this would necessitate teaching the genealogies of scientific racism, incorporating African philosophies of death and personhood, and finally establishing community feedback loops for transparency and trust.

The example of S4J's transdisciplinary approach, where medical scientists, clinicians, social scientists, and non-governmental organizations cooperate, demonstrates how knowledge co-production can disrupt insular policy communities and broaden the epistemic base of forensic governance. By dispersing power and amplifying marginalized voices, ViSUN can help transform forensic identification from a procedural exercise into a relational practice of justice. However, as Monno and Khakee

(2012) caution, such participatory efforts must avoid tokenism. These efforts must, therefore, enable historically silenced communities to exert real influence over bioethical and forensic decisions. Sustained community engagement, transparency in donor and casework processes, and reflexive pedagogical design would collectively strengthen public trust in both the university and the forensic system.

Rationale

As outlined in the preceding case study, ViSUN demonstrates how academic institutions can merge service delivery with education, aligning forensic practice with humanitarian ethics. However, this model still operates within a state-aligned utilitarian framework, where efficiency, compliance, and throughput take precedence over historical accountability and community legitimacy. To align with Vision 2040, Stellenbosch University must, therefore, shift from providing technical assistance to the state toward cultivating epistemic and ethical leadership within South Africa's medico-legal sector. This reframing situates Stellenbosch University not only as a producer of forensic knowledge but also as a public mediator and is, therefore, capable of modeling how universities can restore trust in institutions managing the dead.

Strategic trajectory

Short-term (1–3 years): Institutional anchoring and ethical integration

For the short term, there should be a focus on embedding ethical reflexivity and curricular transformation within its existing forensic and anatomical programs. This can be achieved through a multi-pronged approach that strengthens the moral and educational foundations of ViSUN and its alignment with Vision 2040. First, curricular reform should take priority, and mandatory modules on *Critical Ethics* and *Forensic Histories* should be introduced across relevant programs, drawing inspiration from: (1) African bioethical philosophies such as Ubuntu; (2) the *Features of Science* module at Stellenbosch University; and (3) international examples as offered by the University College London and Bradford University. The relevance of this is that the former emphasizes relationality, dignity, and communal responsibility, whereas the latter explores and critically examines the genealogies of scientific racism within forensic anthropology training. ViSUN should also be formally embedded within Stellenbosch University's Transformation Office and the Public Square initiative, ensuring that identification services are evaluated against both technical and social justice benchmarks.

Finally, building public trust should form part of ViSUN's ethical infrastructure. Here, annual public briefs on ViSUN's operations, such as detailing the number of identifications completed, the nature of student engagements, and key ethical reflections, could be published to enhance institutional transparency and community accountability. Together, these measures should establish a short-term framework that anchors its forensic work in ethical awareness, academic integrity, and public legitimacy.

Medium-term (3–6 years): Structural reform and sectoral leadership

The next step would take place after the consolidation of institutional reforms and the repositioning of ViSUN as the national hub for ethically grounded medico-legal education and innovation. Achieving this requires a deliberate move from internal transformation toward sectoral leadership that reshapes the governance and ethics of forensic science across South Africa. Central to this effort is the establishment of a Forensic Governance Partnership between HEIs, the SAPS, FPS, and the Department of Health. This partnership would co-develop ethical protocols and national guidelines for the use of human remains in education and research, promoting regulatory coherence and accountability across the medico-legal sector.

In tandem, the evolution of S4J may lead to the creation of a Forensic Humanitarian Action Research Cluster that would serve as an interdisciplinary platform for national scholarship and policy development. This cluster would, therefore, possess the necessary expertise to focus on matters of social justice, death governance, and decolonial reform in scientific practice. Through collaborative research, curriculum development, and public engagement, it could generate actionable knowledge that bridges theory and institutional policy. Capacity building should also become a defining feature of this period. HEIs should introduce postgraduate fellowships and continuing-education programs in *Humanitarian Forensics and Policy.* Modeled after the training frameworks of the EAAF and the ACMS/ICRC Pretoria Delegation, these fellowship programs would prepare graduates for roles that extend beyond state employment by equipping them to contribute to non-governmental organizations, regional forensic networks, and international humanitarian initiatives. Finally, transformation must be tracked through measurable outcomes. Ethical performance and community-engagement indicators should be integrated into Stellenbosch University's institutional scorecard, ensuring that progress toward Vision 2040 is evaluated not only through academic outputs but also through social impact and accountability. Collectively, these initiatives will move Stellenbosch University from ethical anchoring to structural leadership, enabling it to redefine the national standards for medico-legal education, practice, and policy.

Long-term (6–10 years): Systemic reframing and global leadership

By this stage, the South African higher education system would have undergone sufficient transformation to extend its influence that would enable the restructuring of national and regional frameworks governing the ethical management of the dead. Central to this vision is the development of a South African Forensic Identification and Taphonomy Act, informed by the evidence base generated through ViSUN's casework and research outputs. Such legislation would standardize national protocols, promote transparent governance of human remains, and ensure that forensic science operates within ethically coherent, culturally sensitive, and legally robust boundaries.

Parallel to policy reform, the establishment of an African Network for Forensic Humanitarian Action would link universities, forensic laboratories, and medico-legal authorities across the continent and beyond. This reform will foster collaboration in data sharing, capacity development, and ethical standard-setting (Czachur et al., 2021). By building regional infrastructure for cooperative learning and policy exchange, South African HEIs could position themselves as convenors for knowledge co-production, echoing the spirit of Stellenbosch University's Vision 2040's emphasis on global engagement and societal impact.

Pedagogy as restorative justice

Forensic pedagogy in South Africa stands at a critical juncture. Although ViSUN is a vital mechanism for state utility, it remains an incomplete project of ethical transformation. The unit addresses a national backlog of unresolved medico-legal investigations by providing anthropological expertise when conventional identification methods fail, while simultaneously serving as a training ground for students. However, this praxis remains deeply entangled with the racial and colonial legacies of South Africa's medico-legal and sociopolitical system.

The use of unclaimed bodies in education is not new. Colonial and apartheid-era legislation, such as the Anatomy Acts of 1911 and 1959, normalized the transfer of the unclaimed dead to medical institutions. This racialized anatomical economy established a legacy where the marginalized in life became resources in death. While democratic South Africa has moved beyond the formal structures of apartheid, the epistemic residues from the past continue to persist. Current forensic protocols still require racial classification, and legal frameworks fail to adequately address consent, community engagement, or the dignity of the deceased. Considering this, ViSUN's model of integrating students into live forensic casework is both innovative and fraught. While it provides essential public service and real-world training in a resource-constrained system; it risks reproducing extractive practices. Therefore, by addressing the necessary ethical, historical, and philosophical reflections that are composite of FHA, curricula will develop agility. Moving from technocratic proceduralism to a pedagogy of care requires confronting uncomfortable questions. In a fractured medico-legal system, forensic pedagogy offers more than procedural solutions. It provides an opportunity for institutions to model care, accountability, and justice. This outlook recognizes the dead as not anonymous specimens but as political subjects whose identification and dignification matter. Thus ViSUN's work can embody this ethos by preparing practitioners who are technically competent, ethically attuned, and socially responsible. Although rooted in a specific national history, these challenges resonate far beyond South Africa, reflecting global tensions in how forensic, medical, and scientific institutions engage the dead amid inequality, conflict, and resource constraints. With the adoption of more inclusive and diverse models of learning and sharing knowledge, the academy may finally reclaim its position in directing the attitudes and norms of humanity - in life, and in death.

References

Adamovic, N., Howes, L.M., White, R., Julian, R., 2023. Understanding the challenges of disaster victim identification: perspectives of Australian forensic practitioners. Forensic Sci. Res. 8 (2), 107–115.

Adams, D.M., Bedard, J.R., Blatt, S.H., Faisal, E., Goliath, J.R., Gregory-Alcock, G., Gruenthal-Rankin, A., Morales Lorenzo, P.N., Smith, A.C., Tallman, S.D., Tegtmeyer Hawke, R., 2024. Speaking truth to power: toward a forensic anthropology of advocacy and activism. Humans 4 (1), 66–90.

Adebayo, J.O., 2023. South Africa's Military Deployment to Northern Mozambique: lessons from the United States of America's (Mis) Adventure in Afghanistan. Sci. Mil. 51 (1), 25–43.

African Center for Strategic Studies, 2025. Africa's Colliding Conflicts Compound Forced Displacement Crisis. Available at: https://africacenter.org/spotlight/africa-conflicts-compound-forced-displacement/ (2025, October 17).

Agarwal, S.C., 2025. The disposability and inclusion of Brown bodies. Am. J. Biol. Anthropol. 186 (1), e25003.

Ajam, T., 2019. Epistemic Redress, Race, Gender, and the South African Academe. Beyond Lip Serv. doi:10.13140/RG.2.2.15683.67361.

Alblas, A., Gibbon, V.E.A., 2026. Custodians of controversy: navigating stewardship challenges of state-donated anatomically derived skeletonised individuals in South African repositories. Am. J. Biol. Anthropol., AJPA–2025–00198.

Albrecht, F. 2024. "It's not a science isolated in a bubble": grave encounters in forensic anthropology in Colombia and Peru (Doctoral dissertation, The University of Western Ontario, Canada).

Australian Facility for Taphonomic Experimental Research (AFTER), n.d. Available at: https://www.uts.edu.au/about/locations-facilities/after (2025, October 21).

Bélisle-Pipon, J.C., Ravitsky, V., 2025. Consent is dead, long live ethical oversight: integrating ethically sourced data into demonstrated consent models. Am. J. Bioeth. 25 (4), 112–115.

Bhutto, F., 2024. Trust in public institutions: causes of decline and ways to restore it. Res. Consort. Arch. 2 (3), 123–131.

Billings, B.K., Kramer, B., Augustine, T.N., Brits, D., Hutchinson, E.F., Libhaber, E., Štrkalj, G., 2024. Leading the transition to ethical human body sourcing in Africa: the South African experience. Ann. Anat. 254, 152263.

Bird, C.E., Shangraw, A., 2024. The Importance of Accounting for the Dead in Migration. J. Migr. Hum. Secur. 12 (3), 310–320.

Bowers Du Toit, N.F., 2014. Gangsterism on the cape flats: a challenge to 'engage the powers'. HTS Theol. Stud. 70 (3), 1–7.

Bowker, G.C., Star, S.L., 2000. Sorting Things Out: Classification and Its Consequences, MIT Press.

Bowleg, L., 2021. The master's tools will never dismantle the master's house: ten critical lessons for Black and other health equity researchers of color. Health Educ. Behav. 48 (3), 237–249.

Breckenridge, K., 2014. The book of life: the South African population register and the invention of racial descent, 1950–1980. Kronos 40 (1), 225–240.

BusinessLive, 2024. State mortuaries are in a disgusting condition, 11 April. Available at: https://www.businesslive.co.za/bd/national/health/2024-04-11-state-mortuaries-are-in-a-disgusting-condition/

Cape Times, 2025. State mortuaries record marked drop in uncollected bodies, June. Available at: https://capetimes.co.za/news/2025-06-23-state-mortuaries-record-marked-drop-in-uncollected-bodies/

Chaussée, A.S., Winter, J., Ayres, P., 2022. Approaches to decolonising forensic curricula. Sci. Justice 62 (6), 795–804.

CISR, J.M.U., 2023. The Journal of Conventional Weapons Destruction. J. Conv. Weapons Destr. 25 (3), Article 1. Available at: https://commons.lib.jmu.edu/cisr-journal/vol25/iss3/1.

City of Cape Town, 2022. The City's response to homelessness and helping people off the streets - Frequently Asked Questions, 20 May. Available at: https://resource.capetown.gov.za/documentcentre/Documents/Procedures,%20guidelines%20and%20regulations/Street_People%20FAQs_English.pdf

Clegg, L., 2021. The International Committee of the Red Cross and the protection of world war dead. Forensic Sci. Int. 319, 110648.

Cooperation in Science and Technology (COST), 2023. Migrant disaster victim identification (MDVI) - CA22106. Available at: https://www.cost.eu/actions/CA22106/ (October 21, 2025)

Czachur, M.V., Todd, M., Gonçalves Loureiro, T., Azam, J.M., Nyeleka, S., Alblas, A., Davies, S.J., 2021. Opening the floor for discussion: a perspective on how scholars perceive attitudes to science in policymaking in South Africa. S. Afr. J. Sci. 117 (1/2), Art.#7948. doi:10.17159/sajs.2021/7948.

Daily Maverick, 2025. Anton Harber's downplaying of modern-day South African race law is naïve, 25 March. Available at: https://www.dailymaverick.co.za/opinionista/2025-03-25-anton-harbers-downplaying-of-modern-day-south-african-race-law-is-naive/

Demetriou, O., 2016. Counter-conduct and the everyday: anthropological engagements with philosophy. Glob. Soc. 30 (2), 218–237.

Demeyere, B., Ojeda, S., Niemeijer, M., Fonseca, S., Rossouw, V., 2025. Mitigating the risk of military personnel becoming unaccounted for on the battlefield: an interview with Stephen Fonseca and Vaughn Rossouw on the ICRC's Military Personnel Identification Project. Int. Rev. Red Cross 107 (929), 440–461. doi:10.1017/S1816383125100672.

Elgerud, L., Miranker, M., Cawley, W., 2025. Necro-forensic spectrum: employing necropolitics in forensic anthropological practice and research. Ann. Anthropol. Pract. 49 (2), e70017.

Essop, F., 2024. Transforming physiology teaching and learning to empower students to navigate their way in an era of doubt: a unique case study. In: Essop, F. (Ed.), Truth Unveiled. Navigating Science and Society in an Era of Doubt. Academic Press, Amsterdam, pp. 161–179.

Essop, M.F., 2025. Biomedicine meets art: a physiologist's reflections on inter-disciplinary liaisons, curriculum renewal and pursuing social justice. Crit. Stud. Teach. Learn. 13. Special Issue 1. doi:10.14426/cristal.v13iSI1.2705.

Fataar, A., 2022. Advancing transformation as the university's strategic heart. University World News. Available from: https://www.universityworldnews.com/post.php?story=20220220201123373 (Accessed 7 June 2026).

Ferri, G., 2022. The master's tools will never dismantle the master's house: decolonising intercultural communication. Lang. Intercult. Commun. 22 (3), 381–390.

Finn, G.M., Danquah, A., Matthan, J., 2022. Colonization, cadavers, and color: considering decolonization of anatomy curricula. Anat. Rec. 305 (4), 938–951.

Foucault, M., 2007. Security, Territory, Population: Lectures at the Collège de France 1977–78, Palgrave Macmillan.
Fuhr, I., de Klerk, N., 2024. The Human Bridge: Racial Healing in South Africa, Jonathan Ball Publishers.
Gaggioli, G., 2018. International Humanitarian Law: the legal framework for humanitarian forensic action. Forensic Sci. Int. 282, 184–194.
García-Mancuso, R., Plischuk, M., Desántolo, B., Garizoain, G., Sardi, M.L., 2020. Ethical considerations in human remains based research in Argentina. In: Squires, K., Errickson, D., Márquez-Grant, N. (Eds.), Ethical Approaches to Human Remains: A Global Challenge in Bioarchaeology and Forensic Anthropology. Springer International Publishing, Cham, pp. 447–463.
Gibbon, V.E., Feris, L., Gretzinger, J., Smith, K., Hall, S., Penn, N., Mutsvangwa, T.E., Heale, M., Finaughty, D.A., Karanja, Y.W., Esterhuyse, J., 2023. Confronting historical legacies of biological anthropology in South Africa - Restitution, redress, and community-centered science: the Sutherland Nine. Plos One 18 (5), e0284785.
Gibbon, V.E., Heathfield, L.J., Smith, K., Sealy, J.C., Martin, L.J., 2025. A transdisciplinary integrated approach to improve identification outcomes for decomposed decedents in medico-legal death investigations. Forensic Sci. Int. Synergy 10, 100579.
Go, M.C., 2025. The decolonial turn in forensic anthropology. Am. J. Biol. Anthropol. 186 (2), e70003.
Great Britain. n.d. Anatomy Act of 1832. https://www.nationalarchives.gov.uk/education/resources/body-snatchers/source-four-the-anatomy-act-1832/
Hicks, D., 2013. Four-field anthropology: charter myths and time warps from St. Louis to Oxford. Curr. Anthropol. 54 (6), 753–763.
Hoole, D., 2018. Dissection of the destitute: the supply of anatomical subjects to the Medical Schools of Aberdeen c. 1832–1902. J. Scott. Hist. Stud. 38 (2), 238–260.
Hutchinson, E.F., Kramer, B., Billings, B.K., Brits, D.M., Pather, N., 2020. The law, ethics, and body donation: a tale of two bequeathal programs. Anat. Sci. Educ. 13 (4), 512–519.
INRAP, 2025. French: institut national de recherches archéologiques préventives. Available at: https://www.inrap.fr/en/application-forensic-archaeology-investigation-cases-political-ethnic-and-12846 (2025, October 17).
Institute for Security Studies, 2023. Analysis of murder in SA's deadliest provinces calls for policing rethink, ISS. Available at: https://issafrica.org/about-us/press-releases/analysis-of-murder-in-sas-deadliest-provinces-calls-for-policing-rethink. 14 November 2023.
International Centre for Humanitarian Forensics (ICHF), 2022. Leading a new frontier in humanitarian forensics. Available at: https://icrcndresourcecentre.org/wp-content/uploads/2022/05/ICHF-Leaf-let-2022-printed.pdf (2025, October 17).
International Committee of the Red Cross (ICRC), 2020b. Commentary on the First Geneva Convention. Geneva: Vol 1. ICRC, Cambridge University Press.
International Committee of the Red Cross, 2018. The Missing and Deceased Migrants Pilot Project in South Africa and Zimbabwe: 2016–2018 (Accessed 17 October 2025).
International Committee of the Red Cross, 2020a. Guiding Principles for the Dignified Management of the Dead in Humanitarian Emergencies and to Prevent them Becoming Missing Persons, September 2020. Available at: https://www.icrc.org/sites/default/files/document/file_list/english_version_-_call_for_comments_-_dignified_management_of_the_dead.pdf

International Committee of the Red Cross, 2024. South Africa: strengthening IHL and IHRL compliance, September 2024. Available at: https://www.icrc.org/en/article/South-Africa-strengthening-complying-ihl-ihrl (2025, October 17].
International Committee of the Red Cross, 2025. Nairobi regional delegation facts and figures, January - December 2024 - April 2024. Available at: Kenya_Facts and Figures Jan - Dec 2024.pdf (2025, October 17).
International Committee of the Red Cross, n.d. South Africa. Available from: https://www.icrc.org/en/where-we-work/south-africa (Accessed 7 June 2026).
International Organization for Migration (IOM), 2025. IOM Spokesperson on the Tragic Migrant Deaths Off Egypt's Coast. Available at: https://www.iom.int/news/iom-spokesperson-tragic-migrant-deaths-egypts-coast (2025, October 17).
Jagganath, G., 2025. Migration trends in South Africa: post COVID-19 perspectives. Afr. Renaiss. 22 (2), 265.
JMU, C., 2023. The Journal of Conventional Weapons Destruction, Issue 25.3. J. Conv. Weapons Destr. 25 (3), 1.
Kehoe, A.B., 2013. Humans: An Introduction to Four-field Anthropology. Routledge. doi:10.4324/9781003226819.
Kenya National Commission on Human Rights, n.d. Available at: https://www.knchr.org/
Keyes, C.A., Hill, L., Gordon, G.M., 2016. An appraisal of decomposition cases received at the johannesburg forensic pathology service medico-legal mortuary during 2010–2011. J. Forensic Sci. 61 (2), 452–457.
Keyes, C.A., Mahon, T.J., Gilbert, A., 2022. Human decedent identification unit: identifying the deceased at a South African medico-legal mortuary. Int. J. Legal Med. 136 (6), 1889–1896.
Khan, N. 2025. Stakeholders demand action to eradicate massive DNA backlog at SAPS forensic lab. IOL. Available from: https://iol.co.za/thepost/news/2025-03-29-stakeholders-demand-action-to-eradicate-massive-dna-backlog-at-saps-forensic-lab/ (Accessed 7 June 2026).
Kim, J.J., Rosenblatt, A., 2023. Whose humanitarianism, whose forensic anthropology? In: Parra, R.C., Ubelaker, D.H. (Eds.), Anthropology of Violent Death: Theoretical Foundations for Forensic Humanitarian Action. Wiley, Hoboken, NJ, pp. 153–176.
Kramer, B., Hutchinson, E.F., Brits, D.M., Billings, B.K., 2019. Making the ethical transition in South Africa: acquiring human bodies for training in anatomy. Anat. Sci. Educ. 12 (3), 264–271.
Kumar, S., 2022. DNA-based human identification in mass-disaster cases. In: Kumar, A., Goswami, G.K., Huffine, E. (Eds.), Handbook of DNA Forensic Applications and Interpretation. Springer Nature Singapore, Singapore, pp. 149–175.
Labuschagne, B., Mathey, B., 2000. Cadaver profile at University of Stellenbosch Medical School, South Africa, 1956–1996. Clin. Anat. 13 (2), 88–93.
Latchanna, A. 2025. Cracking the code of unexplained deaths. Nature Africa. Available from: https://www.nature.com/articles/d44148-025-00243-w (Accessed 7 June 2026). doi:10.1038/d44148-025-00243-w.
Lave, J., Wenger, E. 1991. Situated Learning: Legitimate Peripheral Participation. Cambridge University Press, Cambridge.
M'charek, A., Casartelli, S., 2019. Identifying dead migrants: forensic care work and relational citizenship. Citizensh. Stud. 23 (7), 738–757.

Mabuza, E., 2025. Only dissection section remains closed at Free State government mortuary. Times Live, 08 May 2025. Available at: https://www.timeslive.co.za/news/south-africa/2025-05-08-only-dissection-section-remains-closed-at-free-state-government-mortuary/

Mahlangu, S. 2025. SA universities rely on foreign academics while local scholars remain stuck. The Star, [online, June 2025]. Available at: https://thestar.co.za/news/2025-05-21-sa-universities-rely-on-foreign-academics-while-local-scholars-remain-stuck/

Manyane, M., 2024. Mortuaries in disastrous state due to unclaimed bodies. Sunday Independent. Available from: https://sundayindependent.co.za/news/2024-09-29-mortuaries-in-disastrous-state-due-to-unclaimed-bodies/ (Accessed 7 June 2026).

Maromo, J. 2024. Cold cases: 3,400 unclaimed bodies frozen in state mortuaries, with some kept for six years - should they be cremated now? IOL. Available at: https://iol.co.za/news/crime-and-courts/2024-04-28-cold-cases-3400-unclaimed-bodies-frozen-in-state-mortuaries-with-some-kept-for-six-years-should-they-be-cremated-now/

Masters, N.C., Shapiro, H.A., 1951. Medico-legal section: the medico-legal aspects of interference with dead bodies. S. Afr. Med. J. 25 (22), 376–379.

Mbembe, A., 2020. Necropolitics, Duke University Press.

McManners, J. 1981. Death and the Enlightenment. Changing Attitudes to Death among Christians and Unbelievers in 18th-century France. Clarendon Press/Oxford University Press, Oxford/New York.

Meiring, K.O., Gibbon, V.E., Alblas, A., 2024. Anatomical human body donation in South Africa: inconsistencies of informed consent. Ann. Anat. 255, 152292.

Meyer, D., 2019. Cape Flats violence 'at worst ever' with 1,600 murdered already this year. Times Live, 27 June 2019. Available at: https://www.timeslive.co.za/news/south-africa/2019-06-27-cape-flats-violence-at-worst-ever-with-1600-murdered-already-this-year/

Mezirow, J., 1997. Transformative learning: theory to practice. New Dir. Adult Contin. Educ. 1997 (74), 5–12.

Missing Migrants Project, 2025. Available at: https://missingmigrants.iom.int/region/mediterranean. (2025, October 17)

Monno, V., Khakee, A., 2012. Tokenism or political activism? Some reflections on participatory planning. Int. Plan. Stud. 17 (1), 85–101.

Moon, C., 2014. Human rights, human remains: forensic humanitarianism and the human rights of the dead. Int. Soc. Sci. J. 65 (215-216), 49–63.

Morris, A.G., 2014. Controversies about the study of human remains in post-Apartheid South Africa. Archaeological human remains: global perspectives, pp.189–198.

Morris, A.G., 2022. Bones and Bodies: How South African Scientists Studied Race, NYU Press.

Msimang, P.M., 2021. Revisiting the question of race and biology in the South African social sciences. In: Ludwig, D., Koskinen, I., Mncube, Z., Poliseli, L., Reyes-Galindo, L. (Eds.), Global Epistemologies and Philosophies of Science. Routledge, pp. 243–253.

NBC News, 2024. With rise in border migrant deaths, forensic volunteers and students work to identify remains. Available at: https://www.nbcnews.com/news/latino/rise-border-migrant-deaths-forensic-volunteers-students-work-id

Nelson Mandela Foundation, 2025. The unseen sacrifice: remembering South Africans who died in exile. Available at: https://www.nelsonmandela.org/news/entry/the-unseen-sacrifice-remembering-south-africans-who-died-in-exile (2025, October 17).

Office of the High Commission for Human Rights, 2022. Forensic investigators bring DR Congo victims closer to justice. Available at: https://www.ohchr.org/en/stories/2022/11/forensic-investigators-bring-dr-congo-victims-closer-justice (2025, October 17).

Ozcelik, A., 2026. From emergency relief to development holds: the developmental shift in the ICRC's humanitarian assistance in protracted conflict. J. Confl. Sec. Law 31 (1), 45-66.

Parliament of the Republic of South Africa, 2024. Media statement: health committee learns KZN and EC have highest numbers of unclaimed bodies in mortuaries, 20 September. Available at: https://www.parliament.gov.za/press-releases/media-statement-health-committee-learns-kzn-and-ec-have-highest-numbers-unclaimed-bodies-mortuaries

Parliamentary Monitoring Group, 2019. State of the Forensic Pathology & EMS in the province: Department of Health briefing, 14 August. Available at: https://pmg.org.za/committee-meeting/28692/

Parra, R.C., Zapico, S.C., Ubelaker, D.H., eds. 2020. Forensic Science and Humanitarian Action: Interacting with the Dead and the Living. John Wiley & Sons.

Phetha, H.R., Ndlovu, J., 2026. The contribution of South Africa's humanitarian and development aid efforts to conflict resolution and peace in Africa. J. Sustain. Dev. Law Policy 17 (2).

Pillay, P., McQuoid-Mason, D.J., Satyapal, K., 2017. A study of the role and functions of inspectors of anatomy in South Africa. S. Afr. J. Bioeth. Law 10 (2), 86–92.

Ramgoolam, S., 2017. An investigation into the value of supplementing dissection of the human body with alternative resources: perceptions of students and staff at the University of Cape Town.

Rassool, C., 2015. Re-storing the skeletons of empire: return, reburial and rehumanisation in Southern Africa. J. South. Afr. Stud. 41 (3), 653–670.

Rassool, C., Gibbon, V.E., 2024. Restitution versus repatriation: terminology and concepts matter. Am. J. Biol. Anthropol. 184 (1), e24889.

Republic of South Africa, 1950. The Group Areas Act. [online]. Available at: https://sahistory.org.za/archives.

Republic of South Africa. (1953). The Separate Amenities Act [online]. Available at: https://sahistory.org.za/archives

Republic of South Africa, n.d.a. National Health Act (NHA) No. 61 of 2003. [online]. Available at: https://www.gov.za/about-government/government-system/justice-system/legislation-and-policies.

Republic of South Africa, n.d.b. National Health Act (NHA), Regulation 180 of 2003, Section 10. [online]. Available at: https://www.gov.za/about-government/government-system/justice-system/legislation-and-policies.

Republic of South Africa, n.d.c. South African Constitution (Chapter 2; Section 10). [online]. Available at: https://www.justice.gov.za/constitution/SAConstitution-web-eng.pdf (Accessed: 12 December 2025)

Republic of South Africa, n.d.d. The Anatomy Act 3 of 1911. [online]. Available at: https://sahistory.org.za/archives.

Republic of South Africa, n.d.e. The Population Registration Act (No. 30 of 1950). [online]. Available at: https://sahistory.org.za/archives.

Rhodes, R.A.W., 1997. Understanding, Open University Press.

Rousseau, N., Moosage, R., Rassool, C., 2018. Missing and missed: rehumanisation, the nation and missing-ness. Kronos 44 (1), 10–32.

Saayman, G., 2020. Death investigation and forensic medicine in South Africa: historical perspectives, status quo, and quo vadis? Acad. Forensic Pathol. 10 (3-4), 115–130.

Sanford, V., 2008. On the frontlines: forensic anthropology. In: Poole, D. (Ed.), Companion to Latin American Anthropology. Blackwell Publishing, Malden, MA, pp. 485–501.

Sangweni, M.M., 2023. The failure of conflict early warning systems in preventing the 2017 insurgency in Mozambique's Cabo Delgado Province.

Serra, G. 2024. SA faces crisis of unclaimed bodies: families concerned as missing are buried as paupers. IOL, November 2024. Available at: https://iol.co.za/weekend-argus/news/2024-10-12-sa-faces-crisis-of-unclaimed-bodies-families-concerned-as-missing-are-buried-as-paupers/

Serradinho, J., 2024. The conflict in Cabo Delgado and the relationship between international disaster response law and international humanitarian law treaties. In: Mozambique's Cabo Delgado Conflict, pp. 158-172. Routledge.

Siegert, C.C., Kaplan, M.A., Herrmann, N.P., Spradley, M.K., 2024. The Texas landscape: accounting for migrant mortality and the challenges of a justice of the peace medicolegal system. J. Migr. Hum. Secur. 12 (3), 257–276.

Singh, S., 2022. Disaster Victim Identification. Acad. J. Anthropol. Stud. 7 (01), 21–32.

Smith, J.H., Horne, J.S., 2024. Regulating forensic sciences in Southern Africa. J. Forensic Sci. Crim. Investig. doi:10.19080/JFSCI.2024.18.555987.

Solomons, L., 2024. Rising concerns over hundreds of unclaimed bodies in Western Cape state mortuaries. News24, 04 May. Available at: https://www.news24.com/southafrica/news/rising-concerns-over-hundreds-of-unclaimed-bodies-in-western-cape-state-mortuaries-20240504

South African Criminal Law (Forensic Procedures) Amendment Act, Act No. 37 of 2013, n.d. Available at: https://www.gov.za/sites/default/files/gcis_document/201409/37268act37of2013crimlawamend27jan2014.pdf.

South African History Online, 2020. The Missing Persons Task Team (MPTT) to investigate apartheid missing persons cases. Available at: https://sahistory.org.za/dated-event/missing-persons-task-team-mptt-investigate-apartheid-missing-persons-cases (2025, October 17).

Southern African Development Community, 2024. SADC Mission in the DRC conducts' Training together for peace in the DRC Available at: (2025, October 17).

Spradley, M.K., 2014. Toward estimating geographic origin of migrant remains along the United States–Mexico border. Ann. Anthropol. Pract. 38 (1), 101–110.

Stellenbosch University, 2025b. The public squares teams, science for justice (S4J). Available at: http://www.sun.ac.za/english/research-innovation/Research-Development/Public-Squares-Initiative/Pages/s4j.aspx. Accessed on: 03 July 2025

Stellenbosch University, 2019. Vision 2040 and strategic framework 2019–2024. Available at: https://www.sun.ac.za/english/Documents/Strategic_docs/2018/Vision-2040-Strategic-Framework-2019-2024.pdf

Stellenbosch University, 2025a. Innovative system reduces ID times for unknown decedents, 02 March. Available at: https://www.sun.ac.za/english/Lists/news/DispForm.aspx?ID=11137

Sumathipala, A., Siribaddana, S., Perera, C., 2006. Management of dead bodies as a component of psychosocial interventions after the tsunami: a view from Sri Lanka. Int. Rev. Psychiatry 18 (3), 249–257.

Thompson, D., 2023. New biomedical institute a 'game changer' for healthcare. University World News. Available from: https://www.universityworldnews.com/post.php?story=20230417142839672 (Accessed 7 June 2026).

Thompson, D., 2026. Beyond silos: Rethinking how research gets done. University World News. Available from: https://www.universityworldnews.com/post.php?story=20260420103122551 (Accessed 7 June 2026).

Tidball-Binz, M.V., Cordner, S., 2022. Humanitarian forensic action: a new forensic discipline helping to implement international law and construct peace. Wiley Interdiscip. Rev. Forensic Sci. 4 (1), e1438.

Tobias, P.V., 1980. Apartheid and medical education: the training of black doctors in South Africa. J. Natl. Med. Assoc. 72 (4), 395.

Ubelaker, D.H., Shamlou, A., Kunkle, A.E., 2019. Forensic anthropology in the global investigation of humanitarian and human rights abuse: perspective from the published record. Sci. Justice 59 (2), 203–209.

United Nations Guidelines on the Management of the Dead, 2024. Protection and dignified management of the dead in war and peace are cornerstones for preventing unlawful killings: UN expert, 26 June 2024. Available at: https://www.ohchr.org/en/press-releases/2024/06/protection-and-dignified-management-dead-war-and-peace-are-cornerstones

United Nations High Commissioner for Refugees, n.d. Available at: https://www.unhcr.org/where-we-work/countries/south-africa (2025, October 17).

United Nations Sustainable Development Goals (SDGs) 2030 n.d.. [online]. Available at: https://sdgs.un.org/goals

United States of America. Federal Bureau of Investigations. Unearthing Stories for 20 Years at the 'Body Farm', 2019. Available at: https://www.fbi.gov/news/stories/body-farm-20th-anniversary-032019 (2025, October 21).

University College London, 2022. Cultures of decolonisation at UCL. Available at: https://discovery.ucl.ac.uk/id/eprint/10163919/ (2025, October 17).

University of Bradford, n.d. Forensic Anthropology BSc (Hons) course outline. Available at: https://www.bradford.ac.uk/courses/ug/forensic-anthropology/#:~:text=By%20the%20Chartered%20Society%20of%20Forensic%20Sciences,presentation%20of%20evidence%20in%20anthropology%20and%20archaeology (2025, October 17).

University of the Western Cape, n.d. Department of Historical Studies Projects. Available from: https://www.uwc.ac.za/study/all-areas-of-study/departments/department-of-history/projects (Accessed 7 June 2026).

University of the Witwatersrand, n.d. Human Variation and Identification Research Unit (HVIRU). Available from: https://www.wits.ac.za/anatomicalsciences/hviru/ (Accessed 7 June 2026).

University, S., 2024. "Working outside the box": SU launches public squares initiative to foster transdisciplinary research at SU, March. Available at: https://www.su.ac.za/af/fakulteite/geneeskunde/doktorale-kantoor/nuus/working-outside-box-su-launches-public-squares-initiative-foster-transdisciplinary-research-su (2025, October 17).

Victor, G., Barbu, A., Bélisle-Pipon, J.C., 2024. Moral values in medical AI: a scoping review.

Walters, C., Jansen, J.D., 2022. A troubled body of knowledge: the durability of racial science in human anatomy research in South Africa. Comp. Educ. Rev. 66 (1), 1–18.

Walters, H., 2023. 'The Reconfirmation of the Same': on recent genetic studies of the afrikaner population. In: Jansen, J.D., Auerbach, J. (Eds.), The Politics of Knowledge in the Biomedical Sciences: South/African Perspectives. Springer Nature Switzerland, Cham, pp. 119–144. doi:10.1007/978-3-031-31913-6_6.

Western Cape Government, 2025. The crime stats are shocking – but they show exactly what the solution is too. Available from: https://www.westerncape.gov.za/department-premier/article/crime-stats-are-shocking-they-show-exactly-what-solution-too (Accessed 7 June 2026).

Zavattaro, S.M., Entress, R., Tyler, J., Sadiq, A.A., 2021. When deaths are dehumanized: deathcare during COVID-19 as a public value failure. Adm. Soc. 53 (9), 1443–1462.

CHAPTER

7 Case study: Decolonizing the postgraduate biomedical sciences curriculum

Faadiel Essop[a] and Rhoda Meyer[b]

[a]*Centre for Cardio-metabolic Research in Africa (CARMA), Division of Medical Physiology, Faculty of Medicine and Health Sciences, Stellenbosch University, Cape Town, South Africa*

[b]*Department of Health Professions Education, Faculty of Medicine and Health Sciences, Stellenbosch University, Cape Town, South Africa*

Introduction

Although the colonists were defeated by indigenous resistance struggles and physically withdrew from occupied countries, their remaining effects continue to impact former colonized territories. The lasting influence of colonial remnants on knowledge systems, value judgments, thinking patterns, and structural legacies is referred to as "coloniality," which is distinct from the term "colonialism" (Quijano, 2007; Gonzalez, 2023). As Maldonado-Torres (2007: 243) explains,

> *Coloniality, instead, refers to long-standing patterns of power that emerged as a result of colonialism, but that define culture, labor, intersubjective relations, and knowledge production well beyond the strict limits of colonial administrations. Thus, coloniality survives colonialism.*

Hence, there is a strong need for formerly colonized territories to "decolonize" their thinking and organizational structures, including current education systems (Maldonado-Torres, 2007). Focusing on biomedical and medical education, there is an emerging interest in addressing this issue as part of a broader process of decolonization of tertiary curricula. The process of decolonization is complex and multifaceted, requiring the confrontation and deconstruction of the colonial past (Ratner et al., 2024). This must be accompanied by a reconstructive effort aimed at fostering greater inclusivity, dismantling structural power asymmetries, and advancing social justice and equity.

The historical dovetailing between medical science and colonialism contributed to, and still perpetuates, false beliefs about racial hierarchies (Wolf et al., 2020). Moreover, these so-called physiological race-based differences were instrumental in justifying the conquest and oppression of Indigenous populations. In support, Quijano (2007: 171) states that,

Decolonizing Medical Education, Research and Practice. https://doi.org/10.1016/B978-0-443-44998-7.00016-8

> *Coloniality of power was conceived together with America and Western Europe, and with the social category of 'race' as the key element of the social classification of colonized and colonizers. Unlike in any other previous experience of colonialism, the old ideas of superiority of the dominant, and the inferiority of dominated under European colonialism were mutated in a relationship of biologically and structurally superior and inferior.*

Racial hierarchies deeply shaped medical research, education, and clinical practice, leading to the marginalization of Indigenous knowledge systems and the unethical, discriminatory treatment of both research subjects and patients (Chadha et al., 2020). Such race-based research endeavors included, among other things, assessments of skull sizes, eugenic research, and the pursuit of differences in organ functions. This resulted in the development of race-based clinical algorithms to assess the functionality of various organ systems. For example, for chronic kidney disease, this meant that the disease stage of "Black" patients was classified as less severe than their "White" counterparts, despite displaying identical serum creatinine levels (input biomarker for such algorithmic calculations) (Chadha et al., 2020).

Attempts to address such issues can be challenging, as existing biomedical sciences and medical curricula are rooted in Eurocentric perspectives and historical traditions. Moreover, the decolonization of medical and biomedical sciences curricula has not been tackled by many institutions. Considering this, the current study endeavored to undertake a process of decolonizing the postgraduate biomedical sciences curriculum at our institution. Here, we reasoned that incorporating conversations on medical racism, coloniality, and decolonization into the curriculum would enable the deconstruction of colonial educational vestiges and begin a process of reconstruction. This, in turn, would be expected to promote the personal growth and awareness of students while also creating opportunities for greater inclusivity by fostering an environment of open discussion and learning in the classroom. Moreover, we envisaged that such initiatives would better equip students with the necessary tools to challenge and mitigate coloniality in their future careers and to become advocates for social justice, humanism, inclusivity, and equity. This aligns with the perspectives of Naidu and Abimbola (2022: 556), who stated that,

> *Much of what happens in the world of medicine is shaped by an educational system that socialises students into a hierarchical system dominated by the Western biomedical model. This dominant model of health is insensitive to the experiences of most of the world's people. The denial of majority world's epistemologies by medical schools remains deliberately or by default obscure to students.*

We ground this work in humanizing pedagogy, which moves beyond mere content diversification to fundamentally shifting power relations in the learning environment by valuing students' experiences, cultural knowledge, and ways of being as legitimate sources of understanding. This approach recognizes that decolonizing biomedical sciences and medical education requires not only changing what is taught but also transforming how we relate to each other and whose knowledge counts as valid in the construction of knowledge and understanding in the field.

This study, therefore, introduced a new module into the curriculum of postgraduate biomedical students at Stellenbosch University (Cape Town, South Africa). Thereafter, we performed a thematic analysis of responses to an open-ended questionnaire to gain deeper insight into students' experiences and perspectives on decolonizing the biomedical sciences.

Methods

As mentioned in earlier sections, this study was conducted during the *Features of Science* module that forms part of a 1-year BSc (Hons) course for biomedical students at the Faculty of Medicine and Health Sciences at Stellenbosch University (Cape Town, South Africa). Details of the *Features of Science* module were described in a previous book chapter (Essop, 2025). The module consisted of 8 × 90 min (n = 16 lecture slots) and covered a wide variety of topics, such as the social contract of science with society, the nature of the scientific process (critical examination of all steps), and the influence of the posttruth world on science and medicine. Some of the classes also focused on the decolonization of science and medicine, where students were required to complete prescribed readings before classes to ensure a more active learning approach. At the start of each class, the lecturer shared a brief synopsis of the prescribed readings (15 min) with the class, followed by group discussions for about an hour. The 2024 cohort of 30 students (from the Divisions of Medical Physiology, Clinical Anatomy and Human Genetics, and Human Biology) was divided into five groups (n = 6 per group). Each group was then assigned a staff member to help facilitate such discussions that were guided by several open-ended questions provided by the lecturer, namely:

1. *What is the definition of race, if at all?*
2. *What is the definition of ethnicity?*
3. *What is genetic ancestry and admixing?*
4. *What is "eugenics" and did this occur in an ideological vacuum?*
5. *From a historical perspective, did science assist colonialism and help entrench racism?*
6. *From a clinical perspective, provide two examples of how the adoption of race-based science has resulted in poor treatment of Black patients.*
7. *Regarding the "Colored" women study* (Nieuwoudt et al., 2020), *is the scientific process (or parts of it) flawed, and if so, how can we address this?*
8. *What could a successful and inclusive future science look like?*
9. *Should science consider the cultivation of humanity and empathy when doing research studies?*

Following such critical discussions, each group then shared their unique responses with the entire class. This process stimulated interesting discussions, with further inputs provided by the other students. As the objective of our study was to explore their perceptions and experiences regarding the decolonization of science and medicine, student feedback was gathered (at the end of the module) using an anonymous, open-ended questionnaire, following ethics approval. The questionnaire included the following questions:

1. *What is your understanding of the term "decolonization" in the context of science and medicine?*
2. *In your own words, please write a sentence or two on what you think the term "decolonization" means in the context of science and medicine.*
3. *Do you think it is important for us in South African higher education (focus on science and medicine) to begin such conversations? If so, why?*
4. *Can you provide an example(s) of how colonialism has influenced the development of scientific and medical knowledge?*
5. *What actions or initiatives do you think faculty members and tertiary institutions can take to promote decolonization in the teaching of science and medicine?*
6. *How do you think the curriculum of the course (and at undergraduate level) can be made to be more inclusive and equitable?*
7. *What do you think will be some of the challenges that will be encountered for this process ("decolonization" in our curriculum) to be implemented?*
8. *Do you think the inclusion of decolonization of science and medicine in the BSc (Hons) curriculum will be of value in your future career? Please provide a short motivation for your answer.*

Data collected from the open-ended questionnaires (90% response rate) were subsequently collated and prepared for thematic analysis. After reading the document multiple times, codes were developed by highlighting the key ideas in the text. A second level of coding was then performed to examine these codes more critically, refine them further for clarity and relevance, and group similar codes into categories. This iterative process helped ensure that the codes accurately represented the participants' perceptions and experiences while aligning with the overarching themes of the study. The third phase of data analysis involved developing themes and subthemes that were related to the research question and literature around decolonization in medical education and research. Before data collection, this study received ethical approval from the Research Ethics Committee (Social, Behavioral, and Education Research) at Stellenbosch University (Cape Town, South Africa) (#44378).

Results

Four interconnected themes were eventually developed, reflecting students' perspectives, namely, understanding the colonial legacy in scientific knowledge, rethinking knowledge, pursuing responsive education, and building inclusivity and promoting diversity (refer to Table 7.1 for detailed data analysis).

Table 7.1 Thematic analysis of student responses.

Theme number	Theme name	Quotations
Theme 1	**Understanding colonial legacy in scientific knowledge**	
Subtheme 1	*Critical historical consciousness*	
	Understanding decoloniality	Decolonization means looking closely at how colonial history has affected these fields, and working to make them fairer and more inclusive by listening to and respecting indigenous perspectives (P7). Examining events in our history as case studies (ex. outbreaks, social events influenced by poor health/wellbeing etc.) teaches students several valuable things simultaneously, including thinking about how different factors link to create a bigger picture (P17). I think recognizing historical biases is very important and recognizing how colonialism could have framed the content of the course (P7). Understanding how colonialism has marginalized local and traditional knowledge systems (P10).
	The link between colonialism, scientific racism, and pseudoscience	The concept of scientific racism was used to promote the idea of racial superiority and inferiority, which perpetuated discrimination and oppression against certain ethnic groups (P8). Colonialism perpetuated racial hierarchies and pseudoscientific theories of racial superiority and inferiority. European colonial powers used racialized science to justify the subjugation and exploitation of indigenous peoples and to legitimize discriminatory practices, for example, sterilization of black women (P19). The need for the scientific paradigm to shift, based on the fact that it has historically been and is still significantly dominated by Western repressive and racist processes (P18). It contributed to scientific racism that was used to justify practices that were discriminating. Racialized theories in medicine where it was said that some races are more susceptible to certain diseases (P10).
Subtheme 2	*Recognizing the effects of systemic bias*	
	Bias in health care treatment	Medical diagnoses are not always so accurate for all racial groups (P2). There has been a percentage placed on the development of kidney disease for black patients that allows their pain to not be acknowledged and allow them to be under diagnosed. This allows them to not receive the correct treatment needed that could save them (P23). Skin rashes and issues of darker skinned individuals are mis looked and go undiagnosed as the textbooks doctors used to study these illnesses and problems only used light skin individuals which is super problematic (P20). People also used to think that people's "race" influenced certain physiological mechanisms such as pain tolerance or glomerular filtration rate which meant they were treated differently based on skin color instead of pathology (P24). Colonialism has resulted in Eurocentric biases that has negatively impacted the treatment of black individuals such as the race modifier for eGFR (P26).

Continued

Table 7.1 Thematic analysis of student responses—cont'd

Theme number	Theme name	Quotations
	Bias in research	How focus groups in research are picked and more particularly separated based on race but in the context of certain studies, separating participants based on race is not necessary (P25). Medicine disadvantages people of color, and treats them unfairly. Little data is available as most study participants are white (P3). Colonial powers often conducted medical experiments on indigenous populations without their consent or understanding, leading to exploitation and harm. For example, in the context of European colonialism, indigenous peoples were subjected to unethical medical trials and treatments, such as vaccine experiments, forced sterilizations, and exposure to diseases, contributing to immense suffering and loss of life (P7). Medical experimentation, where the colonial powers conducted medical experiments on colonized population without their consent and not even considering their well-being e.g. Tuskegee study (P14).
	Knowledge suppression	Suppression of traditional knowledge, unequal distribution of resources, exploitation of vulnerable populations, codifying race and racism (P9). Considering that the West colonized vast parts of the world and forced their thought systems while dismissing indigenous systems, indigenous science and philosophy has not been incorporated into modern "accepted" paradigms (P17).
Theme 2	**Rethinking knowledge**	
Subtheme 1	*Challenging existing knowledge*	
	Challenging knowledge production	It's about challenging established norms and power structures within the scientific community (P9). Shaking up and getting rid of old colonial ways of thinking and acting. It's about making research, healthcare, and learning more fair, open, and respectful of different cultures and people (P8).
	Challenging Western worldviews	Challenging the ideas and paradigms established by modern Western thought and philosophy in science and medicine (P17). Challenging historical structures and acknowledging historical injustices (P10).
	Deconstructing harmful views	Decolonization in science and medicine means attempting to correct previous errors made by the industry (P5). South Africa is a very diverse nation and some harmful views are entrenched into certain people based on older generations' opinions. Education is not only about what you study in your specific field. It is also about broadening people's minds. Deconstructing someone's harmful views starts with educating them about why it is harmful and what is factually true (P25).

	Destigmatizing indigenous and local knowledge	Destigmatizing and incorporating indigenous and local knowledge to create an improved and locally relevant science (P17).
	Incorporating modern and indigenous knowledge	It would be important to emphasize an expansion of knowledge rather than a replacement: demonizing Western science could offend individuals of Western descent, increasing racial/cultural tensions and potentially exacerbating the "brain drain" (P17). Principles and foundations in science were developed so many years ago and there has been a great amount of change that needs to be expressed. This would allow more students to learn and allow this knowledge to help us understand further (P24).
	Expanding own knowledge	We need to broaden the knowledge and understanding of others (P13).
	Ensuring local leadership and relevance	South Africans need to make a difference in the science and medical world, as we as scientists are the ones dealing with the firsthand experience of the darker skinned individual patients with various diseases and infections. An example would be TB or HIV, which is more prevalent in South Africa as opposed to the northern hemisphere or Europe, where the "leaders" of the science world are dictating the research (P22).
Theme 3	**Pursuing responsive education**	
Subtheme 1	*Curriculum renewal*	
	Review existing curricula	Decolonization curriculum involves rethinking, and limiting race and looking at the factors like medical or biological factors (P15). Review and revise existing curricula to include perspectives from diverse cultures and regions, highlighting contributions from indigenous knowledge systems and non-Western scientific traditions (P8). Discuss the use of indigenous knowledge (historical perspective) and its beneficial medical applications (P6). Decolonization requires restructuring, rethinking and recreating the current curriculum and research that's based on colonialism (P9).

Continued

Table 7.1 Thematic analysis of student responses—cont'd

Theme number	Theme name	Quotations
	Integrate indigenous perspectives	Improved exposure to the country's rich history, philosophy and biodiversity can inspire students and lead them to reflect on angles they have not considered before, as well as potentially inspire an interest in things they would otherwise not know exist—increasing research in local topics (P17). An example is by including indigenous research and practices (like in healthcare) and this allows the acknowledgement of indigenous people by including their insight and experience into future research and healthcare practices (P15). Integrate indigenous perspectives, knowledge, and practices into the curriculum of science and medicine courses. This could involve including case studies, guest lectures, or readings that highlight indigenous contributions to science and medicine, as well as their traditional healing practices and approaches to healthcare (P7).
	Include contributions of non-European scientists	I think it would be beneficial, if done well. Something as "simple" as introducing a compound used in modern medicine by sharing how a local medicinal plant contains it (or something similar) and how it is/was used, makes the theory feel more relevant (P17). Include contributions of non-European scientists (over the course of history and that is relevant to this course) in terms of the scientific process of discovery and advancement of knowledge (P1).
	Incorporating decolonization into the curriculum	By incorporating decolonization into the curriculum, students can gain a more comprehensive understanding of the social and cultural contexts in which science and medicine operate, and develop the skills to critically evaluate and challenge dominant narratives and practices (P2).
Subtheme 2	*Appropriate pedagogical practices*	
	Engaging in open dialogue	Having these discussions not only allows conversation about the topic but also raises awareness (P16). We are part of a unique hub of bristling diversity, and should take advantage of the fact that we have people from all different backgrounds able to gather and engage in impactful discourse on topics that are experienced so differently by different people (P18). It is important to begin such conversations so that we can redress the injustices of the past that were committed in the medical fraternity on basis of race. We could also curb the possibilities of such taking place (P20). It is essential to start having these discussions about decolonization in South African higher education, especially in the fields of science and health. By addressing historical injustices and promoting a varied viewpoint in research and healthcare practices, it can result in an academic atmosphere that is more inclusive and culturally sensitive (P26). Due to our countries past, it is important to engage in such conversations. Biases and inequalities still exist (P27). More conversations need to be had about the mistakes that the scientific community has made in the past and there needs to be open communication if any behavior or research is making an individual or a group of individuals feel excluded (P4).

	Student involvement	It is important as it helps students explore and engage with decolonial, anticolonial, and postcolonial approaches in their current research (P9). Science and scientists have a huge responsibility now to ensure the past never repeats itself thus it's super important to discuss past mistakes with regards to science with us at this age (futures in the field). We need to be the change and to do so we need to be educated first (P21). One of the best ways to make your curriculum more inclusive is to involve your learners in the process. Ask them about their needs, preferences (P2).
Theme 4	**Building Inclusivity and promoting diversity**	
Subtheme 1	*Challenging norms*	
	Exploring own belief	I think it is important to remind us of what we believe and how much of an impact that the work we do has on the general populations beliefs and values (P11).
	Removing segregation	Kind of removing the racial labels, allowing all people to fall under one category as such (P6).
	Challenging societal norms	Given SA's past of which its social divide still persist, it is important so we can encourage diverse perspectives being valued in the science to avoid incidences such as the "coloured women" study (P19). Yes because how things were done by the previous generations or our ancestors need to be addressed as they were mostly not done in an ethical way and very questionable way (P12).

Continued

Table 7.1 Thematic analysis of student responses—cont'd

Theme number	Theme name	Quotations
Subtheme 2	*Inclusive and equitable practice*	
	Inclusivity of all people	Decolonization involves personalization and inclusivity of all people (P1). Remove any bias or race-based research methods in science and research practices (P16). Ensure a more diverse environment so that all students feel included (P14). That we move away from the narrow thinking and make it more multicultural and inclusive (P11). Through the inclusion of decolonization of science and medicine we are able to adopt an inclusive and culturally competent way of thinking. It will also allow for unity within a work environment. It will promote equality and make one question previous "norms" (P21). Initiating conversations around decolonization in South African higher education is important for promoting equity and inclusion (P7).
	Diverse thinking	Taking into account different opinions and being able to look at it objectively (P13). Holistic education approaches are extremely valuable (in my opinion) for problem-solving, critical thinking and making theory feel relevant—as opposed to a reductionist approach that teaches each "knowledge point" as an isolated credit towards (P17). Incorporation of diverse perspectives and promoting inclusive research practices could all be potential initiatives. Also addressing ethical concerns (P10).
	Equity in treatment	We are a diverse country and need to support everyone equally with treatments or research to help underrepresented groups (P4). Means giving the science understanding about the culture, about the identities and how science is conducted taking care of people's health (P15).
	Responsibility to discuss injustices	We need to reassess aspects that have had an impact on education such as addressing the historical injustices which have occurred and making sure that nothing similar is ever done ever again. The more we speak about the errors and injustices that have been made the more aware we are of our mistakes (P5). Re-evaluate research practices to make sure they are beneficial and respectful to local communities for example (P10). European people came to Africa and used the people living here to do their "dirty" work (P6).
	Empathy	Ensuring research projects are not insensitive to marginalized groups after exploiting data obtained from said groups. Empathy is something everyone could benefit from (P1). Decolonization in science and medicine means moving away from the Western cultural ideals and views that have previously informed science and medicine and becoming more inclusive of diverse perspectives that will drive contextual research (P19).

Theme 1: Understanding the colonial legacy in scientific knowledge

Students acknowledged that an understanding of the colonial legacy in scientific knowledge is a crucial step in the decolonization of medical education and research. Two main subthemes were identified, namely, critical historical consciousness and recognizing systemic bias.

Subtheme 1: Critical historical consciousness

This subtheme captures participants' perceptions of the process of decolonization to develop the abilities of students to critically examine how colonial history has shaped medical knowledge. Here, one student wrote:

> *Decolonization means looking closely at how colonial history has affected these fields, and working to make them fairer and more inclusive by listening to and respecting indigenous perspectives (P7).*

In addition, some perceived decolonization as the opportunity to recognize historical biases in curriculum content and to understand the marginalization of indigenous knowledge systems. It was suggested that such understandings should inform curriculum design. For example, some participants stated,

> *Examining events in our history as case studies (ex., outbreaks, social events influenced by poor health/wellbeing, etc.) teaches students several valuable things simultaneously, including thinking about how different factors link to create a bigger picture (P17).*
> *I think recognizing historical biases is very important, and recognizing how colonialism could have framed the content of the course (P7).*
> *Understanding how colonialism has marginalized local and traditional knowledge systems (P10).*

Participants also shared their insights on how colonial science was not only established but also legitimized racial hierarchies through pseudoscientific theories. This approach resulted in lasting consequences, paving the way for discriminatory practices in medicine and research that continue to affect communities today. For example, some stated that,

> *The concept of scientific racism was used to promote the idea of racial superiority and inferiority, which perpetuated discrimination and oppression against certain ethnic groups (P8).*
> *Colonialism perpetuated racial hierarchies and pseudoscientific theories of racial superiority and inferiority. European colonial powers used racialized science to justify the subjugation and exploitation of indigenous peoples and to legitimize discriminatory practices, for example, sterilization of black women (P19).*

It contributed to scientific racism that was used to justify practices that were discriminating. Racialized theories in medicine, where it was said that some races are more susceptible to certain diseases (P10).

Moreover, there was an urge to challenge Western-dominated scientific paradigms that continue to perpetuate racist paradigms in contemporary medicine:

There is a need for the scientific paradigm to shift, based on the fact that it has historically been and is still significantly dominated by Western repressive and racist processes (P18).

Subtheme 2: Recognizing the effects of systemic bias

This subtheme encompasses the critical awareness and acknowledgment of deeply embedded prejudices and discriminatory practices within medical education. Participants explained how bias in health care treatment can lead to inaccurate diagnoses. Such biases are Eurocentric in nature and based on race and skin color. The participants commented that,

Medical diagnoses are not always so accurate for all racial groups (P2).
There has been a percentage placed on the development of kidney disease for black patients that allows their pain to not be acknowledged and allows them to be underdiagnosed. This allows them to not receive the correct treatment needed that could save them (P23).
Skin rashes and issues of darker-skinned individuals are 'mis-looked' and go undiagnosed as the textbooks doctors used to study these illnesses and problems only used light-skinned individuals, which is super problematic (P20).
People also used to think that people's 'race' influenced certain physiological mechanisms such as pain tolerance or glomerular filtration rate, which meant they were treated differently based on skin color instead of pathology (P24).
Colonialism has resulted in Eurocentric biases that have negatively impacted the treatment of black individuals, such as the race modifier for eGFR [estimated glomerular filtration rate] (P26).

Participants also highlighted the diverse ways in which bias permeated health care research, for example,

How focus groups in research are picked and more particularly separated based on race, but in the context of certain studies, separating participants based on race is not necessary (P25).
Medicine disadvantages people of color, and treats them unfairly. Little data is available as most study participants are white (P3).

There were suggestions that colonial powers supported the undertaking of medical experiments on indigenous populations without their consent or understanding, leading to exploitation and harm. Some stated,

For example, in the context of European colonialism, indigenous peoples were subjected to unethical medical trials and treatments, such as vaccine experiments, forced sterilizations, and exposure to diseases, contributing to immense suffering and loss of life (P7).

Medical experimentation, where the colonial powers conducted medical experiments on the colonized population without their consent and not even considering their well-being, e.g. [the] Tuskegee study (P14).

Participants also suggested that colonialism allowed for the systematic suppression of indigenous knowledge systems while exploiting vulnerable populations and institutionalizing racial categories. For example,

Suppression of traditional knowledge, unequal distribution of resources, exploitation of vulnerable populations, codifying race and racism (P9).

Considering that the West colonized vast parts of the world and forced their thought systems while dismissing indigenous systems, indigenous science and philosophy have not been incorporated into modern 'accepted' paradigms (P17).

Theme 2: Rethinking knowledge

This theme captures participants' perceptions regarding the need to rethink knowledge and confront Western-centered knowledge production and worldviews to deconstruct harmful perspectives that exploit certain groups of people.

Subtheme 1: Challenging existing knowledge

With regards to challenging knowledge production, they referred to the need to identify the power structures within the scientific community, eliminate outdated ways of thinking, and strive to make research and health care fairer and more equitable. Some stated,

It's about challenging established norms and power structures within the scientific community (P9).

Shaking up and getting rid of old colonial ways of thinking and acting. It's about making research, healthcare, and learning more fair, open, and respectful of different cultures and people (P8).

Challenging the ideas and paradigms established by modern Western thought and philosophy in science and medicine (P17).

Part of challenging knowledge is about confronting historical structures that support injustices, such as:

Challenging historical structures and acknowledging historical injustices (P10).

Decolonization in science and medicine means attempting to correct previous errors made by the industry (P5).

Subtheme 2: Knowledge reconstruction

Participants argue for an active process of rebuilding and reimagining medical knowledge systems by dismantling colonial frameworks and integrating previously marginalized perspectives. There was a sense that knowledge construction cannot occur without considering the context within which education takes place. For example, participants argued that because South Africa is a diverse nation, some harmful views may be entrenched in certain individuals based on older generations' opinions. Thus education starts with the "broadening of people's minds." For example,

> *Education is not only about what you study in your specific field, but it is also about broadening people's minds. Deconstructing someone's harmful views starts with educating them about why it is harmful and what is factually true (P25).*

Knowledge reconstruction was also described as the process of destigmatizing indigenous and local knowledge systems:

> *Destigmatizing and incorporating indigenous and local knowledge to create an improved and locally relevant science (P17).*

Conversely, some participants acknowledged the value of current knowledge systems, cautioning against replacing, and emphasizing the need for integration:

> *It would be important to emphasize an expansion of knowledge rather than a replacement: demonizing Western science could offend individuals of Western descent, increasing racial/cultural tensions and potentially exacerbating the brain drain (P17).*
>
> *Principles and foundations in science were developed so many years ago, and there has been a great amount of change that needs to be expressed. This would allow more students to learn and allow this knowledge to help us understand further (P24).*

There were strong affirmations that such changes would require a strong drive and determination from local individuals to ensure relevance. One participant stated that,

> *South Africans need to make a difference in the science and medical world, as we, as scientists, are the ones dealing with the firsthand experience of the darker-skinned individual patients with various diseases and infections. An example would be TB or HIV, which is more prevalent in South Africa as opposed to the northern hemisphere or Europe, where the 'leaders' of the science world are dictating the research (P22).*

Theme 3: Pursuing responsive education

This theme describes participants' views regarding the need to create educational approaches that are sensitive to cultural differences, local health challenges, and community-specific health priorities, rather than applying universal Western-centered models.

Subtheme 1: Curriculum renewal

Participants suggested reviewing the existing curriculum to identify factors that dismiss or exclude indigenous knowledge systems and practices. One student indicated that,

> *Decolonization [of the] curriculum involves rethinking, limiting race and looking at the factors like medical or biological factors (P15).*

Others commented that curriculum renewal should also include perspectives from diverse cultures and contributions from indigenous knowledge systems:

> *Review and revise existing curricula to include perspectives from diverse cultures and regions, highlighting contributions from indigenous knowledge systems and non-Western scientific traditions (P8).*
> *Discuss the use of indigenous knowledge (historical perspective) and its beneficial medical applications (P6).*
> *An example is by including indigenous research and practices (like in healthcare), and this allows the acknowledgement of indigenous people by including their insight and experience into future research and healthcare practices (P15).*
> *Integrate indigenous perspectives, knowledge, and practices into the curriculum of science and medicine courses. This could involve including case studies, guest lectures, or readings that highlight indigenous contributions to science and medicine, as well as their traditional healing practices and approaches to healthcare (P7).*

In addition, there was a request to integrate contextual perspectives into such renewed curricula:

> *Improved exposure to the country's rich history, philosophy and biodiversity can inspire students and lead them to reflect on angles they have not considered before, as well as potentially inspire an interest in things they would otherwise not know exist - increasing research in local topics (P17).*

Furthermore, some participants argued for the incorporation of the concept and process of decolonization into the curriculum to allow students to develop a critical disposition toward dominant narratives and practices. For example,

> *By incorporating decolonization into the curriculum, students can gain a more comprehensive understanding of the social and cultural contexts in which science and medicine operate, and develop the skills to critically evaluate and challenge dominant narratives and practices (P2).*

In addition, participants highlighted the need for active student participation in the decolonization process. They perceived this as an opportunity to empower students to take ownership of their learning, contribute their perspectives and experiences, and participate in curriculum development and reform. Some noted that

> *It is important as it helps students explore and engage with decolonial, anticolonial, and postcolonial approaches in their current research (P9).*
> *Science and scientists have a huge responsibility now to ensure the past never repeats itself, thus it's super important to discuss past mistakes with regards to science with us at this age (futures in the field). We need to be the change, and to do so, we need to be educated first (P21).*
> *One of the best ways to make your curriculum more inclusive is to involve your learners in the process. Ask them about their needs, preferences (P2).*

Subtheme 2: Appropriate pedagogical practices

The participants emphasized the need for teaching methods that are culturally responsive, contextually relevant, and designed to challenge colonial approaches to medical education. This means adopting educational strategies that move beyond traditional Western pedagogical models to incorporate diverse ways of knowing and learning that respect different knowledge systems. Participants indicated the importance of creating spaces for open and honest conversations about decolonization:

> *Having these discussions not only allows conversation about the topic but also raises awareness (P16).*
> *We are part of a unique hub of bristling diversity, and should take advantage of the fact that we have people from all different backgrounds able to gather and engage in impactful discourse on topics that are experienced so differently by different people (P18).*
> *It is important to begin such conversations so that we can redress the injustices of the past that were committed in the medical fraternity on the basis of race. We could also curb the possibilities of such taking place (P20).*
> *It is essential to start having these discussions about decolonization in South African higher education, especially in the fields of science and health. Addressing historical injustices and promoting a varied viewpoint in research and healthcare practices can result in an academic atmosphere that is more inclusive and culturally sensitive (P26).*
> *Due to our country's past, it is important to engage in such conversations. Biases and inequalities still exist (P27).*
> *More conversations need to be had about the mistakes that the scientific community has made in the past, and there needs to be open communication if any behavior or research is making an individual or a group of individuals feel excluded (P4).*

They also proposed that more holistic educational approaches that encourage problem-solving and critical thinking are necessary. For example, one participant noted that,

Holistic education approaches are extremely valuable (in my opinion) for problem-solving, critical thinking and making theory feel relevant, as opposed to a reductionist approach ... (P17).

Theme 4: Building inclusivity and promoting diversity

This theme focuses on the intentional creation of medical education and research environments that welcome, value, and actively include people from all backgrounds, identities, and perspectives.

Subtheme 1: Challenging norms

Participants called for the deliberate questioning and confrontation of established practices, beliefs, and standards within medical education that have been accepted as traditional or "normal" but may perpetuate colonial, racist, or exclusionary frameworks. The idea of challenging the status quo by reflecting on one's own beliefs and values was put forward:

I think it is important to remind us of what we believe and how much of an impact that the work we do has on the general populations' beliefs and values (P11).

This should be followed by the intentional removal of segregation:

Kind of removing the racial labels, allowing all people to fall under one category as such (P6).
Remove any bias or race-based research methods in science and research practices (P16).

Part of this process requires looking at one's own history to identify unethical or questionable practices. For example, some stated,

By addressing historical injustices and promoting a varied viewpoint in research and healthcare practices, it can result in an academic atmosphere that is more inclusive and culturally sensitive (P26).

Subtheme 2: Inclusive and equitable practice

There is a need for a strong commitment to creating medical education and research practices that actively include all individuals while ensuring fair and just treatment across diverse populations. Thus the participants suggested that decolonization should be about inclusivity for all. This also relates to making provisions for students in the learning environment:

Decolonization involves personalization and inclusivity of all people (P1).
Ensure a more diverse environment so that all students feel included (P14).

Decolonization should also, therefore, result in the incorporation of diverse perspectives and the promotion of inclusive research practices:

> *Incorporation of diverse perspectives and promoting inclusive research practices could all be potential initiatives (P10).*
> *Decolonization in science and medicine means moving away from the Western cultural ideals and views that have previously informed science and medicine and becoming more inclusive of diverse perspectives that will drive contextual research (P19).*
> *That we move away from the narrow thinking and make it more multicultural and inclusive (P11).*
> *Through the inclusion of decolonization of science and medicine, we are able to adopt an inclusive and culturally competent way of thinking. It will also allow for unity within a work environment. It will promote equality and make one question previous 'norms' (P21).*

There was a clear call for equity in health care treatment and research:

> *We are a diverse country and need to support everyone equally with treatments or research to help underrepresented groups (P4).*
> *Initiating conversations around decolonization in South African higher education is important for promoting equity and inclusion (P7).*
> *We need to re-evaluate research practices to make sure they are beneficial and respectful to local communities (P10).*

There was also an emphasis on the need for sensitivity and empathy when dealing with marginalized groups, together with increased cultural awareness:

> *Ensuring research projects are not insensitive to marginalized groups after exploiting data obtained from said groups. Empathy is something everyone could benefit from (P1).*
> *Means giving the science understanding about the culture, about the identities and how science is conducted, taking care of people's health (P15).*

Discussion

The Council on Higher Education of South Africa recently called for a reconsideration of decolonization in higher education curricula, aiming to shift from viewing the curriculum as passively received by students to reframing it as a co-constructed process (Council on Higher Education, 2017). This highlights the importance of the student voice in the decolonization process. Our study responded to this call by embarking on a unique curriculum remodeling process that included a focus on the decolonization of medicine and the biomedical sciences. Moreover, we provided students with an opportunity to share their perspectives regarding the decolonization of medical education and research, and how teaching and learning programs should be designed and delivered to meaningfully decolonize the curriculum.

Decolonization was traditionally regarded as removing colonial influences from all domains of higher education (Tight, 2024). However, the term "decolonization" lacks a single definition, and it can therefore be conceptualized in various ways. For example, Fanon (1963) described it as an intense process of unlearning the harms of decolonization, while Monchalin (2016) views it as a catalyst for transitioning from colonized discourses and ideologies. The findings from this study suggest that decolonization is not only about removing colonial vestiges but also about acknowledging the deeply embedded prejudices and discriminatory practices within medical education and research. This acknowledgment and recognition of historical biases align with the suggestions by Saini and Begum (2019: 218), who noted that:

> *Decolonization requires a critical historical lens and a transformative approach to knowledge building in order to expose and dismantle the presence of colonial and imperial practices within the university.*

While most participants elaborated on the nature of decolonization, some also focused on the harms inflicted by colonization. For example, participants pointed out the various ways in which bias permeated health care research and how colonialism suppressed indigenous knowledge systems while exploiting vulnerable populations and institutionalizing discrimination. Such sentiments are shared by Held (2023), who argues that Western science still dominates despite the global prevalence of many different "sciences" and ways of knowing. Thus there was an urgent call by participants to rethink the idea of knowledge and to confront Western-centered worldviews and knowledge production. This strongly aligns with Dawson's (2020: 85) suggestion that a shift is needed from:

> *The establishment of knowledge hierarchies and entrenchment of epistemic hegemonies and internationalization.*

From these findings, it is evident that an active process of rebuilding and reimagining medical knowledge systems is required—in other words, decolonizing the curriculum. There was a sense that the decolonization of the curriculum would provide an opportunity for epistemic justice and equity. This notion is supported by Le Grange (2019), who suggested that decolonizing the curriculum is about achieving epistemic or cognitive justice. However, a rhetorical question posed is what the decolonization of the curriculum would practically involve (Badat, 2017). For example, would it constitute the substitution or replacement of a Western curriculum with an indigenous one? Gopal (2021) warns against this and calls for the recognition of other knowledges rather than a rejection of Western knowledges. This indicates that substituting one knowledge system with another could create other forms of academic or intellectual exclusion. Similarly, while knowledge reconstruction involves destigmatizing indigenous and local knowledge systems, some participants recognized the value of existing knowledge systems and advocated for integration rather than replacement. Thus combining multiple ways of knowing by using the wisdom of indigenous knowledge to enrich scientific inquiry may be more appropriate to address current medical education and research challenges. These findings align with

Le Grange's (2019) view that decolonization does not necessarily entail dismantling traditional approaches but rather involves engagements with current educational challenges in a thoughtful and responsive manner.

There was a strong sense among the participants that curriculum renewal is necessary. In alignment with the literature, participants emphasized the importance of including diverse cultural perspectives and contributions from indigenous knowledge systems to incorporate alternative ways of knowing (Saidi, 2023). In addition, some argued for the incorporation of the concept and process of "decolonization" into the curriculum to allow students to develop a critical disposition toward dominant narratives and practices. There was also an expressed need for students to actively participate in decolonization processes, curriculum development, and reform. As Vandeyar (2020) proposes, the emphasis when decolonizing the curriculum should not only be on the content of what is taught but also on the methods and approaches through which it is delivered. Hence, appropriate pedagogical practices that are inclusive and equitable are needed. The participants also argued for the adoption of educational strategies that move beyond traditional Western pedagogical models to incorporate more diverse ways of knowing and learning. Examples include creating spaces for open and honest conversations about decolonization, active student participation, and empowering students to take ownership of their learning. This was successfully achieved during the *Features of Science* class sessions, where the group discussions created a safe space for students to freely express and share diverse views on coloniality and colonialism. While curriculum renewal is vital, attention to the hidden curriculum is essential, as it often conveys exclusionary practices and messages (Salazar, 2013).

Salazar (2013) argues that colonization is also reflected in the dominant language used within academic spaces, which is characterized by more objective and detached ways of expression. In support of Paulo Freire's (1970) work, she highlights the moral responsibility of educators to adopt more humanizing approaches. This includes a humanizing pedagogy that is a crucial element that can support the social resilience of students. In agreement, a key focus was to include such a cultivation of humanity as a central focus in both the curriculum and student interactions, inside and outside the classroom. Moreover, the intentional creation of environments that welcome, value, and actively include students, researchers, and staff from all backgrounds, identities, and perspectives can also contribute toward maximizing the positive influences of the hidden curriculum (Meyer et al., 2025). While the hidden curriculum often perpetuates hierarchies and exclusionary practices, it can also transmit positive values and institutional practices like equity, belonging, and collaborative learning. When students observe such inclusive behaviors, these implicit lessons become internalized as professional norms and practices.

However, this process is not just about changing the way the curriculum is delivered, but also about developing a critical consciousness in students. Freire (1970: 17) defines critical consciousness as the process of

> *Learning to perceive social, political, and economic contradictions, and to take action against the oppressive elements of reality.*

Thus students become more engaged in their learning while critically reflecting on how they can counteract inequities and injustices, as well as on ways to challenge such oppressive behaviors. Participants emphasized the need to critically challenge established norms in medical education that, though deemed traditional or "normal," may reinforce exclusionary frameworks. In addition, the participants noted that challenging the *status quo* starts with a reflection on personal beliefs and values, aligning with the notion that decolonial learning experiences should be deeply transformative (Tight, 2024).

Summary

This study's findings revealed a consistent emphasis on the idea that decolonization should fundamentally be "about inclusivity for all." This should extend beyond addressing colonial legacies to creating educational and research environments where every individual can participate meaningfully. The call of the students for the incorporation of diverse perspectives and inclusive research practices demonstrates an understanding that meaningful change requires the recognition of multiple ways of knowing beyond Western biomedical paradigms. The emphasis on equity in health care treatment and research reflects an awareness that decolonization must address educational content, research practices, and the clinical management of patients. Students demonstrated critical consciousness by emphasizing the responsibility to explore historical and current injustices in medical education and research, while calling for increased sensitivity, empathy, and cultural awareness when working with marginalized groups. Our findings collectively suggest that decolonization requires systemic changes in educational environments, research practices, health care delivery, and interpersonal relationships. However, it is important to note that it goes beyond just changing structures to include the marginalized. There also needs to be an acknowledgment of the cultural resources that students require to thrive within the medical education and research system (Gonzalez et al., 2005) and how to build and nurture this.

Recommendations

Based on these findings, we recommend that decolonization begins with a critical awareness of the historical influences of colonization on current knowledge systems, medical practice, and research. Furthermore, curriculum planners and educational leaders should develop explicit frameworks for addressing historical injustices in medical education. This should include regular curriculum audits to identify and eliminate colonial biases and establish mentorship programs that support students from marginalized backgrounds in developing culturally responsive research and clinical practices. This could involve collaboration with experts in decolonization and the implementation of feedback mechanisms from students and communities.

Curricula should be redesigned to incorporate multiple knowledge systems and research methodologies, while establishing community partnerships that ensure research benefits reach the populations being studied. Indigenous communities should also be actively engaged as co-creators in medical research (Tight, 2024). Of note, indigenous knowledge and sciences are highly contextual and localized, which need to be considered when developing an agenda for action.

A cultural shift is also needed across medical education institutions. This could be supported by implementing structured diversity and inclusion training that goes beyond cultural competency to address power dynamics and historical context in health care relationships. Ultimately, educational institutions must prioritize the creation of supportive learning environments and invest in continuous professional development for faculty, focusing on inclusive pedagogical practices. These initiatives are essential to ensure that all students undertaking studies in higher education receive an equitable and enriching educational experience.

References

Badat, S., 2017. Trepidation, Longing and Belonging: Liberating the Curriculum at Universities in South Africa [Public Lecture]. University of Pretoria.

Chadha, N., Lim, B., Kane, M., Rowland, B., 2020. *Toward the Abolition of Biological Race in Medicine* [Report]. Institute for Healing and Justice in Medicine. Available at: https://belonging.berkeley.edu/race-edicine. (Accessed 10 August 2021).

Council on Higher Education (CHE), 2017. *Briefly Speaking: Decolonising the Curriculum – Stimulating Debate* [Online]. Council on Higher Education, Pretoria. Available at: https://www.che.ac.za/sites/default/files/publications/BrieflySpeaking%20(3)%20Curriculum%20decolonisation.pdf. (Accessed 9 June. 2025).

Dawson, M., 2020. Rehumanising the university for an alternative future: decolonisation, alternative epistemologies and cognitive justice. Identities 27 (1), 71–90.

Essop, F., 2025. Transforming physiology teaching and learning to empower students to navigate their way in an era of doubt—A unique case study. In: Essop, F. (Ed.), Truth Unveiled: Navigating Science and Society in an Era of Doubt. Academic Press, pp. 161–179.

Fanon, F., 1963. The Wretched of the Earth. Grove Press.

Freire, P., 1970. Pedagogy of the Oppressed. Seabury Press, New York.

González, N., Moll, L.C., Amanti, C. (Eds.), 2005. Funds of Knowledge: Theorizing Practices in Households, Communities, and Classrooms. Lawrence Erlbaum Associates Publishers.

Gonzalez, M., 2023. *Decolonizing Education: What Does It Mean to Decolonize the Curriculum?* [Online]. QUEST. Available at: https://quest-eu.org/publication/decolonizing-education-what-does-it-mean-to-decolonize-the-curriculum/. (Accessed 28 July 2025).

Gopal, P., 2021. On decolonisation and the university. Text. Pract. 35 (6), 873–899.

Held, M., 2023. Decolonizing science: undoing the colonial and racist hegemony of western science. J. Multidiscip. Eval. 19 (44), 88–101. doi:10.56645/jmde.v19i44.785.

Le Grange, L., 2019. The curriculum case for decolonisation. In: Jansen, J., (Ed.), Decolonisation in Universities: The Politics of Knowledge. WITS University Press, Johannesburg, pp. 29–47.

Maldonado-Torres, N., 2007. On the coloniality of being: contributions to the development of a concept. Cult. Stud. 21 (2–3), 240–270.

Meyer, R., Archer, E., Smit, L., 2025. The positive influence of the hidden curriculum in medical education: a scoping review. Med. Sci. Educ. doi:10.1007/s40670-025-02380-1.

Monchalin, L., 2016. The Colonial Problem: An Indigenous Perspective on Crime and Injustice in Canada, University of Toronto Press.

Naidu, T., Abimbola, S., 2022. How medical education holds back health equity. Lancet 400 (10352), 556–557.

Nieuwoudt, S., Dickie, K.E., Coetsee, C., Engelbrecht, L., Terblanche, E., 2020. *Retracted Article*: age- and education-related effects on cognitive functioning in Colored South African women. Aging Neuropsychol. Cogn. 27 (3), 321–337.

Quijano, A., 2007. Coloniality and modernity/rationality. Cult. Stud. 21 (2–3), 168–178.

Ratner, L., Sridhar, S., Owusu, S., Rosman, S.L., Molina, R.L., Kasper, J., 2024. Talk the talk and walk the walk: a novel training for medical students to promote decoloniality in global health. Front. Educ. 9, 1271924. doi:10.3389/feduc.2024.1271924.

Saidi, A., 2023. Reflections on the conceptualisation and practices of community engagement as a core function of universities. S. Afr. J. High. Educ. 37 (1), 1–19. doi:10.20853/37-1-5700.

Saini, R., Begum, N., 2019. Demarcation and definition: explicating the meaning and scope of 'decolonisation' in the social and political sciences. Polit. Q. 91 (1), 217–221.

Salazar, M., 2013. A humanizing pedagogy: reinventing the principles and practice of education as a journey toward liberation. Rev. Res. Educ. 37 (1), 121–148. doi:10.3102/0091732X12464032.

Tight, M., 2024. Decolonization and higher education: a critical review of a contemporary concern, policy and theory. High. Educ. Policy 37, 59–72. doi:10.1057/s41307-022-00296-8.

Vandeyar, S., 2020. Why decolonising the South African university curriculum will fail. Teach. High. Educ. 25 (7), 783–796. doi:10.1080/13562517.2019.1592149.

Wolf, S.T., Jablonski, N.G., Kenney, W.L., 2020. Examining "race" in physiology. Am. J. Physiol. Heart Circ. Physiol. 319 (6), H1409–H1413.

CHAPTER

8

Decolonizing the medical curriculum: A history, a process, and the fleshy remainder

Rachel Niehuus

Departments of Surgery and Social Medicine, University of North Carolina School of Medicine, Chapel Hill, NC, United States

This chapter on decolonizing the United States (US) medical curriculum builds on the robust literature on colonial medicine and medical education. It is also the product of my sustained reflection on the ways inequity inscribes itself on bodies. The form of inscription depends on the historical particularities of a place. In eastern Democratic Republic of the Congo, where I worked as an anthropologist, I argue that centuries of antiblack violence coalesce as women's bodies that leak urine and feces, emaciated children that refuse to eat, and newborn bundles on the hospital floor that exceed counting.[a] In the US where I work as a trauma surgeon, I have wondered if the colon blown apart by gun violence is that which remains in the wake. But that is an exploration for another time. Here, I write as a surgeon who takes care of wounded, infected, broken bodies and who thinks about how they came to be this way. I also write as clinical faculty in a large US university training hospital, where I am tasked with educating medical students and surgical residents. My most durable experience with trainees is hands-on—in the emergency room, the clinic, the operating room, and at our weekly conferences, where my trainees watch me, quiz me, and (sometimes) learn from my surgical practice. My classroom time with medical students is irregular, and my contributions to medical curricular reform are limited to those I have trained with or trained. Hence, I write with the caveat that my understanding of the formal US medical curriculum relies heavily on anthropologists' studies of it and my friendships with those actively engaged in changing it.

The first section of this chapter begins by establishing the coproduction of biomedicine with the historical processes of imperialism, colonialism, and enslavement. Because the racialization of Black bodies, in particular, is so fundamental to biomedical knowledge, I draw largely on histories of the African continent and chattel slavery in the US. Certainly, there are many other imperial and colonial histories, including that of settler colonialism in the US and the resulting genocide of the Native

[a] See Niehuus (2024).

Decolonizing Medical Education, Research and Practice. https://doi.org/10.1016/B978-0-443-44998-7.00018-1

American population. Instead of constituting biomedicine's fundamental structure, as Blackness does, I view these histories more often as being erased by biomedicine.[b] By illustrating how race emerged and came to structure medical knowledge, I aim to trace the construction of the body that I know and teach in the operating room.

The second (and main) section of this chapter then dives into the process of decolonizing the medical curriculum. I understand this to be an iterative, fragmented, and ongoing process, which represents a continual striving toward a more equitable practice of medicine. To make sense of the many layers of activism and scholarship, I articulate five different nodes around which action has centered. Some of the work has been directed at leveling the hierarchies within biomedicine more generally, which improves biomedicine for all people; and some of the activism works more specifically to improve the health of people who have been historically excluded from biomedicine (or included in harmful ways). While this latter category is more likely to be recognized as decolonial work, I believe the scholarship and advocacy devoted to improving the culture of biomedicine have been crucial to advancing equity in the biomedical space, and so they are included here.

In the final section, I describe two directions for continued decolonizing practice. Surely, biomedicine everywhere is laden with coloniality—that is, the persistent patterns of power, knowledge, and being that shap societies and cultures long after the formal end of colonial rule. Where ongoing settler colonialism perpetuates remarkable violence and inequity,[c] the decolonizing process feels especially arduous. And still, even where colonial effects are especially "tenacious,"[d] there are openings. Persistent demands for reparations and a clinical practice of attention feel like two possible directions for the continuing work of decolonizing the medical curriculum.

The coproduction of colonialism and biomedicine

The contemporaneous emergence of European life sciences with imperialist and colonial plunder is well documented and also discussed in detail in this book volume. For example, Comaroff (1993) traces the way that Enlightenment discourse used the metaphor of the (female) human body in descriptions about exploration in Africa. In archival documents, the continent's interior was described as dark and mysterious, with openings that called for exploration at a moment when European biology was adopting more invasive techniques to explore the body.[e] Furthermore, Foucault (1976) describes the emergence of an interest in comparative anatomy and biology that might differentiate man from animal in Europe in the late 18th century, while Stoler (1995) documents the impact of this scientific interest on imperialistic forays,

[b] I find Duana Fullwiley's (2024) description of "ghost DNA" helpful here.

[c] Povinelli's (2021) discussion of ancestral catastrophe is very helpful here.

[d] See Stoler (2016).

[e] See also Comaroff and Comaroff (1991) and Comaroff (1993).

during which human bodies, like land, were measured, mapped, and illustrated.[f] As rubber, timber, and cocoa from the African continent were appropriated into raw materials for European industry, African bodies were also seized as raw material for European science. Through the parading of Africans at circuses and the display of their deconstructed bodies at museums,[g] a taxonomy of humans emerged. Given an unlimited pool of ready research subjects, there were no limits—physical nor ethical—to studying presumed biological differences. Anthropomorphic measurements were but a precursor to much crueler experimentation.

Coincident with imperialism on the African continent was the institutionalization of American chattel slavery.[h] Enslavement began in the US in 1619; a ban on importing Africans to the US went into effect in 1808, and enslavement was abolished in 1863—around the same time that Europe transitioned from imperialism to colonialism.[i] Just as European imperialism arose within and was supported by a particular interest in human taxonomies, so too American chattel slavery depended on biological difference. Hence, because Negroes were considered biologically nearer to animals than to whites, they could be owned and bred, enslaved men used as "studs," and worked like animals in the fields. Similarly, Black flesh was thicker, the Negro felt less pain—and so, enjoyment could be taken in whipping enslaved people, and their bodies could be operated on without anesthesia.[j] Weheliye writes, "Despite having no real basis in biochemistry, the hieroglyphics of the flesh [pigmentation, hair texture, lip thickness] requires grounding in the biological sphere so as to facilitate—even as it conceals and because it masks—the political, economic, social, and cultural disciplining … of the Homo sapiens species into assemblages of the human, not-quite-human, and nonhuman; this is what I am referring to as racialization."[k]

Eventually, in the middle of the 19th century, the scientific focus shifted from defining and characterizing difference to controlling contagion.[l] In the US, Jim Crow laws forcibly segregated public spaces to preserve the purity of the white race, while across the "pestiferous" African continent, missionary healers proliferated to treat diseases that were portrayed as running rampant. Africans were reservoirs of malaria and sexually transmitted infections; they birthed on dirt floors because they did not understand sanitation; and—a consequential slippage here—they walked naked and

---

[f] See also McClintock (1995).

[g] See Chase-Riboud (2003).

[h] It is not my geographic location in the US that occasions my inclusion of American chattel slavery here. Rather, I include enslavement here because of its influence on the creation of racial difference within biomedicine. In this, I join a growing literature of authors who also suture together bodily practices in the colony and on the plantation (Gilroy, 2003; Weheliye, 2014; Sharpe, 2016; Fullwiley, 2024).

[i] See Hunt (2008) and Hochschild (1999) on the human rights movement against King Leopold's Congo that precipitated this transition.

[j] Roberts (2017) and Judy (2020).

[k] Weheliye (2014, p. 45).

[l] In Foucaultian language, in the 19th century, there was this shift from discipline to biopower, a focus on governing conduct, how an individual lives, as a way of governing a population. Foucault never addressed colonialism in depth—and still, his analysis includes this shift. See Stoler (1995) and Weheliye (2014) on Foucault's omission of race.

worshipped idols and thus needed both God and medicine.[m] As filth, disease, and godlessness condensed into Black bodies, medicine became a key site of governance in the colony. The control of the bodies of Africans both protected Europeans from illness and created docile, governable subjects.[n] Thus in the Congo, maternities were built where colonial missionaries evangelized, counted babies, and taught women how to birth, nurse, and reproduce "correctly."[o] Where every new body had income-generating potential, pronatalism was everywhere reinforced, and the bodily practices that emerged during this time durably shifted gender and reproduction across the continent.[p]

After the Second World War, independence movements spread across the continent. Colonial governments were dismantled, European doctors returned to the metropole, and health became an emblem of nation building and modernization across Africa. In the US, hospital wards began to be desegregated.[q] At the same time, the discourse in biomedicine shifted to one of commensurability, emphasizing similarity across bodies.[r] Where race had once appeared so prominently in descriptions of pathology, it gradually became less prominent. But continuities remained, and "bodies remembered."[s]

Certainly, there are particularities in the ways that debris continues to find its way into bodies living in postcolonial ruin;[t] but my argument is more global than that. Because biomedicine emerged out of and in response to the connected processes of imperialism, enslavement, and colonialism, I contend that the differential treatment of racialized bodies is foundational to biomedicine more generally. Imperial and colonial powers produced race to justify their brutal techniques of governance, and the institutionalization of race into biomedicine reinforced the need for vicious techniques to control dangerous bodies. And so, it is not merely that health in the postcolony needs to be decolonized; nor is it just that the course of biomedicine needs to be righted after colonialism, adjusted away from a too-near tie to colonial violence, toward something more just. Rather, where race is at the core of biomedicine, so too are scientific racism, medical neglect, and the denigration of certain bodies. Which is to say that the many violations that biomedicine has committed in the wake of enslavement and colonialism—the forced sterilization of Black women,[u] the Tuskegee

---

[m] See Comaroff (1993), Hunt (1999, 2008), and Vaughan (1991).

[n] Taussig (1991) and Hunt (2016) describe the nervousness that surrounding this control, this boundary making in the colony.

[o] Hunt (1988, 1999).

[p] See also Spillers (1987) on the durable effects of pronatalism on the plantation. It is also crucial to note that the effects of colonial medicine were not limited only to African bodies. Rather, Stoler writes, "Colonialism was not a secure bourgeois project. It was not only about the importation of the middle-class sensibilities to the colonies, but about the *making* of them" (Stoler, 1995, p. 99, emphasis in original).

[q] See Lovasik et al. (2020) for the story of Grady Memorial Hospital's desegregation.

[r] Lock and Nguyen (2018).

[s] Fassin (2007).

[t] See Mbembe (2001), Good et al. (2008), and Stoler (2016).

[u] Roberts (2017).

Study of Untreated Syphilis in the Negro Male,[v] the theft of HeLa cells,[w] the use of the global poor to populate clinical trials[x] to name but a few of its most egregious—are not exceptions to the rule, but rather central to biomedical thought. It is from this vantage point that I now evaluate efforts to decolonize the medical curriculum.

Decolonization: A process

Everywhere, the vestiges of colonialism endure in the present: landscapes, grammars, institutions. Scholars of the postcolony (and the US in the wake of enslavement) have written extensively on that which remains—and the work done every day to live within the duress, to shift under its weight, to forge paths within (for out is not yet an option).[y] As I endeavor to recount the decolonizing work that has begun in biomedicine, I stay near to those who write about the hold. For, in biomedicine, as elsewhere in the postcolony, inequality structures all thought, reasoning, and rationale. And so, efforts toward decolonization, however heroic, are only ever partial. Even when they successfully shift one facet of biomedical practice, many others endure, and some become even more tenacious. The work is necessarily fragmentary; it's also iterative and continuing. In what follows, I describe five foci for decolonizing the medical curriculum. These are by no means the only efforts, but they do seem to be the most cohesive.

1. **The framing of biomedicine as a particular site of cultural production**: In the early 1990s, anthropologist Margaret Lock challenged the idea that menopause was a universal experience.[z] By demonstrating that Japanese women did not experience menopause as hot flashes and night sweats, she introduced the idea into biomedicine that illness is historically and culturally constructed. Psychiatrist-anthropologist Arthur Kleinman had already brought the term "illness narratives" into medicine,[aa] and other medical anthropologists had been pluralizing disease categories for a decade[bb]—but Lock's inclusion of large survey data allowed her work to gain traction in the medical field.

   Culture and cultural diversity had always been the domain of anthropologists. But as the field of medical anthropology blossomed, physicians began talking about culture: the culture of biomedicine, and also other cultures, which

[v] Reverby (2009).
[w] Skloot (2011).
[x] Crane (2010).
[y] On colonial vestiges, see especially Stoler (2016), Hunt (1999), and Niehuus (2024). On that which remains in the wake of enslavement, see Spillers (1987), Sharpe (2016), and Harney and Moten (2013).
[z] Lock (1995).
[aa] See Kleinman (1989).
[bb] See, for example, the 1980 Special Issue of Social Science & Medicine on medical pluralism beginning with (Leslie, 1980). Lock and Nguyen (2018) have a synthesis of this literature starting page 50 of their volume.

produced their own illnesses. Gradually, a transition took place among medical providers: to start, biomedicine became contextualized as one of many healing systems; and further, there emerged a realization that, because biomedicine was a powerful site of cultural production, it produced particular disease categories, particular experiences of illness, and particular normal and abnormal bodies. From here emerged the idea that, in addition to understanding pathophysiology, medical practitioners needed to be conversant in culture: they both needed to understand how they reproduced the culture of biomedicine, and they needed to recognize when (Other) cultures were creating different illnesses. In the 1970s and 1980s, medical anthropologists working in Mexico and Latin America described *susto*, "soul loss," which transpires when a person experiences fear and their soul leaves their body;[cc] in 1994 *susto* became a diagnosis in DSM-4 (*Diagnostic and Statistical Manual of Mental Disorders*). Cultural competence—defined as the ability to understand, appreciate, and interact with people of different cultural backgrounds—became one of the core competencies in medical schools. It had its limitations, but the teaching of culture in medicine was an important step toward removing biomedicine from its position of authority.

2. **The limitations of objectivity and the emergence of patient-centered care**: Alongside medicine's embrace of culture came the questioning of biomedical objectivity. If biomedicine was no longer the preeminent (and only legitimate) science but rather one-of-many systems of healing, then the doctor–patient relationship needed to shift. Biomedicine was built on objectivity and a clearly defined normal,[dd] but where "normal" bodies and processes were now recognized in relation to other systems of knowledge, the physician's prescriptive manner was less tenable. Patients also had valuable insights into their conditions and deserved a role in decisions about how their illnesses might be managed. Illness narratives multiplied in the press, and the physician was displaced from his/her role as the sole bearer of expertise.

   In 2012 the *New England Journal of Medicine* published a call for "patient-centered care" and "shared decision making."[ee] But the leveling of the historical hierarchies that existed between physician and patient did not come naturally; instead, new skills had to be acquired. Medical schools developed courses aimed at teaching good bedside manner, where we learned how to have difficult conversations and give bad news. We read books like *The Spirit Catches You and You Fall Down* so that we might face our patients with more empathy and more understanding. Medical expertise had already become more relative, and

[cc] See the following review article, which nicely synthesizes the large body of literature on *susto*: Jacobo Herrera and Orr (2020). See also Green's (1999) work on fear in Guatemala.

[dd] See especially Lock and Nguyen (2018) for a helpful review of objectivity and the "normal" in biomedicine.

[ee] Barry and Edgman Levitan (2012).

now we were asked to tread even more lightly. As physicians took up less space and patients more, new conversations about equity became possible.

3. **The codification of the rights of patients and research subjects**: During empire, colonialism, and enslavement, the victims of medicine's abuse and unequal treatment had few advocates.[ff] However, the Nazi's atrocious uses of medicine and medical research during the Second World War drew more attention. In 1947 the Nuremberg Code of Ethics was written to codify informed consent and the protection of research subjects from harm.[gg] A few decades later, the Black Panther Party in the US pushed further with their fight against medical discrimination. Here, point six of their ten-point platform declared health to be a right and not a privilege. It further demanded "medical self-defense"—the protection of vulnerable communities from harmful biomedical experimentation—and that the US government, whose historical oppression had made Black people sick, develop research that promoted Black survival.[hh] Over time, these ideas about medicine's responsibility to protect vulnerable patients from harm percolated into mainstream medicine. For example, in 1979 the Belmont Report established that human subjects research should demonstrate the principles of beneficence, respect, and justice;[ii] and in 1980 the American Medical Association added the language of patient rights to its Code of Ethics: "A physician shall respect the rights of patients, colleagues, and other health professionals."[jj]

   In many ways, this codification of the rights of patients and research subjects marked a break from the past, when biomedicine was colonialism (and enslavement's) handmaiden. Internal review boards were set up to examine research protocols and curb unethical practices; ethics committees were established at hospitals to help physicians decide the best ways forward. Certainly, as Dorothy Roberts demonstrates, forced sterilization of women of color—the Mississippi appendectomy—continued in full force into the 1980s.[kk] However, with time (and litigation), the language around patient rights has opened the possibility that biomedicine might break from its troubled past and instead serve as a tool in the fight for more just futures.

4. **The problematization of the biological basis of race and the rise of health advocacy**: Amidst these broader cultural shifts toward justice and equity in biomedicine, conversations have emerged regarding more specific sites of

[ff] The American Medical Association published a Code of Ethics in 1848, which many reference as the beginning of a new conception of medical ethics, but in its early versions, this institutionalized bias instead of protecting vulnerable populations.

[gg] For a (Western) history of medical ethics, see Thobaben (2024).

[hh] Nelson (2011).

[ii] Again, see Thobaben (2024) history.

[jj] Reiser (1991).

[kk] Roberts (2017).

inequity. One such conversation revolves around the biological basis of race. While social scientists have long discussed the problematic use of biomedicine in perpetuating race-based inequality (during imperialism, colonialism, enslavement, and even today), these discussions have only more recently entered the biomedical discourse. In 2003 the National Academies' Institute of Medicine argued that race was not tightly coupled to biology but rather a social construct based on perceived differences.[ll] Shortly thereafter, there was a push to remove race from patient presentations, as it narrowed differential diagnoses in ways that perpetuated bias and contributed to medical error. This was an important change that sought to mitigate provider bias—and gradually, the conversation has widened to include the systematic ways that race has been used and misused in medicine.[mm] In a seminal article in the *New England Journal of Medicine*, a table lists clinical algorithms that operationalize biological race.[nn] There is an entry for nearly every specialty. Some of these algorithms and diagnostic categories had already generated significant scholarly attention. For a decade, clinicians have been discussing how the addition of Black race as a "corrective" to the calculation of kidney function overestimates kidney function in Black patients. Because the estimated glomerular filtration rate (eGFR) has been shown to delay the treatment of kidney disease and eventually kidney transplant in Black patients, task groups have formed to create more equitable calculations for renal function.[oo] Similarly, the use of a race "corrective" factor in pulmonary function tests has garnered criticism for the way that it masks higher exposures to environmental toxins in lower-paying jobs and in poor neighborhoods that Black patients experience.[pp] This limits advocacy efforts and durable change toward less harmful living conditions for people of color; it also likely contributes to disparities in lung cancer outcomes along racial lines. Important efforts to remove race "correctives" from algorithms and create race-neutral algorithms are underway. At the same time, salient conversations about the way that racism adversely affects health—increasing allostatic load, which negatively affects cancer outcomes, for instance—deepen our understanding of the ways that race continues to create differences today.[qq] Taken together, all of this scholarly work on race in biomedicine has transformed the conversation from being about biological difference to focusing on the devastating health effects of racism.

Parallel to this particular conversation about race as a social construct, physician-anthropologist Dr. Paul Farmer began a more global conversation

[ll] See the consensus statement: Institute of Medicine (2003).
[mm] See Braun et al. (2007) for more on racial categories in medicine.
[nn] Vyas et al. (2020).
[oo] See a summary of this work by Powe (2022).
[pp] See, for example, Bhakta et al. (2023).
[qq] See, for example, Moore et al. (2022).

about the ways that structural violence impacts clinical medicine more generally. Using examples of HIV care in Baltimore, Haiti, and Rwanda, he described the ways that health outcomes were dramatically improved via community-based care, including reliable access to affordable medications, transportation, and food.[rr] In many different countries, across a variety of pathologies, he demonstrated that removing structural barriers to care reduced disparities in clinical outcomes. Fundamentally, he argued that equal pathology should receive equal treatment across the globe. This audacious notion, that one's geography, race, or financial ability should not define the medical options available to them, has radically altered the role of biomedicine across the globe. It created anew the field of global health, and it fundamentally shifted the boundaries of biomedicine from the walls of the clinic, which had always excluded so many, into all communities, into all homes. These two conversations—on the appropriate use of race in medicine and on the impact of structural violence on health—opened a new chapter in biomedicine. Here, physicians began to speak about the ways in which social, structural, and systematic inequalities affect health, and advocacy on behalf of patients was institutionalized as a critical part of clinical care.

5. **Most recently, the introduction of structural competency into medical education**: Synthesizing discussions in the social sciences, public health, and race theory, Jonathan Metzl and Helena Hansen proposed a new competency for medical education: structural competency. They defined their intervention thus: "Central to our intervention is the belief that, just as stigma in clinical encounters must be addressed structurally, so too must inequalities in health be conceptualized in relation to the institutions and social conditions that determine health-related resources. We contend that medical education needs to more systematically train healthcare professionals to think about how such variables as race, class, gender, and ethnicity are shaped both by the interactions of two persons in a room, and by the larger structural contexts in which their interactions take place."[ss] Advocating that structural competency replace cultural competency in the medical curriculum, Metzl and Hansen (2014) describe five intersecting skill sets that physicians would need to acquire to move from recognizing structures that shape clinical interactions to intervening on these structures in the name of improving health. Their proposal circulated during a particular time in American politics, in the wake of the Occupy Wall Street and Black Lives Matters movements, after President Obama had campaigned on health care reform and made visible how unequal health care access in the US was, and after Dr. Farmer's argument about health equity irrespective of geography or financial status had become well-known. Because it gave voice to a growing sentiment that health disparities drive disease, structural

[rr] Farmer (2005, 2009) and Farmer et al. (2006).
[ss] Metzl and Hansen (2014).

competency resonated with medical trainees, practitioners, and educators. In 2023 the governing body of US medical training added structural competency as a core competency in undergraduate medical education. At first, much was written about how to teach structural competency.[tt] Now, sandwiched between pathology review and pharmacology, medical educators teach about the ways that systems and practices—American policing, redlining, urban infrastructures, to name a few—sustain inequality in the US and globally.

Since the integration of structural competency into medical education, discussions continue about its applicability and durability. Clinicians wonder what structural competency looks like outside the classroom and in clinical practice. Bourgois et al. (2017) developed a structural vulnerability assessment tool to be used in the outpatient setting;[uu] however, this tool does not easily fit in many clinical contexts, including my own. Despite my formal training in social sciences and a relatively deep understanding of structural violence, I am still learning what it looks like to operationalize structural competency in emergency general surgery. Separately, trainees debate the ways in which training on structural competency changes the positionality of the physician: does knowing more about the structures that constrain their patients' health make trainees more efficacious patient advocates? Or does it contribute to frustration, hopelessness, and burnout because physicians can't also simultaneously do the work of politicians, social workers, and public health workers?[vv]
In addition, educators nervous about the American political climate worry about the politicization of "woke medicine" and what comes after structural competency. At the time of this writing, there is a proposal to eliminate structural competency from required medical education. The future of this decolonizing tactic, which has done so much to advance conversations about health and equity, feels uncertain.

In this section, I describe the foci of decolonizing work on the American medical curriculum. Those teaching in the US will recognize these efforts, as they have iteratively been taken up in medical practice and enshrined in the curricular standards that govern American undergraduate medical education. While their fluency in particular terms used here—"culture" or "patient-centered care" or "structural competency"—might vary, educators across the country are teaching this more ethical version of biomedicine, which works to identify and level relations of power. Truly, we have come very far from the version of biomedicine that was so integral to the colonial project, and I often feel grateful to my predecessors and colleagues for their scholarship, advocacy, and mentorship that have brought us here.

[tt] Neff et al. (2020), Glenn et al. (2024), Hassan and Bui (2022), Castillo et al. (2020), Kolmes et al. (2025), and Neff et al. (2017).

[uu] See Bourgois et al. (2017).

[vv] Goel and Kavanagh (2020) and Burson et al. (2021).

And still, the work must carry on. Despite all of the movement in biomedicine, we still have not taken seriously the appeal (Farmer, 1999) that we take care of all people, "providing pragmatic services to the afflicted."[ww] Most students read his arguments on health as a human right at some point in their preclinical years, but the medical practice that they learn is exclusive, privatized, and financialized.[xx] High-quality care is provided to those who have the money and resources to gain access; many more get sicker in the very system that is supposed to keep them well. Across the globe, people are also turned away from even public hospitals due to lack of funds, and catastrophic health expenditures continue to render the poor even more vulnerable.[yy] Physician organizing, bolstered by the teaching of political advocacy in medical school, is imperative.[zz] We must all work toward a different system that values every life, regardless of their geography or ability to pay. Health is a human right, and our everyday practice, as well as our advocacy, research, and writing, must reflect this.

Reparations and the fleshy remainder

In the first two sections of this chapter, I have established coloniality and biomedicine as coproductive of each other and outlined some foci of work around decolonizing medical education. In this last section, I describe two potential directions for continued decolonizing practice. Here, as always, I write from my vantage point as an American surgeon and educator. Surely, clinicians and educators elsewhere are decolonizing medical education in different, wildly brilliant ways. For example, Wendland (2010) describes heart work in Malawian medical education, while in Brazil the use of Freire's critical pedagogy allows for an "unfinishedness";[aaa] and the list could go on. May the hegemony of American biomedicine not overshadow this work and all the other endeavors elsewhere.

Because of the way that race was inscribed on bodies during imperialism, colonization, and enslavement, there is a certain fixedness in biomedicine's relationship with race (and inequality more generally). On the continuity of anti-Black racism after enslavement, Wynter (2003) writes, "Man will always be defined by race."[bbb] On the dismemberment of the body within a regime of anti-Blackness, Spillers (1987)

[ww] This particular line is taken from an article (Farmer, 1999) that precedes Farmer's larger opus. For more from Farmer on health as a human right, see Farmer (2005, 2009) and Farmer et al. (2006).

[xx] In 1969 leftists decried the "medical industrial complex" (Ehrenreich and Ehrenreich, 1970); this term entered mainstream American medical audience a decade later when Dr Arnold "Bud" Relman, editor of the *New England Journal of Medicine*, decried the growing corporatization of medicine (Relman, 1980), Today, the situation is even more dire.

[yy] Much has been written about catastrophic health expenditures globally. See Garcia-Diaz et al. (2024) for just one example.

[zz] See the argument for politics to become a required part of medical school curriculum here: Goel and Kavanagh (2020).

[aaa] Young (2025).

[bbb] Wynter (2003).

asserts that "Body and flesh are now divided."[ccc] And still, I wonder how a commitment to reparations for past plunder might loosen the grip that race has on biomedicine. Here, I am not talking about a feeble gesture such as corporate America's $50 billion promise after George Floyd's murder.[ddd] Rather, when I speak about the power of reparations, I am referring to a robust commitment to reparations as described by Gene Richardson, Tasseli McKay, Ta-Nehisi Coates, or the Lancet Commission for Reparations and Redistributive Justice.[eee] Such reparations have the potential to transform the otherwise cemented inequalities into social mobility for folks of color; reparations that might transfigure social relationality so completely as to revolutionize our understanding of bodies and our practice of medicine. Because of their capacity to durably and permanently disrupt the hold that race has on biomedicine, reparations seem to be a crucial step in decolonizing medicine.

And, while we wait for reparations to become politically viable, the bodily practice of attending to flesh feels transformative. Weheliye (2014) writes, "In the absence of kin, family, gender, belonging, language, personhood, property, and official records, among many other factors, what remains in the flesh, the living, speaking, thinking, feeling, and imagining flesh: the ether that holds together the world of Man while at the same time forming the condition of possibility for this world's demise."[fff] Where racializing assemblages have transformed so many into flesh—and tasers, bullets, fists, and rot have perpetuated the wounding—attending to flesh constitutes a form of recognition. As a surgeon in a large US public hospital, it is my job to care for wounds: to monitor my incisions, to open infected wounds at the bedside, to go to the operating room and cut off anything that looks dead. Furthermore, if the wound is clean, then to rearrange flesh to close gaping skin, this corner can go to that space, and if we mobilize this area, we can sew this to that. "The biology of wound healing has been drilled into our brains since medical school," I tell the residents, "but taking care of wounds is more physical than that." It's the tactile knowledge of what hidden infection feels like when I probe under a scab; it's the olfactory knowledge of what Group A Streptococcus—in common parlance, flesh-eating bacteria—smells like; it's being able to feel heat through a glove, to scan the grainy black-and-white pictures of computed tomography scans and see fluid that even a radiologist misses, to distinguish between very subtle shades of red skin—the irritated, infected, healing. And it's the sound of the screams, the whimpers, the held breaths as tape is taken down and wounds are examined and debrided without anesthesia, because we're just looking and local anesthesia doesn't work in infected fields anyway; it's the slop of too-wet dressings in a trash can; it's the silence in the room as residents show me and the patient a wound that they're concerned about, my audible exhale while I gather my thoughts about how to tell a patient news that they are not going to want to hear. "There are other ways to take care of people who have been radically excluded,"

[ccc] Spillers (1987).
[ddd] Washington Post (2021).
[eee] Richardson et al. (2021), McKay (2022), and Coates (2014).
[fff] Weheliye (2014, p. 40).

I tell students, but attending to flesh—of the beaten, the rotten, the left for dead, those who were never meant to survive—is a radical transgression. When the work of shifting language, policies, and paradigms feels intractable, teaching physical caretaking is another practice in our decolonizing toolkit.

Conclusion

In this chapter, I have sought to describe the dense entanglement of biomedicine with imperialism, colonialism, and enslavement—and the work that has been done to disentangle these power relations. The last century has witnessed significant shifts in the way we teach and practice medicine: race is now recognized as a powerful social construct that generates poor health; medical pluralism, medical ethics, health inequities, and structural drivers of health have been codified in the US medical curriculum. I believe that because of these changes, students and trainees in their clinical years approach patients differently, more ethically, with an awareness of patients as people with complex emotional, social, and economic landscapes in addition to their physiology. And still, the extreme financialization of health care continues to limit our ability to take care of the sick. In the US, I can operate on someone when they come in shot because it is an emergency, but I often cannot get them primary care, the resources they need to take care of themselves and their families, protection from medical bankruptcy, or anything that resembles equity. The situation across the globe is even more dire, as the effects of imperial and colonial plunder continue to be observed in the widespread (and likely also systematic) underfunding of health care infrastructure. As practitioners, we must keep pushing toward decolonizing medicine and toward a more just version of care. May our patients be able to hold on long enough to feel the effects of our labor.

References

Barry, M.J., Edgman-Levitan, S., 2012. Shared decision making—the pinnacle of patient-centered care. N. Engl. J. Med. 366 (9), 780–781. doi:10.1056/NEJMp1109283.

Bhakta, N.R., et al., 2023. Race and ethnicity in pulmonary function test interpretation: an official American thoracic society statement. Am. J. Respir. Crit. Care Med. 207 (8), 978–995. doi:10.1164/rccm.202302-0310ST.

Bourgois, P., et al., 2017. Structural vulnerability: operationalizing the concept to address health disparities in clinical care. Acad. Med. 92 (3), 299–307. doi:10.1097/ACM.0000000000001294.

Braun, L., et al., 2007. Racial categories in medical practice: how useful are they? PLoS Med. 4 (9), e271. doi:10.1371/journal.pmed.0040271.

Burson, R., Familusi, O., Clapp, J., 2021. Reimagining 'the Social' in education and practice: a longitudinal study of perceptions of structural competency in medicine. Health Serv. Res. 56 (Suppl 2), 79. doi:10.1111/1475-6773.13833.

Castillo, E., et al., 2020. Reconsidering systems-based practice: advancing structural competency, health equity, and social responsibility in graduate medical education. Acad. Med. 95 (12), 1817–1822.

Chase-Riboud, B., 2003. Hottentot Venus: A Novel. Knopf Doubleday Publishing Group, Westminster.

Coates, T.-N., 2014 The case for reparations. The Atlantic. Available at: https://www.theatlantic.com/magazine/archive/2014/06/the-case-for-reparations/361631/. (Accessed 12 November 2025).

Comaroff, J., Comaroff, J.L., 1991. Of Revelation and Revolution, Volume 1: Christianity, Colonialism, and Consciousness in South Africa. University of Chicago Press, Chicago.

Comaroff, J., 1993. The diseased heart of Africa: medicine, colonialism, and the black body. In: Lindenbaum, S., Lock, M.M., Eds., Knowledge, Power, and Practice: The Anthropology of Medicine and Everyday Life. University of California Press (Comparative studies of health systems and medical care), Berkeley, pp. 305–329.

Crane, J., 2010. Adverse events and placebo effects: African scientists, HIV, and ethics in the 'global health sciences' Soc. Stud. Sci. 40 (6), 843–870. doi:10.1177/0306312710371145.

Ehrenreich, J., Ehrenreich, B., 1970. The medical-industrial complex. The New York Review of Books. Available at: https://www.nybooks.com/articles/1970/12/17/the-medical-industrial-complex/. (Accessed 11 December 2025).

Farmer, P., 1999. Pathologies of power: rethinking health and human rights. Am. J. Public Health. 89 (10), 1486–1496. doi:10.2105/ajph.89.10.1486.

Farmer, P., 2005. Pathologies of Power: Health, Human Rights, and the New War on the Poor: With a New Preface by the Author. California Series in Public Anthropology, vol. 4, 2005 ed. University of California Press, Berkeley.

Farmer, P., 2009. Infections and Inequalities: The Modern Plagues, Updated ed. with a new preface, 9. print. Univ. of California Press, Berkeley.

Farmer, P.E., et al., 2006. Structural violence and clinical medicine. PLoS Med. 3 (10), e449. doi:10.1371/journal.pmed.0030449.

Fassin, D., 2007. When Bodies Remember: Experiences and Politics of AIDS in South Africa. University of California Press, Berkeley.

Foucault, M., 1976. The History of Sexuality: An Introduction. Knopf Doubleday Publishing Group, Westminster.

Fullwiley, D., 2024. Tabula Raza: Mapping Race and Human Diversity in American Genome Science. Atelier: Ethnographic Inquiry in the 21st century, vol. 14. University of California Press, Oakland.

Garcia-Diaz, R., Prasad Sapkota, V., Flores, G., 2024. Persistency of catastrophic out-of-pocket health expenditures: measurement with evidence from three African countries—Malawi, Tanzania, and Uganda. Soc. Sci. Med. 357, 117156. doi:10.1016/j.socscimed.2024.117156.

Gilroy, P., 2003. The Black Atlantic: Modernity and Double Consciousness, 8. print. Harvard Univ. Press, Cambridge, Mass.

Glenn, J.E., et al., 2024. Evaluating the impact of an educational intervention on the history of racism in America for teaching structural competency to medical academicians. BMC Med. Educ. 24 (1), 638. doi:10.1186/s12909-024-05626-5.

Goel, R.R., Kavanagh, N.M., 2020. Beyond structural competency: why politics should be taught during medical education. Acad. Med. 95 (11), e10. doi:10.1097/ACM.0000000000003614.

Good, M.-J.D., et al., (Eds.), 2008. Postcolonial Disorders. Ethnographic Studies in Subjectivity, vol. 8. University of California Press, Berkeley. doi:10.1525/9780520941021.

Green, L., 1999. Fear as a Way of Life: Mayan Widows in Rural Guatemala. Columbia University Press, p. 240.

Harney, S., Moten, F., 2013. The Undercommons: Fugitive Planning & Black Study, Minor Compositions.

Hassan, I.F., Bui, T., 2022. The structural analysis: incorporating structurally competent clinical reasoning into case-based presentations. J. Gen. Intern. Med. 37 (13), 3465–3470. doi:10.1007/s11606-022-07751-7.

Hochschild, A., 1999. King Leopold's Ghost: A Story of Greed, Terror, and Heroism in Colonial Africa. Houghton Mifflin Harcourt, Boston.

Hunt, N.R., 1988. 'Le Bebe en Brousse': European women, african birth spacing and colonial intervention in breast feeding in the Belgian Congo. Int. J. Afr. Hist. Stud. 21 (3), 401. doi:10.2307/219448.

Hunt, N.R., 1999. A Colonial Lexicon: Of Birth Ritual, Medicalization, and Mobility in the Congo. Duke University Press, Durham.

Hunt, N.R., 2008. An acoustic register, tenacious images, and congolese scenes of rape and repetition. Cult. Anthropol. 23 (2), 220–253. doi:10.1111/j.1548-1360.2008.00008.x.

Hunt, N.R., 2016. A Nervous State: Violence, Remedies, and Reverie in Colonial Congo. Duke University Press, Durham, NC.

Institute of Medicine, 2003. Unequal Treatment: Confronting Racial and Ethnic Disparities in Health Care (Full Printed Version). National Academies Press, Washington, D.C, p. 10260. doi:10.17226/10260.

Jacobo Herrera, F., Orr, D., 2020. Susto, the anthropology of fear and critical medical anthropology in Mexico and Peru. In: Gamlin, J., Gibbon, S., Sesia, P.M., Berrio, L. (Eds.), Critical Medical Anthropology: Perspectives in and from Latin America. University College London, pp. 69–89. Available at: https://muse.jhu.edu/pub/354/oa_edited_volume/chapter/2778444#fn-ch3_fn9. (Accessed 10 December 2025).

Judy, R.A., 2020. Sentient Flesh: Thinking in Disorder, Poiēsis in Black. Black Outdoors: Innovations in the Poetics of Study. Duke University Press, Durham. doi:10.1515/9781478012559.

Kleinman, A., 1989. The Illness Narratives: Suffering, Healing, and the Human Condition. Basic Books, New York.

Kolmes, S., et al., 2025. Incorporating structural competency into clinical ethics: piloting new bioethics education. J. Clin. Ethics 36 (2), 158–166. doi:10.1086/734776.

Leslie, C., 1980. Medical pluralism in world perspective [1]. Soc. Sci. Med. B Med. Anthropol. 14 (4), 191–195. doi:10.1016/0160-7987(80)90044-7.

Lock, M.M., 1995. Encounters with Aging: Mythologies of Menopause in Japan and North America, 1. pbk. print, 3. print. Univ. of California Press, Berkeley, Calif.

Lock, M.M., Nguyen, V.-K., 2018. An Anthropology of Biomedicine, second ed. Wiley Blackwell, Hoboken, New Jersey Oxford.

Lovasik, B.P., et al., 2020. 'The living monument': the desegregation of Grady Memorial Hospital and the Changing South. Am. Surg. 86 (3), 213–219.

Mbembe, A., 2001. On the Postcolony. Studies on the history of society and culture, vol. 41. University of California Press, Berkeley.

McClintock, A., 1995. Imperial Leather: Race, Gender, and Sexuality in the Colonial Contest. Routledge, New York.

McKay, T., 2022. Stolen Wealth, Hidden Power: The Case for Reparations for Mass Incarceration. University of California Press, Oakland.

Metzl, J.M., Hansen, H., 2014. Structural competency: theorizing a new medical engagement with stigma and inequality. Soc. Sci. Med. 103, 126–133. doi:10.1016/j.socscimed.2013.06.032.

Moore, J.X., et al., 2022. Exploring racial disparities on the association between allostatic load and cancer mortality: a retrospective cohort analysis of NHANES, 1988 through 2019. SSM Popul. Health 19, 101185. doi:10.1016/j.ssmph.2022.101185.
Neff, J., et al., 2017. Teaching structure: a qualitative evaluation of a structural competency training for resident physicians. J. Gen. Intern. Med. 32 (4), 430–433. doi:10.1007/s11606-016-3924-7.
Neff, J., et al., 2020. Structural competency: curriculum for medical students, residents, and interprofessional teams on the structural factors that produce health disparities. MedEdPORTAL. 16, 10888. doi:10.15766/mep_2374-8265.10888.
Nelson, A., 2011. Body and Soul: The Black Panther Party and the Fight against Medical Discrimination. University of Minnesota Press, Minneapolis London.
Niehuus, R.M., 2024. An Archive of Possibilities: Healing and Repair in Democratic Republic of Congo. Critical Global Health: Evidence, Efficacy, Ethnography. Duke University Press, Durham.
Povinelli, E.A., 2021. Between Gaia and Ground: Four Axioms of Existence and the Ancestral Catastrophe of Late Liberalism. Duke University Press, Durham.
Powe, N.R., 2022. Race and kidney function: the facts and fix amidst the fuss, fuzziness, and fiction. Med. 3 (2), 93–97. doi:10.1016/j.medj.2022.01.011.
Reiser, S.J., 1991. Medical ethics reflected in codes of ethics: the Hippocratic Oath and the 1980 AMA code compared. Tex. Med. 87 (1), 77–81.
Relman, A.S., 1980. The new medical-industrial complex. N. Engl. J. Med. 303 (17), 963–970. doi:10.1056/NEJM198010233031703.
Reverby, S.M., 2009. Examining Tuskegee: The Infamous Syphilis Study and Its Legacy. The John Hope Franklin Series in African American History and Culture. University of North Carolina Press, Chapel Hill.
Richardson, E.T., et al., 2021. Reparations for Black American descendants of persons enslaved in the U.S. and their potential impact on SARS-CoV-2 transmission. Soc. Sci. Med. 276, 113741. doi:10.1016/j.socscimed.2021.113741.
Roberts, D.E., 2017. Killing the Black Body: Race, Reproduction, and the Meaning of Liberty, second vintage books ed. Vintage Books, New York.
Sharpe, C., 2016. In the Wake: On Blackness and Being, Duke University Press.
Skloot, R., 2011. The Immortal Life of Henrietta Lacks, 1. pbk. ed. Broadway Paperbacks, New York.
Spillers, H.J., 1987. Mama's baby, papa's maybe: an American grammar book. Diacritics. 17 (2), 65–81. doi:10.2307/464747.
Stoler, A., 1995. Race and the Education of Desire: Foucault's History of Sexuality and the Colonial Order of Things. Duke University Press, Durham, NC. Available at: https://www.thriftbooks.com/w/race-and-the-education-of-desire-foucaults-history-of-sexuality-and-the-colonial-order-of-things_ann-laura-stoler/363350/. Accessed 19 August 2025.
Stoler, A.L., 2016. Duress: Imperial Durabilities in Our Times. A John Hope Franklin Center Book. Duke University Press, Durham.
Taussig, M., 1991. Shamanism, Colonialism, and the Wild Man: A Study in Terror and Healing. University of Chicago Press, Chicago.
Thobaben, J., 2024. The history of medical ethics. In: Wolfe, B.N., et al. (Eds.), St Andrews Encyclopaedia of Theology. Available at: https://www.saet.ac.uk/Christianity/HistoryofMedicalEthics. (Accessed 10 December 2025).
Vaughan, M., 1991. Curing Their Ills: Colonial Power and African Illness, 1. publ. Stanford University Press, Stanford, Calif.

Vyas, D.A., Eisenstein, L.G., Jones, D.S., 2020. Hidden in plain sight—reconsidering the use of race correction in clinical algorithms. N. Engl. J. Med. 383 (9), 874–882. doi:10.1056/NEJMms2004740.

Washington Post, 2021. Big business pledged nearly $50 billion for racial justice after George Floyd's death. Where did the money go? Available at: https://www.washingtonpost.com/business/interactive/2021/george-floyd-corporate-america-racial-justice/. (Accessed 27 August 2025).

Weheliye, A.G., 2014. Habeas Viscus: Racializing Assemblages, Biopolitics, and Black Feminist Theories of the Human, Duke University Press.

Wendland, C.L., 2010. A Heart for the Work: Journeys through an African Medical School. University of Chicago Press, Chicago.

Wynter, S., 2003. Unsettling the coloniality of being/power/truth/freedom: towards the human, after man, its overrepresentation–an argument. CR. 3 (3), 257–337. doi:10.1353/ncr.2004.0015.

Young, J., 2025. Using critical pedagogical for SDOHs. AMA J. Ethics. 27 (9), E700–707.

CHAPTER

9 Toward sustainable change: Embedding decolonial implementation science in medical training to reimagine our world

Patricia Maritim[a], Ramya Kumar[b,c,d], and Muneera Rasheed[e]
[a]*School of Public Health, University of Zambia, Lusaka, Zambia*
[b]*School of Public Health, University of Washington, Seattle, WA, United States*
[c]*University of California, Davis, CA, United States*
[d]*Centre for Infectious Disease Research Zambia (CIDRZ), Lusaka, Zambia*
[e]*Independent Professional, Karachi, Pakistan*

Introduction

Unfair and avoidable differences in access to health care and in the outcomes of health care across social dimensions (such as age or gender) persist in our societies today (Bambra et al., 2025). For instance, individuals in the world's poorest countries are likely to die 12.5 years earlier than their peers in richer countries, while mothers in fragile settings display relatively higher mortality rates (World Health Organization, 2025). Primrose and Loeppky write that this is a result of four decades of neoliberal efforts, which have resulted in the restructuring of the socioecological sphere through: (1) the commodification of health care systems that reorient them from public goods to profit-making entities, thus making them inaccessible and lowering the quality of health care services; (2) diminished public spending, with governments shifting responsibility to private stakeholders (such as pharmaceutical companies) that modify the agenda to curative, rather than preventive care; and (3) heightened population vulnerability to disease through a negative influence on the social determinants of health (Primrose and Loeppky, 2024).

Within medical education training programs, this restructuring manifests through the biomedicalization of health, where the training of medical practitioners is often centered on understanding the etiology, prevention approaches, and treatment strategies of common diseases in their immediate contexts (Brownson et al., 2022). Without an explicit focus on the living and working conditions that shape health, particularly global systems such as the economy, social order, governance structures,

Decolonizing Medical Education, Research and Practice. https://doi.org/10.1016/B978-0-443-44998-7.00006-5

the environment, international security, and transport and communication systems (Krieger, 2024; McNamara and Bambra, 2025; Medvedyuk and Raphael, 2025), medical training may uphold the notion that health problems arise from personal choice (McNamara, 2025). Further reinforcing the neoliberal ideology is the notion that differences across populations can be addressed primarily through individual agency, with a singular focus on health care, rather than the right to health and the right to live in healthy societies (McCartney et al., 2019; Buse, 2025).

This narrowing of vision is itself a colonial legacy: colonial medicine historically reduced illness to biological dysfunction, detaching it from the political economy, cultural knowledge, and living conditions of colonized populations (Naidu, 2021; Naidu and Abimbola, 2022; Razack et al., 2025). Such framings have persisted and have embedded structural ignorance in medical education worldwide. This structural ignorance may lead to downstream drift, where practitioners select public policies and public health interventions that prioritize biomedical solutions. These are selected based on their focus on changing individual behavior and depoliticizing health, consequently creating and recreating inequalities (refer to Box 1) (Lynch, 2020; Raphael and Bryant, 2022; McNamara, 2025).

Medical practitioners, through their everyday work, causally link the upstream determinants of health with routine public health practice and are therefore in a good position to challenge, materially disrupt, and reconfigure these systems of domination that shape health outcomes (Nunes, 2020; Karatekin et al., 2024) (refer to Box 2).

Incorporating structural competency in medical education equips trainees with the ability to identify and respond to the structural causes of health and health inequities by engaging with the systems that determine the generation and distribution of socioeconomic resources and how they are translated into health-relevant resources (Reynolds, 2021; Harvey et al., 2022). As such, medical training ought to strengthen practitioners' capacity to conceive of other ways that these systems can be structured, and, otherwise, through decolonial medical education (Sharpe, 2016; Khasnabish, 2020; Raphael and Bryant, 2022).

Box 1 Key concept: downstream drift

Definition: The shift from addressing structural drivers of health to focusing mainly on individual behaviors or biomedical fixes.

Example: Promoting diet counseling instead of addressing food insecurity through policy.

Why it matters: It depoliticizes health and risks reinforcing inequalities.

Box 2 Key concept: upstream determinants of health

Definition: Structural conditions—like governance, education, and the economy—that shape health outcomes.

Example: Regulating air pollution to lower community-wide respiratory illness.

Why it matters: Acting upstream can prevent disease and reduce inequities.

In keeping with the central premise of this book, decolonization offers a critical framework that challenges extractive, Eurocentric models of knowledge and practice and calls for "an otherwise" health system built around justice, equity, and values grounded in lived realities, rather than imported models (Sharpe, 2016; Khasnabish, 2020; Raphael and Bryant, 2022). The authors draw on experiences from both high- and low-income settings, including decades of work on the African continent and in Pakistan. These perspectives inform our approach to implementation science (IS), while emphasizing the importance of local knowledge and priorities.

This chapter explores how IS can be reimagined through a decolonial lens to enhance the structural competencies of medical trainees. With this approach, medical practitioners are trained not only to apply technical solutions, but also to question the structures that shape health, and hopefully become *agents of transformation*. This training prepares them to address colonial legacies and systemic determinants of health. In this sense, decolonial medical education becomes not only about diagnosing disease, but also about unlearning structural blindness and cultivating the imagination to design systems that serve communities on their own terms.

Decolonization as a framework to redress structural ignorance

Decolonizing medical education means moving beyond the narrow biomedical model to question the structures that sustain health inequities (D'Ambruoso et al., 2025). Structural competency is a step forward; but without a decolonial lens it risks reinforcing the same hierarchies it seeks to challenge. One enduring legacy of colonial medicine is training systems that focus primarily on technical expertise and individual behavior change, while sidelining community knowledge and histories. A decolonial approach unsettles these legacies, values diverse ways of knowing, and positions medical education as a space for transformation, rather than the reproduction of inequity.

Hence, decolonization is a critical aspect of public health practice because it offers a framework through which we can critique the material infrastructure responsible for ill health, while also creating a space to build our capacity for freedom and liberation through alternative, value-driven methods that accommodate the complexity inherent in our social systems (Feukeu, 2024; Bosch, 2025). Beyond the critique of systems, it pushes us to build other ways of "knowing, citing, collaborating, and theorizing that do not extract from the margins to enrich the center" (Bosch, 2025). When envisioned through the lens of training, a key goal of intellectual production is to bring about societal change and repair through our intellectual resources and creative expressions (Sinitiere, 2021). To realize this transformative vision, medical education requires approaches that move beyond critique and actively engage with the processes of change. IS is particularly relevant here, as at its core, it is concerned with change. Its focus is on a myriad of theories, models, and frameworks used to link dynamic contexts, interventions, and the values of the different stakeholders involved in their implementation (Metz et al., 2023).

In this chapter, we explore how IS training for medical practitioners can play a role in establishing more just and healthy futures. We view medical education as preparation for handling the future of public health. Therefore we highlight how several approaches (which are central to the field of IS) can equip medical practitioners with the capability to reshape the social, political, and economic systems in which interventions are being implemented, or support the adaptation of interventions to promote greater compatibility with the systems they are being introduced to (von Thiele Schwarz et al., 2019; Gaias, 2021).

Implementation Science as a discipline and tool for change!

IS refers to the "scientific study of methods to promote the systematic uptake of research findings and other evidence-based practices into routine practice and, hence, to improve the quality and effectiveness of health services and care" (Eccles and Mittman, 2006). IS integrates research aimed at generating generalizable knowledge on implementing innovation within specific settings, while applying this knowledge to guide the processes of change in practice settings (Leppin et al., 2021). Specifically, it uses theory, models, and frameworks to understand and explain implementation processes and outcomes, as well as to develop, apply, and assess the effectiveness of implementation strategies designed to address determinants that can enhance the uptake of evidence-based interventions in practice (Kirchner and Bauer, 2024). Because IS is translational science, its training aims to prepare the implementation workforce (including medical practitioners) to generate knowledge on implementation. This knowledge can be used to enhance the capacity of the systems into which interventions are being introduced, so they can absorb these interventions in practice (Leppin et al., 2021).

Despite their crucial role in implementing new innovations in routine practice and policy settings, training initiatives have often ignored practitioners such as nurses, doctors, and pharmacists, instead focusing more on researchers (Juckett et al., 2022; Maritim et al., 2024). Recent efforts have resulted in increased inclusion of practitioners. For instance, in Zambia, a TB operational research training program demonstrates how context-driven training can shift knowledge production to local practitioners and institutions, supporting a decolonial reorientation of IS (Kumar et al., 2024a). However, academic spaces, where most of the training initiatives are based, are not structurally equipped to adequately prioritize the needs of practitioners (Moore and Khan, 2024).

Decolonizing IS training

Conventional approaches to implementing interventions often assume that there is a set of simple, linear, rational, and uncritical processes one must follow. In reality, implementation is a set of complex, value-driven processes (Bradley et al., 2024;

Engebretsen et al., 2025). When training ignores such complexities, it can reinforce unjust systems. This happens when we ignore the "competing framings, as well as political and epistemic conflicts" associated with the interventions themselves, as well as *how* they are implemented and *whose interests* implementation seeks to serve (Boulton et al., 2020; Bradley et al., 2024; Engebretsen et al., 2025). When translated into training curricula, it leads to the exclusion of the politics of evidence production and the epistemological underpinnings of *why* it is essential to understand implementation (Dada et al., 2025). Effective IS training ought to allow trainees to contest what is knowable and whose voices are included in scientific endeavors. Furthermore, IS training should use the knowledge we generate to guide action (Dada et al., 2025).

To do this, trainees need structured approaches that help them connect individual health problems to the wider social and political forces that shape them. One such approach is Harvey et al.'s structural competency framework (Harvey et al., 2022). The next sections use this framework to highlight how training medical practitioners in critical IS can play a dual role: equipping them with the skills needed to place implementation efforts within their wider social and historic contexts, which would make it possible to prescribe solutions that target patient vulnerabilities while also supporting them in transforming the systems that cause these challenges (Boulton et al., 2020; Snell-Rood, 2021). We provide examples from our own experiences of how this can be realized in real-world settings. The following subheadings correspond to components of Harvey et al.'s (2022) structural competency framework.

Examining the role of social structures in producing and maintaining health inequities

A defining feature of IS is its use of theory to structure our understanding of implementation and appropriate strategies that can lead to its successful execution (Nilsen and Birken, 2020). Value-driven implementation assessments draw our inquiry into tangible events and behaviors, as well as deeper patterns (i.e., social structures) that may need to change to realize better health outcomes (Metz et al., 2023). Social structures within this context refer to policies, institutions, and systems that determine how society is organized, and that which establishes social hierarchies, which are the foundation for health inequities (Harvey et al., 2022).

Training medical professionals in IS can build their capability to conduct critical inquiry, which interrogates pathogenic social structures by concretizing and contextualizing the implementation of interventions (Burgess, 2022; Lynch, 2023). This ensures that, as they envision systems change, their solutions can be appropriate for local communities to disrupt the material infrastructure that causes health inequalities, with the capacity to be sustained (Metz et al., 2023; Bosch, 2025). It should be noted that building structural competency is not limited to clinical encounters; it also involves strengthening the workforce and institutions (Kumar et al., 2024a).

Through its theoretical openness, the field can provide trainees with epistemic freedom that gives them the "right to think, theorize, interpret the world, develop their own methodologies, and write from where one is located, unencumbered by Eurocentrism" (Ndlovu-Gatsheni, 2018). Nevertheless, as they use theory, trainees should consider that theory and methodologies (knowledge) are not neutral and that they carry assumptions. Trainees should, therefore, be aware of the assumptions that underlie the approaches that they opt to use to better understand the realities (conditions, challenges, and contexts) they face in practice (Boulton et al., 2020). Here, they can use theory to reject the existing knowledge canon (traditional academic frameworks) in favor of those that view their everyday practice experiences as sites of knowledge generation (Butler, 2021; Bosch, 2025). In other words, trainees can adopt approaches that treat their own everyday experiences—such as patient care, community engagement, or clinical practice—as legitimate sources of knowledge.

Mapping the pathways through which structural inequalities are naturalized

Once trainees have become familiar with the social structures that shape their practice, it then becomes important for them to understand how these structures trigger social, economic, and political mechanisms that result in poor health and health inequities (Reynolds, 2021). To do so, they require what the critical theorist Amilcar Cabral has termed "new ways of describing social constructions that can alter human experiences and human organization" through an analysis of the causes of social problems, as well as how such causes give rise to the observed effects (Rabaka, 2009, 2014). This inquiry process creates a rupture through which pathogenic social structures can be transformed into health-promoting ones (Harvey et al., 2022). It is important to note that the capacity of trainees to map these conditions may be curtailed by the history of imperialism and the ongoing coloniality of the mind, which accepts prevailing social conditions as natural and immutable via epistemological colonization (Ndlovu-Gatsheni, 2018; Tamale, 2020) rather than as problems that can be addressed using different tools (May, 2023). Through grounding in methodologies such as mechanism mapping and process tracing, they can understand how and why social structures and interventions are operating to produce certain outcomes. This will provide them with entry points through which they can intervene.

Understanding the impact of structures on the practice of global health

IS makes it possible for trainees to not only link what is happening in their local practice to national, regional, and global structures, but also allows them to understand how power and privilege are consolidated. Moreover, it enables them to understand the advantages that arise from the resulting distribution and meaning of material and cultural resources in multiple social systems (Reynolds, 2021; May, 2023). They can then observe how such factors shape their interactions with patients and communities, and how medical providers could actively contribute to health inequalities (Harvey et al., 2022). On a broader scale, local workforce development can contribute to regional and global health security priorities, while also revealing the power dynamics that shape whose expertise is recognized internationally (Kumar et al., 2020).

Identifying structural interventions needed to address health inequities

Beyond identifying and critiquing pathogenic social structures, IS training also equips trainees with the ability to identify and implement structural interventions that can reduce harm, address health inequities, and/or replace harmful social structures with equitable, health-promoting ones (Harvey et al., 2022).

Practicing structural humility in public health practice

Finally, IS training can be one mechanism through which structural humility is cultivated during the training of medical providers (Harvey et al., 2022). Structural humility is a self-reflexive approach to one's practice that is based on cultural humility, which entails an ongoing commitment to unlearning one's own stereotypes and biases that may produce power differentials during patient-provider interactions (Harvey et al., 2022). It also promotes collaboration with patients and communities in mutually beneficial ways (Harvey et al., 2022) that value context-specific ways of knowing, placing a focus on pluralistic epistemological approaches, relationality, and political commitment (Bosch, 2025). To support this reflexive practice, Box 3 introduces an applied learning activity in which students analyze how expertise is

Box 3 In-class activity: Science communication in practice

This activity is designed to reflect on how institutional hierarchies, credentialism, and colonial logics shape who is authorized to speak in health contexts.

Learning goal: to assess how expertise is socially constructed and to practice structural humility in interpreting scientific authority.

1. Watch the TEDx talk: *How do we pick our experts?* by Dr. Ramya Kumar (TEDxLusaka, 2025).

This talk illustrates how questioning authority, interrogating who is seen as an "expert," and practicing humility in science communication are essential skills for decolonial IS. It also serves as a real-world example of making complex ideas accessible to wider audiences.

2. On your own, reflect and journal: where in your training have you been taught to speak *to* communities rather than *with* them?
3. In small groups, discuss the following questions:
   - How do you currently decide who is an "expert" in your own field or practice?
   - Whose expertise is routinely excluded in medical or public health settings?
   - Propose one institutional reform that would operationalize structural humility in how expertise is recognized and rewarded.

constructed and communicated by viewing a public science communication talk, followed by small-group dialog on how structural humility can expand their own definitions of expertise.

Furthermore, Table 9.1 provides a summary of how IS training can enhance structural competency.

A transformative outcome from decolonial IS becomes restored dignity, where (1) equitable systems ensure that structures and policies are fair, responsive, and accountable to the communities they serve; (2) empowered communities activate people as the primary agents of their own wellbeing and transformation; and (3) evolving relationships shift power dynamics, promoting collaboration and mutual respect among all actors. These outcomes collectively realize dignity in its three essential forms, as illustrated in Fig. 9.1:

i. Dignity as *agents* (autonomy): the power to act and make choices, enabled by empowered communities.
ii. Dignity as *contributors* (worth): the recognition of inherent value, affirmed through equitable systems and evolving relationships.
iii. Dignity as *knowers* (knowledge): the respect for one's knowledge and perspective, enabled by just structures and collaborative partnerships.

Table 9.1 Harvey et al.'s (2022) structural competency framework.

Component of structural competency (Harvey et al., 2022)	Description	Application to IS training
Recognizing structures	Identify how social, political, and economic systems shape health outcomes and clinical interactions.	Trainees learn to see how policies, institutions, and inequities influence intervention uptake (e.g., stigma deterring pre-exposure prophylaxis [PrEP] use).
Developing structural humility	Acknowledge limits of one's own perspective and commit to **reflexivity** in practice.	Trainees reflect on their positionality, biases, and how power dynamics shape relationships with communities.
Reframing clinical problems	Move beyond individual-level explanations of illness to situate health within structural determinants.	Trainees understand asthma as linked to housing and pollution, not only to inhaler use; human immunodeficiency virus (HIV) prevention as shaped by stigma and policing.
Intervening structurally	Design and implement interventions that target **upstream determinants**, not just individual behaviors.	Trainees develop strategies such as safe-space PrEP depots or integrated services that address systemic barriers.

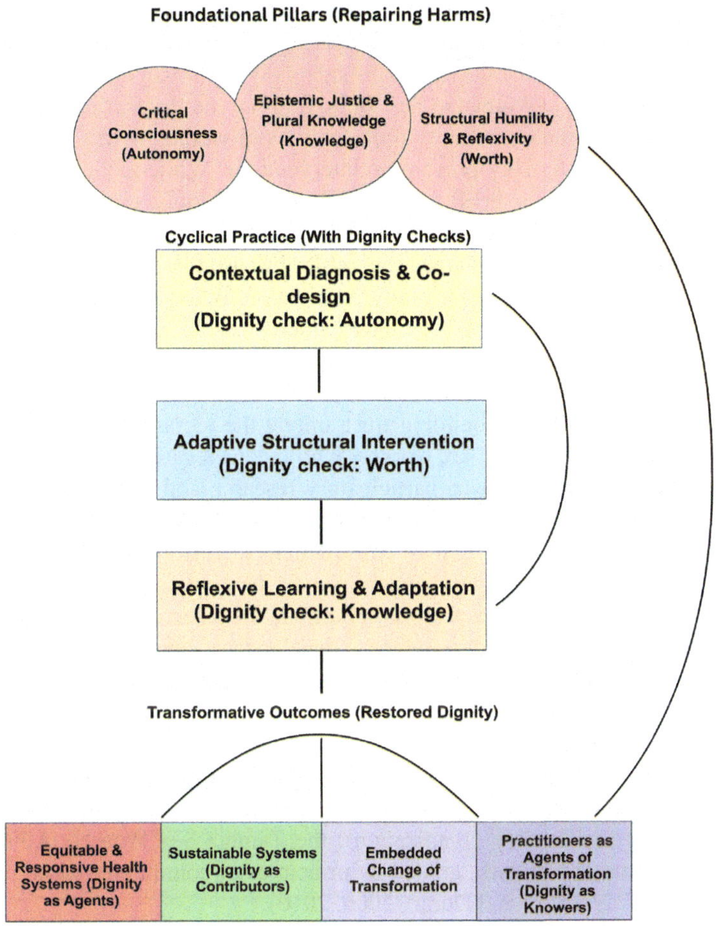

FIG. 9.1

Repairing harm through decolonial implementation science (IS).

Moving from theory to praxis

The abstract principles of IS gain meaning when applied to real-world contexts. Case studies are, therefore, essential because they demonstrate how theories, models, and frameworks of IS interact with lived realities, revealing both the possibilities and limitations of these approaches. By grounding decolonial IS in the specific examples shared in the next section, case studies make visible the ways practitioners negotiate complexity, adapt interventions, and reconfigure relationships of power within health systems. In this chapter, the case studies serve not only as illustrations of methods, but also as evidence of how medical education and practice—when informed by IS and a decolonial lens—can be mobilized to transform practice, and contribute to more just, and healthy futures.

Case study 1: Community-based HIV prevention research with sex workers in Zambia

In Zambia, the WiSSPr (Women in Sex Work, Stigma, and Prevention) research team partnered with multiple community organizations of women in sex work to understand barriers regarding HIV prevention interventions. While global policy emphasizes expanding access to PrEP and other biomedical tools, community members repeatedly pointed to structural barriers (such as police harassment, stigma in clinics, and economic precarity) that shaped their ability to use these interventions. Guided by the COM-B (Capability, Opportunity, Motivation, Behavior) framework (Michie et al., 2008; McDonagh et al., 2018), interventions were tailored to not just increase PrEP knowledge but also to address the structural barriers and everyday realities of women's lives (Kumar et al., 2024a, 2024b).

A critical early task was building trust among the researchers, service delivery teams, and the community of sex workers who received care. Trust development is a construct of community-based participatory research and has also emerged as a synthesizing theory (Belone et al., 2016; Lucero et al., 2020). Trust types are ordered along a relative continuum from least (trust deficit) to most (critical reflective trust), which reflects an ability to discuss and move on after a misstep. Given the historical marginalization of women in sex work in Zambia, the researchers anticipated a trust deficit. They invested time and resources to develop trust along a continuum: starting with role-based trust (from researchers), building proxy trust (through implementing partners and civil society organizations), and ultimately working toward critical-reflective trust, where the community and researchers could acknowledge and repair missteps together. To ground the study in community priorities, the study established a Community Advisory Board (CAB) with the Lusaka District Health Office and two key population civil society organizations: the Zambia Sex Workers Alliance and Tithandizeni Umoyo Network, guided by principles of community-based participatory research (Wallerstein and Duran, 2006, 2010).

The CAB played a central role in co-designing research questions, outcome measures, and implementation strategies (Kumar et al., 2024a). The board provided input on recruitment strategies for sex workers who find clients in different places, i.e., brothels, bars, night clubs, universities, or on the street. The board also provided strategies to minimize risks, including procedures to reduce emotional distress and ensure physical safety, while maintaining women's privacy, even within the community hubs where they receive health care. The board also helped tailor interventions. For instance, the board advised on how and where peer navigators should accompany patients, referral pathways for psychosocial support, and formats for disseminating results back to the community of women engaging in sex work. The CAB's ongoing involvement ensured that interventions were contextually relevant, ethically responsive, and accountable to their community. By embedding the advisory board's processes into IS, medical practitioners can learn how participatory structures strengthen both equity and effectiveness in health research (refer to Box 4).

Box 4 Reflection questions on the community advisory board in case study 1

Q1: How does the board embody principles of structural humility?
Q2: In what ways does the board disrupt conventional researcher–participant hierarchies?
Q3: What trade-offs might arise in balancing advisory board recommendations with donor requirements?
Q4: How could the community advisory board model be adapted for other marginalized communities in different settings?

Rather than importing "best practices" from outside, interventions were designed through participatory workshops with sex workers, in local languages, and built on women's own knowledge and experiences. Recognizing women's daily experiences as valid sources of knowledge to design interventions was transformational: it shifted *expertise* from external stakeholders to the women themselves. By engaging with sex workers as equal partners, the research team connected local practice to global power structures. For example, women's experiences of police violence were not isolated incidents, but reflected broader legal and social systems shaped by colonial-era laws. Training medical practitioners to recognize these connections makes it possible to view HIV risk not only as an individual issue but also as a result of structural injustices. Through CAB discussions and key interviews with women in sex work, participants revealed how health inequities were reinforced by institutions. For instance, clinic staff often required women to disclose their occupation, exposing them to stigma and deterring them from seeking care. This highlighted how institutional policies—which were seemingly neutral on paper—functioned as pathogenic social structures that maintained inequities.

Qualitative interviews with such women further revealed how stigma and rigid clinic systems discouraged PrEP use. For example, women feared being judged if they disclosed alcohol use or pregnancy status. Conventional health education might dismiss these concerns as "misconceptions," but IS reframes them as structural barriers requiring systemic solutions. Drawing on reflections from our "Behind the Paper" essay, the researchers emphasized that mistrust was not a matter of individual reluctance, but the cumulative effect of extractive clinical encounters (Kumar et al., 2025a). This made adherence less dependent on motivation than on whether systems could operate at the *speed of trust.* These findings reinforced that women wanted PrEP—oral and injectable—but that adherence was constrained by stock-outs, punitive provider attitudes, mobility, and anticipated stigma, not by a lack of motivation (Kumar et al., 2025a).

By tracing how stigma worked, researchers identified a damaging pathway:

Clinic judgement → misinformation → discontinuation of PrEP

This mapping helped clinicians to view inequities not as "natural" outcomes of lifestyle, but as products of social and political mechanisms. To counter this, peer-led education programs trained peer health navigators (who were also sex workers) to share accurate information and challenge myths in supportive group discussions. This peer-to-peer structure disrupted stigma by normalizing PrEP use in trusted community spaces. Donor indicators emphasized the number of women initiated on PrEP but did not capture whether women were able to persist with treatment amidst stigma and economic insecurity. By embedding support systems such as buddy networks, Short Message Service reminders, and small follow-up incentives, the intervention challenged the limits of existing monitoring frameworks.

Based on these findings, community leaders and practitioners co-designed interventions that targeted structures and not just behaviors. This included:

- **Convenient service delivery**: Expanding refill depots into safe community spaces, such as the homes of more senior sex workers, brothels, and bars, to make access more practical.
- **Integrated health services**: Training health care providers to deliver PrEP alongside reproductive, substance use, and mental health services.
- **Reducing stigma**: Campaigns to normalize PrEP use that enlist influential members of their neighborhoods.

Together, such interventions shifted the focus from individual responsibility to structural accessibility and acceptance. These interventions were aimed at disrupting harmful systems rather than solely educating individuals. Perhaps most importantly, the project required clinicians, district health leaders, and researchers to practice structural humility. Instead of assuming they held the "best" solutions, they listened to sex workers' experiences and treated those as legitimate knowledge. This reflexive stance reshaped relationships of power, positioning community members not as "subjects" but as co-producers of knowledge (Kumar et al., 2025a).

This case study demonstrates how critical decolonial IS training enables practitioners to: (1) connect global health priorities to community-defined needs; (2) identify and critique structural determinants, such as stigma and rigid service delivery; (3) map inequities to design tailored interventions; (4) implement systemic changes, ranging from safe-space depots to integrated care; and (5) practice humility by honoring women's voices as legitimate sources of knowledge. By grounding theory in lived experience, trainees learned to view interventions not as neutral, technical fixes, but as political and social processes.

This approach intentionally embeds ethical care into its design and delivery process. By shifting decision-making authority to peer navigators and affected women, it disrupts power asymmetrics that typically producc "*mundane ethical harms*" such as misrecognition of local expertise, exclusion from authorship and credit, tokenistic engagement, and extractive evaluation metrics (Vogel et al., 2025). In doing so, it not only expands access to HIV prevention but strengthens trust, responsiveness, community ownership, and the likelihood of meaningful and sustained impact.

We offer a few questions to help with reflection regarding approaches in our case study (refer to Box 5).

Box 5

Reflection questions on case study 1

Q1: How would a conventional, biomedical-only approach to PrEP implementation among sex workers have framed the problem differently? What might it have overlooked?
Q2: In what ways did mapping the pathways of inequity (e.g., clinic judgment → misinformation → discontinuation of PrEP) change how the intervention was designed?
Q3: How do global funding priorities shape what gets measured in HIV programs? What kinds of community priorities might remain invisible under these frameworks?
Q4: What does it mean for clinicians and researchers to practice *structural humility*? How could you apply this in your own training or practice?
Q5: Which of the structural interventions described do you think would be hardest to sustain? Why?

Case study 2: Responsive playful parenting programs in Pakistan

This case study situates responsive play within both outpatient and inpatient pediatric services, resisting the donor-driven logic that frames early childhood development (ECD) as a remedial service for "the poor" and delivered only through undervalued community workers. Instead, it shows how re-embedding interventions into mainstream health services can shift who is regarded as a legitimate recipient of care and who is valued as a provider. The detailed methods and outcomes of this work are reported elsewhere (Rasheed et al., 2021b, 2021c, 2022); here, its relevance lies in demonstrating how a different epistemic starting point—one attentive to equity, legitimacy, and sustainability—opens space for new practices and new imaginaries.

In Pakistan, ECD is often framed as a service for "the poor," typically defined as those in rural communities or urban slums who are struggling to meet basic needs for survival, delivered by public community health workers (CHWs). These CHWs are usually women who are underpaid and undervalued in the community compared to other professions. Here, Granados and Prabhu explain that decisions by consumers are not solely driven by costs and that interventions need a realistic rather than solely an empathetic approach (Granados and Prabhu, 2022). They share examples in India of affordable products like low-cost cars and refrigerators. In this context, such refrigerators failed to scale to more than a few thousand in 15 years, while the production of such cars had to be halted after a few years. Although both products were priced lower than existing alternatives, low-income consumers still chose to pay more for other brands—and even considered second-hand cars or refrigerators preferable. Postmortem analyses revealed that this failure stemmed, in part, from the fact that such products signaled poverty, and many consumers actively avoided purchases that reinforced this perception.

Our approach aligned with this insight, and to encourage organic uptake, we designed an ECD intervention that reflected the aspirations of low-income families.

We piloted an intervention that embedded play-based practices into a private, elite pediatric hospital, spanning both routine inpatient and outpatient department (OPD) services. Rather than creating a parallel program, we integrated developmental play sessions into existing clinical routines. Psychology trainees from a local, well-reputed university were trained to deliver structured, developmentally appropriate interactions with children and caregivers, thus transforming routine care encounters into opportunities for responsive engagement. The intervention was co-designed with pediatricians, psychologists, and families—by aligning with their schedules and embedding activities within their usual care routines. Psychology trainees underwent structured training on responsive caregiving, developmental milestones, and play facilitation, followed by supervised practice. The program was implemented in two hospital contexts:

- **Inpatient pediatric wards**: The inpatient component was more extensive, as children often spent several days admitted for routine management of acute physical health issues. Trainees conducted structured play sessions at the hospital bedside—primarily to ease children's emotional stress—with simple activities like blowing bubbles or filling and emptying containers. This type of play provided comfort and engagement. Activities were adjusted to each child's developmental level and medical condition. Sessions typically lasted 20–30 minutes and were repeated daily during hospitalization. Caregivers were guided on how to continue play and encourage stimulation outside sessions. Developmental screenings were integrated, and children at risk were flagged for follow-up. Psychology trainees also provided psychoeducation to parents on stress, bonding, and responsive caregiving in the context of illness.
- **Outpatient departments**: The goal was to transform otherwise idle waiting time into opportunities for learning through playful activities and parent coaching. Parents completed a short milestones and risk survey (interview or self-administered), which was scored and added to the child's medical record. Trainees then engaged children in brief play sessions (5–10 minutes) using play materials like a ball or a book, focusing on age-appropriate play, language, and enhancing caregiver–child warmth during their interaction. Data collection included the number of families reached (OPD vs. inpatient), developmental risks identified through standardized screening, family satisfaction surveys, trainee reflections, and supervisor evaluations of competence.

We found that our study led to shifts in different dimensions. In terms of family experiences (of 223 inpatient families), only 5% reported stress about their child's illness postintervention, while 45% ranked this program among the top three factors improving their hospital experience. Over 90% felt respected, listened to, and understood by therapists, identifying the program as the strongest driver of family satisfaction and recommendation likelihood. From 49 outpatient families, 86% valued learning about developmental and emotional needs, 78%

wished to discuss results with the pediatrician, and 73% requested a booklet to guide play-based learning. Parents highlighted increased awareness of milestones and their role in child growth. Furthermore, of 231 inpatients screened, 76 (32.9%) were at risk of developmental delays. Poor health nearly doubled this risk (OR = 1.95; 95% CI 1.36–2.79; $p < 0.001$), particularly among less-engaged parents. Children of parents using fewer than three stimulation practices were almost twice as likely to be at risk (mothers: OR = 1.90, CI = 1.27–2.81; fathers: OR = 1.96, CI = 1.35–2.84). Outpatient screening of 182 families revealed 54% of children being at risk, highlighting the need for integrating developmental surveillance into routine services.

From a training perspective, 33 psychology trainees (32 females and one male) engaged in inpatient services, reporting gains in counseling, developmental knowledge, and addressing parental stress. Outpatient involvement by four trainees enhanced skills in research presentation, early detection of developmental delay (a challenge in the context), and understanding the parent–child relationship. Importantly, the intervention was sustained within existing hospital structures and required minimal external resources, demonstrating both feasibility and scalability. These outcomes extended beyond families and trainees to the overall service experience, with play-based care identified as a significant driver of family experience (Rasheed et al., 2021a). Based on the reading of case study 2, the prompts in Box 6 can be used to seek answers.

Box 6 Responsive playful parenting programs

What is being resisted/refused? The framing of ECD services as only for "the poor" and the reliance on undervalued community health workers.

How does this case reveal the influence of donor-driven global agendas on what counts as "legitimate" ECD interventions? In what ways does embedding play into mainstream pediatric services to disrupt those global hierarchies?

Why does this matter for decolonial IS? It expands who is regarded as a legitimate recipient of interventions, challenges donor-driven targeting logics, and builds sustainable systems from within.

Donor logics often assume that only "the poor" deserve early care. How does this targeting naturalize inequities, and how does including middle-class families in this intervention unsettle that assumption?

What structural determinants (e.g., class assumptions, donor conditionalities, undervaluing of local expertise) shaped both the problem and the solution? How do these determinants reproduce inequities in global health?

What is being reimagined? Integration of play-based interventions into mainstream pediatric OPD and inpatient services, engaging the middle class, and positioning psychology trainees as valued contributors.

The intervention didn't just add activities—it altered service structures (training psychology trainees, embedding care in pediatrics, prioritizing caregiver dignity). How can this model inform the design of structural interventions elsewhere?

If you were part of the team implementing this program, what blind spots or privileges might shape your approach? What practices of reflexivity would help ensure accountability to families rather than donors?

A comparison between "business-as-usual" ECD funding and this approach highlights how donor-driven ECD programs often prioritize visibility, scalability, and cost-efficiency over a long-term system fit. In contrast, integrating pediatric psychology within routine OPD and inpatient care demonstrates feasibility, high acceptability, and sustainability by embedding services into existing structures and strengthening local capacity. While such approaches are frequently dismissed as "unfundable" because they do not align with narrow deficit-based frameworks (that caregiving is not sufficient, hence children have developmental deficits), the evidence here shows that they expand equity, reduce parental stress, and build professional skills that endure beyond project cycles (Scheidecker et al., 2025).

From a decolonial IS lens, this shift matters as it resists the framing of ECD as charity for the poor and instead reimagines it as a universal service embedded within health systems. It opens possibilities for valuing human relationships, professional growth, and middle-class engagement in ways that donor logics often overlook. In doing so, it shows how IS—freed from colonial assumptions—can help us determine not just what is fundable, but what is truly transformative (Table 9.2).

Table 9.2 Comparing "business-as-usual" ECD vs. integrated pediatric psychology approach.

Dimension	Business-as-usual ECD	Integrated pediatric psychology approach	Why it's deemed 'unfundable'	Strength of this approach
Target population	Poor, rural, marginalized families (donor-driven targeting).	All families in hospital OPD and inpatient care, including middle-class.	Middle-class not seen as "at risk".	Expands reach and equity → >70% uptake by middle-class families who rarely engage in ECD services.
Delivery agent	Community health workers, undervalued and underpaid.	Psychology trainees, supervised by clinical psychologists.	Seen as "too costly/professionalized".	Builds professional workforce; 25 trainees gained clinical skills; high feasibility and low dropout.
Service location	Public health systems (preferred for service funding).	Integrated into existing well-reputed hospital system (preferred for research funding).	Channeling resources through under-resourced public services affords them greater influence over systems and policy.	Embedding in well-reputed health system → >80% families returned for follow-up, showing trust and continuity.

Dimension	Business-as-usual ECD	Integrated pediatric psychology approach	Why it's deemed 'unfundable'	Strength of this approach
Framing of need	"Deficit model" → lack of stimulation = poor GDP outcomes.	Holistic: addresses parental stress, family well-being, developmental growth.	Doesn't fit narrow economic logic.	Parents reported reduced stress and stronger bonding; families asked for continued services.
Intervention type	Standardized play kits, group sessions.	Individualized counseling, screening, bedside play sessions (inpatients more intensive).	Not easily branded/ scaled.	Flexible to family needs; clinicians noted better parent–child interactions in wards.
Evaluation focus	Cognitive test gains, cost-effectiveness (Rasheed et al., 2023).	Feasibility, family satisfaction, system fit and feasibility, family satisfaction, system fit.	Funders demand "hard outcomes" i.e., quantifiable and measurable (Cook-Lundgren and Girei, 2025).	90% satisfaction; >60% families recommended service to peers; pediatricians requested expansion.
Sustain-ability	Ends when donor project ends.	Builds local workforce and integrates into pediatric care.	Workforce development seen as slow.	Sustainable skills transfer; trainees continue service beyond project period.

Practitioners as agents of just futures

Medical education is future-oriented training as it equips trainees with the necessary knowledge and skills to address broader societal challenges associated with the promotion of health and wellbeing (Kligyte et al., 2025). Decolonial IS allows us to operationalize equity, not just in intervention delivery, but in how we relate, collaborate, and build trust among those driving changes. To move toward decolonization, IS must ask harder and more humane questions, for example, how does an individual lead in a way that builds the types of communities that others, especially emerging leaders, can trust? What traits, emotional skills, and commitments are required to lead without dominating? How do we inspire others to hold on to a long-term vision of genuine change? And how do we help people resist trading their values for short-term rewards?

Through the transdisciplinarity embodied in decolonial IS, practitioners can be trained to ground their practice in "critical reflection, intellectual humility, and integrative thinking," as they navigate emerging challenges such as pandemics and climate change, while seeking to transform our societies (Kligyte et al., 2025). If trained to observe the link between the patients they manage in clinical settings, the diseases commonly found in their communities, and the social drivers shaping health (local realities vs. large structures) (Piñones-Rivera et al., 2024), they would then be able to provide novel solutions (Krishnamurthy et al., 2023). Decolonial IS training, therefore, becomes a justice-making tool, a way through which the capability to resist social, economic, and political systems that are responsible for poor health and well-being can be strengthened (Bradley et al., 2024).

Conclusions

Placed in context with the other chapters in this book volume, the argument becomes clearer, namely that coloniality in global health is not only sustained through structures and systems, but also through what we allow ourselves to imagine as being possible. Building on the call to reclaim imagination in global health through systems rooted in dignity and sovereignty, this chapter extends that call into the domain of training and practice. We believe decolonial IS equips future practitioners with the tools and reflexivity needed to translate imagination into architecture, into interventions, systems, and futures that refuse to reproduce colonial logic. The case studies from Zambia and Pakistan demonstrate what this refusal looks like in practice—reimagining HIV work through community-based leadership and rethinking child development interventions by engaging middle-class families rather than defaulting to "the vulnerable poor." Each, in its own way, disrupts dominant global health scripts and insists on different possibilities. Taken together, we assert that dignity and justice must be cultivated simultaneously, i.e., both in the way we design systems and in how we educate those who shape and transform them.

Acknowledgments

Dr. Ramya Kumar was supported by the Fogarty International Center of the National Institutes of Health (NIH) under Award Number D43TW009343 and the University of California Global Health Institute (UCGHI). The content is solely the responsibility of the authors and does not necessarily represent the official views of the NIH or UCGHI.

Glossary

Decolonial implementation science (IS) An approach to IS that centers local knowledge, histories, and priorities, while challenging colonial legacies in health interventions. Example: designing HIV prevention strategies with community leaders instead of importing external models.

Downstream drift The tendency to focus on individual behaviors or biomedical fixes rather than addressing the structural drivers of health. Example: Offering diet counseling instead of addressing food insecurity through policy.

Epistemic justice Fair recognition of diverse forms of knowledge, especially those marginalized by dominant systems. Example: Treating the lived experiences of patients and communities as valid sources of evidence.

Pathogenic social structures Social arrangements that systematically produce illness or inequities. Example: Housing policies that lead to overcrowding and higher rates of tuberculosis.

Plural knowledge The recognition that multiple forms of knowledge (scientific, experiential, cultural) are valuable in shaping health interventions. Example: Integrating biomedical data on PrEP with sex workers' strategies for safe daily use.

Reflexivity The practice of critically examining one's own assumptions, biases, and positionality in relation to communities and knowledge. Example: A clinician reflecting on how their social class or professional authority shapes patient trust.

Structural competency The skill of recognizing and responding to how social, political, and economic systems affect health. Example: Understanding high asthma rates as being linked to housing and pollution, not only genetics.

Structural humility A self-reflexive approach that involves unlearning stereotypes, addressing power imbalances, and engaging with communities as partners. Example: Providers listening to community members as co-producers of knowledge rather than as passive recipients of care.

Structural ignorance The neglect of systemic drivers of health when training and practice focus only on biology or behavior. Example: Treating obesity as a dietary failure while ignoring poverty and food deserts.

Upstream determinants of health The structural conditions that shape health outcomes, such as governance, education, housing, and the economy. Example: Regulating air pollution to reduce community-wide respiratory illness.

References

Bambra, C., Lynch, J., Smith, K.E., 2025. Introduction: the holy grail of reducing health inequalities. In: Getting Better. Policy Press. pp. 1–14. http://Users/patriciamaritim/Downloads/978-1-4473-7288-2-9781447372882%20(1).pdf.

Belone, L., et al., 2016. Community-based participatory research conceptual model: community partner consultation and face validity. Qual. Health Res. 26 (1), 117–135.

Bosch, T., 2025. Decolonisation is not a vibe: on anti-capitalist praxis, citation politics and epistemic refusal. Media Cult. Soc. 47 (8), 1714–1724. doi:10.1177/01634437251360382.

Boulton, R., Sandall, J., Sevdalis, N., 2020. The cultural politics of "Implementation Science". J. Med. Humanit. 41, 379–394.

Bradley, C.D., Irie, W.C., Geng, E.H., 2024. Situating implementation science (IS) in res (IS) tance: a conceptual frame toward the integration of scholarship from the black radical tradition. Front. Public Health. 11, 1286156.

Brownson, R.C., et al., 2022. Revisiting concepts of evidence in implementation science. Implement. Sci. 17 (1), 1–25.

Burgess, R.A., 2022. Working in the wake: transformative global health in an imperfect world. BMJ Glob. Health 7 (9), e010520.

Buse, K., 2025. Thinking and acting politically for effective policies for physical activity. J. Phys. Act. Health 22 (5), 525–530.
Butler, P., 2021. Critical Black Futures: Speculative Theories and Explorations, Springer Nature.
Cook-Lundgren, E., Girei, E., 2025. Ethics of quantification and randomised control trials in international development: a decolonial analysis. J. Bus. Ethics 196 (2), 241–254.
D'Ambruoso, L., et al., 2025. Redefining implementation science for global health decolonisation. Lancet. 405, 439.
Dada, D., et al., 2025. Epistemic resistance in global health implementation science: considerations for trainees.
Eccles, M.P., Mittman, B.S., 2006. Welcome to implementation science. Implement. Sci. 1 (1), doi:10.1186/1748-5908-1-1.
Engebretsen, E., Greenhalgh, T., Han, P.K. , 2025. Educating healthcare students in the Sustainable Development Goals: from translational science to translational humanities. Med. Humanit. 51 (4), 428–434. doi:10.1136/medhum-2025-013292.
Feukeu, K.E., 2024. A pluriversal definition of Futures Studies: critical futures studies from margin to centre. In: Poli, R. (Ed.), Handbook of Futures Studies. Edward Elgar Publishing. pp. 80–97.
Gaias, L.M., et al., 2021. Adapting strategies to promote implementation reach and equity (ASPIRE) in school mental health services. Psychol. Sch. 59 (12), 2471–2485.
Granados, C., Prabhu, J., 2022. Making the affordable aspirational: increasing the adoption of frugal innovations at the base of the pyramid. Stanford Soc. Innov. Rev. 15.
Harvey, M., et al., 2022. Structural competency and global health education. Glob. Public Health. 17 (3), 341–362. doi:10.1080/17441692.2020.1864751.
Juckett, L.A., et al., 2022. Leveraging academic initiatives to advance implementation practice: a scoping review of capacity building interventions. Implement. Sci. 17 (1), 49.
Karatekin, C., et al., 2024. Re-politicizing the WHO's social determinants of health framework. Health Promot. Int. 39 (5), daae122.
Khasnabish, A., 2020. Ecologies of the radical imagination. Inform. Commun. Soc. 23 (12), 1718–1727.
Kirchner, J.E., Bauer, M.S., 2024. Fundamentals of implementation science. In: Nilsen, P. (Ed.), Implementation Science: Theory and Application. Routledge. pp. 22–32.
Kligyte, G., Melvold, J., Pratt, S., Bowdler, B., 2025. Educating for sustainable and equitable futures: A transdisciplinary future-making capability framework. Futures 173, 103665. doi:10.1016/j.futures.2025.103665.
Krieger, N., 2024. Epidemiology and the People's Health: Theory and Context, Oxford University Press.
Krishnamurthy, S., Soltany, K.A., Montez, K., 2023. Incorporating health policy and advocacy curricula into undergraduate medical education in the United States. J. Med. Educ. Curric. Dev. 10. doi:10.1177/23821205231191601.
Kumar, R., et al., 2020. Zambia field epidemiology training program: strengthening health security through workforce development. Pan Afr. Med. J. 36, 323. doi:10.11604/pamj.2020.36.323.20917.
Kumar, R., et al., 2024a. Designing an operational research TB training program in Zambia. Public Health Action. 14 (1), 26–29.
Kumar, R., et al., 2024b. Mixed-methods protocol for the WiSSPr study: women in Sex work, Stigma and psychosocial barriers to pre-exposure prophylaxis in Zambia. BMJ Open 14 (9), e080218.

Kumar, R., et al., 2025a. Beyond the pill: understanding barriers and enablers to oral and long-acting injectable PrEP among women in sex work in Zambia. PLOS Glob. Public Health. 5 (6), e0004461.

Kumar, R., et al, 2025b. Behind the paper: beyond the pill: understanding barriers and enablers to oral and long-acting injectable PrEP among women in sex work in Zambia. PLOS Blogs Speak. Med. Health.

Leppin, A.L., et al., 2021. Teaching for implementation: a framework for building implementation research and practice capacity within the translational science workforce. J. Clin. Transl. Sci. 5 (1), e147. doi:10.1017/cts.2021.809.

Lucero, J.E., et al., 2020. Engage for equity: the role of trust and synergy in community-based participatory research. Health Educ. Behav. 47 (3), 372–379.

Lynch, J., 2020. Regimes of Inequality: The Political Economy of Health and Wealth, Cambridge University Press.

Lynch, J., 2023. The political economy of health: bringing political science in. Annu. Rev. Polit. Sci. 26, 389–410.

Maritim, P., et al., 2024. Lost in translation: key lessons from conducting dissemination and implementation science in Zambia. Implement. Sci. Commun. 5 (1), 121.

May, C.R., et al., 2023. Complex interventions and service innovations: development and application of the COMPLETE framework for patient-centered and justice-oriented design.

McCartney, G., et al., 2019. Impact of political economy on population health: a systematic review of reviews. Am. J. Public Health. 109 (6), e1–e12.

McDonagh, L.K., et al., 2018. Application of the COM-B model to barriers and facilitators to chlamydia testing in general practice for young people and primary care practitioners: a systematic review. Implement. Sci. 13 (1), 130.

McNamara, C., Bambra, C., 2025. The Global Polycrisis and Health Inequalities. Int. J. Soc. Determ. Health Health Serv. 55 (3), 238–248.

McNamara, C., 2025. Conceptualizing the social determinants of health: the value of a global political economy of health perspective. In: Bryant, T. (Ed.), Handbook on the Social Determinants of Health .Edward Elgar Publishing. pp. 27–34.

Medvedyuk, S., Raphael, D., 2025. Historical perspectives. In: Bryant, T. (ed.), Handbook on the Social Determinants of Health. Edward Elgar Publishing. pp. 13–26.

Metz, A., Kainz, K., Boaz, A., 2023. Intervening for sustainable change: tailoring strategies to align with values and principles of communities. Front. Health Serv. 2, 959386.

Michie, S., et al., 2008. From theory to intervention: mapping theoretically derived behavioural determinants to behaviour change techniques. Appl. Psychol. 57 (4), 660–680.

Moore, J., Khan, S., 2024. A critique of implementation science. In: Nilsen, P. (Ed.), Implementation Science: Theory and Application. Routledge. pp. 251–260.

Naidu, T., 2021. Modern medicine is a colonial artifact: introducing decoloniality to medical education research. Acad. Med. 96 (11S), S9–S12.

Naidu, T., Abimbola, S., 2022. How medical education holds back health equity. Lancet. 400 (10352), 556–557.

Ndlovu-Gatsheni, S., 2018. Epistemic Freedom in Africa: Deprovincialization and Decolonization, Routledge.

Nilsen, P., Birken, S., 2020. Overview of theories, models and frameworks in implementation science. In: Nilsen, P., Birken, S.A. (Eds.), Handbook of Implementation Science. pp. 8–31.

Nunes, J., 2020. The everyday political economy of health: community health workers and the response to the 2015 Zika outbreak in Brazil. Rev. Int. Polit. Econ. 27 (1), 146–166.

Piñones-Rivera, C., Holmes, S.M., Morse, M., Ferrall, J., Nambiar, K., Martínez-Hernáez, Á., 2024. Structural competency in global perspective. Glob. Public Health 19 (1). doi:10.1080/17441692.2024.2326631.

Primrose, D., Loeppky, R., 2024. Revitalizing the political economy of health and healthcare in a context of crisis. In: Primrose, D., Loeppky, R., Chang, R. (Eds.), The Routledge Handbook of the Political Economy of Health and Healthcare. Taylor & Francis. pp. 1–16.

Rabaka, R., 2009. Africana Critical Theory: Reconstructing the Black Radical Tradition, from WEB Du Bois and CLR James to Frantz Fanon and Amilcar Cabral, Lexington Books.

Rabaka, R., 2014. Concepts of Cabralism: Amilcar Cabral and Africana Critical Theory, Lexington Books.

Raphael, D., Bryant, T., 2022. Emerging themes in social determinants of health theory and research. Int. J. Health Serv. 52 (4), 428–432.

Rasheed, M.A., 2022. Scaling nurturing care interventions in the health sector: a theory of change perspective. Front. Public Health. 10, 903342.

Rasheed, M.A., et al., 2023. The predictive validity of bayley scales of infant and toddler development-III at 2 years for later general abilities: findings from a rural, disadvantaged cohort in Pakistan. PLOS Glob. Public Health 3 (1), e0001485.

Rasheed, M.A., Bharuchi, V., et al., 2021b. Development and feasibility testing of a play-based psychosocial intervention for reduced patient stress in a pediatric care setting: experiences from Pakistan. Pilot Feasibility Stud. 7 (1), 63.

Rasheed, M.A., Kedzierski, J.T., Hasan, B.S., 2021a. Improved family experience outcomes in a pediatric hospital in Pakistan: mentoring, human-centered practice, and theory of change. NEJM Catal. Innov. Care Deliv. 2 (7).

Rasheed, M.A., Mughis, W., et al., 2021c. A family-centered intervention to monitor children's development in a pediatric outpatient setting: design and feasibility testing. Front. Health Serv. 1, 739655.

Razack, S., Richardson, L., Pillay, S.R., 2025. The violence of curriculum: dismantling systemic racism, colonisation and indigenous erasure within medical education. Med. Educ. 59 (1), 114–123.

Reynolds, M.M., 2021. Health power resources theory: a relational approach to the study of health inequalities. J. Health Soc. Behav. 62 (4), 493–511.

Scheidecker, G., et al., 2025. Beyond deficit models in early childhood development science. Lancet 405 (10495), 2127.

Sharpe, C., 2016. In the Wake: On Blackness and Being, Duke University Press.

Sinitiere, P.L., 2021. The black futures of WEB Du Bois. In: Butler, P. (Ed.), Critical Black Futures: Speculative Theories and Explorations. Springer. pp. 19–36.

Snell-Rood, C., et al., 2021. Advancing health equity through a theoretically critical implementation science. Transl. Behav. Med. 11 (8), 1617–1625. doi:10.1093/tbm/ibab008.

Tamale, S., 2020. Decolonization and Afro-Feminism. Daraja Press, Ottawa. p. 278.

TEDxLusaka, 2025. https://www.youtube.com/watch?v=KoIXWFtHjhc&t=21s.

Vogel, B., Dhungana, N., Džuverović, N., Khanal, N.A., Zapata, M.L., Fast, L., 2025. Caught in the middle: local researchers' experience of mundane ethical harms in crisis settings. Res. Ethics. doi:10.1177/17470161251386531.

von Thiele Schwarz, U., Aarons, G.A., Hasson, H., 2019. The value equation: three complementary propositions for reconciling fidelity and adaptation in evidence-based practice implementation. BMC Health Serv. Res. 19, 1–10.

Wallerstein, N.B., Duran, B., 2006. Using community-based participatory research to address health disparities. Health Promot. Pract. 7 (3), 312–323.

Wallerstein, N., Duran, B., 2010. Community-based participatory research contributions to intervention research: the intersection of science and practice to improve health equity. Am. J. Public Health. 100 (S1), S40–S46.

World Health Organization, 2025. World Report on Social Determinants of Health Equity. World Health Organization, Geneva.

CHAPTER 10

Integrating indigenous knowledge into the health sciences curriculum: A path toward decolonial healing and holistic care

Mabitja Moeta

Faculty of Health Sciences, University of Johannesburg, Johannesburg, South Africa

Introduction

The health sciences curriculum across Africa and globally has been profoundly shaped by colonial histories that privileged Euro-Western biomedical paradigms while systematically excluding, marginalizing, or delegitimizing indigenous knowledge systems (IKS). This exclusion is not an innocent oversight; it is a manifestation of epistemic violence, which involves the silencing of other ways of knowing, of alternative ontologies of health and healing, and of lived practices embedded in communities. Hence, the dominant curricula were designed to sustain biomedical hegemony, thereby producing graduates whose training is largely disconnected from the social, cultural, and historical contexts in which they will practice. The aim of this chapter is to interrogate this marginalization, critically examine the importance of integrating IKS into health sciences education, and propose pathways for curriculum transformation that move beyond tokenistic inclusion toward genuine epistemic justice and decolonial healing.

The colonial legacy of medical education

Colonialism was not only a political and economic project; it was also an epistemic one. Western biomedical science, framed as universal and objective, was imposed as the singular valid approach to health, while African and other indigenous systems of medicine were dismissed as unscientific, superstitious, or primitive (Harding, 2023). Such framings have persisted in curricula long after the end of formal colonial rule. The reproduction of biomedical dominance in health sciences education perpetuates a system where medical students and health professionals are trained to ignore,

Decolonizing Medical Education, Research and Practice. https://doi.org/10.1016/B978-0-443-44998-7.00012-0

undervalue, or actively oppose community-based forms of healing that continue to hold deep significance for the populations that they serve.

This legacy has modern-day implications as many patients in African contexts engage in plural health-seeking behaviors, drawing simultaneously on biomedical and indigenous healing systems. However, health professionals are often ill-prepared to navigate this complexity due to their training and curricula that delegitimize non-biomedical epistemologies (Chinouya and Reynolds, 2022). The result is not only a gap in clinical competence but also an erosion of trust between communities and health care providers.

Benefits of integration

Integrating IKS into health sciences curricula is not merely about adding cultural sensitivity modules or acknowledging traditional health practitioners (THPs) in passing. It is more complex and includes restructuring knowledge hierarchies to ensure that students are exposed to diverse epistemologies of health and illness, better preparing them to deliver care that is both scientifically rigorous and culturally responsive. This is especially critical in the African context, where indigenous healing practices continue to play a central role in community health.

Integration is also a matter of justice. As scholars of epistemic justice argue, excluding indigenous knowers from knowledge production and validation processes wrongs them in their capacity as knowers (Fricker, 2007; Asea, 2022). By failing to recognize the legitimacy of IKS, medical education perpetuates historical injustices and compromises its own relevance. A transformed curriculum that honors diverse ways of knowing can foster a new generation of health professionals who bridge divides between biomedical and indigenous systems in ways that respect the dignity and agency of all knowledge holders.

Key concepts

To ground this discussion, it is important to clarify some of the key concepts that underpin the chapter. *Epistemology* refers to the study of knowledge: how it is defined, validated, and transmitted (Steup et al., 2021). In health sciences, the epistemological dominance of biomedicine has led to the exclusion of alternative frameworks of healing, such as African IKS. In contrast, *ontology* (the study of being and existence) shapes how health and illness are understood (Smith, 2020). For example, biomedical frameworks often adopt a mechanistic ontology of the body, whereas indigenous systems may view health holistically as a balance of spiritual, social, and physical forces (Ntshoe, 2020).

Another important concept is curriculum transformation, which refers to the deliberate process of reshaping educational content and pedagogy to reflect inclusivity, relevance, and social justice (Luckett, 2016). Here, cultural competence becomes an essential skill for students and practitioners, enabling them to engage effectively with patients from diverse cultural backgrounds (Msomi and Barath, 2025). In addition, we draw attention to the notion of *scientific racism,* which refers to the misuse of scientific claims to justify racial hierarchies, a clear reminder of the dangers of allowing curricula to reproduce discriminatory assumptions under the guise of objectivity (Dordunoo et al., 2022). Other fundamental concepts are added in the glossary, which should help the reader gain a clearer understanding of this chapter.

Positioning this chapter

This chapter positions itself at the intersection of history, philosophy, and pedagogy. It seeks not only to critique but also to offer constructive models for integration. By drawing on African philosophies such as Ubuntu (focus: becoming human through others) and Botho (focus: being a good human being), as well as decolonial frameworks of epistemic justice, the chapter aims to articulate a vision of health sciences education that is ethically grounded and contextually relevant. In addition, case studies of collaborative teaching ventures between THPs and medical educators are presented to strengthen our arguments. Such examples also reveal challenges encountered in terms of such integration. Ethical considerations, such as intellectual property and cultural protocols, are also addressed, highlighting the need for respectful and equitable co-production of knowledge.

This introduction thus sets the stage for a deeper exploration of how historical legacies of colonialism continue to shape medical education, the barriers to integrating IKS into existing curricula, and practical strategies for curriculum transformation. The chapter argues that genuine integration is not optional but imperative for justice, cultural responsiveness, and the future of health care in Africa and beyond.

Theoretical and historical background

Colonial epistemologies in health sciences

The dominance of Western biomedicine in health sciences education has deep historical roots in colonialism. Colonial authorities did not simply govern politically and economically; they also shaped epistemic frameworks by positioning Western science as the only legitimate form of knowledge. Thus indigenous forms of healing, long practiced in Africa, Asia, and the Americas, were delegitimized, criminalized, or relegated to the margins. For example, missionary medicine, introduced in the 19th and early 20th centuries, often framed traditional health practices as obstacles to Christian "civilization" and biomedical progress (Comaroff and Comaroff, 1991).

Focusing on South Africa, the 1928 Medical, Dental and Pharmacy Act and subsequent legislation, such as the Witchcraft Suppression Act (1957), entrenched biomedicine as the dominant health care system, excluding THPs from formal recognition for decades (Edwards, 2020). Such regulatory frameworks were underpinned by epistemological hierarchies that equated "science" with objectivity and universality, while framing indigenous knowledge as anecdotal, spiritual, or irrational. These hierarchies persist in contemporary curricula, where medical students are trained almost exclusively within biomedical paradigms.

Scientific racism and medical education

The rise of scientific racism during the 18th and 19th centuries (as is discussed in the previous chapters in this book volume) entrenched the marginalization of IKS, with pseudoscientific practices such as craniometry, eugenics, and racialized clinical practices used to justify colonial domination and slavery by promoting false claims of inherent biological differences between so-called races (Gould, 1996; Dordunoo et al., 2022). These ideas infiltrated medical curricula, shaping research agendas and clinical practices in ways that persist today. For example, African bodies were historically portrayed as biologically inferior, resistant to certain diseases, or predisposed to others. Such narratives have now been widely discredited, although some still echo in medical education and practice.

Fragments of scientific racism can still manifest (more subtly) in curricula and research. For example, race-adjusted diagnostic algorithms/tools previously adjusted for "Black" patients demonstrate how biomedical science historically incorporated race as a biological category rather than a sociopolitical construct (Vyas et al., 2020). Such practices perpetuate inequities and undermine trust in medical institutions.

African philosophies as alternative epistemologies

In contrast to Western biomedical epistemology, African philosophies provide rich alternative frameworks for understanding health and healing. Ubuntu, often expressed as *umuntu ng'umuntu ngabantu* ("a person is a person through other people"), emphasizes relationality, interdependence, and collective wellbeing (Ramose, 2002). Within this worldview, health is not merely the absence of disease but the presence of harmony between individuals, communities, ancestors, and the environment.

Ubuntu/Botho highlights ethical responsibility and respect as essential to human flourishing. These philosophies underpin indigenous healing practices, where illness is understood not only in physical terms but also as a disturbance of social or spiritual balance (Mokgoro, 1998). Healing practices thus involve rituals, herbs, divination, and counseling aimed at restoring equilibrium rather than simply treating symptoms. Framing IKS within African philosophies demonstrates their epistemological coherence. Far from being "unscientific," these systems are grounded in ontologies and epistemologies that are internally consistent and pragmatically effective in their contexts. Hence, recognizing them as legitimate knowledge systems is the first step toward successful integration into curricula.

Decoloniality and epistemic justice as frameworks

Decolonial theory provides critical tools for analyzing and challenging the persistence of colonial epistemologies. Decoloniality, as articulated by scholars such as Mignolo (2011) and Ndlovu-Gatsheni (2018), involves dismantling the coloniality of power, knowledge, and being that continue to shape postcolonial societies. In health sciences education, decoloniality involves a critical examination of whose knowledge is taught and privileged, whose bodies are studied, and whose voices are allowed to shape the narrative.

Epistemic justice is a concept developed by Fricker (2007) and was expanded in African contexts by Asea (2022), emphasizing the ethical dimension of knowledge production. It recognizes that marginalizing certain groups as knowers is not only intellectually limiting but also morally wrong. Integrating IKS into curricula is therefore not an act of charity or cultural sensitivity, but instead a demand for justice.

Critical pedagogy as inspired by Freire (1970) also adds to this framework. Freire argued that education should not reproduce oppressive structures but should empower learners to question and transform their realities. In the context of medical education, critical pedagogy suggests that students should be equipped to interrogate biomedical dominance and engage meaningfully with diverse epistemologies.

Persistent effects in contemporary health education

Despite policy shifts in many African countries regarding the recognition of THPs (e.g., South Africa's Traditional Health Practitioners Act [2007]), the integration of IKS into health sciences curricula remains minimal (Chitindingu et al., 2014). Most medical and nursing students complete their degrees without structured exposure to indigenous healing systems, leaving them ill-equipped to engage with patients who rely on both systems. The consequences are substantial, as patients may conceal their use of traditional medicine from biomedical practitioners, fearing judgment or dismissal, which can result in harmful drug-herb interactions and/or delayed diagnoses (Chinouya and Reynolds, 2022). Conversely, biomedical practitioners may fail to build trust with communities if they disregard culturally embedded healing practices. This epistemic gap undermines both clinical outcomes and the ethical responsibility of health care to be contextually relevant.

Towards a pluralist knowledge framework

Theoretical debates increasingly call for a pluralist framework that does not replace biomedicine with IKS but situates them in dialog. Such a framework acknowledges the strengths and limitations of each system. Biomedicine offers advanced diagnostic and surgical technologies, while IKS provides holistic, culturally resonant approaches to healing that address psychosocial and spiritual dimensions often neglected in clinical care. *Pluralism* requires more than coexistence; it demands structured engagement in curricula. Students should learn not only the pharmacological properties of medicinal plants but also the cultural contexts of their use. They should study the ethical principles governing traditional healing, such as respect for ancestors and communal responsibility, alongside biomedical ethics. This pluralist orientation fosters *cultural humility*, which is an ongoing process of self-reflection and openness to other worldviews (Ogunyemi et al., 2024).

Challenges in integrating indigenous knowledge

While the recognition of IKS as legitimate forms of healing and knowledge has gained momentum, translating this recognition into practical integration within health sciences curricula faces significant challenges. These challenges are not merely logistical but are deeply rooted in epistemological, institutional, cultural, and political dynamics. This section critically examines the barriers to integration, highlighting how entrenched hierarchies of knowledge, institutional resistance, ethical concerns, and systemic inequities continue to impede meaningful transformation.

Epistemological hierarchies

At the heart of the challenge is the persistence of epistemological hierarchies that privilege Western biomedical knowledge over indigenous systems. These hierarchies stem from a positivist epistemology that defines valid knowledge as that which can be empirically tested and generalized. Indigenous systems, which often include spiritual, relational, and experiential dimensions, are therefore dismissed as unreliable.

This binary framing ignores the fact that all knowledge systems are shaped by cultural contexts and ontological assumptions. As Ntshoe (2020) argues, biomedical science itself is underpinned by a mechanistic ontology of the body, privileging reductionism and objectivity, yet it is presented as universal. By contrast, IKS often adopts a relational ontology, regarding health as a balance between physical, social, and spiritual dimensions. The inability of curricula to accommodate such plurality reflects a rigid epistemological gatekeeping that impedes integration.

Institutional resistance and structural barriers

Even where policymakers acknowledge the value of IKS, academic institutions often resist transformation. Universities, especially those historically modeled on Western traditions, remain structured around disciplinary silos, accreditation requirements, and performance metrics tied to global biomedical standards (Moeta et al., 2019). Faculty members trained exclusively in biomedical traditions may lack exposure to or an appreciation for indigenous systems, leading to resistance or skepticism when proposals for integration arise (Asea, 2022). Furthermore, institutional disinterest is compounded by accreditation and regulation frameworks. Health professions councils that are responsible for maintaining "standards" often default to Western benchmarks. As a result, even when educators are willing to innovate, they may face obstacles in terms of securing approval and recognition by such accreditation bodies. The risk of being perceived as lowering standards or compromising scientific rigor discourages experimentation with the integration of IKS into contemporary biomedical and medical curricula.

Lack of champions and capacity constraints

Another barrier is the limited number of champions advocating for IKS within academia (Moeta et al., 2019). Integration efforts often rely on a small group of dedicated scholars and practitioners who face institutional pushback. Without broader support, these initiatives remain isolated and vulnerable to discontinuation when leadership changes.

Capacity is also a concern, whereby indigenous practitioners are often excluded from academic spaces due to credentialing requirements that privilege formal university qualifications over lived expertise. This exclusion deprives students of the opportunity to learn directly from knowledge holders. Moreover, there is a shortage of structured training programs that could equip academics with the necessary tools to teach indigenous content responsibly and rigorously (Chinouya and Reynolds, 2022).

Ethical and intellectual property concerns

Ethical concerns present another layer of difficulty. Indigenous knowledge is often considered communal property that is orally transmitted and protected by cultural protocols. When universities seek to integrate such knowledge, communities may fear appropriation, misrepresentation, or exploitation. The long history of colonial extraction, ranging from bioprospecting to anthropological research, has left legacies of mistrust. For example, the global commercialization of the Hoodia plant, traditionally used by the San people to suppress appetite, illustrates how indigenous knowledge can be extracted without fair benefit-sharing (Wynberg, 2017). The increased use of traditional herbs such as Moringa and Aloe vera during and after the COVID-19 pandemic illustrates the ongoing appropriation of indigenous medicinal knowledge, often without proper recognition or benefit-sharing with communities that have a longstanding tradition of using such remedies (Mpofu et al., 2021; WHO, 2022). These cases highlight the importance of suitable protocols and agreements that ensure communities retain agency over their knowledge and benefit from its use.

Intellectual property law is grounded in Western notions of individual ownership and is therefore ill-suited to protecting communal knowledge systems. This creates tension between academic imperatives to publish and the ethical need to respect indigenous ownership. Without clear frameworks for co-authorship, acknowledgment, and benefit-sharing, integration risks perpetuating epistemic exploitation rather than redressing it.

Marginalization of indigenous practitioners

Even in contexts where IKS are recognized at the policy level (e.g., South Africa's Traditional Health Practitioners Act [2007]), indigenous practitioners often remain marginalized in practice. Their roles are frequently limited to community health, with little formal collaboration with hospitals or academic teaching, and such marginalization reflects both structural inequities and professional prejudices. Medical professionals may perceive indigenous practitioners as competitors or threats to their professional authority. Students trained in biomedical curricula may internalize dismissive attitudes toward THPs, reinforcing stereotypes of irrationality or danger (Msomi and Barath, 2025). Such prejudices are not only epistemically unjust but also detrimental to patient care, as they discourage open dialog between patients and providers about plural health-seeking behaviors.

Curriculum design limitations

Another challenge lies in curriculum design itself, as most contemporary curricula are already overloaded with biomedical content, leaving minimal room for the addition of new course content. Faculty often argue that including IKS would require removing "essential" biomedical topics, reflecting implicit assumptions regarding what counts as indispensable knowledge. In addition, pedagogical models in medical education emphasize evidence-based medicine, clinical skills laboratories, and problem-based learning. While these are valuable, they leave limited room for narrative-based, experiential, or community-centered education and training that are central to teaching indigenous perspectives. Without rethinking pedagogy, attempts to integrate IKS risk becoming superficial by arranging an odd guest lecture or a token module, as opposed to a systemic transformation of the curriculum.

Case studies on the integration of IKS into health sciences

Policy momentum and programmatic models

At a global level, the World Health Organization (WHO) (2013, 2023) has consistently advocated for the integration of traditional and complementary medicine into national health systems and education. Its Traditional Medicine Strategy emphasizes training, regulation, and evidence generation as enablers for safe integration. Moreover, India's AYUSH system provides a notable example, where Ayurveda, Yoga, Unani, Siddha, and Homeopathy are formally integrated into universities and hospitals, supported by collaborative research (Patwardhan, 2014). Similarly, China has long maintained dual medical systems, where students are trained in both Traditional Chinese Medicine and biomedicine (Li et al., 2025). These global precedents demonstrate that integration is feasible when backed by strong policy frameworks and institutional commitment.

African universities and health systems have piloted different forms of collaboration. For example, Okyere-Manu and Ntlapo (2018) described how community dialogs with THPs in Lesotho created mutual respect and facilitated referrals for maternal health. Moreover, Kayombo et al. (2013) highlighted co-training models where biomedical staff taught THPs in Tanzania about hygiene and referral systems, while, in turn, THPs shared plant knowledge and cultural frameworks. Chatora (2022) argues that African universities should formalize such engagements into curricula to produce graduates who are competent in plural health systems. South African universities have also initiated models of co-creation in training and practice, for example:

University of Pretoria

We previously described an Imbizo approach (traditional public gathering in African societies) for curriculum integration, where THPs, students, and academics co-designed modules and assessment criteria (Moeta et al., 2019). The process was rooted in dialog and cultural protocol, ensuring THPs were treated as knowledge partners rather than guests. The University of Pretoria's research also shows that competency frameworks for African Traditional Health Knowledge can be developed collaboratively, balancing academic rigor with cultural respect. Furthermore, the university facilitated student visits to Indumba (healing spaces) and engagements with royal houses, where learners directly observed healing practices. These immersive visits enabled students to experience cultural rituals first-hand, demystifying practices often imitated in textbooks. Such experiential pedagogy reduced stigma and improved openness to plural health-seeking behaviors.

University of Limpopo

The University of Limpopo has convened conferences, bringing together THPs, biomedical researchers, and policymakers to advocate for integration. Nkuna and van der Merwe (2021) note that these forums have produced research collaborations, policy dialogs, and momentum for curriculum inclusion. Conferences serve as safe spaces for relationship building, allowing THPs and academics to co-identify research priorities and teaching content.

University of KwaZulu-Natal

The University of KwaZulu-Natal's (UKZN's) TradMed initiative, which was supported by the US President's Emergency Plan for AIDS Relief (PEPFAR), piloted modules where THPs co-teach medical students about human immunodeficiency virus (HIV) and cultural approaches to healing. Here, Leclerc-Madlala (2006) documented that early iterations faced mistrust, yet evaluations showed that students valued hearing from THPs alongside biomedical lecturers. Similarly, Gandugade (2016) established that co-teaching fosters respect, with students reporting increased confidence in engaging with patients using traditional medicines.

Sefako Makgatho Health Sciences University

Sefako Makgatho Health Sciences University (SMU) has invested in research exploring IKS integration into the spheres of child health care and pharmacology. Mokgobi (2014) noted that SMU collaborated with THP organizations in the northwestern parts of South Africa to assess the nature of herbal medicine usage. More recent examples include THP training programs where biomedical lecturers taught first aid and referral criteria, while THPs explained the culturally grounded ways of illness. Similarly, Masola and Maotoana (2025) initiated programs to facilitate cooperation between THPs and biomedical practitioners regarding mental health issues. These programs demonstrated that reciprocal training builds mutual trust and improves referral systems. Through these initiatives, SMU is contributing to the development of health care professionals who are knowledgeable and respectful of traditional medicine, facilitating a more inclusive and holistic approach to patient care.

Thus collectively these efforts reveal several practical mechanisms for the co-creation of curriculum remodeling in this context:

- *Mutual curriculum design and co-teaching:* Documented at the University of Pretoria (Moeta et al., 2019) and UKZN's TradMed project, where THPs and academics co-design learning outcomes, ensuring epistemic equity.
- *Student visits to Indumba*: At the University of Pretoria, immersive visits reduce stigma and broaden students' cultural competence.
- *Training for THPs:* The University of Limpopo and SMU examples (Nkuna and van der Merwe, 2021; Masola and Maotoana, 2025) show how biomedical skills (first aid, referral pathways) are taught, while THPs share ritual practices and herbal knowledge.
- *Joint research and ethical frameworks:* At the University of Pretoria (Peu et al., 2020), memoranda of understanding ensure co-authorship and prevent appropriation.
- *Referral pathways:* Evaluations in the KwaZulu-Natal region of South Africa (Leclerc-Madlala, 2006) show that structured THP–clinic collaboration increases adherence to HIV care.

Synthesis of case studies

These examples demonstrate that integration succeeds when it is institutional, relational, and ethical. Curriculum transformation must, therefore, include shared module design, immersive learning, reciprocal training, ethical frameworks, and referral linkages. Where these mechanisms are implemented, outcomes include improved cultural competence among students, stronger trust between communities and health professionals, and more contextually responsive health systems.

Power dynamics and tokenism

A recurring risk is that integration becomes tokenistic rather than transformative. Thus inviting a traditional healer for a guest lecture once a semester may create the appearance of inclusion without fundamentally shifting epistemological hierarchies. Similarly, reducing IKS to pharmacological discussions of medicinal plants risks stripping practices of their cultural and spiritual dimensions, reproducing colonial extraction. True integration requires shifting power relations in curriculum design and knowledge production. This means not only including IKS content but also recognizing indigenous practitioners as co-educators, reshaping pedagogy to value narrative and relational learning, and challenging students to critically reflect on their own epistemological assumptions. Without such shifts, integration risks reinforcing, rather than dismantling, epistemic hierarchies.

Political and funding contexts

Political and funding contexts also influence the feasibility of integration, as national health policies may rhetorically support IKS but fail to allocate sufficient resources for training, research, and collaboration. Donor-driven health programs often prioritize biomedical interventions aligned with global agendas (e.g., HIV, TB, malaria) over locally embedded practices. Universities dependent on international rankings may hesitate to invest in indigenous scholarship, fearing it will be underrated in metrics that privilege Western outputs.

Inference about the complex (but not impossible) task of integration

The challenges to integrating IKS in health sciences curricula are numerous and deeply rooted, as they reflect broader struggles around epistemology, power, and justice in higher education. Yet, recognizing these barriers is the first step toward addressing them, whilst overcoming them requires critical reflection, institutional commitment, and ethical engagement with indigenous communities.

As the next section will argue, curriculum transformation is not only possible but necessary. By developing frameworks that promote inclusivity, cultural humility, and co-production of knowledge, health sciences education can begin to dismantle epistemic hierarchies and equip future practitioners to deliver care that is both scientifically rigorous and culturally responsive.

Curriculum transformation—a framework for integration

Curriculum transformation is more than the superficial inclusion of indigenous concepts in isolated modules. It requires a systematic rethinking of what counts as valid knowledge, how knowledge is taught, and who teaches it. The goal is to create curricula that are epistemically just, contextually relevant, and ethically grounded. In formerly colonized countries and other postcolonial contexts, curriculum transformation is inseparable from broader debates on decolonization, social justice, and institutional reform. This section develops a framework for integration that highlights guiding principles, curricular analysis, models of integration, pedagogical strategies, and the role of policy and accreditation bodies.

Guiding principles for inclusive curriculum design

- *Epistemic plurality*
  Recognizing that no single knowledge system has a monopoly on truth, biomedicine and IKS should, therefore, be viewed as complementary rather than competitive.
- *Cultural humility*
  Unlike cultural competence which implies a level of mastery, cultural humility is an ongoing process of self-reflection and openness to learning from others (Tervalon and Murray-García, 1998).
- *Dialogic pedagogy*
  Knowledge should be co-constructed through dialog between educators, students, and indigenous practitioners (Moeta et al., 2019).
- *Ethical reciprocity*
  Integration must avoid appropriation by ensuring benefit-sharing, acknowledgment, and protection of indigenous intellectual property.
- *Policy alignment*
  Transformation efforts must align with national frameworks and global initiatives such as the WHO's Traditional Medicine Strategy (World Health Organization, 2013, 2023).

Curriculum analysis: Identifying gaps and silences

The first step in transformation is analyzing existing curricula to identify what is missing. Most South African medical and nursing curricula prioritize biomedical content and global evidence-based medicine, with little attention to cultural, spiritual, or community health practices. As Luckett (2016) argues, curriculum analysis should move beyond content audits to interrogate underlying epistemological assumptions: Whose voices are included? Whose are excluded? What knowledge is deemed essential, and why? For example, while pharmacology modules may cover drug–herb interactions, they rarely contextualize why patients use traditional medicines or how THPs understand disease causality. Likewise, ethics courses may focus on Western bioethics (autonomy, beneficence, nonmaleficence, justice) while ignoring African philosophies like Ubuntu and Botho, which focus on relationality and communal responsibility. Recognizing these silences allows educators to map areas for integration.

Models of integration

Several models can be considered for integrating IKS into health sciences curricula:

- *Additive model*—adding standalone modules on traditional medicine. While this raises awareness, it risks tokenism and marginalization.
- *Integrative model*—embedding indigenous perspectives across existing modules (e.g., teaching indigenous concepts of mental health in psychiatry,

herbal pharmacology in pharmacology, and rituals in medical anthropology). This normalizes plurality.
- *Transformative model*—reimagining curricula holistically by restructuring outcomes, pedagogy, and assessment to value multiple epistemologies. This requires co-teaching, community engagement, and systemic reform.

International examples suggest that integrative and transformative models are the most sustainable. For example, in China, dual training in Traditional Chinese Medicine and biomedicine is fully institutionalized (Xu and Yang, 2009), while in South Africa, UKZN's TradMed project offers a practical blueprint for embedding co-teaching and experiential learning across modules.

Pedagogical strategies

Pedagogy is where integration becomes real, as it focuses on the practical application. Here, various strategies become noteworthy in this regard:

- *Storytelling and narrative learning*: THPs often teach through stories that convey moral, spiritual, and practical lessons. Incorporating storytelling into medical training develops cultural sensitivity and narrative competence.
- *Case-based learning*: Case studies can illustrate plural health-seeking behaviors. For example, a patient with HIV may use antiretroviral therapy together with herbal remedies, prompting discussion of pharmacological, cultural, and ethical considerations.
- *Experiential learning and immersion*: Student visits to THP's practice sites allow them to observe indigenous practices in context. Such exposure fosters cultural humility and reduces stigma.
- *Co-teaching with THPs*: UKZN's TradMed program (Lawrence et al., 2021) demonstrates how co-teaching between academic staff and THPs allows students to access information directly from both knowledge traditions.
- *Simulation and role-play*: Simulated consultations, where students must navigate patients' varied health care choices, can develop communication and clinical reasoning skills.
- *Reciprocal training*: Providing THPs with biomedical training (first aid, referral pathways) while students learn about indigenous cosmologies creates mutual respect and shared learning.

The role of accreditation bodies and policy reform

Curriculum transformation cannot succeed without support from professional councils and accreditation bodies, as such entities ultimately determine what counts as a legitimate qualification. If such councils equate "quality" with the sole adherence to Western biomedical standards, the integration of IKS into curricula will remain marginal. Hence, the Health Professions Council of South Africa, the South African Nursing Council, and other councils in health care must develop policies that explicitly endorse pluralism in terms of curricula. This could involve requiring cultural competence and indigenous health knowledge as core competencies. National policy frameworks should also incentivize institutions to implement transformative models, while ensuring safeguards against exploitation and/or pseudoscience. This scenario, of course, should not only apply to South Africa but be considered at a global level. For example, accreditation bodies in India and China provide useful models to consider as they mandate dual training in traditional and biomedical systems, thereby institutionalizing integration. We argue that African accreditation councils could adapt these models to their specific contexts.

Recommendations for curriculum design

- Embed IKS across modules rather than confining it to electives.
- Institutionalize co-teaching arrangements with THPs, ensuring clear agreements on authorship, remuneration, and recognition.
- Develop immersive placements (e.g., Indumba visits) with cultural protocols and reciprocity agreements.
- Train educators in cultural humility and decolonial pedagogy.
- Engage accreditation bodies to ensure integration is endorsed at the national level.
- Evaluate outcomes rigorously to demonstrate their impact on student competence, community trust, and health outcomes.

Taken together, curriculum transformation is a moral, pedagogical, and institutional imperative. A framework for integration must rest on principles of epistemic plurality, cultural humility, dialogic pedagogy, and ethical reciprocity. By analyzing gaps, adopting integrative and transformative models, and implementing innovative pedagogies, health sciences education can move beyond tokenism toward genuine decolonial reform. Significantly, transformation must be institutionalized through policy and accreditation reform. Some of the case studies discussed here demonstrate that integration is not only possible but beneficial, yielding more culturally responsive graduates and stronger community trust.

Future directions and recommendations

The integration of IKS into health sciences curricula is no longer a question of "if" but of "how." The previous sections demonstrated both the moral imperative and the practical feasibility of such integration. Moving forward, the challenge lies in consolidating lessons from pilot initiatives, scaling them across institutions, and embedding them into national and international frameworks. This requires coordinated action at multiple levels: policy, pedagogy, research, and institutional culture.

National policy frameworks

Integration must be supported by clear national policy that results in successful implementation and practices at the ground level. For example, while South Africa enacted the Traditional Health Practitioners Act (2007) and the Protection, Promotion, Development and Management of Indigenous Knowledge Act (2019), these laws remain underutilized in shaping higher education. Thus we argue that the Ministries of Health and Higher Education should issue joint directives mandating the inclusion of IKS into existing curricula. Such directives should not prescribe a one-size-fits-all model but should set minimum standards; for example, all health sciences graduates should demonstrate competence in cultural humility, knowledge of local healing practices, and awareness of ethical issues regarding knowledge use. Such competencies could be benchmarked against global policies such as the WHO's Traditional Medicine Strategy (2023), which emphasizes education and capacity building.

Accreditation and professional councils

Although accreditation bodies are powerful levers of change, their current curricular focus reflects only biomedical priorities. Future directions must, therefore, include unambiguous recognition of IKS as a legitimate domain of health knowledge and could take the form of accreditation standards requiring curricula to demonstrate integration of indigenous perspectives. Furthermore, the development of competency frameworks that include indigenous epistemologies alongside biomedical outcomes could go a long way in the recognition of IKS. Lastly, the establishment of joint panels should be strongly encouraged to enable, for example, the inclusion of THPs in accreditation site visits and/or to manage curriculum reviews. By embedding indigenous content in accreditation, councils would move integration from the margins into the mainstream of higher education.

Training: Educators and THPs

A transformed curriculum requires transformed educators. Many university faculty members feel unprepared to teach indigenous perspectives, either due to a lack of exposure or fear of misrepresentation (Moeta et al., 2019). Professional development programs should therefore train educators in decolonial pedagogy, cultural humility, and ethical engagement with communities. In parallel, THPs should be supported to teach within universities. This does not mean coercion into Western credentialing systems but rather providing bridging programs to help orient them to academic contexts while respecting their epistemologies. The University of Pretoria's Imbizo approach (Moeta et al., 2019) demonstrates how such collaborations can be formalized without compromising cultural integrity.

Institutional reforms

Future directions should also involve addressing institutional cultures. Here, universities must recognize THPs as colleagues and not as outsiders. This requires formal contracts and/or honorary appointments for THPs involved in teaching, together with suitable remuneration that reflects their expertise. Shared governance mechanisms, such as curriculum committees (that include THP representatives), should ensure that reforms are not merely lip service but legitimate ones. Thus we argue that institutions should also invest in centers of excellence for indigenous health knowledge, housing archives, training spaces, and research hubs co-managed with healer associations. Such centers could serve as repositories of both oral and written knowledge that are protected under IKS legislation.

Research and knowledge production

Future research must move beyond documenting traditional practices toward evaluating models of integration and their impact. Areas of focus could include:

- *Pedagogical outcomes*: How do students exposed to co-teaching or Indumba visits perform in cultural competence compared to those without such exposure?
- *Health outcomes*: Does integration improve patient trust, adherence, or health-seeking behaviors?
- *Ethical governance:* Which models of co-authorship and benefit-sharing are most effective in preventing appropriation?

Collaborative research teams should include THPs as co-investigators, with funding allocated for community-based research infrastructure. Journals and funders must also adapt to recognize oral and experiential knowledge as legitimate contributions.

International collaboration and benchmarking

While integration must be locally grounded, international collaboration can provide valuable benchmarks; for example, countries such as China and India have demonstrated how dual systems can indeed be institutionalized (Xu and Yang, 2020; Patwardhan, 2014). African institutions could establish exchange programs that allow students and faculty to observe these systems in practice. South Africa could position itself as a continental leader in indigenous health education, while collaborations with other African universities could generate comparative case studies, thereby fostering African-owned models of integration.

Recommendations

Drawing together these threads, the following recommendations emerge:

- **Policy alignment**
  Ministries should issue joint guidelines mandating IKS integration across health sciences curricula.
- **Accreditation reform**
  Professional councils must include indigenous competencies in accreditation standards.
- **Educator development**
  Universities should train academics in decolonial pedagogy and cultural humility while equipping THPs to teach within academic settings.
- **Institutional commitment**
  Create formal roles, remuneration, and governance structures for THPs within universities.
- **Research agenda**
  Fund collaborative research on pedagogical and health outcomes of integration.
- **International benchmarking**
  Establish exchange programs with countries that have successfully institutionalized dual medical systems.
- **Community engagement**
  Ensure curricula are co-designed with THPs and governed by community advisory boards.

Implications for the future

Future directions for integrating IKS into health sciences education must move beyond fragmented initiatives and require systemic reform that includes policy, accreditation, pedagogy, institutional culture, and research domains. The case studies demonstrate that integration is possible and beneficial, but sustainability depends on embedding such practices into national frameworks and institutional structures. By adopting these recommendations, universities and councils can lead genuine transformation that not only redresses historical injustices but also equips health professionals to provide care that is scientifically rigorous, culturally responsive, and ethically grounded.

Conclusion

The exclusion of IKS from health sciences curricula is not accidental but instead the product of colonial histories that entrenched epistemological hierarchies.

By privileging Western biomedical frameworks as the only legitimate form of health knowledge, universities and professional councils have historically silenced, delegitimized, or criminalized indigenous epistemologies. This has created generations of health professionals ill-equipped to serve communities that continue to rely on plural health-seeking behaviors. The consequences are visible in mistrust between medical practitioners and patients, fragmented care, and the perpetuation of epistemic injustice.

This chapter argued that integration is both a moral imperative and a practical necessity. Drawing on theoretical frameworks of decoloniality, epistemic justice, and critical pedagogy, it revealed that integration requires not only the addition of indigenous content, but also the restructuring of curricula to truly value epistemic plurality. African philosophies such as Ubuntu and Botho provide guiding principles for this transformation, as they emphasize relationality, community, and ethical responsibility.

The case studies discussed demonstrate that IKS integration is not aspirational but already underway. These initiatives highlight practical strategies such as co-teaching between traditional healers and biomedical lecturers, immersive visits to Indumba, reciprocal training programs, and curriculum design through Imbizo dialogs. International examples from China and India further show that dual medical systems can be institutionalized without compromising quality. Together, these cases offer both inspiration and evidence that such integration is indeed possible.

However, as discussed, IKS integration faces formidable challenges such as epistemological gatekeeping, institutional resistance, lack of champions, ethical risks of appropriation, and rigid accreditation standards. Without systemic reform, integration efforts risk becoming tokenistic, reinforcing hierarchies rather than dismantling them. We argue that overcoming such barriers requires embedding ethical engagement and co-production at the heart of curriculum design. Respecting cultural protocols, ensuring co-authorship and benefit-sharing, and aligning with legal frameworks are legitimate safeguards.

Ultimately, the integration of IKS into health sciences curricula is not only about inclusion; it is about justice, healing, and relevance. A curriculum that embraces epistemic plurality produces graduates who are not only technically competent but also culturally humble and ethically attuned. Such graduates are better prepared to serve communities holistically, bridging divides between biomedical and indigenous systems. In the broader struggle for decolonial transformation, curriculum reform is both a symbolic and substantive step. It symbolizes the recognition of indigenous knowledge as legitimate, and it substantively equips health professionals to build a health care system that is equitable, inclusive, and contextually responsive. The task ahead is challenging, but the evidence presented here demonstrates that it is achievable. What is required is the collective will of policymakers, educators, students, and communities to move from marginalization to integration, from epistemic injustice to epistemic justice, and from fragmented care to holistic healing.

Glossary

Cultural competence The ability of health care professionals to provide care that is respectful of, and responsive to, the cultural and linguistic needs of diverse patients.

Curriculum transformation A deliberate process of reforming academic content, pedagogy, and policy to make education inclusive, relevant, and socially just.

Decolonization The process of undoing colonial legacies, especially in knowledge, power, and culture, and replacing them with inclusive, contextually relevant frameworks.

Epistemic justice The fair recognition and valuation of all ways of knowing, preventing the silencing and/or marginalization of some groups as knowers (Fricker, 2007).

Epistemic violence The silencing, erasure, or delegitimization of alternative knowledge systems through dominant discourses and practices.

Epistemology The branch of philosophy concerned with the theory of knowledge, examining how knowledge is created, validated, and legitimized.

Indigenous knowledge systems (IKS) Bodies of knowledge, skills, and practices developed by indigenous peoples over generations, adapted to local culture and environment, often transmitted orally and through lived practice.

Indumba It is a sacred healing space used by traditional health practitioners in some African cultures. It is often a small hut or room where spiritual rituals, consultations, and ceremonies take place to connect with ancestors and seek guidance or healing.

Ontology The philosophical study of being and existence, shaping how health, illness, and healing are defined within different paradigms.

Pedagogy The way teachers help students learn. It includes the methods, activities, and approaches used to enhance the ease of learning and to make it more meaningful.

Scientific racism The misuse of scientific methods to justify racial hierarchies and discrimination, evident in colonial medicine and persisting subtly in contemporary knowledge systems.

Transformation A process of systemic change aimed at making institutions and practices more equitable and inclusive; in South Africa it is often linked to redress and the decolonization of higher education (Council on Higher Education (CHE), 2013).

Ubuntu/Botho An African philosophy emphasizing interconnectedness, community, compassion, and shared humanity, often summarized as "I am because we are."

References

Asea, T.W., 2022. Epistemic decoloniality of westernised higher education: a discourse on curriculum justice and knowledge integration at historically white universities in South Africa. Policy Futures Educ. 20 (7), 746–762.

Chatora, R., 2022. Integrating traditional medicine into African universities: the case for curriculum reform. Afr. J. Health Prof. Educ. 14 (1), 45–52.

Chinouya, M., Reynolds, R., 2022. Health pluralism and HIV care in Southern Africa: engaging with traditional medicine. Glob. Public Health. 17 (9), 1665–1678.

Chitindingu, E., George, G., Gow, J., 2014. A review of the integration of traditional, complementary and alternative medicine into the curriculum of South African Medical Schools. BMC Med. Educ. 14, 40. doi:10.1186/1472-6920-14-40.

Comaroff, J., Comaroff, J., 1991. Of Revelation and Revolution: Christianity, Colonialism and Consciousness in South Africa. University of Chicago Press, Chicago.

Council on Higher Education (CHE), 2013. A Proposal for Undergraduate Curriculum Reform in South Africa: The Case for A Flexible Curriculum Structure. CHE, Pretoria.
Dordunoo, D., Etowa, J., Bagnall, A., Smith, T., 2022. Dismantling "race" in health research: the misuse of race as a biological variable. Can. J. Public Health 113 (1), 1–5.
Edwards, S.D., 2020. Traditional healers and the law in South Africa: a historical perspective. Indilinga Afr. J. Indig. Knowl. Syst. 19 (1), 35–47.
Freire, P., 1970. Pedagogy of the Oppressed. Continuum, New York.
Fricker, M., 2007. Epistemic Injustice: Power and the Ethics of Knowing. Oxford University Press, Oxford.
Gandugade, P.V., 2016. Collaboration with and integration of African traditional healers into the South African health care system (Master of Medical Science (Pharmacology) dissertation), University of KwaZulu-Natal. Available at: https://researchspace.ukzn.ac.za/bitstreams/91a04681-af68-4a53-867f-8a4d35274ad0/download.
Gould, S.J., 1996. The Mismeasure of Man, Rev. ed. W.W. Norton, New York.
Harding, S., 2023. Revisiting colonial science: legacies and futures. Sci. Technol. Human Values 48 (2), 215–229.
Kayombo, E.J., Uiso, F.C., Mahunnah, R.L.A., 2013. Experience on healthcare utilization in seven administrative regions of Tanzania. J. Ethnobiol. Ethnomed. 9 (1), 1–13.
Lawrence, C., Bollinger, J., Stewart, K.A., Moshabela, M., 2021. Improving South African medical curricula related to traditional health systems'. Afr. J. Health Prof. Educ. 13 (2), 145–149. doi:10.7196/ajhpe.2021.v13i2.1246.
Leclerc-Madlala, S., 2006. Traditional healers and the integration of HIV/AIDS education into medical training: lessons from KwaZulu-Natal. Soc. Dyn. 32 (2), 57–78.
Li, Y., Yang, Z., Kong, W., Zhao, L., Liu, H., Chen, C., 2025. Drug innovation via integration of traditional Chinese and Western medicine. Sci. Tradit. Chin. Med. 3 (2), 97–112. doi:10.1097/STC.0000000000000033.
Luckett, K., 2016. Curriculum contestation in a post-colonial context: a view from the South. Teach. High. Educ. 21 (4), 415–428.
Masola, N.J., Maotoana, M., 2025. Managing collaboration between traditional and western health practitioners in mental health care, Limpopo province, South Africa. Health SA Gesondheid. 30, 1–9. doi:10.4102/hsag.v30i0.2905.
Mignolo, W.D., 2011. The Darker Side of Western Modernity: Global Futures, Decolonial Options. Duke University Press, Durham.
Moeta, M.E., Mooa, R.S., Mgolozeli, S., Bhana, V., Moagi, M., 2019. Integrating African traditional health knowledge and practices into health sciences curricula in higher education: an imbizo approach. Int. J. Afr. Renaiss. Stud. - Multi- Inter- Transdiscip. 14, 67–82. doi:10.1080/18186874.2019.1635505.
Mokgobi, M.G., 2014. Understanding traditional African healing. Afr. J. Phys. Health Edu. Recreat. Dance 20 (2), 24–34.
Mokgoro, Y., 1998. Ubuntu and the law in South Africa. Buff. Hum. Rights Law Rev. 4 (1), 15–23.
Mpofu, S., Moyo, M., Dube, B., 2021. Traditional medicine and the COVID-19 pandemic: the case of Moringa and Aloe Vera in Southern Africa. J. Indig. Knowl. Dev. Stud. 3 (2), 45–58.
Msomi, N., Barath, D., 2025. Cultural competence of paediatric doctors: a qualitative study in a rural setting. J. Child Health Care South. Afr. 7 (1), 77–89.
Ndlovu-Gatsheni, S.J., 2018. Epistemic Freedom in Africa: Deprovincialization and Decolonization. Routledge, London.

Nkuna, E., van der Merwe, W., 2021. Conferences as spaces for curriculum transformation: traditional healers and scientists at the University of Limpopo. J. High. Educ. Afr. 19 (3), 133–150.

Ntshoe, I.M., 2020. Ontologising social justice in decolonised and post-apartheid settings. S. Afr. J. High. Educ. 34 (3), 263–280

Ogunyemi, D., Thind, B.S., Teixeira, A., Sams, C.M., Ojo, M., Dinkins, G.A.E., Serseni, D., 2024. Integrating cultural humility into medical education using a structured and interactive workshop. Adv. Med. Educ. Pract. 15, 575–583.

Okyere-Manu, B.D., Ntlapo, N., 2018. Traditional healers and maternal health: lessons from Lesotho. HTS Teol. Stud./Theol. Stud. 74 (1), a4973.

Patwardhan, B., 2014. Ayurveda and integrative medicine: global opportunities. Chin. J. Integr. Med. 20 (1), 3–8.

Peu, M.R., Mulaudzi, F.M., Rikhotso, R.S., Ngunyulu, R.N., Rasweswe, M.M., 2020. Reflections on accessing indigenous research settings: encounters with traditional health practitioners and leaders in Vhembe district, South Africa. Cult. Psychol. 27, 1354067X2097124. doi:10.1177/1354067X20971249.

Ramose, M.B., 2002. African Philosophy through Ubuntu. Mond Books, Harare.

Smith, B., 2020. The Ontology of Medicine: What Concepts of Disease and Health Mean for Medical Practice. Cambridge University Press, Cambridge.

Steup, M., Turri, J., Sosa, E., (Eds.), 2021. Contemporary Debates in Epistemology, third ed. Wiley-Blackwell, Malden, MA.

Tervalon, M., Murray-García, J., 1998. Cultural humility versus cultural competence: A critical distinction in defining physician training outcomes in multicultural education. J. Health Care Poor Underserved 9 (2), 117–125.

Vyas, D.A., Eisenstein, L.G., Jones, D.S., 2020. Hidden in plain sight—reconsidering the use of race correction in clinical algorithms. N. Engl. J. Med. 383 (9), 874–882.

World Health Organization, 2022. WHO expert meeting on evaluation of traditional Chinese medicine in the treatment of COVID-19. WHO, Geneva.

World Health Organization, 2013. WHO Traditional Medicine Strategy: 2014–2023. WHO, Geneva.

World Health Organization, 2023. Draft Global Strategy on Traditional Medicine 2025–2034. WHO, Geneva.

Wynberg, R., 2017. The commercialization of Hoodia: a cautionary tale for bioprospecting. Dev. South. Afr. 34 (3), 282–296.

Xu, J., Yang, Y., 2009. Traditional Chinese medicine in the Chinese health care system: Health policy, practice, and research. Health Policy 90 (2–3), 133–139.

CHAPTER

11 Not asking for a seat, but building a new table: Case for a Global South–centered model in medicine

Muneera Rasheed
Independent Professional, Karachi, Pakistan

Introduction beyond inclusion: A call to rebuild

> **What it entails:**
> Establish the core argument: global health is not just structurally unjust; it is imaginatively constrained, and transformation must be grounded in Southern values, dignity, and autonomy.
> **Transition to Section 2:**
> To understand what needs transformation, we must first examine the architecture we've inherited.

"We are not asking for a seat at your table. We are building our own."
(Anonymous, Global South scholar)

Calls for transforming global health and medical education have intensified in recent years. Terms like equity, inclusion, and decolonization now routinely appear in institutional vision statements, journal editorials, and funding calls (Rahman-Shepherd et al., 2023; Kumar et al., 2024; Khan et al., 2022a; Krugman, 2023). Most of these calls are directed toward actors in the Global North (Hafez et al., 2025; Krugman and Bayingana, 2025; McAllister et al., 2025; Obasi et al., 2022), with far less written about what those in the Global South can or should do (Clarke et al., 2024). The prevailing expectation is often framed as creating more opportunities and inclusion for those from the Global South (Pai et al., 2024).

For those of us rooted in the Global South not only geographically, but through lived marginalization, these calls often feel disconnected from our realities (Sharma and Sam-Agudu, 2023; Contractor and Dasgupta, 2022). This is because, while in the Global North discourse has shifted, even if structures remain (Hirsch and Martin, 2022; Koum Besson et al., 2022a; Koum Besson, 2022b), for us neither has moved. It is from this space of dissonance that I write this chapter intentionally for those in the Global South. The literature that speaks to us, about our institutions, systems, and

Decolonizing Medical Education, Research and Practice. https://doi.org/10.1016/B978-0-443-44998-7.10011-0

experiences, is scarce, even within the decolonization discourse. My gaze and my pose are for those rarely addressed in these spaces. I share my observations and experiences not to instruct or prescribe, but to offer insight, to tell our stories, to make visible what we live, witness, and endure.

This chapter addresses the actors in the Global South directly. It asks questions that move beyond the pursuit of power through a seat at the table: What does a seat at the table actually lead to? What do you think you will be able to achieve with this gained power? What if inclusion is not the ultimate goal? What if the table itself the foundational site of knowledge-making, funding, and recognition is exclusionary by design? And what if justice demands not reforming that table, but dismantling and rebuilding it?

The premise is not simply that inclusion falls short; it is that inclusion without transformation is an illusion. Global South actors may be added to institutions, journals, and partnerships, but often without real power to change the architecture. Representation may increase, yet the terms of participation remain fixed: what counts as evidence, who decides priorities, and how legitimacy is conferred (Mehjabeen et al., 2025).

Why this chapter, why now?

This moment is an inflection point. The COVID-19 pandemic made visible what many had long experienced: deep asymmetries in whose lives are protected, whose expertise is heard, and whose knowledge is funded (Abimbola et al., 2021; Büyüm et al., 2020; Kumar et al., 2024). From vaccine apartheid to extractive research, the crisis exposed what global health often conceals. But the reckoning did not stop with COVID. The genocide in Gaza unfolding in real time, one of the cruelest in human history, has shown the limits of institutional moral clarity (Albanese, 2025). Speaking up comes with consequences: backlash, job loss, silencing. Global health is not shielded from systems of power and violence; it is implicated in them (Krugman et al., 2024).

Meanwhile, geopolitical shifts have further destabilized the field. Since the Trump presidency, the United States (US) has turned inward, cutting United States Agency for International Development (USAID) (Sablich and Ainsworth, 2025; Green, 2025), abandoning multilateral commitments, and reducing funding for structural and equity-focused work (Flannery, 2025; Cornwall, 2025). These shifts have hit Global South institutions and systems the hardest (Green, 2025; Sablich and Ainsworth, 2025). In parallel, the #DecolonizeGlobalHealth movement gained momentum, but it remains fragmented. Despite powerful critique, it has not translated into sustained institutional transformation. Much of the work has been individual—important, but insufficient (Stewart et al., 2025; Mehjabeen et al., 2025; Kwete et al., 2022).

Hence, this moment demands that we stop relying on the Global North for validation or assistance (Kyobutungi et al., 2025). If engagement comes, it must be approached strategically (Onyes et al., 2025; Adeyi, 2025): how do we protect our agency, set our terms, and ensure that partnerships do not reproduce the same inequities we seek to dismantle? The time is now for Global South actors to claim their space, articulate their priorities, and act with both vision and vigilance.

This chapter's proposition

This chapter issues a different call. It does not seek more panels, better seats, or softer language. It demands a reimagining of global health, not in opposition to the North, but grounded in the values, priorities, and epistemologies of the South, not as a retreat into regionalism; but as a call for justice, for leadership, and for dignity as a foundation.

But before we build new systems, we must understand what we've inherited. Many reforms have been superficial. The illusion of progress—where gestures are mistaken for structural change—makes it harder to challenge the status quo. Recent evidence on localization confirms this, showing how Northern nongovernment organization (NGO) centrism continues to reproduce power inequities even in efforts framed as reform (van Selm et al., 2025).

Aim and structure of the chapter

This chapter opens by confronting the imaginative constraints at the heart of global health. It argues that the issue is not only structural injustice but also imaginative, leading to a deeper erosion of dignity, autonomy, and epistemic possibility, particularly for those in the Global South. It then turns to the architecture we've inherited, exposing how current systems were built to serve imperial priorities and continue to centralize power, even under the language of inclusion. What often appears as progress—greater visibility, new funding streams, increased representation—can, in fact, mask a deeper bind: the illusion of agency within donor-defined boundaries. This logic not only limits what can be imagined or pursued but also inflicts real harms on knowledge systems, ethical integrity, and the social fabric of local institutions. Against this backdrop, the chapter presents grounded examples from the Global South where dignity, local leadership, and structural autonomy have taken root not as exceptions but as models of what's possible. It reflects on the human qualities that have enabled such alternatives to emerge: integrity, community trust, and principled resistance. Finally, it considers how these values might scale across journals, grants, partnerships, and institutions, and ends with an invitation to stop asking for inclusion and start building with courage, on our own terms.

The table we inherited: Origins and current architecture

What it entails:

Unpacks the colonial origins of global health's power structures: who built the "table," who controls it, and why inclusion alone reinforces the status quo.

Transition to Section 3:

But even when we are included, we are often still inside a system designed to constrain us and what looks like progress can be a carefully managed illusion.

Medicine was never neutral

Modern medicine is often framed as a triumph of rationality, science, and humanitarian progress. But this narrative conceals a more complex legacy. Although Anderson (2014) does not use the term, his account exemplifies what might be described as a *Whiggish history*—a linear, celebratory tale of progress that overlooks how deeply medicine was entangled with empire. From its inception, biomedicine functioned as a tool of governance, used to control and discipline colonized populations. Hospitals in colonial contexts were not merely spaces of healing (Wasino et al., 2025). They served as instruments of surveillance and moral regulation, prioritizing the care of European settlers and troops, while tying care for local populations to political compliance or labor productivity (Anderson, 2014; Washington, 2006). These institutions did not exist outside the political project of empire; they were shaped by it (Naidu and Abimbola, 2022).

Medical education mirrored this dynamic. In colonized nations, schools were often designed to train technical assistants—not independent thinkers or intellectual leaders (Saini, 2016). Curricula privileged European epistemologies, while Indigenous systems of healing were dismissed as unscientific, anecdotal, or outdated (Naidu, 2021a, 2021b). Shahjahan et al. (2022) describe this process as dominance due to *epistemic privilege*—where Western ways of knowing are positioned as objective and universal, and all others are rendered inferior or invisible. These silences were not incidental; they were structural. As Razack et al. (2025) argue, medical education legitimizes whiteness, settler colonialism, and the biomedical model while marginalizing other ways of knowing, and actively erasing them (Kumar, 2025; Naidu and Abimbola, 2022).

Beyond erasure, colonial medicine was also extractive. Black and Indigenous bodies were frequently used as material for scientific experimentation and training. As Washington, 2006 documents, while enslaved Black women endured surgery without anesthesia, the bodies of Black people were taken, often without consent, for medical dissection and study. These practices were foundational to a racialized knowledge economy that dehumanized marginalized communities to serve scientific "progress."

This logic persists in global health. Global South researchers often remain data collectors, while authorship, prestige, and recognition flow North (Liverpool, 2021; Odjidja, 2021; Demeter, 2020; Crane, 2013). Scientific capital, grants, citations, and awards continue to accrue to institutions that have historically benefited from these extractive dynamics (Rucavado Rojas and Postigo, 2025).

The table we inherited

From empire to global health: How the system persists

The origins of global health lie in the structures of empire. Early international health campaigns were often less about improving population wellbeing than protecting colonial economies (Anderson, 2014). The rhetoric has shifted from civilizing missions to capacity-building—but many of the underlying structures remain. As Naidu (2021a) observes, medicine and medical education are "colonial artifacts," their universalist claims masking a profoundly Eurocentric genealogy. The architecture of authority—who funds, who publishes, who sets the rules—remains anchored in the Global North (Horton, 2021). Even when institutions are based in the South, their legitimacy is often measured by their proximity to Northern norms (Hamid and Rasheed, 2022).

In other words, the colonial architecture of the global health "table" has not been dismantled; it has been overlaid with the language of "assistance" and "generosity" (Büyüm et al., 2020) Medical schools in former colonies still rely on Western curricula (Wong et al., 2021; Naidu, 2021a; Naidu and Abimbola, 2022). Research agendas are shaped by donor priorities rather than community needs (Charani et al., 2022). Journals privilege English-language, euro-centric methodologies (Khan et al., 2024), while metrics such as citation counts and impact factors concentrate recognition and resources in the Global North. Curricular inertia entrenches rote memorization and detachment rather than empathy and contextual understanding (Razack et al., 2025; Naidu, 2021b).

The illusion of progress

Symbolic gestures, such as renaming buildings and removing statues, risk becoming illusions of action that distract from deeper reform (Horton, 2021). As Hirsch (2021) notes, even the most progressive institutions struggle to dismantle the structures they have inherited. These interlocking mechanisms—donor logics (Stirling, 2024), publication norms (Abimbola, 2019; Bhakuni and Abimbola, 2021; Lawrence and Hirsch, 2020), prestige metrics (Mahajan, 2019), and outdated curricula (Naidu, 2021a; Razack et al., 2025)—work together to preserve control, even as the rhetoric of equity and decolonization gains prominence.

Within this system, Global South scholars may be included in international collaborations yet remain positioned primarily as data collectors or local implementers, while agenda-setting, analysis, and authorship continue to be concentrated in Northern institutions (Hedt-Gauthier et al., 2019; Crane, 2013; Santos, 2014). Success depends on navigating rather than challenging dominant frameworks (Smith, 2012; Bhambra, 2014; Abimbola, 2023b), creating a constrained form of participation that reinforces epistemic hierarchies (Spivak, 1988; Ndlovu-Gatsheni, 2014).

Inclusion, therefore, comes at a cost. To be accepted, many Southern researchers must be strategically diplomatic—not burn their bridges—reframe questions to meet the inclusion criteria of recent requests for proposals, and adopt "standard" Western thresholds and humanity (Santos, 2018; Ahmed, 2012; Naidu and Abimbola, 2022). Their work becomes legible to donors but increasingly disconnected from community realities (Smith, 2012). As Kalinga (2019) observes, visibility may increase, but autonomy diminishes. Shahjahan et al. (2022) and Ahmed (2012) warn that diversity initiatives which fail to alter structural conditions become "empty signifiers"—gestures that comfort but do not liberate.

In this context, the call for "a seat at the table" has become a dominant metaphor in global health. But what we need to remind ourselves is that this table is not neutral. It was built by institutions rooted in empire—elite universities, philanthropic foundations, and multilateral agencies—and continues to function as a mechanism of control. As Packard (2016) notes, these actors have historically defined what counts as evidence, whose expertise matters, and which problems deserve attention. Inclusion may offer visibility, but rarely sovereignty. Often, the reward is representation, not power. "Slavery is imprinted on the DNA of epidemiology" (Downs, 2020) is a reminder that the epistemic foundations of the field are entangled with systems of domination and dehumanization—exactly what has been unfolding in Gaza (Amnesty International, 2024).

These dynamics help explain why many well-intentioned reforms feel unsatisfying. Even when "included," we operate within systems designed to limit autonomy (Santos, 2018). What appears to be progress can be a carefully managed illusion, maintaining the status quo under the guise of reform (Ahmed, 2012). "Inclusion," thus, becomes a badge of legitimacy—but according to whose standards?

The next section examines this illusion through the Oliver Twist dilemma: how bounded autonomy, token recognition, and donor logics shape decisions and constrain the ability to imagine differently.

Inside the maze—The illusion of progress

> **What it entails:**
>
> Explores the *Oliver Twist dilemma*: token inclusion, bounded autonomy, and how donor logics shape our imagination and decisions.
>
> **Transition to Section 4:**
>
> These structural constraints are not benign; they inflict real harms on knowledge, practice, and people.

Inside the maze: The illusion of progress

Much of the reform language we now hear—"inclusion," "capacity building," "co-creation"—creates the appearance of openness without redistribution of power, reinforcing the very hierarchies it claims to dismantle (Chaudhuri et al., 2021; Affun-Adegbulu and Adegbulu, 2020), operating as a form of performative inclusion (Khan et al., 2022b; Crane, 2010). To understand why such solutions fall short, we must interrogate not just who is present, but the design of the system itself. Until we confront these deeper logics, we risk mistaking symbolic gestures for structural transformation—not a seat at the table, but the need to build an entirely new one (Kwete et al., 2022).

Why inclusion without autonomy fails

> *"Western thinking tells us we must earn our place through work, achievement, and purpose. But many Indigenous teachings say we belong simply because we are. Like rivers flow and stars shine, we exist to be, to feel, to grow, to wonder. We don't need to prove our worth. Being alive is already enough"*
>
> **(Indigenous, 2025).**

Global health is built on this Western logic of conditional belonging (Abimbola and Pai, 2020). Our "place" at the table is never given freely; it must be earned through compliance (Kumar et al., 2024), credibility as defined by others (Bhakuni and Abimbola, 2021), and constant performance (Chaudhuri et al., 2021; Opara, 2021). "But when visibility comes on the terms of those who have displaced you from your lands and sought to destroy your cultures, it is another form of control packaged as 'global solidarity'" (Sahibzada, 2025). This is the root of the problem. The harms we witness today are symptoms of a deeper system designed not to share control, but to centralize it (Crane, 2013). You can be included in a system that still harms you (Ahmed, 2012). You can be welcomed to the table, only to find that the menu and the rules were decided in your absence (Hirsch, 2021; Büyüm et al., 2020).

Belonging in such a system is not necessarily a sign of progress—it can be a form of complicity (Ong'era et al., 2021; Opara, 2021). When we operate within colonial residues without challenging their foundations, we risk legitimizing them through our presence (Chaudhuri et al., 2021; Equinox Racial Justice Initiative, 2023). What is lost in this process is not only autonomy but dignity—the most fundamental measure of human worth (Abimbola, 2023a). When knowledge is extracted without recognition, when compliance is rewarded over competence, and when agency is stripped away in the name of science, dignity suffers.

I have come to believe that inclusion without autonomy is meaningless. Armenteras (2025) exposes this "extractive collaboration"—invitations offering no real power, revealing a system designed for containment, not liberation. Inclusion can work only when we are allowed to define what matters: which problems to solve, which diseases to prioritize, which partnerships to enter, and how to design funding calls, always centering dignity (Achana et al., 2023; Hirsch, 2021). Without this, a seat at the table leads nowhere; at best, it is performative, at worst, complicit. And so, a different table is critical.

I wish I could be more hopeful—some may even think I am being ungrateful—but what I have seen over the past 33 months makes that impossible: genocide being livestreamed, children bombed, torn apart, burned alive, forced into mass starvation (Amnesty International, 2024). A population of two million is slowly dying in the most inhuman ways, while being funded by those very governments that provide developmental assistance for health. I do not see hope in the current global health system that holds us hostage with its money—and, when it chooses to, just takes it away, crippling us—or, when it likes to, can exterminate us (Oxfam, 2025).

Inside the maze: The architecture of bounded agency

If belonging without autonomy is complicity, then we must ask: what structures make autonomy impossible? The answer lies in the architecture of global health—a maze built to contain us, not to free us. Margaret Atwood's (1985) words capture this reality: "A rat in a maze is free to go anywhere, as long as it stays inside the maze."

In global health, that maze is donor-driven philanthropy (Wilson, 2021), which paradoxically can cause harm even if meant to be generous—by shifting priorities (Arumugasamy and Sekaran, 2025; Barwise and Liebow, 2019). Our institutions are told they have agency—but only within boundaries defined by Global North funders:

what gets funded, what counts as innovation, and which methods are legitimate (Birn, 2014; Chaudhuri et al., 2021). This creates what I call the Oliver Twist dilemma (Fig. 11.1)—as long as we silently keep accepting what is being offered at the table and remain within the maze, we are included. The moment we ask for more than what is allowed—question the fairness in the systems—we risk exclusion, and at times "violence" (Rasheed, 2021, 2023a). Faced with this trade-off—visibility or voice—we learn to play by the rules of the maze, mistaking constrained gains for transformation.

The illusion of progress may come from reforms like more grants to lead, direct awards, more first and last authorship for us, invitations to panels, roles as advisors or reviewers, greater visibility (though still conditional), while the underlying ethos of power and exclusion remains unchanged (Kumar, 2025; Horton, 2021). As Abimbola, 2019 cautions, these reforms may merely grant *moral license* rather than dismantle the maze of power "without addressing the problem of gaze." Worse still, local actors operating under a foreign gaze risk losing their own sense of reality, shaping their scholarship for external validation rather than local impact. Despite the reforms, our agency is bounded. We are rarely free to pursue what we believe is best for our communities—because priorities are already scripted (Crane, 2013; Storeng and Béhague, 2014).

Take the Gates Foundation's recent $2.5 billion pledge to women's health innovation. The announcement radiates ambition, signaling the "largest investment yet" in women's health (Gates Foundation, 2025). But money, too, can be an illusion of progress. The deeper question is: who decides what women need? Currently, this decision rests not with the women most affected but with donors in Seattle. The investment targets five disease areas—maternal health, contraception, menstrual health,

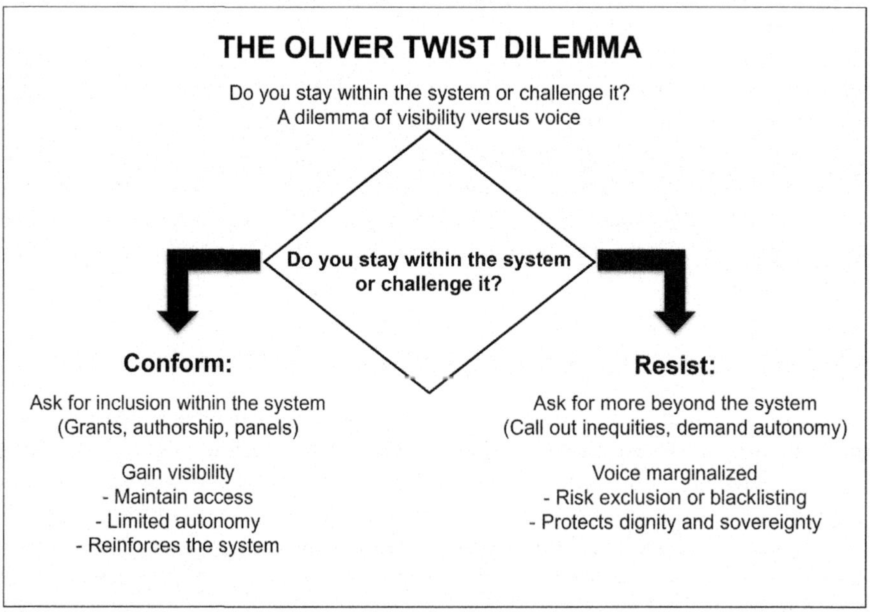

FIG 11.1

The Oliver Twist dilemma.

sexually transmitted infections, and obstetric care and maternal immunizations—as though these categories exhaust the spectrum of women's health concerns. Where in this framework is safety, dignity, freedom from violence—issues repeatedly identified by women themselves as urgent? When the sole focus on women's health is reproductive, it reduces women to objects—of producing babies.

Around two decades in early childhood development, I have seen Northern funding shape what counts as "innovation," often regardless of local need. Play-based stimulation, promoted by Grand Challenges Canada (2025) and the LEGO Foundation, dominates—not necessarily due to proven effectiveness, but because it is fundable: measurable, marketable, and exportable (Scheidecker et al., 2023). Rooted in white, middle-class American parenting norms, these interventions are adapted across the Global South while structural barriers like poverty and poor-quality schooling remain unaddressed (Scheidecker et al., 2024). Programs often create a narrative of parenting in the target countries as "deficient" (Scheidecker et al., 2025). They position the lack of early stimulation in the first three years as a key explanation for poor school achievement, leading to low national productivity, drawing on a "landmark" study by Grantham-McGregor et al. (2007)—now cited >5,000 times—laying the ground for a multibillion-dollar enterprise that further individualizes systemic inequities (Scheidecker et al., 2023).

Similarly, the Wellcome Trust's call for digital mental health solutions champions technological fixes, narrows innovation to app-based interventions, and excludes systemic reforms or culturally embedded care models (Wellcome, 2024). This framing sidelines community-based care and reinforces Northern expertise, while measurable outputs like app downloads eclipse systemic reform. Importantly, it represents a crucial missed opportunity to harness the social capital of close-knit family structures in many countries—resources that could offer far more sustainable and contextually meaningful care. In effect, digital tools become gatekeepers of what mental health can look like—anchoring priorities in the Global North.

In the current global health architecture, interventions prioritize data generation over contextually grounded solutions, benefiting funders more than communities. Much of our work remains applied science; instead of generating locally relevant research questions (Rasheed and Holding, 2024), we often ask questions we already know the answers to, wasting resources in the process (Chattopadhyay, 2024)—but this spending is far from accidental. It sustains a system in which billionaire donors, well-versed in advancing their own interests, reap reputational, political, and economic returns (Birn, 2014). The result is the repeated replication of these models, reshuffled across sites and repackaged as novel, reflecting a deeper power asymmetry—decision-making and priorities remain dictated by the Global North. So, while one may celebrate winning grants, at what expense are we achieving it?

This logic does more than dictate priorities; it reshapes how we think. Over time, what begins as external control becomes internalized (Rasheed, 2023a; Naidu, 2021c; Quijano, 2000). We start to believe that what donors value is what matters most. Even those who have lived in or been raised in these contexts often fail to see—or knowingly choose to unsee—the quiet erosion of dignity embedded in this system (Oppong, 2023). While donors sound progressive, their frameworks control more than funds—they control imagination (Birn, 2014; Moyo, 2009). We internalize the belief that our communities deserve less because implementing effective interventions is deemed a cost we cannot afford.

This is what has disillusioned me most—the lack of imagination. Even well-meaning colleagues, who are brilliant, committed people, often struggle to envision interventions beyond donor logics. Our creativity bends toward what looks fundable, not what works for our communities. Calls for proposals dictate what problems matter, and the system rewards compliance, not contextual innovation (Koum Besson, 2022b). I would like more of us to ask: how does having an innovation guarantee its uptake? Doesn't the underlying assumption that all we need is a technological fix ignore the structural, social, and political determinants that make or break real change (Bardosh et al., 2022; Niang et al., 2021)? And who will invest in these determinants, the slow, less visible work of building systems, trust, and equity? And so, without these difficult conversations, we keep moving, writing proposals, joining advisory boards, accepting grants, believing we are advancing. But like Atwood's rat, we are free only inside the maze that narrows our imagination. Until we dismantle its walls, progress will remain an illusion sustained through interlocking mechanisms.

Coloniality of imagination operates not only at the epistemic level but also at the neurocognitive level. Neuroscience demonstrates that the brain recruits similar neural pathways when imagining a possibility as it does when experiencing it in reality (Dijkstra et al., 2025). Imagination, therefore, functions as a rehearsal space for creation: the more expansive the imagined future, the greater the neural capacity to enact it. When colonial logics determine whose visions are legitimate, they suppress not only knowledge but also the neurocognitive scaffolding of alternative futures. Thus the coloniality of imagination is not merely symbolic—it materially constrains our collective capacity to innovate and create new realities (Felix, 2025).

Consequently, donor priorities—not community needs—shape what gets studied (Crane, 2013; Abimbola, 2023a ; Stuckler et al., 2011). Journals reinforce this hierarchy by upholding standards such as randomized controlled trials (RCTs) (Khan et al., 2024; Hedt-Gauthier et al., 2019), which are feasible only with large donor backing; without publication, further funding becomes impossible. In this cycle, donor logics—not local realities—also determine what research survives. Peer-reviewed journals then marginalize alternative epistemologies (Smith, 2012; Mbaye et al., 2019), while citation counts and impact factors act as gatekeepers that privilege the North (Sinha et al., 2021). Medical training sustains the same logic, valuing memorization over empathy, detachment over context (Farmer, 2013; Naidu and Abimbola, 2022).

Therefore inclusion without autonomy is not neutral—it is a system of containment (Abimbola, 2021; Crane, 2010; Sharma and Sam-Agudu, 2023). These aren't abstract constraints; they shape how knowledge is produced, how care is delivered, and who gets to decide (Hirsch, 2021; Naidu, 2021b). And they are not benign. I have lived this harm—being told to stay quiet, not question, align myself with Western colleagues, accept their superior credibility as knowers simply because of their power, translate dignity into a language that doesn't upset the donors, invalidate my own experience, and lose myself in the process. More importantly, it felt like an insult to my intellect and to my people.

In what follows, I trace these harms—ethical, social, and epistemic—not as conceptual categories—but as deeply lived realities. Because the table didn't just exclude—it hurt.

The harms of the current system

What it entails:

Outlines three intersecting harms: structural, ethical, and social, that result from donor-driven, extractive global health systems.

Transition to Section 5:

If these harms are systemic, then what does it look like when systems are built differently, grounded in dignity and local leadership?

The harms of the current system

The legacies of colonial medicine are a living architecture that continues to shape what knowledge is produced, who benefits from care, and how ethical decisions are made (Razack et al., 2025; Naidu and Abimbola, 2022), perpetuating systemic violence across three interlocking domains: epistemic, social, and ethical. Understanding them is critical for those operating in the system and advocating for inclusion, as without naming how coloniality persists, we risk mistaking symbolic inclusion for transformation and, worse, reproducing these harms in the new table we design.

These harms are not accidental; they are structural. Decolonial scholars describe this as the colonial matrix of power, which operates through three interlocking logics: power, being, and knowledge (Quijano, 2000; Maldonado-Torres, 2007).

- Coloniality of power organizes the global hierarchy of who governs resources, sets priorities, and decides what matters, while casting the Global South as dependent.
- Coloniality of being dehumanizes those at the margins, reducing them to passive recipients of charity rather than rights-bearing subjects.
- Coloniality of knowledge dictates what counts as truth, privileging Northern epistemologies while erasing local ways of knowing.

These logics translate into global health in three kinds of harm: ethical and power harm (loss of sovereignty), social fabric harm (loss of trust), and epistemic harm (loss of truth).

Power and ethical harms: Architectures of control

Power and ethical harm in global health governance arise from asymmetries in decision-making, where control over resources translates into control over priorities, governance, and national agendas (Lambin and Surender, 2023; McCoy et al., 2023).

Donor-driven systems are not neutral assistance mechanisms; they are architectures of power (de Bengy Puyvallée et al., 2025). As Sheikh and Schneider (2025) show, abrupt USAID withdrawals have left countries dependent rather than resilient. Decades of vertical, disease-specific programs, often justified in the name of "efficiency," have steadily eroded public systems and outsourced sovereignty, placing agenda-setting firmly in donors' hands.

This concentration of power is further reinforced by philanthropic foundations. Schwab (2023) illustrated how billionaire philanthropies, like the Gates Foundation, govern without consent, using tax-sheltered wealth to promote technology-centric fixes that bypass democratic accountability. The result is the "Bill Chill," the chilling effect of the extent of influence maintained through institutions that choose silence to safeguard their funding streams. Mahajan (2019) demonstrated how the Gates-funded Institute for Health Metrics and Evaluation (IHME) has eclipsed the WHO in setting global priorities—ensuring that "what gets measured gets funded," while nonquantifiable health issues quietly disappear.

These dynamics are underpinned by cultural logics that normalize Northern dominance. Martin's (2016) critique of the "reductive seduction" of other people's problems captures how powerful actors privilege quick, technical fixes abroad over confronting the structural injustices within their own borders. Such logics, as Adeyi (2023) and Storeng and Béhague (2014) argue, prioritize measurable wins over systemic reform, narrowing the very horizon of what counts as progress.

The consequences are now impossible to ignore. The termination of 90% of USAID's 6,300 projects—described as "geopolitical vandalism" (Kyobutungi et al., 2025)—reveals the fragility of health architectures built on external scaffolding. Likewise, analyses warn of "fiscal and programmatic cliffs" that follow donor exits, particularly in contexts where vertical programs have crowded out investment in primary care and governance (Sheikh and Schneider, 2025). These are not accidental failures but predictable outcomes of a model that systematically weakened local capacity.

Addressing this legacy requires more than managed transitions or technocratic adjustments. Transformation can only emerge from building systems with, not for, and anchoring governance, research, and investment in models that center sovereignty, equity, and epistemic justice. Anything less amounts to abandoning the duty of care donors claim to uphold and violating the very principles of fairness, reciprocity, and accountability they demand within their own borders (Adeyi, 2023).

Social fabric harms: The erosion we don't count

One of the least examined yet most corrosive effects of donor-driven systems in global health is the harm done not only to local capacity but also to the broader social fabric—the trust, prestige, and value systems that hold societies together. This harm is subtle and gradual, reshaping norms, incentives, and aspirations under the guise of assistance.

Indeed, foreign aid is never a neutral transfer; it intervenes in social, political, and intellectual ecosystems (Nadim, 2025; Moyo, 2009). Built around donor priorities and the security–development nexus, aid reorders local hierarchies of power, trust, and legitimacy. As Nadim, 2023 shows in US–Pakistan relations, aid entrenched asymmetrical power, created elite competition, and eroded institutional integrity. What is framed as partnership, then, often functions as control—producing dependencies that persist long after projects end.

Global health initiatives reproduce these same dynamics. Rather than dismantling paternalism, they frequently entrench it. Northern institutions control funding, timelines, and agendas, determining who is deemed "worthy" of investment (Hamid and Rasheed, 2022). Credibility becomes tethered to proximity to whiteness, both epistemically and behaviorally (Bhakuni and Abimbola, 2021). Those who adopt Northern frameworks, cite Western scholarship, and perform deference are rewarded; those who resist or challenge priorities risk exclusion (Naidu, 2021c). Local institutions are left to implement external visions, with their staff often reduced to "glorified data collectors" (Odjidja, 2021).

The consequences extend well beyond the institutions themselves. Donor salary structures distort local labor markets, creating perverse incentives. As a former United Nations (UN) employee observed, a mid-level NGO administrator can out-earn a senior surgeon, drawing talent away from ministries, universities, and hospitals (Tahir, 2025). Wade (2023) offers a similar assessment from West Africa: "UN and NGO contracts way out-price the local market, diverting local talent away from productive, entrepreneurial activity to useless, bureaucratic activity instead… Their budgets don't make sense anywhere except where you're dealing with 'free money'.'"

Such distortions undermine public systems, draining resilience and stifling entrepreneurial energy. Worse, they normalize a culture of compliance. Speaking up about inequities can cost opportunities; silence buys survival. As I wrote previously (Rasheed, 2023b): "Local talent is killed when mediocrity is rewarded to serve the agenda of the few. Money is powerful in poor economies; even an international trip can buy loyalty."

These dynamics also erode the integrity of "partnerships." An example is the Wellcome Trust's 26-year partnership in Malawi, where—despite repeated claims of "local capacity building"—no Malawian was ever appointed as director. When staff reported bullying, the Trust allegedly breached confidentiality and sidelined the local institution (Nordling, 2021). Such cases illustrate the limitations of partnership rhetoric, where Southern actors may remain structurally subordinate and the institutional voice is constrained by donor dependence (Birn, 2014; Odjidja, 2021; Büyüm et al., 2020).

Over time, compliance is rewarded over competence, and silence over dissent. What is framed as diplomacy erodes self-worth (Tahir, 2025). Survival often means self-censorship, trading autonomy for acceptance. The deepest harm is aspirational: dependency becomes a dream rather than a disgrace.

Education and training pipelines further reinforce these inequities. Global health fellowships and scholarships disproportionately benefit elite, English-speaking professionals, effectively pricing out the majority (Svadzian et al., 2020). Far from leveling the playing field, opportunities touted as equity pathways reinforce hierarchies within and between countries.

Academic global health illustrates this vividly. Within local institutions, managers tied to Western grants may wield more authority and earn more than some (marginalized) principal investigators—not because of merit, but because donor logics dictate hierarchies. Leadership, wary of disrupting funding streams, rarely intervenes, even as trust corrodes. Grant agendas follow donor priorities, and "capacity building" often means producing compliant labor rather than autonomous thinkers (Table 11.1). If power and ethical harm ask who decides, then social harm asks: whose careers and futures are reordered to keep this system running?

As Wade (2023) concludes, "That might be worth it if the 'work' they did actually made a difference. But it doesn't. Instead of helping the continent rise, it has a culture of dependency and loss of dignity for the local people."

Knowledge and methodological harms: Whose knowledge is legitimate?

Colonial residues are most visible in knowledge production (Abimbola et al., 2024; Bhakuni and Abimbola, 2021). Global health still clings to a rigid hierarchy of evidence one that privileges RCTs, statistical abstraction, and biomedical outcomes over community knowledge, lived experience, and plural epistemologies. What can't

Table 11.1 Social fabric harm: Key mechanisms and their consequences.

Mechanism	Description	Social consequences
Distorted labor markets	Donor salaries outpace public wages; UN/NGO pay scales restructure incentives (Tahir, 2025; Wade, 2023).	Talent siphoned from public service; weakening of ministries, universities, and health systems.
Elite pipelines and conditional inclusion	Fellowships and leadership programs favor urban, Western-aligned professionals (Global Health 50/50, 2023).	Reinforces internal inequities; rural and non-English-speaking professionals excluded from access.
Proxy power and gatekeeping	Credibility tied to donor-defined behavior; proximity to whiteness determines funding and visibility (Odjidja, 2021).	Conformity rewarded; dissent punished. Inclusion becomes performance, not equity.
Aspiration shift	Youth see NGO contracts as success; innovation in public systems devalued (Tahir, 2025).	Career dreams recalibrated around dependency and donor alignment, not local needs or systemic reform.
Internalized chaos/ethical silence	Speaking up risks opportunity; silence becomes survival. Values diluted to fit donor logics (Rasheed, 2023b).	Identity erosion; the psychological cost of constant self-censorship.
Capacity extraction	Programs branded as capacity-building become pipelines for data extraction and donor-led implementation (Nadim, 2025).	Local actors relegated to implementers; long-term sovereignty undermined.

be randomized is sidelined, and what doesn't fit Northern academic frameworks is rendered illegible (Suleiman, 2025)

A decolonial analysis of the dominance of RCTs in global health suggested that it has shifted from methodological rigor to epistemic violence (Cook-Lundgren and Girei, 2025). Designed for testing drugs, they were transplanted into communities facing complex social realities, where need was randomized not because of resource constraints, but to protect the purity of experimental design. Families received or were denied help to preserve "clean" data, turning human suffering into a variable (Khan et al., 2024; Suleiman, 2025).

As Suleiman (2025) warns, when a tool determines which questions can be asked, what gets funded, and what counts as real knowledge, it becomes a "sieve filtering out anything that cannot be randomized." This bias rewards what is small, measurable, and politically safe, while ignoring the structural drivers: land grabs, debt traps, and extractive trade rules that perpetuate inequity.

For local researchers, donor logics force alignment with externally approved designs, reframing evidence as performance and legitimacy as gatekeeping (Cook-Lundgren and Girei, 2025). Communities become sites of extraction, local scholars are reduced to data collectors, and entire research agendas orbit donor priorities (Crane, 2010). The casualty is not methodological purity; it is dignity: the ability of people and places to define their own realities and shape their own solutions.

Dignity: The hidden casualty

The three harms—epistemic, social, and ethical—converge on one fundamental loss: dignity (Table 11.2). The dignity of being a knower, an agent, a contributor. The dignity of mattering, not as a beneficiary, but as a sovereign actor in one's own health and future.

- When interventions strip agency, dignity vanishes.
- When systems reward compliance over courage, dignity erodes.
- When knowledge is extracted without recognition, dignity suffers.

Table 11.2 Violation of dignity across the three harms.

Harm domain	Core violation	Manifestation	How dignity is violated
Power and ethical	Denial of autonomy	Donor-driven agendas, imposed interventions	Dignity as agents, stripped of autonomy and choice
Social fabric	Conditional worth	Distorted labor markets, elite pipelines, conditional inclusion	Dignity as contributors' worth tied to subservience
Knowledge and methodology	Epistemic injustice	Methodological gatekeeping, exclusion from agenda-setting	Dignity as knowers denied legitimacy as knowledge holders

Other frameworks of dignity in global health, including (Bayingana et al., 2025) epistemic dignity, align with these lines of thinking. While they articulate different expectations—transparency, nonextraction, democratization, and transformation—they converge on the principle that people must be recognized as knowers, agents, and leaders rather than passive subjects of interventions. These frameworks help us see that dignity is not only about preventing harm but also about centering agency, legitimacy, and structural transformation.

Dignity is not an ideal; it is, instead, the foundation of justice. Without it, global health becomes a machine that dehumanizes even as it claims to heal (Fig. 11.2). In the next section, we explore what it looks like to build differently to design systems not from scarcity or control, but from dignity and local leadership.

Beyond the maze: Toward a new architecture

Power without true agency means nothing if it does not allow us to solve problems in ways our contexts demand. This is why we need an architectural redesign of not a bigger maze, but an open landscape, a new table grounded in dignity, sovereignty, and trust (Abimbola, 2019; Erondu et al., 2021). Until then, global health's rhetoric of inclusion will remain a carefully managed illusion a slightly bigger maze, but a maze nonetheless (Armenteras, 2025; Chaudhuri et al., 2021).

If the goal is liberation, then we must reject the architecture of the maze altogether. We need systems anchored in values that prioritize reciprocity, epistemic diversity, and justice (Ndlovu-Gatsheni, 2014; Santos, 2014; Escobar, 2018). If the new table merely reproduces the same extractive logic, then it is no different from the old one. Transformation must begin with values that honor lived knowledge and collective wisdom.

That is where principles matter, not as decorative mission statements, but as structural blueprints to answer foundational questions:

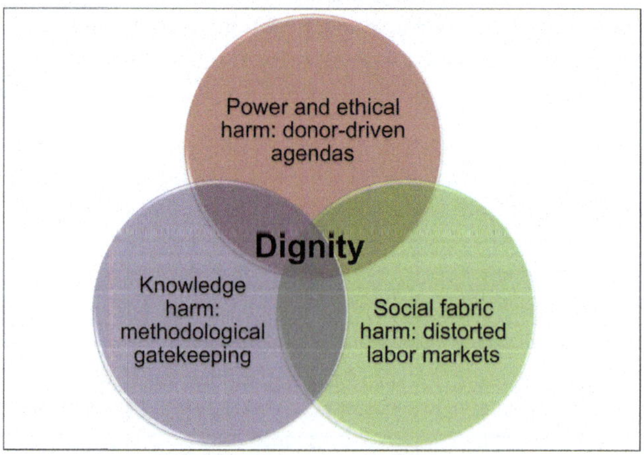

FIG 11.2

Dignity as the key loss across the three harms.

- What harms must never be repeated?
- What values must be protected at every level?
- Who defines the problem and who owns the solution?

Without a shift in these principles, even the most "inclusive" reforms risk reproducing colonial logics under a façade of progress (Anderson, 2014; Ahmed, 2012). In the next section, I move from metaphor to method and propose a set of principles grounded in Southern-led models that respond directly to the three core harms: ethical, social, and epistemic. These principles are seeds of a new space rooted in transparency, dignity, and empathy, rather than dominance. The question is no longer how to move through the maze, but why the maze exists at all.

What can be done differently? Lessons from the Global South

What it entails:

Presents case studies (SIUT, Narayana, Dhulikhel) as real-world examples of dignity-centered, sovereign system-building.

Transition to Section 6:

These systems were built by individuals with the courage to lead from their values—often quietly but powerfully.

Principles for a Global South–centered model

The call to decolonize global health is not just about who designs the systems; it is about the values they are built on (Pant et al., 2022). In the Global South, where survival is too often framed as the highest attainable good, what's most urgently needed is a model rooted not in discounted fixes but in dignity. The vignette that follows makes this argument personal and political.

When low cost becomes low value

Vignette: Silence in the maternity ward

Several years ago, on a visit to our rural field site Naushahro Feroze (which is located five hours from Karachi), I met a female physician at a maternity health center. She had moved there from Karachi. I asked if the job conditions in such a remote location were difficult. To my surprise, she replied, "No, it's much easier."

When I asked why, her response stayed with me: "In the city, when a mother or child dies during delivery or soon after, the family becomes really upset, and it can be hard to explain our point of view. Here, when a child dies, the family doesn't question. The woman just gets up and returns home silently."

That silence revealed a truth about the underlying cause of high maternal and child mortality: the value placed on human life, particularly women and children in disadvantaged settings. Women are seen as replaceable; men can remarry, and women can bear another child the next year.

Complex sociocultural norms cannot be dismantled by low-cost, quick-fix interventions that dominate global health agendas. As a woman and mother in Pakistan, I find these notions offensive and detrimental to our struggle for equal value. Equity should not be sold as cost-efficiency. Programs that claim to advance justice must measure success in quality and dignity, not merely in affordability. If we lack resources to deliver that, let us not settle for low-quality solutions. If an intervention is truly equitable, it should be good for all women everywhere and not only for "the other."

If equity becomes about cost alone, dignity is the first casualty.

What Pakistan needs: low cost, high courage.

For countries like Pakistan, the challenge is not only financial; it is cultural and systemic. No "cheap fix" can substitute for interventions that shift the value of women's lives from objects of gratification to full human beings with dignity. Yes, we need low-cost interventions, but they must also be high-courage intervention strategies that confront social norms head-on, engage religious leaders, and dismantle narratives justifying a woman's disposability. Change will not come through distributing more kits or running donor-friendly campaigns; it will come by elevating dignity as the core outcome of health systems, not just survival.

This shift from price to principles is not optional; it's the only way to make equity real.

Beyond cost: The real measure of equity

Global health often responds to inequity by asking: how can we make this cheaper? This lens reduces dignity to a budget line and displaces the real question: how can we build systems that treat all human life as equally valuable?

True equity demands more than affordability. It demands systems that do not trade quality for cost or impose low-expectation fixes in the name of innovation. As Martin (2016) warns, the "reductive seduction" of solving other people's problems lies in their imagined simplicity while avoiding structural reform at home. We must resist this illusion.

Principles: Toward structural, moral, and practical redesign

The harms detailed earlier, i.e., ethical, social, and epistemic, cannot be addressed through minor corrections to a broken system. What is needed is a fundamental reimagining of health systems from the ground up, rooted in Global South experiences and values. The following principles build on what I had envisioned in a previous piece (Rasheed, 2023a), asking: what would it mean to build from dignity rather than dependency?

Addressing power and ethical harm: Coloniality of power

Dare to dream

To dream differently is a political act. It affirms that we are not merely surviving but are capable of shaping sovereign futures. Krugman, 2025 writes, "Africa will not simply roll over and die without USAID." That belief exposes the narrative of dependency that underpins current systems. The so-called fiscal cliff is not a death sentence; it is a moral opportunity. Dreaming differently means refusing donor logics and building regional alternatives

Power is also in receiving

Colonial legacies frame power as giving money, knowledge, and legitimacy. But true power includes the right to receive without subservience. Sovereignty means saying no; it means rejecting extractive projects and demanding accountability from donors. The Global North must learn to listen and follow, not dictate. As long as we beg, we remain petitioners. Dignity demands the right to shape, question, and redefine the terms of engagement.

Addressing social fabric harm: Coloniality of being

Beyond charity: Toward value-based models

Charity frames the Global South as a passive recipient. Justice demands that we position ourselves as equal partners, setting agendas, shaping methods, and leading the conversations that define our futures. A value-based model moves beyond benevolence to solidarity, acknowledging histories of extraction and insisting on systems that center our dignity rather than donor approval (Rasheed, 2023; Cook-Lundgren and Girei, 2025). We must rise above being sites of data extraction and claim our role as originators of theory, methodology, and evidence, adding value to global knowledge production on our own terms.

Horizontal and vertical justice

We must challenge not just inequity between North and South (horizontal), but also within our own systems (vertical). Local elites often replicate colonial power's centralizing voice, rewarding compliance and excluding the marginalized. Justice must descend and travel (Opara, 2021) .

Addressing epistemic harm: Coloniality of knowledge

From representation to leadership

Token inclusion is not justice (Pant et al., 2022). True decolonization requires Southern leadership at the table, not as a formality or a box-ticking exercise. We must lead agenda-setting, funding, authorship, and governance. We must uphold dignity as the foremost value—refusing to compromise it for the sake of representation or visibility. Presence without dignity is not progress; it is assimilation. Leadership must mean shaping the terms of engagement, not merely accepting a seat at a table that was never built for us.

Reclaiming the right to define systems

We must reclaim the authority to determine what counts as valid evidence, which methodologies are legitimate, and what outcomes truly matter. This is not just about challenging rigid methodologies like RCTs, but about dismantling the broader colonial and Eurocentric structures that dictate how knowledge is produced, validated, and valued. True equity in research demands centering local epistemologies, lived experiences, and community-driven priorities in defining what constitutes meaningful evidence (Hafez et al., 2026; Smith, 2012; *Scheidecker* et al.*, 2023*).

The power of rejecting external validation

The demand to validate knowledge through Western journals, rankings, and metrics is a form of epistemic colonialism. The right to refuse that validation without sacrificing rigor or impact is part of intellectual sovereignty. We must stop seeking approval for truths already lived (Naidu, 2021c).

Building the table—Case studies from the Global South

These principles are alive in the Global South not as lofty ideals but as daily practice. The Sindh Institute of Urology and Transplantation (SIUT) in Pakistan, Dhulikhel Hospital in Nepal, and Narayana Health in India are dignity-driven models born out of their own contexts. If such models could have emerged with the Global North assistance, they would have by now. The fact that they haven't underscores why the Global South must stop waiting for inclusion and start building on its own terms. These homegrown models embody an approach where universal, high-quality care is delivered without charity optics or donor-driven compromise. Their strength lies not in an abundance of capital, but in an abundance of vision dignity as design.

Sindh Institute of Urology and Transplantation (SIUT), Pakistan
Ethos: Health care is a birthright.

Founded in the early 1970s by Dr. Adeebul Hasan Rizvi, SIUT was born from a radical commitment: no one should be denied care due to cost. "*We cannot let them die because they cannot afford to live.*" Starting with just eight beds, it has grown into one of South Asia's largest public-sector institutions, offering dialysis, transplant, oncology, and critical care for kidney health, all free of cost (Memon, 2015), and has now turned into a health care network offering state-of-the-art services like robotic surgery. Crucially, SIUT refuses the visual cues of charity; patients are not "beneficiaries" but citizens with rights. Funded through the government, local philanthropy, and diaspora support, SIUT's independence protects it from external conditionalities. Its integrated model of care, research, and training shows how equity can scale when dignity leads and not donor logic.

Dhulikhel Hospital, Nepal
Ethos: Quality health service to the poor.

Founded in 1996 by Dr. Ram Kantha Makaju Shrestha, Dhulikhel Hospital evolved from a rural initiative into a major teaching and referral center serving over 2.5 million people. Rejecting one-size-fits-all solutions, its model centers on community needs, cultural relevance, and dignity (Shrestha, 2016). Shrestha chose to return to Nepal over a career abroad to "start something that matters" (Shrestha, 2022). Today, Dhulikhel is sustained by local partnerships and diaspora philanthropy, and not by external donors, thereby allowing it to stay anchored in context. Its outreach, telemedicine, and education arms all reinforce health as a right, not a favor.

Narayana Health, India
Ethos: Compassionate care at scale.

Dr. Devi Shetty, trained in the United Kingdom, posed a bold question: "If industries can achieve efficiency through scale, why can't health care?" Narayana Health answers this with a network of 30 hospitals performing thousands of cardiac surgeries at a fraction of global costs (Shetty et al., 2022). Its frugal innovations like assembly-line surgery, task-shifting, and bulk procurement, challenge the false trade-off between quality and affordability, providing heart surgery for <10% of the average cost in the US (Bhatti et al., 2017). The model is sustained through cross-subsidization: paying patients offset costs for those who cannot pay. Narayana proves that scale does not have to erode dignity when innovation is guided by empathy.

A shared ethos: Dignity as an operational reality

What inspires me is that these institutions did not wait for external permission before initiating action. They commenced with limited resources, yet expanded by embedding dignity as a central organizing principle across all operational domains, from financing structures to service delivery mechanisms. Rejecting the prevailing charity-based paradigm, each draws on community trust and safeguards autonomy through financial independence (Crisp, 2010; Sen, 1999). Collectively, these cases contest the dominant assumption that high-quality health care provision in the Global South must originate from externally designed models (Abimbola and Pai, 2020). Within their operational and ethical frameworks, dignity functions beyond a rhetorical statement, serving as a structuring logic that shapes decision-making and priorities (Farmer, 2013; Marmot, 2015). Their trajectories illustrate that equity is not achieved through donor-imposed frameworks but rather authored through courage, contextual responsiveness, and an enduring commitment to justice (Grassi et al., 2024). While the future of health justice may not lie in replicating these models in their entirety, it may depend on adopting their underlying ethos one in which dignity is not merely an outcome but a foundational design principle (Farmer, 2003, 2013; Sen, 1999).

Why these models matter

The emergence of such models in South Asia carries particular significance. The region, home to nearly a quarter of the global population (World Bank, 2024), bears a disproportionate share of the worldwide disease burden, encompassing high maternal and child mortality, undernutrition, infectious diseases, and a rising prevalence of noncommunicable conditions (WHO, 2023; UNICEF, 2023). Public health systems are frequently underresourced, with out-of-pocket expenditures accounting for more than half of health spending in some countries, thereby pushing millions into poverty each year (World Bank, 2023). Within this context, South Asian institutions that provide affordable, high-quality, and dignified care offer not merely local solutions but concrete evidence that equity and quality health outcomes can be realized under conditions of constraint.

Contrast with Global North–centric innovation

Global North approaches have framed equity as a technical challenge, solvable through "low-cost innovations" and advocacy events.

This mindset, while well-intentioned, reproduces three deep harms:

- Power and ethical harm—reinforcing dependency through donor-driven agendas.
- Social fabric harm—eroding trust and institutional integrity.
- Epistemic harm—erasing local knowledge and leadership.

This is the logic that dominates high-profile global health spaces: Goalkeepers Awards (Gates Foundation, 2025), Skoll Awards (Skoll Foundation, n.d.), XPRIZE challenges, and Grand Challenges, where "innovation" is celebrated as the currency of equity. Yet this term itself is rarely interrogated: who defines what counts as innovation, and for whom? In practice, it often functions as a buzzword, privileging novel technical fixes over the more complex work of strengthening and sustaining effective systems. Within these award ecosystems, innovation frequently translates into cost-minimizing solutions that can be easily packaged and measured, rather than structural redesigns aimed at maximizing dignity. Such framing reflects a tendency to reward what is easily quantified and fundable, rather than what is contextually grounded and transformational (Chaudhuri et al., 2021; Birn, 2014; Adams, 2016).

In contrast, Southern-led institutions, such as those discussed above, embody a different logic of change (Table 11.3). They do not seek permission to enter the donor maze; they build new architecture grounded in local values (Abimbola, 2023b; Santos, 2014).

- Power is redefined: Sovereignty through local government and philanthropy, diaspora support, and frugal innovation, not conditional grants (Moyo, 2009).
- Social fabric is restored: Merit is tied to community impact, not proximity to Northern networks. Care is framed as a right and not as charity (Abimbola, 2023a; Bayingana et al., 2025).
- Knowledge is liberated: Success is measured by dignity, universality, and equity, not Western key performance indicators (Mignolo and Walsh, 2018).

They prove a radical truth: equity cannot be imported. It must be authored locally, anchored in principles that prioritize justice over optics, autonomy over dependency (Tuck and Yang, 2012), the redistribution of power over symbolic inclusion, and dignity over cost-efficiency (Sen, 1999; Farmer, 2003).

What the Global North can learn—If it is willing to listen

If global health is serious about equity, it must learn what India's Tata Nano vehicles and ChotuKool cooling boxes have already taught business: affordability without aspiration strips people of dignity (Granados and Prabhu, 2022). These "innovations for the poor" failed because they reduced human needs to price tags, ignoring pride, meaning, and care.

Global health often makes the same mistake, prioritizing top-down interventions in the name of efficiency and measuring success in costs saved per dollar, while neglecting whether individuals are treated as persons of worth or merely as data points

Table 11.3 Comparison of Global North models versus selected case studies in the Global South.

Principle	Global North innovation-centric approach	SIUT (Pakistan)	Dhulikhel (Nepal)	Narayana Health (India)
Daring to dream (*power and ethical harm*)	Dependency reinforced through donor-driven agendas and conditional funding	Public–private partnership, local philanthropy, and diaspora funding sustain independence	Operates largely through Nepali partnerships, avoiding dependency	Leverages cross-subsidization and self-sustaining financial models
Restoring social value (*social fabric harm*)	Merit is defined by proximity to donor systems; loyalty to Global North agendas often outweighs community impact	Key performance indicators correspond to service to the community; dignity, not hierarchy, structures care	Merit is defined by community trust, responsiveness, and culturally grounded care	Value rooted in outcomes for the underserved, not donor validation; rewards frugal excellence
Reclaiming the right to define systems (*epistemic harm*)	Global North models dictate metrics, validation, and priorities; prestige and inclusion are measured by alignment with Western standards	Defines priorities locally; measures success in patient dignity, not donor nor key performance indicators	Community-driven governance; metrics rooted in access and quality	Frames healthcare as a right, not a luxury; adapts processes to Indian realities

(Farmer, 2013; Marmot, 2015). The case studies of SIUT, Dhulikhel Hospital, and Narayana Health exemplify a dignity-centered paradigm:

- Power redefined: sovereignty through local leadership and autonomous financing.
- Trust restored: legitimacy grounded in community impact.
- Truth reclaimed: evidence standards rooted in dignity, not externally imposed performance indicators.

As Mignolo (2009) argues, these models enact epistemic disobedience—rejecting colonial evaluative frameworks and asserting the Global South's right to define health systems on its own terms. They are not abstract ideals but *living proof* of what Felix (2025) envisions: a decolonial imagination grounded not in individual gain but in collective purpose. Their legitimacy emerges from outcomes and trust (Hirsch, 2021), not from visibility or market validation. In contrast to the capitalist imagination that, as Felix notes, "fuels individual innovation, which is to be sold on the market, deemed valuable only in how it competes with other ideas," these models demonstrate that real innovation lies in solidarity, in the shared pursuit of health equity as a communal act of creation.

The challenge to Northern platforms, such as the Goalkeepers Awards, is not to selectively showcase these models but to support them without altering their core principles (Crisp, 2010; Abimbola and Pai, 2020). The ethos "We cannot let them die because they cannot afford to live" contrasts sharply with Global North narratives of "low-cost interventions," where scarcity is used to justify reduced standards of care. Here, scarcity demands creativity, courage, and an uncompromising commitment to dignity.

Martin (2016) warns that the reductive seduction of "other people's problems" has been institutionalized in global health. Genuine transformation is complex, relational, and grounded in justice. It cannot be designed from Geneva or Seattle; it must be rooted in Karachi, Dhulikhel, and Bengaluru. The task is not replication but resourcing models that already work, building infrastructures of justice rather than charity.

Beyond vision to systemic change

What it entails:

What must Global South researchers learn from the case studies and then implement?

Transition to Section 7:

But for systemic change, this quiet power must evolve into collective action across institutions and platforms.

Building beyond individuals

The stories of SIUT, Dhulikhel Hospital, and Narayana Health are proof of concept that transformation is possible when health systems are rooted in dignity, autonomy, and contextual imagination. These models dismantle the myth that equity requires billion-dollar campaigns or Northern-led designs. But their strength reveals a vulnerability: they depend on the vision of a few. Charisma can spark change—but sustainability needs systems. This is the critical lesson: what begins as personal conviction must now evolve into institutional architecture.

What can guide us forward?

First, there is a need to transition from personality-driven change to principle-based systems (Quinn and Thakor, 2018). While leadership will always matter, enduring progress depends on embedding equity and empathy into governance, financing, and delivery mechanisms. This requires investment in systems thinking and implementation science (Kruk et al., 2018), grounded in real-world examples rather than abstract theory.

Second, we must design systems that prioritize wellbeing for both patients and providers. A human-centered approach cannot stop at the point of care; it must encompass the workforce whose resilience and engagement determine system quality (Rasheed et al., 2021). Structures that value, support, and empower health workers create environments where dignity becomes a shared experience.

Third, future leaders require more than technical competence—they need moral imagination and contextual intelligence (Bhakuni and Abimbola, 2021). The pioneers behind these models were not only clinicians—they were architects of possibility. Education and professional development must instigate critical thinking—the capacity to question dominant paradigms, navigate complexity, and align innovation with justice.

Finally, equity begins internally. Institutions committed to global fairness must reflect these values in their internal practices, promoting transparency, inclusion, and fairness in leadership, remuneration, and decision-making (Hosmer and Kiewitz, 2005). Building systems that honor people outwardly requires practicing equity inwardly.

The task ahead is not to replicate SIUT, Dhulikhel, or Narayana in any form—but to distill their principles and operationalize them in diverse contexts:

- Autonomy over dependency.
- Reciprocity over extraction.
- Dignity over charity.

If these principles guide leadership training, funding priorities, and evaluation frameworks, the work of these pioneers will no longer remain exceptional; it will become systemic.

From the quiet power of character to collective action

What it entails:

Emphasizes the role of personal integrity and relational leadership in resisting extractive norms and nurturing alternative systems. Maps practical strategies for shifting institutional norms, incentives, authorship, partnerships, and journals toward dignity and equity.

Transition to Section 8:

These shifts are possible now, if we choose to act. What we need is not more evidence, but more courage.

The quiet power of character and community

Very often, after my talks on decolonizing global health, the first question I hear is: "So what can I do as an individual?". For a long time, I have been reflecting on that question, and through this reflection, I have come to see that the influence of individual character is often underestimated in systems that value status (Prasad, 2023). In global health, especially within decolonization discourse, structural inequities dominate the conversation, as they should (Kumar et al., 2024; Abimbola, 2023b; Hirsch, 2021). But there is something quieter, more personal: the *soft power of integrity*.

Powerlessness is often assumed because we do not hold institutional titles or control funding streams. Yet, some of the most unsettling and transformative forces in any system are those individuals who live and act from deeply rooted values (Burgess, 2024). When you act from a place of integrity, detached from external validation, you become a mirror reflecting what systems often try to hide (Lachman, 2007). And mirrors can be uncomfortable. Systems built on prestige and performativity often do not know how to respond to those who refuse to detach ethics from action (Packard, 2016).

But character alone is not enough. Its true strength lies in connection—using that quiet power to build community (Aultman, 2008). When we link across geographies and silos with others who lead from the same values, we create a tribe of solidarity. It is within these trust-based networks, rather than isolated resistance, that collective transformation becomes possible. This is where the work of decolonization shifts from rhetoric to reality not by dismantling systems alone, but by rehumanizing them through relational resilience (Abimbola, 2023a).

Communities of dignity-centered leaders can reimagine the architecture of health systems from within, without waiting for permission or external validation. What does this look like in practice? It means moving from critique to construction, i.e., building our own tables rather than waiting for a seat at someone else's, which, if it comes at all, often comes with conditions (Burgess, 2024; Hirsch, 2021). The most sustainable shift will come from leveraging spaces we already control and creating new ones where necessary, aligning them with our values rather than external validation.

The stories of SIUT, Dhulikhel, and Narayana are health care models, but their lessons reach far beyond clinical systems. They teach us that transformation is not born of abundance—it is born of vision, values, and local leadership (Memon, 2015; Shrestha, 2022; Bhatti et al., 2017). If health care can be reimagined through autonomy and dignity, so can global health academia. The question is: how do we translate these principles, i.e., financial independence, cultural rootedness, and systemic equity into the platforms we control: journals, grant agreements, ethics committees, conferences, and institutional metrics?

Table 11.4 outlines practical entry points to embed dignity, shift power, and create new norms that reflect Global South priorities.

This is how character scales when the quiet power of individuals is woven into collective action that reshapes systems. The goal was never to navigate the maze more efficiently; it was to design spaces where no maze exists. So, if you feel powerless, start where you are. Lead from your values and use them to build relationships of trust. That is how we decolonize not just institutions, but ourselves.

A call to action—We cannot afford to wait

What it entails:

Closes with an invitation to act with urgency and moral clarity, especially for those in the Global South, to stop seeking permission and to start building differently.

Table 11.4 Platforms for action: building and owning our tables.

Platform we can control	Action for dignity
Medical training institutes	Redesign medical education to embed dignity, cultural humility, and equity as core competencies. Integrate Southern health priorities, local epidemiology, and community-engaged practice into clinical training. Assess students on technical knowledge and relational skills, ethical leadership, and the ability to work in resource-constrained settings with respect and reciprocity.
Journals	Start and strengthen Global South (GS)-led journals. Value them as primary outlets for thought leadership rather than as fallback options. Build citation ecosystems rooted in Southern priorities.
Regional conferences and symposia	Host GS-owned conferences to showcase locally rooted innovations and create agenda-setting spaces independent of Northern gatekeeping (Mbembe, 2016).
Institutional metrics & incentives	Reform promotion and evaluation systems in GS institutions to reward community impact, mentorship, and collaborative leadership, rather than focusing solely on publications in Northern journals.
ERCs	Expand ERC mandates to include dignity and equity. Require power-sharing and community accountability in research proposals.
Professional networks and alliances	Form South-South networks for researchers, clinicians, and policymakers. Use these to share resources, create collaborative grants, and amplify voices (Abouzeid et al., 2022)
Social media and digital platforms	Use GS-controlled online spaces to frame narratives around dignity and equity. Promote Southern scholarship through coordinated campaigns and knowledge hubs.
Curricula training programs	Develop training programs that center on equity and dignity. Include Southern case studies as core content in medical and public health education.
Institution-led funding mechanisms	Pool resources regionally to create GS-managed research funds (Araújo et al., 2021). Prioritize dignity-centered proposals and invest in long-term system-building rather than donor-driven projects.
Community advisory boards	Institutionalize community voices within GS health systems and research agendas. Ensure participatory decision-making at every stage of project design and delivery.

Abbreviations: ERC, Ethics Research Committee; GS, Global South.

I close with an invitation to act with urgency and moral clarity, especially for those in the Global South, to stop seeking permission and to start building differently. I wrote this chapter for those in the Global South engaged in medical research, teaching, or practice, and particularly for young professionals who hold high aspirations and envision transformative change. I did not intend to leave you in despair, but neither will I offer the false comfort that liberation is ever easy. As the late Palestinian leader Ghassan Kanafani asserted in his final interview with Carleton (1970), "To us, to liberate our country, to have dignity, to have respect, to have our mere human rights;

these are something as essential as life itself." Let us make dignity as essential to our careers as it is to life itself.

The transformation of global health will not be achieved through the refinement of metrics alone, nor through tokenistic authorship practices or grants awarded to the Global South whose foci were externally designed. It is no longer enough to "center Southern voices" while keeping Northern control intact. Structural reform demands investment in early-stage local innovation, acceptance of contextual evidence, and the dismantling of symbolic inclusion (Okumu, 2025; Chaudhuri et al., 2021, Naidu, 2021c). Structural change will occur when our institutions and scholars prioritize building independent knowledge and practice infrastructures, grounded in their own realities, priorities, and epistemologies (Mbembe, 2016).

I realize this means navigating a paradox: career advancement often depends on recognition from Northern gatekeepers, yet uncritical pursuit of such validation risks eroding identities. The persistence of "soft money" academic models, often contingent on Northern approval, continues to reproduce colonial relations, calling instead for what Krugman and Bayingana (2025) describe as "ruinous solidarity": the deliberate redistribution of resources and relinquishing of elite prestige to ethically rebalance power. The task, therefore, is not to reject engagement with global platforms outright, but to engage on terms that preserve autonomy and epistemic integrity (Mignolo, 2009; Naidu, 2021c; Rasheed, 2021). This requires investing in Global South-led journals, conferences, and funding mechanisms, while using existing transnational spaces strategically rather than aspirationally.

The imperative for Northern actors is to listen without defensiveness and to relinquish control where Global South leadership holds greater contextual competence (Bain et al., 2024). For funders, this means shifting investment toward long-term system strengthening and relational infrastructure. A welcome example is the Global Fund's 2025 Community Engagement Strategic Initiative, which supports the meaningful engagement and leadership of communities affected by HIV, TB, and malaria through a combination of short-term technical assistance, long-term capacity strengthening, and regional learning hubs. This reflects a broader institutional shift toward sustained community systems strengthening and community-led monitoring (The Global Fund, 2025). However, real impact depends on *how* funds are allocated and tracked, ensuring they genuinely empower local systems rather than reinforce existing inequities.

Ultimately, the central question is not only what must be done, but how to pursue change without losing the values that define us. Power gained at the expense of identity is not liberation, but assimilation (Bhakuni and Abimbola, 2021). Sustainable transformation will come from building systems that are locally legitimate, globally relevant, and resistant to epistemic erosion—a principle echoed in Indigenous approaches to solidarity, which root collective action in epistemologies, offering alternatives to Western-dominated frameworks (Fayemi et al., 2025; Smith, 2012; Hafez et al., 2026). Readiness for this work is not a fixed state; it is a collective decision to act even when conditions are imperfect, resources are scarce, and recognition is uncertain.

Concluding reflection

Liberation will rarely be comfortable. There will be times of isolation, despair, and feelings of being left behind. Yet, as my spouse reminds me, the privilege of living with your head held high is one that few can afford. I now realize this is the life alluded to in the verse often attributed to Rumi (Barks, 1997): "Every soul will taste death, but only some will taste life"—a life lived fully aligned with what you truly believe in. And it is this understanding that has shaped how I share my experiences with aspiring researchers and practitioners in the Global South: not to prescribe, but to be honest about how the system works, how we must navigate it, and the importance of understanding it before you find yourself in a place where progress means compromising your values. People do not need manuals; they know what is best for themselves if they have true knowledge. I trust you—just keep your heart in the right place.

Moral clarity in compromised systems

Remaining principled within compromised systems is not a naïve ideal, but a calculated choice. It requires refusing the "free lunch" when it costs your dignity. In global health, many of the most generously funded programs come with invisible strings: agendas prewritten elsewhere, metrics that erase meaning, and partnerships that dilute autonomy. Law 40 of Power, "Despise free lunch" (Greene, 2000) reminds us: what is offered freely often extracts quietly. In this case, it is dignity.

This does not mean we reject all collaboration. It means we learn to recognize the price of acceptance, and when the cost is too high, we walk away or build alternatives. We must invest in systems where dignity is the currency. The future will be shaped by those who refuse the terms of the maze and have the courage to set their own.

AI disclosure

The author used an AI-based language model (ChatGPT) to improve the clarity and language of her manuscript draft. The conception, content, and interpretation remain the author's own.

References

Abimbola, S., 2019. The foreign gaze: authorship in academic global health. BMJ Glob. Health 4 (5), e002068. doi:10.1136/bmjgh-2019-002068.

Abimbola, S., 2021. The uses of knowledge in global health. BMJ Global Health 6 (4), e005802. doi:10.1136/bmjgh-2021-005802.

Abimbola, S., 2023a. When dignity meets evidence. Lancet 401 (10374), 340–341. doi:10.1016/S0140-6736(23)00079-4.

Abimbola, S., 2023b. Knowledge from the global South is in the global South. J. Med. Ethics 49 (5), 337–338.

Abimbola, S., Asthana, S., Montenegro, C., Guinto, R., Louskieter, L., Rasanathan, K., et al., 2021. Addressing power asymmetries in global health: imperatives in the wake of the COVID-19 pandemic. BMJ Glob. Health 6 (4), e006779. doi:10.1136/bmjgh-2021-006779.

Abimbola, S., Pai, M., 2020. Will global health survive its decolonisation? Lancet 396 (10263), 1627–1628. doi:10.1016/S0140-6736(20)32417-X.

Abimbola, S., van de Kamp, J., Lariat, J., Rathod, L., Klipstein-Grobusch, K., van der Graaf, R., Bhakuni, H., 2024. Unfair knowledge practices in global health: a realist synthesis. Health Policy Plan. 39 (6), 636–650.

Abouzeid, M., Muthanna, A., Nuwayhid, I., El-Jardali, F., Connors, P., Habib, R.R., Akbarzadeh, S., Jabbour, S., 2022. Barriers to sustainable health research leadership in the Global South: time for a Grand Bargain on localization of research leadership? Health Res. Policy Syst. 20 (1), 136. doi:10.1186/s12961-022-00910-6.

Achana, F., Boah, M., Chirawurah, D., Owusu-Marfo, J., Ndago, J.A., Santah, C., Mohammed, A.-R., Adokiya, M.N., 2023. Equitable global health research collaborations with a mind of human dignity. Front. Educ. 8, 1233018. doi:10.3389/feduc.2023.1233018.

Adams, V., 2016. Metrics: What Counts in Global Health. Duke University Press, Durham, NC. doi:10.1215/9780822374480.

Adeyi, O., 2025. Africa without foreign aid for health: free at last?. *Development Today*. Available at: https://www.development-today.com/archive/2025/dt-2–2025/africa-without-foreign-aid-for-health-free-at-last (Accessed 14 August 2025).

Adeyi, O.O., 2023. Misguided charity: the bane of global health. BMJ Glob. Health 8 (9), e013322. doi:10.1136/bmjgh-2023-013322.

Affun-Adegbulu, C., Adegbulu, O., 2020. Decolonising global (public) health: from Western universalism to global pluriversalities. BMJ Glob. Health 5 (8), e002947. doi:10.1136/bmjgh-2020-002947.

Ahmed, S., 2012. On Being Included: Racism and Diversity in Institutional Life. Duke University Press, Durham, NC. doi:10.1515/9780822395324.

Albanese, F., 2025. UN expert calls on world to end trade with Israel's "economy of genocide". *Al Jazeera*, Available at: https://www.aljazeera.com/news/2025/7/3/un-expert-calls-on-world-to-end-trade-with-israels-economy-of-genocide (Accessed 14 August 2025).

Amnesty International, 2024. *You feel like you are subhuman': israel's genocide against Palestinians in Gaza*. Available at: https://www.amnesty.org/en/documents/mde15/8668/2024/en/ (Accessed 14 August 2025).

Anderson, W., 2014. Making global health history: the postcolonial worldliness of biomedicine. Soc. Hist. Med. 27 (2), 372–384. doi:10.1093/shm/hkt126.

Araújo, L.F., Pilecco, F.B., Correia, F.G.S., Ferreira, M.J.M., 2021. Challenges for breaking down the old colonial order in global health research: the role of research funding. Lancet Glob. Health 9 (8), e1057–e1058. doi:10.1016/S2214-109X(21)00196-0.

Armenteras, D., 2025. Equity in science is a beautiful lie - and I'm done pretending. Nature 645 (8081), 561. doi:10.1038/d41586-025-02916-w.

Arumugasamy, H., Sekaran, S., 2025. Decolonising global health funding: local leadership for global impact. Lancet 405 (10484), 1052–1053. doi:10.1016/S0140-6736(25)00591-7.

Atwood, M., 1985. The Handmaid's Tale, McClelland and Stewart, Toronto.

Aultman, J., 2008. Moral courage through a collective voice. Am. J. Bioeth. 8 (4), 67–69. doi:10.1080/15265160802147149.

Bain, L.E., Adeagbo, O.A., Avoka, C.K., Amu, H., Memiah, P., Ebuenyi, I.D., 2024. Identifying the conundrums of "global health" in the Global North and Global South: a case for Sub-Saharan Africa. Front. Public Health 12, 1168505. doi:10.3389/fpubh.2024.1168505.

Bardosh, K., de Figueiredo, A., Gur-Arie, R., Jamrozik, E., Doidge, J., Lemmens, T., Keshavjee, S., Graham, J.E., Baral, S., 2022. The unintended consequences of COVID-19 vaccine policy: why mandates, passports and restrictions may cause more harm than good. BMJ Glob. Health 7 (5), e008684. doi:10.1136/bmjgh-2022-008684.

Barks, C., 1997. '*The Essential Rumi*', Castlebook, New Jersey.

Barwise, A., Liebow, M., 2019. When generosity harms health care and public health. Am. J. Public Health 109 (7), 997–998. doi:10.2105/AJPH.2019.305073.

Bayingana, A., Bhakuni, H., van de Kamp, J., Lariat, J., Rathod, L., van der Graaf, R., Abimbola, S., 2025. Dignity-based practice in global health research: a framework of expectations. Lancet Glob. Health 9 (3), e013456. Available from: https://www.thelancet.com/pdfs/journals/langlo/PIIS2214-109X(25)00242-6.pdf. doi:10.1016/S2214-109X(25)00242-6.

Bhakuni, H., Abimbola, S., 2021. Epistemic injustice in academic global health. Lancet Glob. Health 9 (10), e1465–e1470. doi:10.1016/S2214-109X(21)00301-6.

Bhambra, G.K., 2014. Connected Sociologies. Bloomsbury Academic, London. doi:10.5040/9781472544372.

Bhatti, Y., Taylor, A., Harris, M., Wadge, H., Escobar, E., Prime, M., Patel, H., Carter, A.W., Parston, G., Darzi, A.W., Udayakumar, K., 2017. Global lessons in frugal innovation to improve health care delivery in the United States. Health Aff. 36 (11), 1912–1919. doi:10.1377/hlthaff.2017.0380.

Birn, A.E., 2014. Philanthrocapitalism, past and present: the Rockefeller Foundation, the Gates Foundation, and the setting(s) of the international/global health agenda. Hypothesis 12 (1), e8.

Burgess, R.A., 2024. Rethinking Global Health: Frameworks of Power, Taylor & Francis, London.

Büyüm, A.M., Kenney, C., Koris, A., 2020. Decolonising global health: if not now, when? BMJ Glob. Health 5, e003394. doi:10.1136/bmjgh-2020-003394.

Carleton, R., 1970. ABC's Richard Carleton interviewing Ghassan Kanafani. [Television broadcast]. ABC News, Available at: https://www.abc.net.au/news/2024-09-19/abc-richard-carleton-interviewing-ghassa/104368218 (Accessed 30 Oct, 2025)

Charani, E., Abimbola, S., Pai, M., Adeyi, O., Mendelson, M., Laxminarayan, R., Rasheed, M.A., 2022. Funders: the missing link in equitable global health research? PLOS Glob. Public Health 2 (6), e0000583. doi:10.1371/journal.pgph.0000583.

Chattopadhyay, S., 2024. On not studying (and stating) the obvious. *BMJ Global Health Blogs*. Available at: https://blogs.bmj.com/bmjgh/2024/08/09/on-not-studying-and-stating-the-obvious/ (Accessed 14 August 2025).

Chaudhuri, M.M., Mkumba, L., Raveendran, Y., Smith, R.D., 2021. Decolonising global health: beyond "reformative" roadmaps and towards decolonial thought. BMJ Glob. Health 6 (6), e006371.

Clarke, A., Richter, K., Lokot, M., Rivas, A.M., Hafez, S., Singh, N.S., 2024. Decolonising humanitarian health: a scoping review of practical guidance. PLOS Glob. Public Health 4 (10), e0003566. doi:10.1371/journal.pgph.0003566.

Contractor, S.Q., Dasgupta, J., 2022. Is decolonisation sufficient? BMJ Glob. Health 7 (12), e009716. doi:10.1136/bmjgh-2022-009716.

Cook-Lundgren, E., Girei, E., 2025. Ethics of quantification and randomised control trials in international development: a decolonial analysis. J. Bus. Ethics 196 (2), 241–254. doi:10.1007/s10551-024-05684-1.

Cornwall, W., 2025. Trump's "fear factor": scientists go silent as funding cuts escalate–many worry about retribution. But for others, speaking out is worth the risk. *Science*. Available at: https://www.science.org/content/article/trump-s-fear-factor-scientists-go-silent-funding-cuts-escalate (Accessed 14 August 2025).

Crane, J.T., 2010. Unequal "partners": AIDS, academia, and the rise of global health. Behemoth 3 (3), 78–97. doi:10.6094/behemoth.2010.3.3.678.

Crane, J.T., 2013. Scrambling for Africa: AIDS, Expertise, and the Rise of American Global Health Science, Cornell University Press, Ithaca, NY.

Crisp, N., 2010. Turning the World Upside Down: The Search for Global Health in the 21st Century, CRC Press, Boca Raton, FL.

de Bengy Puyvallée, A., Storeng, K.T., Rushton, S., 2025. The Gates Foundation's network diplomacy in European donor countries. Glob. Health 21, 22. doi:10.1186/s12992-025-01112-9.

Demeter, M., 2020. Academic Knowledge Production and the Global South: Questioning Inequality and Under-representation, Palgrave Macmillan, Cham.

Dijkstra, N., von Rein, T., Kok, P., Fleming, S.M., 2025. A neural basis for distinguishing imagination from reality. Neuron 113 (15), 2536–2542.e4. doi:10.1016/j.neuron.2025.05.015.

Downs, J., 2020. Maladies of Empire: How Colonialism, Slavery, and War Transformed Medicine, Harvard University Press, Cambridge, MA.

Equinox Racial Justice Initiative, 2023. "NGOs have tamed a generation of activists ... ", *X*. Available at: https://x.com/Equinoxrji/status/1660585452783403008 (Accessed 13 August 2025).

Erondu, N.A., Aniebo, I., Kyobutungi, C., Mafaune, H.W., Jamison, D., 2021. Open letter to international funders of science and development in Africa. Nat. Med. 27 (5), 742–744. doi:10.1038/s41591-021-01307-8.

Escobar, A., 2018. Designs for the Pluriverse: Radical Interdependence, Autonomy, and the Making of Worlds, Duke University Press, Durham, NC.

Farmer, P., 2003. Pathologies of Power: Health, Human Rights, and the New War on the Poor, University of California Press, Berkeley.

Farmer, P., 2013. To Repair the World: Paul Farmer Speaks to the Next Generation, Vol. 29. University of California Press, Berkeley.

Fayemi, A.K., Kirchhoffer, D.G., Pratt, B., 2025. Solidarity and its decoloniality in global health ethics. Int. J. Equity Health 24, 13. doi:10.1186/s12939-025-02380-y.

Felix, M., 2025. Towards a theory of decolonial imaginations. Glob. Policy J. Available at: https://www.globalpolicyjournal.com/blog/11/02/2025/theory-decolonial-imaginations (Accessed 30 October 2025).

Flannery, M.E., 2025. Trump cancels federal research grants. What are the consequences?, *NEA Today*. Available at: https://www.nea.org/nea-today/all-news-articles/trump-cancels-federal-research-grants-what-are-consequences (Accessed 14 August 2025).

Gates Foundation, 2025. *Gates Foundation commits $2.5 billion to advance women's health and wellbeing globally*. [Press release]. Available at: https://www.gatesfoundation.org/ideas/media-center/press-releases/2025/08/womens-health-funding-commitment (Accessed 18 July 2025).

Global Health 50/50, 2023. *Annual report*. Available at: https://globalhealth5050.org (Accessed 1 August 2025).

Granados, C., Prabhu, J., 2022. Making the affordable aspirational: increasing the adoption of frugal innovations at the base of the pyramid, *Stanford Social Innovation Review*. Available at: https://ssir.org/articles/entry/making_the_affordable_aspirational_increasing_the_adoption_of_frugal_innovations_at_the_base_of_the_pyramid (Accessed 12 August 2025).

Grand Challenges Canada, 2025. *Play Learn Thrive funding call*. Available at: https://www.grandchallenges.ca/2025/play-learn-thrive-funding-call/ (Accessed 13 August 2025).

Grantham-McGregor, S., Cheung, Y.B., Cueto, S., Glewwe, P., Richter, L., Strupp, B., 2007. Developmental potential in the first 5 years for children in developing countries. Lancet 369 (9555), 60–70.

Grassi, L., Nanni, M.G., Riba, M., Folesani, F., 2024. Dignity in medicine: definition, assessment and therapy. Curr. Psychiatry Rep. 26 (6), 273–293. doi:10.1007/s11920-024-01506-3.

Green, A., 2025. The emotional fallout of mass USAID and NGO layoffs, *Devex*. Available at: https://www.devex.com/news/the-emotional-fallout-of-mass-usaid-and-ngo-layoffs-110295 (Accessed 13 August 2025).

Greene, R., 2000. The 48 Laws of Power, Penguin Books, New York.

Hafez, S., Clarke, A., Richter, K., Lokot, M., Rivas, A.M., Singh, N.S., 2025. Dismantling colonial legacies: decolonising research and teaching at the Health in Humanitarian Crises Centre, London School of Hygiene and Tropical Medicine. PLOS Glob. Public Health 5 (7), e0004833. doi:10.1371/journal.pgph.0004833.

Hafez, S., Pacho, A., Ponsford, R., Ranganathan, M., Gafos, M., Soremekun, S., 2026. Decolonising implementation science: a call for methodological pluralism. Lancet Glob. Health 14 (2), e281–e285. doi:10.1016/S2214-109X(25)00477-2.

Hamid, M., Rasheed, M.A., 2022. A new path to mentorship for emerging global health leaders in low-income and middle-income countries. Lancet Glob. Health 10 (7), e946–e948. doi:10.1016/S2214-109X(22)00230-3. PMID: 35714639.

Hedt-Gauthier, B.L., Jeufack, H.M., Neufeld, N.H., Alem, A., Sauer, S., Odhiambo, J., Boum, Y., Shuchman, M., Volmink, J., 2019. Stuck in the middle: a systematic review of authorship in collaborative health research in Africa, 2014–2016. BMJ Glob. Health 4 (5), e001853. doi:10.1136/bmjgh-2019-001853.

Hirsch, L.A., 2021. Is it possible to decolonise global health institutions? Lancet 397 (10270), 189–190. doi:10.1016/S0140-6736(20)32763-X. PMID: 33453772.

Hirsch, L.A., Martin, R., 2022. LSHTM and Colonialism: A Report on the Colonial History of the London School of Hygiene & Tropical Medicine (1899–c.1960). London School of Hygiene & Tropical Medicine, London. doi:10.17037/PUBS.04666958.

Horton, R., 2021. Offline: the myth of "decolonising global health". Lancet 398 (10312), 1673. doi:10.1016/S0140-6736(21)02428-4. PMID: 34742372.

Hosmer, L.T., Kiewitz, C., 2005. Organizational justice: a behavioral science concept with critical implications for business ethics and stakeholder theory. Bus. Ethics Q. 15 (1), 67–91. doi:10.5840/beq20051513.

Indigenous [@AmericanIndian8], 2025. [Tweet with photo]. X. https://x.com/AmericanIndian8/status/1952436878164037840/photo/1.

Kalinga, C., 2019. Caught between a rock and a hard place: navigating global research partnerships in the global South as an indigenous researcher. J. Afr. Cult. Stud. 31 (3), 270–272. doi:10.1080/13696815.2019.1630261.

Khan, M.S., Guinto, R.R., Boro, E., Rahman-Shepherd, A., Erondu, N.A., 2022a. The need for metrics to measure progress on racial equity in global public health and medicine. Lancet 400 (10368), 2019–2021. doi:10.1016/S0140-6736(22)02464-3. PMID: 36502829.

Khan, M.S., Naidu, T., Torres, I., Noor, M.N., Bump, J.B., Abimbola, S., 2024. The Lancet and colonialism: past, present, and future. Lancet 403 (10433), 1304–1308. doi:10.1016/S0140-6736(24)00102-8. PMID: 38555135.

Khan, M.S., Obasi, A., Deb, R., Ceesay, S.J., 2022b. Committing to anti-racism reforms? Three critical building blocks for global health organizations. PLOS Glob. Public Health 2 (7), e0000653. doi:10.1371/journal.pgph.0000653.

Koum Besson, E.S., Clarke, A., Kedebe, M.A., Ismail, Z., Pensa, S., 2022a. Introduction to Decoloniality and Anti-racism in Global Health: Student Toolkit, London School of Hygiene and Tropical Medicine. Available at:, London. https://researchonline.lshtm.ac.uk/id/eprint/4664957. (Accessed 12 August 2025) (Unpublished).

Koum Besson, E.S, 2022b. How to identify epistemic injustice in global health research funding practices: a decolonial guide. BMJ Glob. Health 7 (4), e008950.

Krugman, D.W., 2023. Global health and the elite capture of decolonization: on reformism and the possibilities of alternate paths. PLOS Glob. Public Health 3 (6), e0002103. doi:10.1371/journal.pgph.0002103.

Krugman, D.W., 2025. I know nuance breaks our brains these days. *X*. Available at: https://x.com/d_krugman/status/1940123443976196395 (Accessed 4 August 2025).

Krugman, D.W., Bayingana, A., 2025. Soft money, hard power: mapping the material contingencies of change in global health academic structures. PLOS Glob. Public Health 5 (5), e0004622. doi:10.1371/journal.pgph.0004622.

Krugman, D.W., Schor, M., Cipriano, G., Dhlamini, B., Nasser Eddine, G., Salam, Z., Smith, J., 2024. We do not deserve to be called "global health". *Medium*. Available at: https://medium.com/@daniel_krugman/we-do-not-deserve-to-be-called-global-health-d894c73cd36b (Accessed 12 August 2025).

Kruk, M.E., Gage, A.D., Arsenault, C., Jordan, K., Leslie, H.H., Roder-DeWan, S., Adeyi, O., Barker, P., Daelmans, B., Doubova, S.V., English, M., 2018. High-quality health systems in the sustainable development goals era: time for a revolution. Lancet Glob. Health 6 (11), e1196–e1252. doi:10.1016/S2214-109X(18)30386-3.

Kumar, R., 2025. How do we pick our experts?. *TEDxLusaka*, [Video]. Available at: https://www.youtube.com/watch?v=KoIXWFtHjhc (Accessed 10 July 2025).

Kumar, R., Khosla, R., McCoy, D., 2024. Decolonising global health research: shifting power for transformative change. PLOS Glob. Public Health 4 (4), e0003141. doi:10.1371/journal.pgph.0003141.

Kwete, X., Tang, K., Chen, L., Ren, R., Chen, Q., Wu, Z., Cai, Y., Li, H., 2022. Decolonizing global health: what should be the target of this movement and where does it lead us? Global Health Res. Policy 7 (1). doi:10.1186/s41256-022-00237-3.

Kyobutungi, C., Okereke, E., Abimbola, S., 2025. After USAID: what now for aid and Africa? BMJ 388, r479. doi:10.1136/bmj.r479.

Lachman, V.D., 2007. Moral courage: a virtue in need of development? Medsurg Nurs. 16 (2), 131–133.

Lambin, R., Surender, R., 2023. The rise of big philanthropy in global social policy: implications for policy transfer and analysis. J. Soc. Policy 52 (3), 602–619. doi:10.1017/S0047279421000775.

Lawrence, D.S., Hirsch, L.A., 2020. Decolonising global health: transnational research partnerships under the spotlight. Int. Health 12 (6), 518–523. doi:10.1093/inthealth/ihaa073.

Liverpool, L., 2021. Researchers from global south under-represented in development research, *Nature*. Available at: https://www.nature.com/articles/d41586-021-02549-9 (Accessed 15 June, 2025).

Mahajan, M., 2019. The IHME in the shifting landscape of global health metrics. Glob. Policy 10 (S1), 110–120. doi:10.1111/1758-5899.12605.

Maldonado-Torres, N., 2007. On the coloniality of being: contributions to the development of a concept. Cult. Stud. 21 (2–3), 240–270. doi:10.1080/09502380601162548.

Marmot, M., 2015. The health gap: the challenge of an unequal world. Lancet 386 (10011), 2442–2444. doi:10.1016/S0140-6736(15)00150-6.

Martin, C.. 2016. The reductive seduction of other people's problems, *BRIGHT Magazine*, Available at: https://brightthemag.com/the-reductive-seduction-of-other-people-s-problems-3c07b307732d (Accessed 12 August 2025).

Mbaye, R., Gebeyehu, R., Hossmann, S., et al., 2019. Who is telling the story? A systematic review of authorship for infectious disease research conducted in Africa, 1980–2016. BMJ Glob. Health 4 (5), e001855. doi:10.1136/bmjgh-2019-001855.

Mbembe, A., 2016. Decolonizing Knowledge and the Question of the Archive, Wits Institute for Social and Economic Research, Johannesburg.

McAllister, T.G., Walker, L.A., Smiles, N.D., Jennings, L., Moggridge, B., Naepi, S., Kamai, B., Milligan-McClellan, K., 2025. Decolonize scientific institutions, don't just diversify them. Nature 644 (8076), 331–334. doi:10.1038/d41586-025-00442-0.

McCoy, D., Kapilashrami, A., Kumar, R., Rhule, E., Khosla, R., 2023. Developing an agenda for the decolonization of global health. Bull. World Health Organ. 102 (2), 130. doi:10.2471/BLT.23.289807.

Mehjabeen, D., Patel, K., Jindal, R.M., 2025. Decolonizing global health: a scoping review. BMC Health Serv. Res. 25 (1), 828. doi:10.1186/s12913-025-12890-8.

Memon, A., 2015. SIUT: a model of philanthropy in Pakistan. Pak. J. Med. Sci. 31 (6), 1539–1541. doi:10.12669/pjms.316.8915.

Mignolo, W.D., 2009. Epistemic disobedience, independent thought and decolonial freedom. Theory, Culture & Society 26 (7–8), 159–181. doi:10.1177/0263276409349275.

Mignolo, W.D., Walsh, C.E., 2018. On Decoloniality: Concepts, Analytics, Praxis. Duke University Press, Durham, NC. doi:10.2307/j.ctv11g9616.

Moyo, D., 2009. Dead Aid: Why Aid Is Not Working and How There Is a Better Way for Africa, Farrar, Straus and Giroux, New York.

Nadim, H., 2023. Aid, Politics and the War of Narratives in the US-Pakistan Relationa: A Case Study of Kerry Lugar Berman Act, first ed. Routledge, New York, p. 140. doi:10.4324/9781003193210.

Nadim, H., 2025. A single audit of these democracy promotion programs', X (formerly Twitter), @HNadim87. Available from: https://x.com/HNadim87/status/1891141302919549233. (Accessed 30 May 2025).

Naidu, T., 2021a. Modern medicine is a colonial artifact: introducing decoloniality to medical education research. Acad. Med. 96 (11S), S9–S12.

Naidu, T., 2021b. Southern exposure: levelling the Northern tilt in global medical and medical humanities education. Adv. Health Sci. Educ. 26 (2), 739–752.

Naidu, T., 2021c. Says who? Northern ventriloquism, or epistemic disobedience in global health scholarship. Lancet Glob. Health 9 (9), e1332–e1335.

Naidu, T., Abimbola, S., 2022. How medical education holds back health equity. Lancet 400 (10352), 556–557.

Ndlovu-Gatsheni, S.J., 2014. Global coloniality and the challenges of creating African futures. Strategic Rev. South. Afr. 36(2). 181. https://upjournals.up.ac.za/index.php/strategic_review/article/view/189/141 (Accessed 4 August 2025).

Niang, M., Dupéré, S., Alami, H., Gagnon M., 2021. Why is repositioning public health innovation towards a social paradigm necessary? A reflection on the field of public health through the examples of Ebola and Covid-19. Global Health, 17 (1), 46. doi:10.1186/s12992-021-00695-3.

Nordling, L., 2021. Head of Wellcome-funded Malawi health project investigated for bullying, *Nature News*. Available at: https://www.nature.com/articles/d41586-021-02662-9 (Accessed 4 August 2025).

Obasi, A.I., Barley, A., Khan, M., 2022. Racism in two UK global health institutions. Lancet 399 (10332), 1287–1288. doi:10.1016/S0140-6736(22)00441-8.

Odjidja, E.N., 2021. What is wrong with global health? So-called glorified data collectors in low-income regions. Lancet Glob. Health 9 (10), e1365. doi:10.1016/S2214-109X(21)00331-4.

Okumu, N., 2025. Epistemic injustice in global health and development: the best ideas in global health and development often come from the margins. So why do we still ignore them?, *The Elephant*. Available at: https://www.theelephant.info/reflections/2025/08/06/epistemic-injustice-in-global-health-and-development/ (Accessed 14 August 2025).

Ong'era, F.M., Stewart, J., Bukusi, E., 2021. Why and for whom are we decolonising global health? Lancet Glob. Health 9 (10), e1359–e1360. doi:10.1016/S2214-109X(21)00317-9.

Onyes, U.M., Tapsoba, Y., Shellaby, L.S., Chen, N., 2025. The end of USAID: a catalyst for Africa's sustainable community health financing future, *Africa Frontline First*. Available at: https://africafrontlinefirst.org/2025/03/24/the-end-of-usaid-a-catalyst-for-africas-sustainable-community-health-financing-future/ (Accessed 13 August 2025).

Opara, I.N., 2021. It's time to decolonize the decolonization movement. Speaking of Medicine and Health (PLOS Blogs). Available from: https://speakingofmedicine.plos.org/2021/07/29/its-time-to-decolonize-the-decolonization-movement/ (Accessed 30 May 2026).

Oppong, S., 2023. Promoting global ECD top-down and bottom-up. Ethos 51 (3), 321–325.

Oxfam, 2025. What's happening in Gaza and Israel?, *Oxfam in Action*. Available at: https://www.oxfam.org.uk/oxfam-in-action/current-emergencies/whats-happening-in-gaza-and-israel/ (Accessed 13 August 2025).

Packard, R.M., 2016. A History of Global Health: Interventions into the Lives of Other Peoples, Johns Hopkins University Press, Baltimore.

Pai, M., Bandara, S., Kyobutungi, C., 2024. Shifting power in global health will require leadership by the Global South and allyship by the Global North. Lancet 404 (10464), 1711–1713. doi:10.1016/S0140-6736(24)01752-6.

Pant, I., Khosla, S., Lama, J.T., et al., 2022. Decolonising global health evaluation: synthesis from a scoping review. PLOS Glob. Public Health 2 (11), e0000306. doi:10.1371/journal.pgph.0000306.

Prasad, A., 2023. Rajat Khosla: shifting power structures in global health. Lancet 401 (10377), 635. doi:10.1016/S0140-6736(23)00266-6.

Quijano, A., 2000. Coloniality of power and Eurocentrism in Latin America. Int. Sociol. 15 (2), 215–232. doi:10.1177/0268580900015002005.

Quinn, R.E., Thakor, A.V., 2018. Creating a purpose-driven organization. Harv. Bus. Rev. 96 (4), 78–85.

Rahman-Shepherd, A., Erondu, N.A., Anwar, B., Boro, E., Chau, T. D., Guinto, R.R., Hollmann, L., Mejarito, J.A., Rasheed, M.A., Khan, M., 2023. Antiracism in leading public health universities, journals and funders: commitments, accountability and the decision-makers. BMJ Glob. Health 8 (3), e010376. doi:10.1136/bmjgh-2022-010376.

Rasheed, M.A., 2021. Navigating the violent process of decolonisation in global health research: a guideline. Lancet Glob. Health 9 (12), e1640–e1641. doi:10.1016/S2214-109X(21)00412-6.

Rasheed, M.A., 2023a. The case for a Global South centred model in global health. BMJ 383. doi:10.1136/bmj.p2282.

Rasheed, M.A., 2023b. *The same story with global health research money even with local staff as I saw in Pakistan*. [Tweet]. X (formerly Twitter). Available at: https://x.com/muneera_rasheed/status/1651813740889935872 (Accessed 14 August 2025)

Rasheed, M.A., Holding, P., 2024. Whose child is it? A psychological perspective on responsibility and accountability in decision making on nurturing care in early childhood. Ethos 52 (2).

Rasheed, M.A., Kedzierski, J.T., Hasan, B.S., 2021. Improved family experience outcomes in a pediatric hospital in Pakistan: mentoring, human-centered practice, and theory of change. NEJM Catalyst Innov. Care Deliv. 2 (7). doi:10.1056/CAT.21.0099.

Razack, S., Richardson, A., Pillay, N., 2025. Medical education and the coloniality of power. Med. Educ. Q. 29 (1), 112–119.

Rucavado Rojas, D., Postigo, J.C., 2025. The cycle of underrepresentation: structural and institutional factors limiting the representation of Global South authors and knowledge in the IPCC. Clim. Change 178, 19. doi:10.1007/s10584-025-03857-z.

Sablich, J., Ainsworth, D., 2025. US aid tracker: following Trump's cuts to international development, *Devex*. Available at: https://www.devex.com/news/us-aid-tracker-following-trump-s-cuts-to-international-development-109177 (Accessed 13 August 2025).

Sahibzada, M., 2025. *Visibility is not the same as dignity* ... [Post]. LinkedIn. https://www.linkedin.com/posts/activity-7362133743337164800-pIwk/?utm_source=share&utm_medium=member_android&rcm=ACoAACLIJjABQ9QtVqsK_SfwAzfeiJEc79pMePE.

Saini, A., 2016. Physicians of colonial India (1757-1900). J. Fam. Med. Prim. Care 5 (3), 528–532. doi:10.4103/2249-4863.197257.

Santos, B.D.S., 2014. Epistemologies of the South: Justice against Epistemicide, Paradigm Publishers, Boulder.

Santos, B.D.S., 2018. The End of the Cognitive Empire: The Coming of Age of Epistemologies of the South. Duke University Press, Durham, NC. doi:10.1515/9781478002005.

Scheidecker, G., Abebe, T., Abimbola, S., Chaudhary, N., Frankenhuis, W.E., Keller, H., Munakata, Y., Olusanya, B.O., de Oliveira Ramos, D., Rasheed, M.A., 2025. Beyond deficit models in early childhood development science. Lancet 405 (10495), 2127.

Scheidecker, G., Chaudhary, N., Keller, H., Mezzenzana, F., Lancy, D.F., 2023. 'Poor brain development'' in the global South? Challenging the science of early childhood interventions. Ethos 51 (1), 3–26.

Scheidecker, G., Tekola, B., Rasheed, M., Oppong, S., Mezzenzana, F., Keller, H., Chaudhary, N., 2024. Ending epistemic exclusion: toward a truly global science and practice of early childhood development. Lancet Child Adolesc. Health 8 (1), 3–5.

Schwab, T., 2023. The Bill Gates Problem: Reckoning with the Myth of the Good Billionaire, Metropolitan Books, New York.

Sen, A., 1999. Development as freedom. Alfred A. Knopf, New York, p. 366.

Shahjahan, R.A., Estera, A.L., Surla, K.L., Edwards, K.T., 2022. "Decolonizing" curriculum and pedagogy: a comparative review across disciplines and global higher education contexts. Rev. Educ. Res. 92 (1), 73–113.

Sharma, D., Sam-Agudu, N.A., 2023. Decolonising global health in the Global South by the Global South: turning the lens inward. BMJ Glob. Health 8 (9), e013696. doi:10.1136/bmjgh-2023-013696.

Sheikh, K., Schneider, H., 2025. Donor transitions and programmatic cliffs: a critical review. Health Policy Plan. 40 (1), 17–25.

Shetty, V., Shetty, A., Shetty, D.P., 2022. Developing a sustainable high-end cardiovascular surgery program in emerging economies: the Narayana Health (NH) model structured on affordable, accessible, tertiary level care. In: Kpodonu, J. (Ed.), Global Cardiac Surgery Capacity Development in Low and Middle Income Countries. In: Sustainable Development Goals Series. Springer, Cham, pp. 499–507. doi:10.1007/978-3-030-83864-5_37.

Shrestha, R.K.M., 2016. Saving lives was our only goal. Available from: https://www.youtubecom/watch?v=yR0je4Uu4-4.

Shrestha, R.K.M., 2022. A Life of Purpose. Business 360°. Available at: https://www.b360nepal.com/detail/6131/a-life-of-purpose (Accessed 10 June 2025).

Sinha, S., et al., 2021. Inequities in global health research authorship: a bibliometric analysis. BMJ Glob. Health 6 (5), e006621. doi:10.1136/bmjgh-2021-006621.

Smith, L.T., 2012, Decolonizing Methodologies: Research and Indigenous Peoples, second ed. Zed Books, London.

Spivak, G.C., 1988. Can the subaltern speak? In: Nelson, C., Grossberg, L. (Eds.), Marxism and the Interpretation of Culture. University of Illinois Press, pp. 271–313. Available at: https://jan.ucc.nau.edu/~sj6/Spivak%20CanTheSubalternSpeak.pdf.

Stewart, D.E., Amsalu, T., Fairfoot, E., Keen, D., Keenan, J., Butcher, F., Miles, K., Razavi, A., 2025. Decolonising global health: a scoping review. medRxiv 2025–03. doi:10.1101/2025.03.26.25324588.

Stirling, A., 2024. The "Bill Gates problem": do billionaire philanthropists skew global health research? Personal priorities are often trumping real needs and skewing where charitable funding goes. Nature 626, 477–479.

Storeng, K.T., Béhague, D.P., 2014. "Playing the numbers game": evidence-based advocacy and the technocratic narrowing of the Safe Motherhood Initiative. Med. Anthropol. Q. 28 (2), 260–279. doi:10.1111/maq.12072. Epub 2014 Mar 6. PMID: 24599672; PMCID: PMC4314706.

Stuckler, D., Basu, S., McKee, M., 2011. Global health philanthropy and institutional relationships: how should conflicts of interest be addressed? PLOS Med. 8 (4), e1001020. doi:10.1371/journal.pmed.1001020.

Suleiman, A., 2025. Randomized controlled trials (RCTs) were … [LinkedIn post]. Available at: https://www.linkedin.com/posts/suleiman-a-bb945b298_randomizedcontrolledtrials-rcts-were-activity-7318233488023805952-sM9_ (Accessed 13 August 2025).

Svadzian, A., Vasquez, N.A., Abimbola, S., Pai, M., 2020. Global health degrees: at what cost? BMJ Glob. Health 5 (8), e003310. doi:10.1136/bmjgh-2020-003310. PMID: 32759185; PMCID: PMC7410003.

Tahir, N., 2025. The social cost of UN pay structures [LinkedIn post]. Available at: https://www.linkedin.com/feed/update/urn:li:activity:7347786460688777216/ (Accessed 13 July 2025).

The Global Fund, 2025. Community engagement strategic initiative – update June 2025 [Online]. Available at: https://resources.theglobalfund.org/en/more-topics/technical-cooperation/community-engagement/ (Accessed 14 August 2025).

Tuck, E., Yang, K.W., 2012. Decolonization is not a metaphor. Decolonization: Indigeneity, Education & Society 1 (1), 1–40.

UNICEF, 2023. *UNICEF calls for more investment as South Asia remains a global epicentre of undernourished and wasted children.* Retrieved from https://www.unicef.org/rosa/press-releases/unicef-calls-more-investment-south-asia-remains-global-epicentre-undernourished-and.

van Selm, G., Bukenya, B., Kamya, I., Kumi, E., Yeboah, T., Banks, N., Elbers, W., Schulpen, L., van Wessel, M., 2025. Northern NGO-centrism in localisation processes: reproducing power inequities in the aid field. Dev. Pract. doi:10.1080/09614524.2025.2543349.

Wade, M., 2023. *Come to Dakar and you'll see very clearly where your aid money goes.* [Tweet]. X (formerly Twitter). Available at: https://x.com/magattew/status/1648068946656997384 (Accessed 14 August 2025).

Washington, H.A., 2006. Medical Apartheid: The Dark History of Medical Experimentation on Black Americans from Colonial Times to the Present, first ed., vol. 6. Random House Digital Inc., New York.

Wasino, W., Hartatik, E.S., Shintasiwi, F.A., Wijanarko, N.B., Putri, L.O., 2025. A historical analysis of health services in Medan and the tobacco plantation areas of East Sumatra during the Dutch colonial period. Cogent Arts Humanit. 12 (1). doi:10.1080/23311983.2025.2482399.

Wellcome, 2024. Mental Health Award: accelerating scalable digital mental health interventions. Wellcome, London. Available from: https://wellcome.org/grant-funding/schemes/digital-mental-health-interventions (Accessed 4 June 2025).

Wilson, J., 2021. Philanthrocapitalism and global healthIn: Benatar, S., Brock, G. (Eds.), Global Health: Ethical Challenges, Cambridge University Press, Cambridge, pp. 416–428. doi:10.1017/9781108767111.028.

Wong, B.O., Blythe, J.A., Batten, B.E., Turner, J.N., Lau, J.N., Hosamani, P., Hanks, W.F., Magnus, D., 2021. Recognizing the role of language in the hidden curriculum of undergraduate medical education: implications for equity in medical training. Acad. Med. 96 (6), 842–847. doi:10.1097/ACM.0000000000003657.

World Bank, 2023. *Out-of-pocket expenditure (% of current health expenditure)*. Retrieved from https://data.worldbank.org/indicator/SH.XPD.OOPC.CH.ZS.

World Bank, 2024. *Population, total – South Asia*. Retrieved from https://data.worldbank.org/indicator/SP.POP.TOTL?locations=8S.

World Health Organization (WHO), 2023. *Experts, officials meet to accelerate reduction of maternal, newborn and child deaths in WHO South-East Asia Region*. Retrieved from https://www.who.int/southeastasia/news/detail/19-09-2023-experts–officials-meet-to-accelerate-reduction-of-maternal–newborn-and-child-deaths-in-who-south-east-asia-region.

CHAPTER

Toward inclusive medical education, research, and practice: Removing race from genetics

12

Pedro W. Fernandez[a], Phila M. Msimang[b], Caitlin Uren[c], and Desiree C. Petersen[c]

[a]*Division of Urology, Faculty of Medicine and Health Sciences, Stellenbosch University, Cape Town, South Africa*

[b]*Department of Philosophy, Stellenbosch University, Stellenbosch, South Africa*

[c]*South African Medical Research Council Centre for Tuberculosis Research, Division of Molecular Biology and Human Genetics, Faculty of Medicine and Health Sciences, Stellenbosch University, Cape Town, South Africa*

Introduction

"The past is always with us, for it feeds the present"—

Ruskin Bond

Race in education and genetics research—Past and present

Racial classification based on physical appearance and assumed biological relatedness gained a foothold during the colonial expansionism era of the 19th century (Msimang and Sarkar, 2026). The racist ideology that accompanied it formed the basis of injustices like the Trans-Atlantic slave trade, in which darker skin was associated with inferiority. Indigenous peoples across the world subjected to colonization were seen through this same racial ideology, which affected how they were treated and classified by colonists. The persistent use of race as a means of subjugation and classification became pervasive across society, including scientific and educational institutions. The legacies of these ideas and practices in educational institutions are still with us today, although more surreptitiously concealed. Dealing with these legacy issues and how they persist in present practices within institutions remains of paramount importance in improving education and creating a more equitable field of genetics and genomics.

There have been calls to reform medical and health science education as a prerequisite to addressing inequalities in clinical and biomedical research (Donovan, 2022;

Decolonizing Medical Education, Research and Practice. https://doi.org/10.1016/B978-0-443-44998-7.00010-7

Gingell and Bergemann, 2022). Past failures, whether intentional or unintentional, to implement these reforms have contributed to perpetuating genetic essentialism—the idea that we are destined to be a certain way because of our genes, or, more willfully, the false belief that people of the same race share the same set of genes that make them physically, cognitively, and behaviorally uniform, but different from "other" races (Dar-Nimrod and Heine, 2011).

The American Society of Human Genetics in 2018 described race as a social construct that has evolved over time and stated that using genetics to rank populations demonstrates a fundamental misunderstanding of genetics (Box 12.1). More recently, the *Committee on the Use of Race, Ethnicity, and Ancestry as Population Descriptors in Genomics Research,* under the banner of the National Academies of Sciences, Engineering, and Medicine (NASEM), released a seminal document entitled "Using Population Descriptors in Genetic and Genomics Research: A New Framework for an Evolving Field" (NASEM, 2023) with the view of addressing the continued and contentious use of race in genomics—some of the key recommendations of this document are discussed in the latter part of this chapter.

Box 12.1 Why are race and ethnicity not relevant biological variables?

Race, as applied to humans, has most typically referred to a group of people thought to be related by similar origins who share physical characteristics such as skin color, bone structure, or hair type. This idea of race is rooted in early attempts to understand and classify biological variation, where races were defined as natural divisions within the species. Today, it is understood that humans are a single species without natural divisions below the level of the species, meaning that the groups labeled as races in humans are social inventions rather than biological realities.

Ethnicity relates to a cultural identity made up of factors like nationality, language, religion, and the shared traditions of a community. Ethnic groups, as cultural groups, should not be confused with ancestry groups, just as race groups have also been confused in the past. Often, the use of the terms race and ethnicity is conflated, but these are separate concepts despite some of the ways they are related and some of their historical overlap in conception and use. Race is most often connected to physical traits, whereas ethnicity is strongly connected to a cultural identity and shared heritage. Each is an important aspect of self-identity that can have notable differences, one being that a single "race" group is often made up of various ethnicities, and people of the same ethnicity can belong to different so-called races. In most genetic studies, self-reported ethnicity is requested from participants and recorded. In the majority of cases, however, there is no correlation between self-reported ethnicity and genetic ancestry represented by various proxy ancestral proportions.

Ancestry refers to the study of an individual's personal lineage or genetic heritage, focusing on exploring the multidimensional historical and geographical origins of their ancestral roots by measuring genetic variation in DNA. Race, ethnicity, and ancestry are often incorrectly used interchangeably in the literature because they have historically been thought to be the same or refer to similar concepts. However, these concepts are significantly different from one another. The most important difference between them in biomedical research is that race and ethnicity are social rather than biological concepts, while ancestry is a biological concept grounded in genetics, or more precisely, genetic variation, which in turn may be related to biological processes (Thomas and Peters, 2025).

Complexities within the race–ethnicity–ancestry paradigm

In the early 1970s, Dr. Richard Lewontin first proposed that the proportion of genetic variation among socially constructed race groups is low and that racial subdivision is biologically meaningless (Lewontin, 1972). This concept was later supported with the completion of the Human Genome Project, which showed that 93% of human genetic variation occurs within populations, only 4% of variation occurs among populations on different continents, and the remaining 3% occurs within populations on the same continent. In evolutionary biology terminology, population groups are broadly equivalent to a "subdivision" or "subspecies" and can be quantified by the fixation index (FST), a statistic that measures the degree of variation within and between populations. The FST ranges from 0 (no population subdivision) to 1 (complete population subdivision), with an FST greater than 0.25 potentially indicating the existence of genetic subgroups. Interestingly, estimates from the 1000 Genomes Project, using analysis of African ancestry genetic studies, established that the global human FST is low (ranging between 0.052 and 0.083), indicating that the human population exhibits little genetic subdivision (Auton et al., 2015; Lala and Feldman, 2024) and thus falls far below the conventional threshold in biology for being considered to have biological races. Others have asserted that "genetic isolation, sharp boundaries, and distinct evolutionary lineages of 'races' do not exist" (Duello et al., 2021).

Population genetics—Classifications and misclassifications

Genetic variation and genetic ancestry

With humans adapting and changing over time to survive, genetic variation through various processes, including mutation, recombination, and gene flow, serves as the basis for natural selection. Genetic variation is, therefore, driven by the process of evolution and is the result of differences observed at the DNA level in either an individual or a group of individuals, representing what is referred to as a population group. Common human genetic variation data can, therefore, be interpreted to reconstruct and trace an individual's origins and subsequent ancestral migrations, which are linked to their genetic ancestry (Petersen and Croock, 2025).

Specific genetic variants are found at a higher frequency in some populations, and when occurring at an allele frequency greater than 1%, they are referred to as single nucleotide polymorphisms within that population group. The role of genetic variation in determining traits and disease susceptibility is what contributes to certain populations being more genetically diverse and, therefore, unique when compared to other groups. This has been extremely useful when establishing genetic risk profiles for certain health-related traits found at a higher prevalence or an increased susceptibility to specific diseases (Table 12.1).

Estimating various proportions of the human genome from different ancestral populations (e.g., 40% East African, 30% South Asian, 20% European, etc.) using genetic variation is known as global ancestry analysis. This entails the analysis of genome-wide genetic markers and compares genetic variation patterns to reference individuals from

Table 12.1 Defining genetic variation versus genetic ancestry.

	Genetic variation	Genetic ancestry
Classification	Differences due to alternate forms of DNA	Origins of DNA variation and estimated ancestral proportions
Spectrum	All DNA variation across the genome	Population origins inferred from DNA variation
Purpose	Provides a basis for ancestry inference	Can be inferred using patterns of DNA variation

different populations across the world (see Fig. 12.1). The estimation of global ancestry proportions can be calculated with moderate to high accuracy by using software tools such as ADMIXTURE and STRUCTURE (Alexander and Lange, 2011; Alexander et al., 2009; Pritchard et al., 2000). This differs from local ancestry, which infers ancestral contributions at specific chromosomal locations (called loci that contain one or more genes) along the genome rather than an overall estimation of ancestral proportions (refer to Fig. 12.1). Accuracy for local ancestry analysis tools ranges from moderate to high using programs such as RFMix, ELAI, and FLARE to identify and assign the most closely related ancestral population to each chromosomal segment (Maples et al., 2013; Guan, 2014; Browning et al., 2023). Both global and local ancestry analyses, however, are used to trace common ancestors from different geographical sites at different time points, revealing the journey of particular DNA changes (Uren et al., 2020).

Admixture and admixture mapping

Admixture occurs when individuals from genetically distinct populations reproduce, resulting in offspring with ancestry from multiple ancestral populations. This process has been fundamental to human demographic history, particularly following major population migrations and cultural exchanges. In Africa, for example, admixture between European colonizers, enslaved Indigenous, and Asian populations created genetically diverse populations with varying ancestral proportions. This includes numerous southern African populations representing various degrees of admixture, ranging generally from two to five major contributing ancestral populations (Campbell and Tishkoff, 2010; Patterson et al., 2010; Petersen et al., 2013; Uren et al., 2016, 2017). Likewise, admixture is commonly observed in populations from North, West, and East Africa (Caro-Consuegra et al., 2023; Dobon et al., 2015; Bryc et al., 2010). Admixed populations are not confined to Africa and have been characterized globally, including North and South America (Montinaro et al., 2015; Jordan et al., 2019), Europe (Gouveia et al., 2023), Oceania and Australia (Wollstein et al., 2010), and Asia (Wang et al., 2021).

Admixture mapping leverages this genomic mosaic to identify trait-associated loci by comparing local ancestry patterns between defined cases and controls. The method assumes that genomic regions showing ancestry differences between cases and control individuals harbor variants that differ in frequency across ancestral populations and contribute to trait risk. This approach (as further demonstrated in Fig. 12.2) has successfully identified genetic factors underlying conditions like prostate cancer, kidney disease, and asthma in admixed populations.

FIG. 12.1

Global ancestry analysis estimates overall proportions of ancestry as varying percentages across the genome by investigating the amount of variation compared to reference populations worldwide. Local ancestry identifies ancestry at specific chromosome positions called loci. Together, both types of analyses trace ancestral origins and the movement of genetic changes over time across continents.

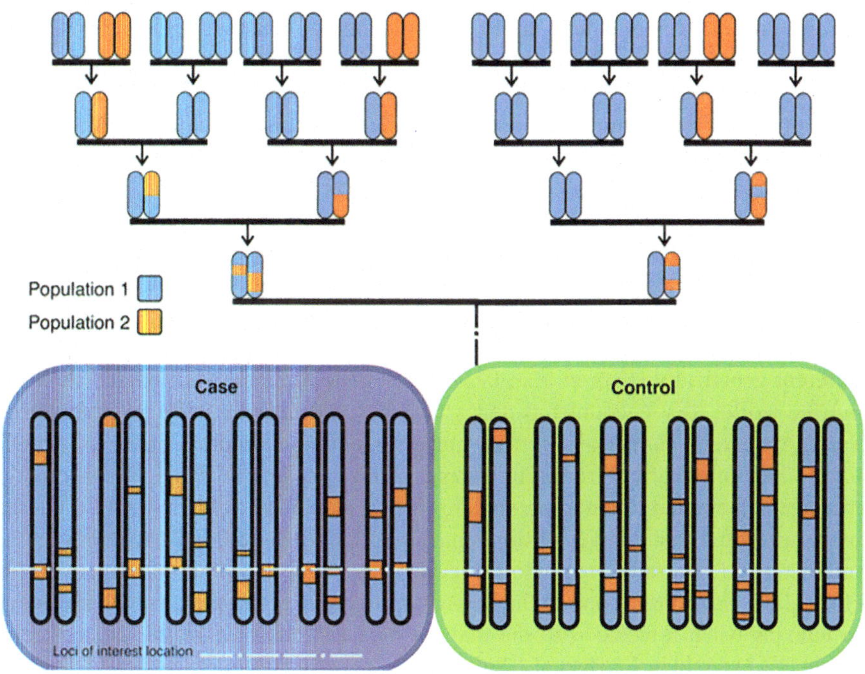

FIG. 12.2

Case–control admixture mapping showing the eventual location of the loci of interest. Admixed populations have segments of DNA from different ancestral populations. If a disease is more common in one ancestral population, regions in the genome inherited from that ancestry will be overrepresented in affected individuals (cases) compared to unaffected individuals (controls). The loci of interest (indicated by the dashed line) are found across genomic regions where local ancestry patterns correlate strongly with disease status.

However, admixture mapping requires careful interpretation. The technique identifies genomic regions where ancestry correlates with phenotype, but this correlation may reflect complex interactions between genetic variants, environmental factors, and social determinants of health rather than simple genetic causation.

Inappropriate use of race in genetics research and clinical practice

Scientific and medical communities continually use racial categories as biologically meaningful disease-risk criteria, as well as to define differences in incidence between racial groups for certain medical conditions. In clinical trials, self-reported race or skin color is frequently used to identify individuals, particularly those of African ancestry (Duello et al., 2021), and responses to medical treatments or drugs are often compared among socially defined racial groups.

Using race as a confounder

A confounder is an unrelated variable that distorts the relationship between the exposure and the outcome, leading to incorrect inferences about causality. A common oversight in biomedical research is the reporting of race and ethnicity as a confounder without providing the appropriate context to clarify the association with the disease. Humans may be allocated into groups using genetic data; however, these groups may only crudely correspond, or not at all, to groupings by race and ethnicity based on physical traits. Evolutionary processes, such as natural selection (i.e., genetic adaptations that may confer survival advantages and the ability to reproduce in a given environment), should be taken into consideration. Although specific traits, including facial features and skin pigmentation, are routinely used to group individuals, populations that share similar physical characteristics due to natural selection can be very different genetically (Kim et al., 2021). As human populations dispersed throughout the world, they encountered new pathogens that exerted selective pressure on them. A classic example of inferring human population structure based on genetic data, which often differs from inferences based on phenotypic characteristics, is the monogenic hereditary hemoglobinopathy, sickle cell disease (SCD). SCD is caused by a mutation in both copies of the human beta-globin gene (*HBB*), resulting in the production of abnormal hemoglobin termed hemoglobin S (Hb S). However, the Hb S allele, as a single copy in the heterozygous state, confers a survival advantage to humans when infected with the malaria parasite *Plasmodium falciparum* (refer to Fig. 12.3) (Ashley-Koch et al., 2000).

SCD is commonly categorized as an African disease despite its prevalence worldwide, spanning Africa, South and Central America, the Caribbean, Saudi Arabia, India, Europe (including Turkey, Greece, and Italy), and the United States (US) (Campbell et al., 2021). Although about 90% of all SCD populations reside in malaria-endemic regions, when research is performed on populations in Western countries,

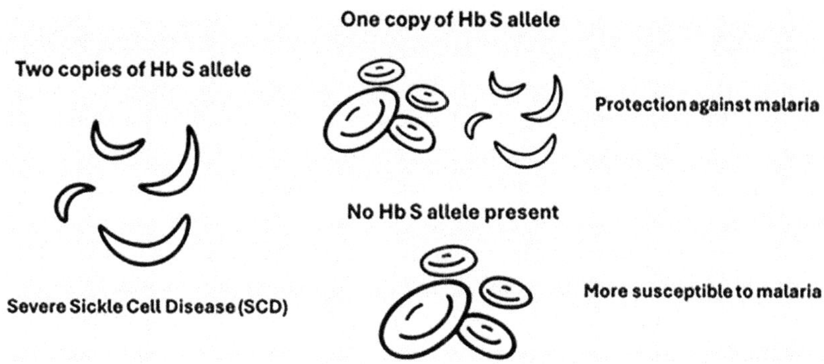

FIG. 12.3

Sickle cell disease caused by a mutation in the human beta-globin gene (HBB).

a Westernized context is often used, describing those with SCD as exclusively "non-White" (Campbell et al., 2021), even though SCD can affect all population groups. This characterization likely stems from the pervasive, scientifically racist practices in the 1920s in the US; i.e., whenever SCD was reported in White Americans, the doctors would reject the diagnosis and conclude that the individuals were not truly "White" (Tapper, 1999). A recent systematic review of 1,105 published SCD studies sought to analyze patterns of using race and ethnicity as a confounder in SCD research and whether the inclusion of race and ethnicity as a confounder was appropriately contextualized (Kidane Gebremeskel et al., 2025). Here, such analyses found that nearly one-third of studies corrected for racial or ethnic categorization, with the highest corrections for race and ethnicity found in studies from the US and Europe, while corrections in studies from the sub-Saharan African region were significantly less. Most notably, though, in 76% of all studies the context or classification criteria for the adjustment of race and ethnicity were not described. The authors concluded that the lack of standardization on the use of race and ethnicity as a confounder in SCD raised the potential for bias and misinterpretations of research outcomes.

Diagnostic and prognostic risk prediction bias

A drawback of identifying race and ethnicity as confounders is that they are often converted into correction factors and applied in clinical algorithms and clinical practice. In the past, the use of race and ethnicity as a proxy for genetic variation has led to incorrect assumptions about disease risks and treatment outcomes. Race and ethnicity are increasingly being used in diagnosis, clinical decision-making, and disease management without consideration of flawed social, environmental, biological, and genetic assumptions. Vyas and colleagues reported that race-based cardiology, nephrology, urology, and obstetrics diagnostic algorithms and risk scores made incorrect adjustments that favored White patients and were biased against non-White patients (Vyas et al., 2020). To illustrate, the estimated glomerular filtration rate (eGFR) Modification of Diet in Renal Disease and the Chronic Kidney Disease Epidemiology Collaboration equations generate higher eGFR values (suggesting better kidney function) for patients who self-identify as Black versus White, even though they display the same circulating creatinine measurements (used as an input value in such equations). Consequently, Black patients with algorithm-ascribed "better kidney functioning" could falsely be delayed specialist care and/or registering for kidney transplantation. This incorrect algorithm adjustment is applied even though kidney failure incidence is three times higher in US Black individuals compared to Whites and is attributed to social inequalities and a higher percentage of genetic risk variants in the apolipoprotein L1 gene (*APOL1*). However, a review of the literature demonstrated that coupling *APOL1* risk variants with Black race and ethnicity supported the misleading impression that racial disparities are genetically determined and that kidney disease is homogeneously distributed across all "Black" individuals (Cerdena et al., 2021). In addition, the authors suggested that this bias limited the use of conventionally applied clinical interventions, further emphasizing the effect of structural racism as a significant determinant of the disparities observed in kidney disease.

Similarly, the Breast Cancer Surveillance Consortium Risk Calculator ranks the highest risk for White patients, while lower risks are ascribed (in descending order) to Native American, Black, Hispanic, and Asian patients, respectively. The lower risk estimate for non-White patients reduces the likelihood that they will receive active surveillance follow-up (Vyas et al., 2020). This skewed risk prediction is used even though, on average, Black women in the US are younger at diagnosis, are more likely to be diagnosed with aggressive or advanced forms of breast cancer, and have the highest disease-associated mortality.

Questions about the use of race adjustment algorithms in South Africa are a growing issue of concern. Similar to the use of race adjustment in the US, the United Kingdom, and other places around the world, race adjustment algorithms in South Africa are still used across medical genetics and clinical decision-making. For instance, recent guidance published by a group of scientists led by the President of the South African Thoracic Society has "strongly" recommended that for Black South African and mixed-ethnicity populations, the Global Lung Initiative's (GLI) "Other reference equation should be utilized when performing spirometry" and that the "GLI-White equation should be utilized for White South Africans, and the GLI-South East Asian equation for Indian [South Africans]" (Masekela et al., 2020). This position is reflected in the South African Thoracic Society's most recent guidelines released in 2022, which state that a patient's "self-identified ethnic group" should be recorded alongside certain biological variables as part of what is "required for reference purposes" (Maree et al., 2022).

Race and pharmacogenomics

Pharmacogenomics is the study of the relationship between genes and the response to drugs, based on the premise that it could lead to "personalized medical treatment." However, costs and feasibility limit the availability of genetic tests for everyone, and even more so in low-resourced countries, which often bear the highest burden of disease. Because of these limitations, pharmacogenomics relies on broader genetic stratification by population group, assuming the genetic differences between populations are sufficient to predict drug response (Chang et al., 2020). However, research has shown that there is greater genetic diversity within so-called race and ethnic groups than between such groups. Several studies reported a high frequency of the human leukocyte antigen B 5801 allele (*HLA-B*5801*) in Southeast Asian and US Black men, and a low frequency in other populations (Chang et al., 2020; Fitzgerald et al., 2020). The *HLA-B*5801* is known to be associated with causing life-threatening severe cutaneous adverse reactions in patients treated with allopurinol, used for gout, kidney stones, and uric acid nephrolithiasis. Consequently, the American College of Rheumatology recommended that all individuals of Southeast Asian and US Black African descent be tested for the *HLA-B*5801*. Subsequent data, however, showed that it was incorrect to broadly assume potential allopurinol effects are confined to individuals of Southeast Asian and Black African descent (Goodman and Brett, 2021). The *HLA-B*5801* frequency in Switzerland, which has a largely homogeneous White population, was shown to vary widely by city; for example, in Basel, the *HLA-B*5801* frequency was found to be comparable to the high allele frequencies that are observed in US Black men (Goodman and Brett, 2021).

Epigenetics and the environment

A limiting factor in genetics-based diagnosis, risk predictions, and pharmacogenomics is the challenge of accounting for the complex effects that physical, social, chemical, and biotic factors (environmental interactions) have on individuals or communities. Assessing environmental exposures has been difficult without physical quantitative measurements, and comprehensive lifestyle interviews of research participants/patients are further compounded by the effects of recall bias. However, the measurement of nongenetic factors as a proxy for environmental exposures has gained impetus through the identification of biological mechanisms such as DNA methylation and histone modification (Cavalli and Heard, 2019). Epigenetics broadly encompasses molecular modifications along the human genome without altering the DNA sequence. Genes are not static; rather, epigenetic modifications permit genes to become fluid and able to respond to environmental stimuli (e.g., cigarette smoking, toxins and carcinogens, infections, radiation, physical activity, etc.) and adapt biological processes in response thereto (Cavalli and Heard, 2019). A recent study demonstrated greater epigenetics-associated molecular aging in older adults living in US neighborhoods with larger concentrations of socioeconomic disadvantage; however, there were no differences per reported race and ethnicity, and the associations were only attenuated when cigarette smoking was included in their analysis models (Robinette and Smith, 2025).

Studies have identified underlying epigenetic differences among population groups, primarily for diseases that show observable disparities in risk, presentation, and clinical outcomes. However, some have suggested that the performance of epigenetic predictors is still not fully understood. A recent analysis of epigenetic predictors of cell proportion estimates, plasma protein levels, telomere length, and chronological age in relation to the self-reported race/ethnicity and sex of adults from the US National Health and Nutrition Examination Survey (NHANES) found significant variation in performance by reported race or ethnicity and sex in terms of correlation and median difference between observations and their model predictions (i.e., median absolute error) (Khodasevich et al., 2025). The variation was suggested to stem primarily from the fact that participants of European ancestry are historically overrepresented in epigenetic research, resulting in limited diversity of populations to appropriately train epigenetic prediction models—broader inclusivity of diverse population groups in biomedical research is discussed in a later section.

Race in addressing health disparities

The National Institutes of Health defines health disparities as preventable health differences that adversely affect populations who experience greater challenges to optimal health and are closely linked with intergenerational social, economic, and/or environmental disadvantages. It is primarily based on identification as an individual from a racial and/or ethnic group and/or by low socioeconomic status. Socioeconomic status is defined as a social construct dependent on variables such as educational attainment, employment, and income. Race and ethnicity are also noted as social constructs that should not be applied as a proxy for human genetic variation. Notwithstanding, there are defined guidelines for selecting an appropriate reference population or group in racial and ethnic health disparities research (Dankwa-Mullan et al., 2021) (Box 12.2). The recent NASEM report notes that even though race and ethnicity may capture some shared exposure and therefore may be suitable for some health disparities studies, it is a poor proxy for other environmental exposures. Therefore, the committee recommended that race should not be used outside of a specific subset of health disparities research.

Box 12.2 Considerations in health disparities research

Race and ethnicity have historically been misused as biological variables; however, there is now broader consensus regarding how racial and ethnic classifications are influenced by social, economic, and political contingencies. Race and ethnicity are social constructs based on self-ascribed identities or socially ascribed classifications that have important social meanings to a particular group but may not have direct biological associations. When used in reporting disease rates (e.g., cancer), race and ethnicity can point to important differences experienced by diverse groups affected by factors such as physical and social determinants of health and environmental exposures. Neglecting to report race and ethnicity in health and medical research can disregard the "lived experiences" and structural factors that produce health inequities along the dimensions of race and ethnicity. Collecting data and reporting on race and ethnicity can provide invaluable insights into the realities of societal injustices and inequalities in how they affect population health. This kind of data can assist in crafting appropriate interventions that are sensitive to these issues of injustice and inequality across such groups.

A study that proposed structural racism as a potential driver of health disparities developed a measure that showed an association between structural racism and breast cancer treatment delays among Black, but not White, patients (Reeder-Hayes et al., 2023). Addressing health disparities remains complex, given the myriad of physical, community-level barriers that may influence socioeconomic status or access to health care. Moreover, individual-level biological factors (age, metabolic and hormonal stressors, pathogens), genetic factors (inherited conditions, epigenetics), and lifestyle/behavioral factors (tobacco use, physical activity, diet) may further contribute to disparities. For certain diseases such as prostate cancer, there appears to be continued use of the term "racial disparities," largely due to the significant differences in incidence, clinical presentation, and disease outcomes (morbidity and mortality) observed between men of African ancestry compared to men of European or Asian ancestry (Roach et al., 2023).

A recent investigation sought to determine whether genetic ancestry was associated with prostate cancer risk and mortality in the context of neighborhood deprivation (surrounding environment) influences (Pichardo et al., 2024). The case–control study of US Black versus White men assessed the percentage of West African ancestry in each participant. In men of African ancestry, they found that West African ancestry was higher in cases than in controls (70% vs. 67%); interestingly, West African ancestry was also detected in cases (9%) and controls (9%) classified as being of European ancestry. Furthermore, they observed that West African genetic ancestry was associated with increased odds of prostate cancer in neighborhoods with high deprivation, while the odds of getting prostate cancer were lower in more affluent neighborhoods. The study demonstrated that surrounding environments may play a key role in defining how genetic ancestry modulates prostate cancer risk.

The use of terminology such as race in genetics research has persisted in spite of its controversial subtext. However, for studies investigating disease disparities, particularly those that include admixed populations, there appears to be a realization that genetic ancestry may potentially be a more accurate measure than self-reported race in capturing differences in clinical, molecular, and health characteristics (Petersen et al., 2019; Kotecha et al., 2024; Telonis et al., 2024; Nunes et al., 2025). These investigations recognize that population differences do exist at genetic, environmental, and economic levels and are crucial toward addressing health disparities, but without an overemphasis on race. In fact, for two of these studies (Petersen et al., 2019; Nunes et al., 2025), race is not mentioned at all.

Deconstructing the race-in-genetics paradigm

Multilevel roles and responsibilities

Genetic and genomic studies are predominantly undertaken in European and US populations, with increasing inclusion of Asian populations. It is widely accepted that there is a lack of global diversity and population representation in genomic studies, especially from populations residing in low-income countries who face the highest disproportionality in terms of disease burden, morbidity, and mortality (Mulder et al., 2021). Populations from the African continent account for only around 1% of genomic studies despite representing about one-fifth of the world's population. To this end, several initiatives have been launched to tackle the underrepresentation of diverse populations in genetic and genomic studies. These include, in Africa, the Human Heredity and Health in Africa (H3Africa) consortium, the Uganda Genome Resource, and the Africa Wits-INDEPTH Partnership for Genomic Studies; in Asia, the Pakistan Alliance on Genetic Risk Factors for Health; in South America, the Latin American Genomics Consortium; and in Australia, the Tiwi Island Aboriginal Population (Mulder et al., 2021; Fatumo et al., 2022).

In the past five years, there has been an increasing awareness in genetics research and education of the intentional or unintentional exclusion and misrepresentation of certain populations, perpetuating structural racism and biases. Numerous frameworks, proposals, and guidelines have been developed to openly and honestly start addressing these issues. Rebbeck and colleagues developed a framework based on four key values: (1) inclusivity—efforts should be inclusive of a broad population, (2) equity—should include diverse perspectives, starting at the research development (design) stage and following through to implementation, (3) usability—the research material should cover a range of literacy/numeracy levels and accommodate different language and cultural perspectives, and (4) bidirectionality—researchers should learn from participants, who should be engaged with and respected throughout the research process (Rebbeck et al., 2022). Similarly, Fatumo and colleagues proposed a forward-thinking framework to diversify genomics research, involving (1) research institutions—promoting workforce diversity representative of the community they serve, (2) researchers—forming partnerships with the communities they serve, (3) research participants—ensuring ethically informed involvement in the research, thereby building trust, (4) funders—being willing to fund research on underrepresented population groups, and (5) governments—instituting policies conducive to sustainable, diverse genomics research (Fatumo et al., 2022).

In 2023, a NASEM-convened committee comprising experts in population genetics, human and clinical genetics, genetic epidemiology, statistical and computational genetics and genomics, anthropology, sociology, social epidemiology, demography, and population statistics, as well as historical, ethical, legal, and social implications research, released a consensus report with the best practices on the use of race, ethnicity, genetic ancestry, and population descriptors in genetics and genomics research (NASEM, 2023). The report included approaches to improve clinician and medical student education in human genetics and made 13 recommendations for the use of population descriptors in genetics and genomics research:

1. Race and ethnicity should not be used as a proxy for human genetic variation.
2. When grouping people in studies of human genetic variation, avoid the assumption and implication of hierarchy, homogeneity, distinct categories, or the stability of the groups over time.

3. Attention should be given to the connotations and impacts of the terminology they use to label groups.
4. Environmental factors or exposures and their potential relevance to studies should be directly evaluated.
5. Ongoing partnerships with the study participants and community experts should be developed to integrate the perspectives of the relevant communities and to inform the selection and use of population descriptors.
6. Tailor the use of population descriptors to the type and purpose of the study.
7. Population descriptor labels should be applied consistently to all participants.
8. The process by which descriptors are selected and assigned, and the rationale for any grouping of samples, should be disclosed.
9. Funding agencies, research institutions, research journals, and professional societies should offer tools to communities to facilitate the implementation of the recommendations.
10. Research institutions and funding agencies should offer incentives to encourage interdisciplinary collaboration among genomics, social sciences, epidemiology, and community-based research, facilitate the inclusion of environmental measures, and promote the engagement of diverse communities in genomics research.
11. Funders and research institutions should advance and develop best practices for population descriptor usage.
12. Funding agencies, research institutions, and scientific journals should ensure that their policies and procedures are aligned with the recommendations.
13. There should be periodic re-evaluation of the recommendations, and the re-evaluation should be overseen by multidisciplinary advisory groups.

Beyond institutions—Biomedical publication responsibility

The NASEM report and other frameworks clearly mention the responsibility of scientific and biomedical research journals. Although contentious, it has been suggested that journals may have contributed to promoting structural racism in their published content (Merchant et al., 2021). Of note, an anonymous survey conducted on the Editorial Board memberships of 151 psychiatry and neuroscience journals found that 26.9% self-reported as Black, 9.9% as Asian, 16.6% as Hispanic/Latinx, and 48.1% as White; with only one respondent who self-reported as Black indicated having ever served as Editor-in-Chief (Shim et al., 2021). The authors noted that individuals from marginalized groups should have greater representation on editorial boards because they possess specific expertise that includes "lived experiences." To this end, Feero and colleagues recommended that, for biomedical journal authors and reviewers of manuscripts that include population descriptors as proxies for genetic ancestry groups, the following ethical principles should be incorporated (Feero et al., 2024):

1. Terminology used for population descriptors should be accurate, respectful, and adhere to current guidance in biomedical science.
2. Self-identified race and ethnicity should not be used as proxies for genetic ancestry or to represent the genetic diversity of study participants, and they are not recommended for use as analytic variables.
3. The rationale for using population descriptors should be provided and justified in the Methods section.
4. Description of how participants were classified using population descriptors, including self-identification for racial and ethnic categories, should be included in the Methods section.
5. The methods section should clearly explain how the selected descriptors were operationalized in participant accrual, study design, data analysis, and data interpretation.
6. Wherever possible, all population groups should be included in analyses; if any groups are excluded, this should be adequately justified scientifically.
7. The discussion and conclusions sections should explain how the use of populations categorized by population descriptors influenced the interpretation of the study data and the conclusions drawn, including whether this is unknown.
8. The discussion and conclusions sections should explain how stratified analyses, using the selected population descriptors, affect the generalizability of the study findings to other populations.
9. When using legacy datasets, it may not be possible to derive and use more accurate measures of population groups. In these instances, authors should explain the limitations of the dataset and the effects these limitations have on data analysis and interpretation.
10. Generally, titles and conclusion sections should avoid the inclusion of race and ethnicity as descriptors for population groups. Such usage may be appropriate for studies in historically isolated populations or those about health care disparities, where the social constructs of race and ethnicity are important contributors to outcomes.

Overlooked responsibility—Participant and community inclusivity

The principled view of clinical and biomedical researchers is to undertake genetics and genomics research in communities, aiming to develop diagnostic, prognostic, and clinical management strategies that will benefit those residing in such communities. Yet, when conceptualizing and designing a research protocol, the reality is often that communities become involved only once funding has been secured and institutional review board approval obtained, and then only in the capacity of "study or research participants." The frameworks and reports discussed in the previous section all note the importance of inclusivity and engagement with research participants, communities, and community representatives (nominated by those in the community). They emphasize that researchers should be open to learning from the communities they serve; this can only be achieved if participants/communities are involved at the conceptualization and design phases, allowing a critical and continued rethink and revision of the study if needed. In addition, participants and communities should be involved in discussing the potential study outcomes and, importantly, how these outcomes will be properly disseminated into the scientific and public domain. A recent guide for researchers working in genomics research, medicine, and health care included instructions on how to incorporate participant and community engagement (Murtagh et al., 2021); the guide was based on four key themes: (1) fairness, (2) context, (3) heterogeneity, and (4) recognizing tensions and conflict.

Lastly, researchers should avoid procedures that privilege their scientific ("expert") viewpoints to the exclusion of those who have been historically marginalized and disempowered. Participant and community voices contributing to appropriate population descriptors and culturally sensitive matters should continually be taken into consideration and implemented. Community perspectives should be included when formulating a genetic study. Interestingly, when investigators from several South African research institutions, including authors of this chapter, undertook to develop a population descriptor decision model, they first sought a community's perspective (unpublished). The community perspective differed from the scientists' "academic expert viewpoint." This should not have been unexpected because none of the scientists were members of that specific community. Consequently, the community-informed perspective provided a previously unheard voice for a more in-depth context, which would otherwise have been missed by the "experts."

The way forward

It is crucial to acknowledge that language matters, especially in genetics and genomics, and that terms for either race, ethnicity, or genetic ancestry are used appropriately to ensure accurate interpretation of findings. There should be an increasing awareness of the need for greater inclusion in the research process of communities that are affected by the research (Herd et al., 2021). Moreover, the continued use or more precisely the inappropriate use of race and ethnicity "labels" to describe genetic ancestry determined by an individual's ancestral contributions may prolong the history of stigmatization and marginalization of communities.

Research on germline (inherited) genetic variation will benefit from more studies across different population groups from various geographical locations, with the potential to improve our understanding of underlying biology, identify new drug targets, and personalize disease risk assessment. Large studies using new high-throughput technologies to generate genetic data have primarily been conducted in specific population groups in the US, Europe, and Asia, limiting the transferability of potential benefits to other population groups and thereby potentially further exacerbating health disparities. Some groups, having realized that present-day categorizations used in research were built on policies of social and political disenfranchisement, have proceeded to move beyond the use of race and ethnicity representing a single ancestry in genomics research. For example, a recent study demonstrated a dynamic clustering model that focused on trait-based genetic similarity rather than using geographically-based ancestry categories (Mohsen et al., 2025).

Population-based studies focusing on somatic (acquired) changes in tissue or tumors should identify features that are related to disease progression, mortality, or other outcomes, rather than continually focusing on racial disparities that may have little to no clinical utility (Thomas and Peters, 2025). Tissue biobanks should be expanded with greater inclusivity and multiomics profiling of samples from underrepresented populations to improve analytical statistical power, thereby avoiding racial biases.

Precision medicine and pharmacogenomics, often driven by large research institutions, pharmaceutical corporations, and/or biotechnology companies, have hitherto relied on race and ethnicity as a proxy for genetics without due consideration of the level of genetic variation within and between population groups. There has been recent recognition that race is often presented in health education as the focal point for pharmacotherapy (Lipscomb et al., 2023). These practices, which are introduced early at educational institutions, have been proposed to contribute to reinforcing ideas that mislabel genetic ancestry and culture; these taught concepts could exacerbate unconscious bias that may be held beyond the educational institution, leading to misinformed research hypotheses, inaccurate diagnoses, and skewed risk predictions.

Redressing race in genetics—A call for education and curriculum transformation

Modern scientific and medical education remains a continuation of past Western/Eurocentric paradigms—curricula and teaching are overwhelmingly geared toward a "Westernized" standard. Images in textbooks and clinical reference ranges are largely based on "Western norms." Unconscious racial biases are reinforced among educators, faculty, and students at scientific and medical institutions within a "blinkered" knowledge-system, which defines race as an independent risk factor for disease without considering the underlying causes of health disparities. Educators misleadingly equate race with genetic ancestry, and students mistakenly learn (and are taught!) to associate race with certain diseases. There have been calls in some education sectors in the US to adopt a race-conscious teaching policy (Cerdena et al., 2020). To this end, numerous US universities and medical schools have abandoned the use of race correction for eGFR (refer to the section "Diagnostic and prognostic risk prediction bias"). Furthermore, there should be a move away from racially tailored training that propagates inequity. Students should be taught that most health disparities are a consequence of past sociopolitical practices. Moreover, institutional leadership should denounce race-based teaching, and research studies should appropriately investigate the structural barriers faced by historically classified "non-Whites" in the context of health care, rather than focusing on race as a proxy for biological differences (Cerdena et al., 2020). Others have sought to promote social justice curricula as a standardization and best practice for teaching, prioritizing health equity and preparing students to address the consequences of health disparities (Draper et al., 2022). There is currently a dearth of transformational scientific and medical education programs, including ones that specifically address race in genetics. This likely reflects hesitancy by leadership, educators, and faculty within the constraints of broader restrictive policies, for example, being fearful of institutional reputation, the response of institutional funders, and political pressure. Notwithstanding, one unique student-led program at the RCSI University of Medicine and Health Sciences in Dublin (Ireland) has successfully improved awareness and advocacy against race-based medical practices, shifting students' incorrect perceptions of race (with their emphasis on genetic, biological, geographical, and cultural factors) to a more inclusive sociopolitical view regarding the nature of "race" (Lynn et al., 2023). Of note, the overwhelming majority of students disagreed with the use of race in clinical practice after completing this program. More importantly, the adoption and implementation of the program were strengthened by it being student-led and in partnership with a diverse group of educators and faculty across various disciplines such as biochemistry, health psychology, clinical microbiology, clinical practice, physiology, and medical education.

For countries on the African continent, with a colonial past mired by racial injustices and intergenerational inequality, there have been attempts to address colonial misrepresentations in scientific and medical curricula. For example, in South Africa, it has been suggested that a program that introduces "decoloniality to medical education" should be adopted, focusing on conscious awareness, deliberation, and action by medical educators and health professionals to begin dismantling colonial artifacts and practices (Naidu, 2021). At this stage, information on the actual implementation of structured programs at educational institutions is limited, given the sociohistorical colonial past on the African continent and the overemphasis on the "African race" in biomedical research and publications. Therefore it might not be surprising that the strongest calls for transforming genetics curricula, in addition to adopting new genomics policy frameworks, will come from Africa (Jongeneel et al., 2022; Lukhele et al., 2025). Key to this success will be the buy-in and coordination between educational institutions, health care providers, governments, and policy makers.

Conclusion

Current frameworks and reports collectively point to the difficulty in completely removing the use of race in biomedical education, research, and clinical practice, as it may still prove useful in terms of understanding societal determinants of health and disease, for example, racism and discrimination that, in turn, have exacerbated socioeconomic inequalities and disparate environmental exposures. However, it should be clear that the use of race as a proxy for social and biological variables is incorrect and misleading. Moreover, the outdated concept of race fails to consider the genetic exchange between populations through migrations across continents over generations throughout history. However, as genetic studies become more inclusive of diverse population groups, there is increasing and clear evidence of genetic admixture and, therefore, the realization that the levels of genetic variation between and within populations cannot be apportioned in a racial fashion. In view of this knowledge, we believe it is now an opportune time to recognize that genetic ancestry is a more appropriate descriptor of human biological variation. Moreover, we argue that the outdated concept of race in genetics should be confined only to past and present reminders of discriminatory practices in education, health, and biomedical research. However, the task of dismantling race-based genetics will not be easy unless there is an epistemological shift in scientific and medical education. This can only be advanced if there is an acknowledgment and willingness by higher educational institutions and relevant stakeholders to adopt practices, theories, and actions that redress past colonial injustices in pursuit of a more equitable society (refer to central illustration) (Fig. 12.4).

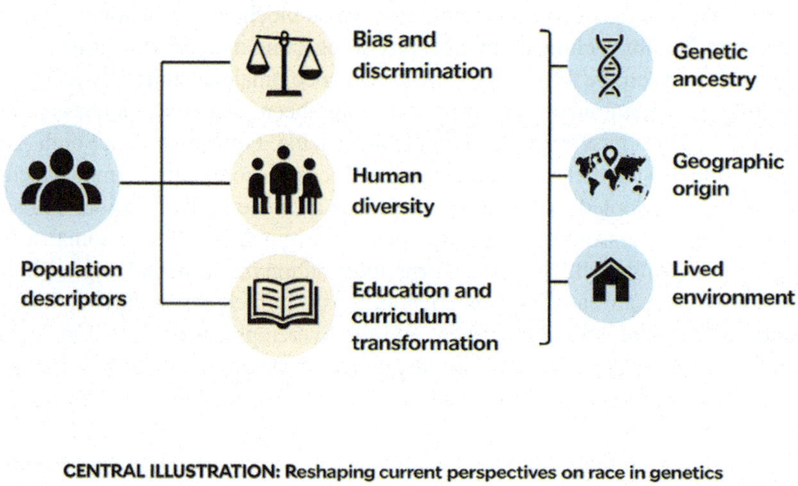

FIG. 12.4

A paradigm shift from the use of race as a biological variable toward ethically and scientifically robust population descriptors in genetics research. The figure highlights the roles of bias and discrimination, human diversity, as well as education and curriculum transformation in shaping more appropriate population descriptors. Recognizing the continuous nature of human genetic diversity supports the use of genetic ancestry, geographic origin, and lived environment as more meaningful and contextually accurate descriptors.

Glossary

Allele A variant form of a deoxyribonucleic acid (DNA) sequence, composed of a specific order of nucleotides at a specific position (locus) on a chromosome.

Gene flow The introduction of genetic material from one population to another, which alters the gene pool of the receiving population.

Genetic variation The differences in the genetic makeup or DNA sequence among individuals within a population.

Genome The entire set of DNA sequences or genetic instructions found in a cell or organism.

Heterozygous An individual who has inherited different versions (alleles) of a genetic marker from each biological parent, i.e., the individual has two different alleles for a specific gene.

Loci Specific positions or locations on a chromosome for a particular gene or genetic marker.

Molecular aging The time-dependent accumulation of damage to cellular components like DNA and proteins, leading to a progressive loss of function and increased vulnerability to disease.

Mutation An alteration in the genetic material of a cell or a living organism that is permanent and can be transmitted to descendants.

Natural selection The process whereby organisms better adapted to their environment tend to survive, pass on genes, and produce more offspring.

Population descriptors ways of describing or distinguishing people from each other based on perceived or actual differences.

Recombination The exchange of genetic material between chromosomes or between different regions within the same chromosome, resulting in the formation of new combinations of DNA sequences.

Selectivepressure The force that drives the evolution of a population by favoring the survival and reproduction of individuals with advantageous traits or genetic variations.

Telomere A region of repetitive DNA sequences at the end of a chromosome.

References

Alexander, D.H., Lange, K., 2011. Enhancements to the ADMIXTURE algorithm for individual ancestry estimation. BMC Bioinformatics 12, 246.

Alexander, D.H., Novembre, J., Lange, K., 2009. Fast model-based estimation of ancestry in unrelated individuals. Genome Res. 19, 1655–1664.

Ashley-Koch, A., Yang, Q., Olney, R.S., 2000. Sickle hemoglobin (HbS) allele and sickle cell disease a huge review. Am. J. Epidemiol. 15, 839–845.

Auton, A., Brooks, L.D., Durbin, R.M., Garrison, E.P., Kang, H.M., Korbel, J.O., Marchini, J.L., Mccarthy, S., Mcvean, G.A., Abecasis, G.R., Genomes Project, C., 2015. A global reference for human genetic variation. Nature 526, 68–74.

Browning, S.R., Waples, R.K., Browning, B.L., 2023. Fast, accurate local ancestry inference with FLARE. Am. J. Hum. Genet. 110, 326–335.

Bryc, K., Auton, A., Nelson, M.R., Oksenberg, J.R., Hauser, S.L., Williams, S., Froment, A., Bodo, J.M., Wambebe, C., Tishkoff, S.A., Bustamante, C.D., 2010. Genome-wide patterns of population structure and admixture in West Africans and African Americans. Proc. Natl. Acad. Sci. U. S. A. 107, 786–791.

Campbell, A.D., Colombatti, R., Andemariam, B., Strunk, C., Tartaglione, I., Piccone, C.M., Manwani, D., Asare, E.V., Boruchov, D., Farooq, F., Urbonya, R., Boatemaa, G.D.,

Perrotta, S., Sainati, L., Rivers, A., Rao, S., Zempsky, W., Sey, F., Segbefia, C., Inusa, B., Antwi-Boasiako, C., 2021. An analysis of racial and ethnic backgrounds within the CASiRe international cohort of sickle cell disease patients: implications for disease phenotype and clinical research. J. Racial Ethn. Health Disparities. 8, 99–106.

Campbell, M.C., Tishkoff, S.A., 2010. The evolution of human genetic and phenotypic variation in Africa. Curr. Biol. 20, R166–73.

Caro-Consuegra, R., Lucas-Sanchez, M., Comas, D., Bosch, E., 2023. Identifying signatures of positive selection in human populations from North Africa. Sci. Rep. 13, 8166.

Cavalli, G., Heard, E., 2019. Advances in epigenetics link genetics to the environment and disease. Nature 571, 489–499.

Cerdena, J.P., Plaisime, M.V., TSAI, J., 2020. From race-based to race-conscious medicine: how anti-racist uprisings call us to act. Lancet 396, 1125–1128.

Cerdena, J.P., Tsai, J., Grubbs, V., 2021. APOL1, black race, and kidney disease: turning attention to structural racism. Am. J. Kidney Dis. 77, 857–860.

Chang, C.J., Chen, C.B., Hung, S.I., Ji, C., Chung, W.H., 2020. Pharmacogenetic testing for prevention of severe cutaneous adverse drug reactions. Front. Pharmacol. 11, 969.

Dankwa-Mullan, I., Pérez-Stable, E.J., Gradner, K.L., Zhang, X., Rosario, A.M., 2021. The Science of Health Disparities Research, Wiley Online Library.

Dar-Nimrod, I., Heine, S.J., 2011. Genetic essentialism: on the deceptive determinism of DNA. Psychol. Bull. 137, 800–818.

Dobon, B., Hassan, H.Y., Laayouni, H., Luisi, P., Ricano-Ponce, I., Zhernakova, A., Wijmenga, C., Tahir, H., Comas, D., Netea, M.G., Bertranpetit, J., 2015. The genetics of East African populations: a Nilo-Saharan component in the African genetic landscape. Sci. Rep. 5, 9996.

Donovan, B.M., 2022. Ending genetic essentialism through genetics education. HGG Adv. 3, 100058.

Draper, J.K., Feltner, C., Vander Schaaf, E.B., Malchuk, A.M., 2022. Preparing medical students to address health disparities through longitudinally integrated social justice curricula: A systematic review. Acad Med. 97, 1226–1235.

Duello, T.M., Rivedal, S., Wickland, C., Weller, A., 2021. Race and genetics versus 'race' in genetics: a systematic review of the use of African ancestry in genetic studies. Evol. Med. Public Health. 9, 232–245.

Fatumo, S., Chikowore, T., Choudhury, A., Ayub, M., Martin, A.R., Kuchenbaecker, K., 2022. A roadmap to increase diversity in genomic studies. Nat. Med. 28, 243–250.

Feero, W.G., Steiner, R.D., Slavotinek, A., Faial, T., Bamshad, M.J., Austin, J., Korf, B.R., Flanagin, A., Bibbins-Domingo, K., 2024. Guidance on use of race, ethnicity, and geographic origin as proxies for genetic ancestry groups in biomedical publications. JAMA 331, 1276–1278.

Fitzgerald, J.D., Dalbeth, N., Mikuls, T., Brignardello-Petersen, R., Guyatt, G., Abeles, A.M., Gelber, A.C., Harrold, L.R., Khanna, D., King, C., Levy, G., Libbey, C., Mount, D., Pillinger, M.H., Rosenthal, A., Singh, J.A., Sims, J.E., Smith, B.J., Wenger, N.S., Bae, S.S., Danve, A., Khanna, P.P., Kim, S.C., Lenert, A., Poon, S., Qasim, A., Sehra, S.T., Sharma, T.S.K., Toprover, M., Turgunbaev, M., Zeng, L., Zhang, M.A., Turner, A.S., Neogi, T., 2020. 2020 American College of Rheumatology guideline for the management of gout. Arthritis Rheumatol. 72, 879–895.

Gingell, G., Bergemann, A.D., 2022. Disrupting essentialism in medical genetics education. Med. Sci. Educ. 32, 255–262.

Goodman, C.W., Brett, A.S., 2021. Race and pharmacogenomics-personalized medicine or misguided practice? JAMA 325, 625–626.

Gouveia, M.H., Bentley, A.R., Leal, T.P., Tarazona-Santos, E., Bustamante, C.D., Adeyemo, A.A., Rotimi, C.N., Shriner, D., 2023. Unappreciated subcontinental admixture in Europeans and European Americans and implications for genetic epidemiology studies. Nat. Commun. 14, 6802.

Guan, Y., 2014. Detecting structure of haplotypes and local ancestry. Genetics 196, 625–642.

Herd, P., Mills, M.C., Dowd, J.B., 2021. Reconstructing sociogenomics research: dismantling biological race and genetic essentialism narratives. J. Health Soc. Behav. 62, 419–435.

Jongeneel, C.V., Kotze, M.J., Bhaw-Luximon, A., Fadelmola, F.M., Fakim, Y.J., Hamdi, Y., Kassim, S.K., Kumuthini, J., Nembaware, V., Radouani, F., Tiffin, N., Mulder, N., 2022. A view on genomic medicine activities in Africa: implications for policy. Front. Genet. 13, 769919.

Jordan, I.K., Rishishwar, L., Conley, A.B., 2019. Native American admixture recapitulates population-specific migration and settlement of the continental United States. PLoS Genet. 15, e1008225.

Khodasevich, D., Gladish, N., Daredia, S., Bozack, A.K., Shen, H., Nwanaji-Enwerem, J.C., Needham, B.L., Rehkopf, D.H., Cardenas, A., 2025. Influence of race, ethnicity, and sex on the performance of epigenetic predictors of phenotypic traits. Clin. Epigenetics. 17, 59.

Kidane Gebremeskel, A.S., Rab, M.A., Van Werkhoven, E.D., Petersen, T.B., Cnossen, M.H., M'Charek, A., Meeks, K.A.C., Rijneveld, A.W., 2025. The use of race and ethnicity in sickle cell disease research. BMC Med. Res. Methodol. 25, 63.

Kim, J., Edge, M.D., Goldberg, A., Rosenberg, N.A., 2021. Skin deep: the decoupling of genetic admixture levels from phenotypes that differed between source populations. Am. J. Phys. Anthropol. 175, 406–421.

Kotecha, R.R., Knezevic, A., Arora, K., Bandlamudi, C., Kuo, F., Carlo, M.I., Fitzgerald, K.N., Feldman, D.R., Shah, N.J., Reznik, E., Hakimi, A.A., Carrot-Zhang, J., Mandelker, D., Berger, M., Lee, C.H., Motzer, R.J., Voss, M.H., 2024. Genomic ancestry in kidney cancer: correlations with clinical and molecular features. Cancer 130, 692–701.

Lala, K.N., Feldman, M.W., 2024. Genes, culture, and scientific racism. Proc. Natl. Acad. Sci. U. S. A. 121, e2322874121.

Lewontin, R.C., 1972. The apportionment of human diversity. In: Dobzhansky, T., Hecht, M.K., Steere, W.C. (Eds.), Evolutionary Biology. Springer, New York. pp. 381–398.

Lipscomb, J., Galvez-Peralta, M., Cropp, C.D., Delgado, E., Crutchley, R., CALINSKI, D., Iwuchukwu, O., 2023. A genetics-focused lens on social constructs in pharmacy education. Am. J. Pharm. Educ. 87, 100077.

Lukhele, S.T., Ras, V., Mulder, N., 2025. Workforce development in geneomic data science for health: a worldview. Annu. Rev. Genom. Hum. Genet. 26, 449–471.

Lynn, T.M., D'Urzo, K.A., Vaughan-Ogunlusi, O., Wiesendanger, K., Colbert-Kaip, S., Capcara, A., Chen, S., Sreenan, S., Brennan, M.P., 2023. The impact of a student-led anti-racism programme on medical students' perceptions and awareness of racial bias in medicine and confidence to advocate against racism. Med. Educ. Online. 28, 2176802.

Maples, B.K., Gravel, S., Kenny, E.E., Bustamante, C.D., 2013. RFMix: a discriminative modeling approach for rapid and robust local-ancestry inference. Am. J. Hum. Genet. 93, 278–288.

Maree, D.M., Swanepoel, R.A., Swart, F., Gray, D.M., Masekela, R., Allwood, B.W., Van Zyl-Smit, R.N., Koegelenberg, C.F.N., 2022. Position statement for adult and paediatric spirometry in South Africa: 2022 update. Afr. J. Thorac. Crit. Care Med. 28, 181–192.

Masekela, R., Koegelenberg, C.F.N., Gray, D.M., 2020. Guidance to the applicability of the Global Lung Initiative spirometry reference equations for South African populations. S. Afr. Med. J. 111, 13186.

Merchant, R.M., Del Rio, C., Boulware, L.E., 2021. Structural racism and scientific journals-a teachable moment. Jama 326, 607–608.

Mohsen, H., Blenman, K., Emani, P.S., Morris, Q., Carrot-Zhang, J., Pusztai, L., 2025. Dynamic clustering of genomics cohorts beyond race, ethnicity-and ancestry. BMC Med Genomics. 18, 87.

Montinaro, F., Busby, G.B., Pascali, V.L., Myers, S., Hellenthal, G., Capelli, C., 2015. Unravelling the hidden ancestry of American admixed populations. Nat. Commun. 6, 6596.

Msimang, P.M., Sarkar, S., 2026. Philosophy and Race Science. Tuboly, A., Padovani, F. (Eds.), In: Routledge Handbook of the History of Philosophy of Science. Oxfordshire, Routledge. pp. 388–397.

Mulder, N., Zass, L., Hamdi, Y., Othman, H., Panji, S., Allali, I., Fakim, Y.J., 2021. African global representation in biomedical sciences. Annu. Rev. Biomed. Data Sci. 4, 57–81.

Murtagh, M.J., Machirori, M., Gaff, C.L., Blell, M.T., De Vries, J., Doerr, M., Dove, E.S., Duncanson, A., Hastings Ward, J., Hendricks-Sturrup, R., Ho, C.W.L., Johns, A., Joly, Y., Kato, K., Katsui, K., Kumuthini, J., Maleady-Crowe, F., Middleton, A., Milne, R., Minion, J.T., Matshaba, M., Mulrine, S., Patch, C., Ryan, R., Viney, W., 2021. Engaged genomic science produces better and fairer outcomes: an engagement framework for engaging and involving participants, patients and publics in genomics research and healthcare implementation. Wellcome Open Res. 6, 311.

Naidu, T., 2021. Modern medicine is a colonial artifact: introducing decoloniality to medical education research. Acad. Med. 96, S9–S12.

NASEM, 2023. Using Population Descriptors in Genetics and Genomics Research: A New Framework for an Evolving Field. The National Academic Press, Washington, DC.

Nunes, K., Araujo Castro, E.S.M., Rodrigues, M.R., Lemes, R.B., Pezo-Valderrama, P., Kimura, L., De Sena, L.S., Krieger, J.E., Catoia Varela, M., De Azevedo, L.O., Aranha Camargo, L.M., Ferreira, R.G.M., Krieger, H., Bortolini, M.C., Mill, J.G., Sacuena, P., Guerreiro, J.F., De Souza, C.M.B., Veronese, F.V., Vianna, F.S.L., Comas, D., Pereira, A.C., Pereira, L.V., Hunemeier, T., 2025. Admixture's impact on Brazilian population evolution and health. Science 388, eadl3564.

Patterson, N., Petersen, D.C., Van Der Ross, R.E., Sudoyo, H., Glashoff, R.H., Marzuki, S., Reich, D., Hayes, V.M., 2010. Genetic structure of a unique admixed population: implications for medical research. Hum. Mol. Genet. 19, 411–419.

Petersen, D.C., Croock, D.A., 2025. Population genomics-the fundamentals. In: Patrinos, G.P., Moller, M., Uren, C. (Eds.), Population Genomics in the Developing World. Elsevier. pp. 1–30.

Petersen, D.C., Jaratlerdsiri, W., Van Wyk, A., Chan, E.K.F., Fernandez, P., Lyons, R.J., Mutambirw, S.B.A., Van Der Merwe, A., Venter, P.A., Bates, W., Bornman, M.S.R., Hayes, V.M., 2019. African KhoeSan ancestry linked to high-risk prostate cancer. BMC Med Genomics. 12, 82.

Petersen, D.C., Libiger, O., Tindall, E.A., Hardie, R.A., Hannick, L.I., Glashoff, R.H., Mukerji, M., Indian Genome Variation, C., Fernandez, P., Haacke, W., Schork, N.J., Hayes, V.M., 2013. Complex patterns of genomic admixture within Southern Africa. PLoS Genet. 9, e1003309.

Pichardo, C.M., Ezeani, A., Acevedo, A.M., Agurs-Collins, T., Bailey-Whyte, M., Dorsey, T.H., Harris, A.R., Franklin, J., Kittles, R.A., Lawrence, W.R., Loffredo, C.A., Minas,

T.Z., Pichardo, M.S., Ryan, B.M., Tang, W., Wooten, W., Liu, J., Ambs, S., 2024. West African genetic ancestry, neighborhood deprivation, and prostate cancer. JAMA Netw Open. 7, e2433546.

Pritchard, J.K., Stephens, M., Donnelly, P., 2000. Inference of population structure using multilocus genotype data. Genetics 155, 945–959.

Rebbeck, T.R., Bridges, J.F.P., Mack, J.W., Gray, S.W., Trent, J.M., George, S., Crossnohere, N.L., Paskett, E.D., Painter, C.A., Wagle, N., Kano, M., Nez Henderson, P., Henderson, J.A., Mishra, S.I., Willman, C.L., Sussman, A.L., 2022. A framework for promoting diversity, equity, and inclusion in genetics and genomics research. JAMA Health Forum. 3, 1–18.

Reeder-Hayes, K., Roberson, M.L., Wheeler, S.B., Abdou, Y., Troester, M.A., 2023. From race to racism and disparities to equity: an actionable biopsychosocial approach to breast cancer outcomes. Cancer J. 29, 316–322.

Roach, M., Coleman, P.W., Kittles, R., 2023. Prostate cancer, race, and health disparity. Cancer J. 29, 328–337.

Robinette, J.W., Smith, J.A., 2025. Accelerated molecular aging in socioeconomically disadvantaged neighborhoods: a racial/ethnic comparison. Health Place. 93, 103446.

Shim, R.S., Tully, L.M., Yu, G., Monterozza, E.C., Blendermann, M., 2021. Race and ethnicity of editorial board members and editors as an indicator of structural racism in psychiatry and neuroscience journals. JAMA Psychiatry 78, 1161–1163.

Tapper, M., 1999. Sickle cell anemia and the politics of race. In: Ludden, D. (Ed.), the Blood. University of Pennsylvania Press, pp. 1–163.

Telonis, A.G., Rodriguez, D.A., Spanheimer, P.M., Figueroa, M.E., Goel, N., 2024. Genetic ancestry-specific molecular and survival differences in admixed patients with breast cancer. Ann. Surg. 279, 866–873.

Thomas, C.E., Peters, U., 2025. Genomic landscape of cancer in racially and ethnically diverse populations. Nat. Rev. Genet. 26, 336–349.

Uren, C., Hoal, E.G., Moller, M., 2020. Putting RFMix and ADMIXTURE to the test in a complex admixed population. BMC Genet. 21, 40.

Uren, C., Kim, M., Martin, A.R., Bobo, D., Gignoux, C.R., Van Helden, P.D., Moller, M., Hoal, E.G., Henn, B.M., 2016. Fine-scale human population structure in southern Africa reflects ecogeographic boundaries. Genetics 204, 303–314.

Uren, C., Moller, M., Van Helden, P.D., Henn, B.M., Hoal, E.G., 2017. Population structure and infectious disease risk in Southern Africa. Mol. Genet. Genomics. 292, 499–509.

Vyas, D.A., Eisentein, L.G., Jone, D.S., 2020. Hidden in plain sight — reconsidering the use of race correction in clinical algorithms. N. Engl. J. Med. 383, 874–882.

Wang, T., Wang, W., Xie, G., Li, Z., Fan, X., Yang, Q., Wu, X., Cao, P., Liu, Y., Yang, R., Liu, F., Dai, Q., Feng, X., Wu, X., Qin, L., Li, F., Ping, W., Zhang, L., Zhang, M., Liu, Y., Chen, X., Zhang, D., Zhou, Z., Wu, Y., Shafiey, H., Gao, X., Curnoe, D., Mao, X., Bennett, E.A., Ji, X., Yang, M.A., Fu, Q., 2021. Human population history at the crossroads of East and Southeast Asia since 11,000 years ago. Cell 184, 3829–3841e21.

Wollstein, A., Lao, O., Becker, C., Brauer, S., Trent, R.J., Nurnberg, P., Stoneking, M., Kayser, M., 2010. Demographic history of Oceania inferred from genome-wide data. Curr. Biol. 20, 1983–1992.

CHAPTER

Shaping an inclusive health care ethics framework in the 21st century: Lessons from African bioethics and relational approaches

13

Berna Gerber, Juan Bornman, and Alida Naudé

Division of Speech-Language and Hearing Therapy, Department of Health and Rehabilitation Sciences, Faculty of Medicine and Health Sciences, Stellenbosch University, Cape Town, South Africa

Introduction

Bioethics education in the health care professions has historically been dominated by Western paradigms, most notably principlism, which emphasizes autonomy, beneficence, nonmaleficence, and justice (Beauchamp and Childress, 2019). While these principles have provided a structured starting point for ethical reasoning, they often marginalize indigenous, communal, and relational traditions that are central to African moral life. This marginalization risks perpetuating a form of ethical imperialism in which Western frameworks are treated as universal, while African philosophies such as Ubuntu are acknowledged only superficially. It also shapes how bioethics is taught, with curricula privileging abstract principles over lived communal realities and relationships, leading to dissonance in learners' experiences of ethics education.

The emergence of bioethics as a formal field of inquiry in sub-Saharan Africa is deeply entangled with both Western Enlightenment traditions and indigenous moral philosophies (Jecker and Atuire, 2022). Enlightenment thought, with its valorization of reason, technology, and progress, promised emancipation from nature's constraints. Yet critics such as D.H. Lawrence (1885–1930) highlighted its darker legacy, arguing that the competitive rationality of advanced industrial and imperialist powers contributed to the devastation of World War I (Wallace, 1999). The Industrial Revolutions that followed reshaped health care alongside warfare, producing both medical advances and new moral dilemmas. The current Fifth Industrial Revolution, marked by artificial intelligence, continues this trajectory, offering both improved diagnostics and access to care and unprecedented ethical challenges (Ziatdinov et al., 2024).

Decolonizing Medical Education, Research and Practice. https://doi.org/10.1016/B978-0-443-44998-7.00002-8

These global transformations reveal how ethical reflection in health care is always context dependent. Health care professionals often turn to professional codes of conduct, legal frameworks, and justice principles as a "safety net" for ethical decision-making. However, as health care becomes increasingly globalized and culturally diverse, reliance on abstract, seemingly universalist approaches is inadequate. A critical task for bioethics education is therefore to interrogate the historical roots of Western dominance, recognize its limitations, and make space for relational, community-based, and indigenous ethical traditions—including, but not limited to, African approaches—that may provide more contextually relevant guidance (Ashby, 2019).

The purpose of this chapter is to contribute to the ongoing reorientation of bioethics education by tracing the emergence of dominant frameworks and exploring how they can be transformed. We begin with a historical overview of Western bioethics, tracing its Enlightenment foundations, its codification in key documents such as the Nuremberg Code, the Declaration of Helsinki, and the Belmont Report, and its subsequent consolidation in principlism. We then examine how colonial legacies shaped bioethics and education in Africa, frequently marginalizing or silencing local moral traditions and knowledge systems. Finally, it turns to relational African bioethics, showing how frameworks such as Ubuntu can enrich ethics education by emphasizing interdependence, harmony, and community-based justice—values that offer a robust alternative to individualistic approaches and open pathways toward decolonizing curricula, research, and practice.

Historical context and the rise of Western bioethics

Enlightenment philosophy and the foundation of Western bioethics

The philosophical foundations of Western bioethics can be traced to Enlightenment-era ideals, particularly the emphasis on individualism and rationalism. During this period, individuals were encouraged to reason independently, signaling a shift from unquestioned religious dogma to an equally rigid faith in reason and scientific rationality (Kurakina, 2023). Thinkers such as Immanuel Kant (1724–1804) profoundly influenced modern conceptions of personhood, rights, and moral agency, and he championed the idea that "theory" and "theorizing" are the best paths to understanding reality (Vest, 2025). The emphasis on the rational, autonomous individual—capable of making independent moral choices—became a cornerstone of Western moral philosophy. These ideas later crystallized into the four guiding principles of biomedical ethics—autonomy, beneficence, nonmaleficence, and justice—which continue to shape the ethical landscape of modern medicine, particularly in contexts where informed consent, individual rights, and personal liberty are privileged over communal or relational values.

With the rapid expansion of modern medicine and biomedical research in the mid-to-late 20th century, the need for ethical oversight became increasingly apparent. Yet, the dominant frameworks were shaped by Western liberal ideologies and often failed to account for cultural diversity and historical injustices.

The rise of principlism

Perhaps the most influential framework in Western bioethics is the four principles approach, developed by Tom Beauchamp and James Childress in their seminal work *Principles of Biomedical Ethics*, now in its eighth edition (Beauchamp and Childress, 2019). This framework is grounded in a concept of a common or universal morality among citizens and proposes four guiding principles for ethical deliberation: autonomy, beneficence, nonmaleficence, and justice.

An important precursor to this approach was the Belmont Report, commissioned by the United States (US) government to establish ethical principles for research involving human subjects. The report articulated the principles of respect for persons, beneficence, and justice, forming a foundational ethical framework in the context of biomedical research. While co-authoring the Belmont Report, Tom Beauchamp was simultaneously developing the principlist approach with James Childress. Although the principles in each work are distinct, their conceptual overlap is evident, as the two projects clearly influenced one another (Trimble, 2024).

Principlism, with its origins in the field of medicine, should, however, not be seen as a coherent ethical or moral theory, but rather as a set of principles that blend different philosophical perspectives (Halse, 2011). As such, it provides a helpful starting point for structured ethical reasoning. However, it has been critiqued for being overly individualistic, privileging autonomy at the expense of more relational or communitarian perspectives. Some scholars argue that solidarity should be regarded as a more fundamental guiding principle for bioethical discourses and practices in Africa and beyond (Fayemi, 2021). In many African cultures, the concept of the self is not isolated but embedded within a network of social, familial, and communal relations. As such, ethical models that prioritize individual decision-making may not fully resonate with African values or the lived realities of African communities.

Whose values? The Declaration of Helsinki, Nuremberg Code, and Belmont Report

The key documents in the field of human research ethics—all products of specific historical and geopolitical contexts—that form the backbone of modern research ethics include, among others, the Nuremberg Code (1947). Here its 10-point set of rules emphasizes consent provided voluntarily, which emerged in response to the horrific medical experiments conducted by Nazi doctors during the Second World War. This code, drafted at the end of the Doctors' Trial in Nuremberg, has been hailed as a landmark document in medical and research ethics (Ten Have and Patrão Neves, 2021). Its primary concern was voluntary informed consent, once again reflecting an emphasis on individual rights.

The Nuremberg Code was later eclipsed by the World Medical Association's Declaration of Helsinki in 1964 (Moreno et al., 2017), which sought to expand these protections, particularly in the context of international human research. Similarly, the Belmont Report was published in 1979 in the US following the Tuskegee Syphilis

Study (1932–1972) that was conducted by the US Public Health Service (USPHS) to investigate the effects of untreated syphilis. Here they withheld treatment from African American men without their informed consent, even after penicillin became the standard treatment. In fact, the USPHS sought to prevent access to treatment from outside the research study (Brandt, 1978). The Tuskegee Study exemplifies how race, power, and structural inequality intersect in research ethics failures (Tobin, 2022).

These milestones also became embedded in medical and health sciences curricula worldwide, forming the backbone of ethics education. However, their universalizing assumptions did not always translate effectively into contexts where literacy levels, communal decision-making, and socioeconomic inequalities shaped and complicated health care interactions (Tobin, 2022).

Taken together, the Declaration of Helsinki, the Nuremberg Code, and the Belmont Report were milestones in codifying ethical standards, yet they implicitly reflect the Western philosophical and legal traditions from which they emerged. Their assumption of a literate, autonomous individual capable of engaging with complex medical information may not hold true in all cultural or socioeconomic contexts (Tobin, 2022).

Colonial legacies and the ethics of medical education

A critical dimension of bioethics that has received increasing attention is the colonial legacy embedded in medical practice and education in Africa. The curricula and ethical frameworks taught in many universities in Africa—particularly in Faculties of Health Sciences and Medicine—have been mainly imported from former colonial powers, often without sufficient adaptation to local cultures, health systems, or epistemologies. This has led to a form of ethical imperialism, where Western norms are treated as universal, and African moral traditions are either marginalized or ignored.

The colonial legacy also continues to shape the ethics curricula and regulatory structures within medical and health sciences education. Many researchers are now encountering what Haggerty (2004, p. 391) refers to as "ethics creep," which describes the expanding reach of ethical regulation beyond the biomedical sciences into health care, the humanities, and social sciences—a trend that has influenced how ethics is taught and governed in universities. As Le Grange (2020) explains, ethics creep involves both the extension of ethics oversight into new domains and the tightening of control over practices already under its purview. Within educational contexts, this has often meant that ethical instruction remains highly procedural and compliance-focused, prioritizing regulatory adherence over critical reflection on power, context, and cultural diversity.

This dynamic is deeply intertwined with the colonial history of medical education itself, in which African bodies were frequently framed as objects of study or experimentation rather than as partners in ethically grounded care. Such pedagogical legacies—for example, the apartheid-era training of health care professionals who treated Black patients as experimental subjects without meaningful consent—have left enduring marks on the values and assumptions embedded in today's curricula. These historical practices modeled a paternalistic view of health care that still echoes in some teaching approaches, reinforcing hierarchies and shaping professional identities in ways that can perpetuate structural inequities.

This exposes the limitations of formal ethical frameworks when power dynamics and systemic discrimination are not addressed. It also highlights the urgent need for a context-sensitive, historically aware bioethics that foregrounds justice, redress, and relational accountability. Bioethics in Africa should involve a critical examination of the social, political, and economic conditions shaping health care, research, and science within the evolving global landscape, where diverse value systems converge (Fayemi and Macaulay-Adeyelure, 2016).

Toward relational African bioethics

Recognizing these limitations, scholars across the continent have called for an African bioethics that is relational, community-oriented, and culturally grounded (Behrens, 2017; Fayemi, 2021; Fayemi and Macauley-Adeyelure, 2016; Jecker and Atuire, 2022). This approach draws from indigenous philosophies such as Ubuntu which emphasize interconnectedness, mutual respect, and the idea that a person is a person through others. In contrast to the Western emphasis on autonomy, African bioethics may give greater weight to relational autonomy, family involvement, and communal wellbeing.

Behrens (2017) suggests that when African researchers base their ethical responsibilities toward participants and society on indigenous values such as harmony, communal solidarity, and Ubuntu, they are more likely to embrace and fulfill these duties. Relational approaches do not reject the foundational principles of Western bioethics; instead, they reinterpret them through a culturally resonant lens. For example, autonomy is not understood as independence but as the ability to make decisions in dialog with one's community and in alignment with shared values. Justice is not merely procedural but also encompasses historical justice and restorative accountability. In especially the southern African region, there is a growing scholarship exploring Ubuntu as a moral framework, highlighting several key ethical concepts common across many African communities that can inform and enrich ethical debates, as well as ethics education in medicine and the health sciences (Sambala et al., 2020; Behrens, 2017). Thus there are vital and neglected aspects of African cultures that can humanize and enrich Western culture (Tangwa, 1996).

The epistemic critique: Whose ethics?

The problem of epistemic injustice and the illusion of universality

Epistemic injustice occurs when knowledge traditions (or ways of knowing), especially those that developed in non-Western or indigenous contexts, are marginalized or treated as less credible. It involves the unfair treatment of individuals as knowers, often based on aspects of their identity such as race, class, gender, and disability status, which can be unintentional yet systemic in nature (Nielsen et al., 2025). The problem of epistemic injustice in health sciences curricula is closely linked to the dominance of Western ethics, which often presents itself under the illusion of universality. This is pointedly indicated in the title of Alasdair MacIntyre's seminal work, "Whose Justice? Which Rationality?" (1988). Within health sciences curricula this injustice is perpetuated by the dominance of Western ethical frameworks which often position themselves as universally applicable—an assumption that obscures their cultural specificity and marginalizes alternative epistemologies.

In the context of health sciences, the prevailing Western ethical frameworks tend to universalize particular moral and epistemic norms, which may not adequately account for diverse cultural, social, and epistemic standpoints. This universality claim risks silencing or disregarding alternative epistemologies and lived experiences, especially those of marginalized groups, thus reinforcing testimonial and hermeneutical injustices. Testimonial injustice occurs when "a speaker receives an unfair deficit of credibility from a hearer owing to prejudice on the hearer's part" (Fricker, 2007, p. 9). According to Fricker (2007), hermeneutical injustice occurs at an earlier stage than testimonial injustice, when marginalized groups lack the concepts or language within the dominant culture that enable them to make sense of their experiences. Testimonial injustice may occur when patients from marginalized backgrounds find their accounts of pain or other symptoms dismissed. Hermeneutical injustice may arise when someone suffers "sexual harassment in a culture that still lacks that critical concept" (Fricker, 2007, p. 1).

Marks (1999) argues that doctors have strong reasons to take patients' own accounts of their symptoms seriously, even when there is no clear physical evidence to explain what they are experiencing. She illustrates this with sickle cell disease, i.e., an inherited blood disorder in which abnormal hemoglobin causes red blood cells to become rigid and "sickle" shaped. These changes can block blood flow, leading to severe episodes of pain that may be triggered by illness, stress, dehydration, exhaustion, or exposure to cold. Because this pain is largely invisible and cannot be directly measured, health professionals sometimes doubt its severity and may undertreat it. This would be an example of hermeneutical injustice. Marks (1999) also points out that racism can worsen this problem, since sickle cell anemia often affects people of African-Caribbean, Mediterranean, Middle Eastern, and Indian heritage. In this case, it would also be a form of testimonial injustice.

However, the conceptualization of epistemic injustice in health care is still developing. There is a need for more explicit criteria and empirical evidence to understand how these injustices manifest in medical education and practice. The limited empirical grounding and the reliance on anecdotal cases make it difficult to generalize or systematically address these issues, which may contribute to the persistence of Western-centric curricula that overlook epistemic diversity (Nielsen et al., 2025).

The dominance of Western ethics in health sciences curricula functions as a form of epistemic power, determining what is recognized as legitimate knowledge and ethical reasoning. Challenging this requires critical reflection on the epistemic structures underpinning medical education and deliberate inclusion of diverse knowledge systems to promote epistemic justice (Fricker, 2007).

Philosophical critiques of autonomy and individualism

Philosophical critiques of autonomy and individualism within African bioethics emphasize the limitations of Western autonomy-based frameworks, particularly their view of the individual as an isolated, self-sufficient decision-maker (the idea of atomistic self-determination). In many African cultures, the self is understood relationally, where social relationships and interdependence form the basis for moral decisions, rather than relying solely on individualism. This relational self contrasts with the Western principle of autonomy, which often privileges individual choice and self-sufficiency as central ethical values (Onuoha, 2007).

Critiques highlight that bioethical principlism, grounded in autonomy, nonmaleficence, beneficence, and justice, does not adequately reflect African moral realities. African bioethics emphasizes communal relationships that extend beyond the living to include ancestors and future generations, thereby challenging the individualistic and anthropocentric (or human-centered) assumptions of Western bioethics. This relational approach consists of a threefold sense of community that embraces the living, the ancestors whose histories continue to shape the present, and future generations yet to be born. It promotes a communitarian ethic that is more attuned to African social contexts (Buyondo, 2024). Communitarianism refers to the idea that people's identities are shaped mainly through the communities and social relationships they belong to, and that this understanding of human nature should guide ethical reasoning, political decisions, as well as the design of policies and institutions (Bell, 2024). Such a perspective calls for ethical frameworks that reflect collective wellbeing and responsibility rather than focusing solely on individual rights.

The conventional emphasis on individual autonomy and informed consent may overlook the fundamental role of interpersonal relationships in shaping identity. Illness, disability, and suffering can disrupt a person's identity. A relational conception of autonomy acknowledges that supporting a patient's relationships is essential to fostering their capacity for autonomous decision-making in a more holistic sense. This approach argues for care practices that respect and restore identity through relationality, rather than focusing narrowly on individual choice (Stanfield, 2019).

Philosophical critiques of autonomy and individualism in African bioethics thus advocate for a shift from individualistic frameworks toward relational and communitarian ethics. This shift has significant implications for ethics education in health care professions, highlighting the need to incorporate cultural realities and relational understandings of personhood to more effectively address moral issues in African health care contexts (Buyondo, 2024; Stanfield, 2019; Onuoha, 2007).

Ethical pluralism and why a single lens fails diverse realities

Ethical pluralism recognizes the coexistence of diverse moral outlooks within societies, challenging the notion that a single ethical framework can adequately guide decision-making in health care and bioethics. In the context of African bioethics, this pluralism is particularly salient because Western bioethics, often grounded in autonomy-based principles, does not universally apply across different cultural realities. This fundamental difference highlights why a single ethical lens, particularly one rooted in Western individualism, is insufficient to capture the moral complexities inherent in diverse cultural contexts (Onuoha, 2007). An acceptable bioethical framework must balance sensitivity to cultural contexts with recognition of common moral features of human experience. The common moral features of human life refer to the ethical aspects of life that most people share, such as the interaction of emotions and reason in moral decision-making, the experience of moral emotions like regret, and the social dynamics that influence moral judgments (Tosam, 2020). A "global bioethics" (Tosam, 2020, p. 611) requires negotiating norms through democratic dialog rather than imposing dominant cultural values, thereby avoiding both moral relativism and moral imperialism. To engage with pluralistic realities effectively, it is necessary to expand the dominant bioethical discourse of principlism to include collective values such as solidarity and to recognize the sociopolitical context of moral agency (Sutrop, 2024). This pluralistic stance supports a more inclusive and context-sensitive ethics education in health care professions (Sutrop, 2024; Tosam, 2020; Onuoha, 2007). For instance, fostering students' capacity to navigate multiple moral frameworks rather than privileging one as superior.

Ubuntu and African bioethics: A relational alternative

Core tenets

We have already alluded to Ubuntu as an African philosophy and moral theory. In this section, its core tenets will be explored by mainly drawing on Metz's (2011) account. The word "Ubuntu" comes from the Nguni languages and has related forms in Sotho-Tswana and Shona. The word literally means "humanness." The moral theory of Ubuntu follows the maxim, well-known in South Africa, "A person is a person through other persons." In Nguni, "Umuntu ngumuntu ngabantu." It can be interpreted as: "One becomes a moral person insofar as one honors communal relationships," or "A human being lives a genuinely human way of life to the extent that she prizes identity and solidarity with other human beings" (Metz, 2011, p. 540). Ubuntu is a multidimensional concept that represents the fundamental values of African worldviews: "respect for any human being, for human dignity and for human life, collective sharedness, obedience, humility, solidarity, caring, hospitality, interdependence, communalism, to list but a few" (Kamwangamalu, 1999, p. 26).

Metz (2011) demonstrates how a moral theory grounded in Ubuntu can credibly explain the human rights principles embedded in the South African Constitution, presuming that human rights are grounded in human dignity—that is, that individuals have a “superlative non-instrumental value” (p. 542). He draws upon resources in southern African moral thought to construct a conception of human dignity. He suggests that “one is to develop one’s humanness by communing with those who have a dignity in virtue of their capacity for communing. That is, individuals have a dignity insofar as they have a communal nature, that is, the inherent capacity to exhibit identity and solidarity with others” (Metz, 2011, p. 544). Identifying with others involves seeing oneself as part of the same group, thinking in terms of “we,” sharing in the group’s pride or shame, and working together on common goals by coordinating actions. Solidarity means offering mutual support and behaving in ways that can reasonably be expected to benefit one another. It also includes holding positive attitudes toward others, such as feeling empathy and helping them for their own sake (Metz, 2011). Individuals are thus viewed less as independent and more as interdependent (Sambala et al., 2020).

A third central tenet of Ubuntu is social consciousness. This principle acknowledges the impact of racial and socioeconomic inequalities on health care, which is a result of historical discrimination (Maqutu, 2023; Moll, 2021). Maqutu (2023) argues that social consciousness expands on Beauchamp and Childress’s (2019) bioethical principle of justice by ensuring equitable distribution of goods and services in health care, as well as equal treatment. It also enables health care professionals to humanize and empathize with all patients and to become aware of implicit biases that can negatively impact patient care.

Contrasting Ubuntu with Western bioethical concepts

Ubuntu thus stands in contrast to much of Western bioethics due to its strong focus on relationships, community, and collective wellbeing, rather than on individual autonomy. Whereas Western approaches to bioethics tend to emphasize personal rights and self-determination, Ubuntu places the individual firmly within a broader social, communal, and spiritual context, recognizing that a person’s identity and moral worth are inseparable from their connections to others.

In Western bioethics, principles such as autonomy, beneficence, and justice generally focus on the individual as the key moral agent. By comparison, an Ubuntu-based ethic understands personhood as fundamentally relational, extending moral concern beyond the individual to include social duties and communal solidarity. This perspective influences how moral rights and human dignity are understood, framing dignity as both a collective and an individual attribute, rooted in shared humanity and mutual care. In this way, Ubuntu offers an alternative ethical framework that both challenges and enriches dominant Western models.

Ethics education in the health professions: Current challenges

The Western canon and its dominance

Bioethics education across health professions globally has been heavily shaped by a specific intellectual tradition rooted in Western liberal thought. Central to this are the works of Immanuel Kant, John Stuart Mill, and John Rawls—philosophers whose ethical theories emphasize reason, individual rights, autonomy, and justice as fairness. These theories form the bedrock of what is often referred to as principlism, encapsulated in the four widely taught principles of biomedical ethics: autonomy, beneficence, nonmaleficence, and justice (Beauchamp and Childress, 2019).

Ethics curricula, professional codes of conduct, and international declarations such as the Belmont Report and the Declaration of Helsinki (Cohen-Almagor, 2017; Fischer, 2006) have entrenched these perspectives in medical and health sciences training worldwide. Yet, in African contexts, their adoption has often occurred without adaptation to local traditions, epistemologies, or communal values (Fraser-Burgess et al., 2024; Onuoha, 2007). This uncritical transfer produces a form of ethical imperialism, in which students are taught to privilege autonomy and individual rights even in contexts where relationality and communal decision-making are central.

Absence and tokenization of indigenous ethical traditions

The dominance of Western paradigms in ethics education has led to the near-complete absence—or, at best, token inclusion—of indigenous ethical systems (Bayuo, 2025). In most health science programs across Africa, the curriculum includes either no reference to African bioethics or only a cursory mention as a supplementary viewpoint. These inclusions are often presented as cultural anecdotes rather than as valid and systematic ethical frameworks.

This tokenism reinforces the notion that indigenous worldviews lack philosophical rigor or are merely context-specific customs, rather than structured approaches to moral reasoning. For example, Ubuntu is rarely treated with the same depth and analytical engagement as Kantian or utilitarian theory (Iyioke, 2019). Its relational foundation, communal accountability, and spiritual dimensions are often deemed unscientific or overly subjective within the dominant Western academic paradigm (Makanjee et al., 2018).

The result is a curriculum that implicitly teaches students that ethical reflection is the domain of Western philosophy (Matahela and Ngwenya, 2025), while African or indigenous approaches are peripheral or "soft." This perpetuates epistemic injustice by marginalizing ways of knowing and valuing that are central to the lived experiences of many students, educators, and patients.

Impact on students: Dissonance, alienation, and lack of relevance

Epistemic bias carries significant implications for students, particularly in African contexts. Students are often taught to approach ethical dilemmas through frameworks that may feel alien or even incompatible with their own values and cultural norms (Iyioke, 2019; Matahela and Ngwenya, 2025). For example, the principle of individual autonomy may clash with communal decision-making practices in many

African societies, where family or community elders often play an active role in health decisions (Ujewe, 2023).

Students report experiencing cognitive dissonance when the moral reasoning taught in the classroom, typically grounded in Western frameworks, does not align with the moral traditions and frameworks they have grown up with in their communities (Schrepel et al., 2019; van Zyl et al., 2023). This dissonance is not merely philosophical; it can erode the perceived relevance of ethics education altogether. When students cannot see their values and realities reflected in what they are taught, they may disengage or resort to the uncritical application of ethical rules, leading to superficial moral reasoning (van Zyl et al., 2023; Shih, 2024).

Moreover, the absence of African ethical voices can contribute to a form of internalized inferiority among students, subtly implying that their inherited values are less legitimate than the Western norms institutionalized in textbooks and lectures (Musitha and Mafukata, 2018). This form of curricular colonization not only hampers identity formation and professional confidence but also limits the moral repertoire of future health care professionals operating in diverse and complex environments.

Empirical evidence: South African student experiences

Emerging studies in South Africa illustrate these tensions. In a qualitative study conducted by Makola and Ntoyanto-Tyatyantsi (2023), postgraduate students reported a lack of engagement with ethical content that reflected their cultural contexts. Students expressed frustration at the emphasis on abstract Western theories and described a need for more practical, grounded ethical tools relevant to their future workplaces—often rural, underresourced, and communally structured.

Similarly, a metasynthesis by Matahela and Ngwenya (2025) highlights how Ubuntu philosophy is frequently acknowledged in nursing education but rarely embedded in ways that allow students to engage with it as a legitimate ethical framework. Instead, Western moral theories continue to dominate formal teaching, leaving Ubuntu positioned as peripheral or secondary. The study notes that this lack of integration can create tensions for students, as their culturally grounded values of relationality and communal decision-making are sidelined in favor of abstract, individualist approaches.

These findings reinforce concerns that ethics curricula in South Africa remain Eurocentric in both content and pedagogy. Even when African perspectives, such as Ubuntu, are acknowledged, they are often treated as add-ons rather than as legitimate frameworks for systematic moral reasoning. This not only limits students' ability to engage deeply with ethical complexities but also risks eroding the relevance of ethics education to their professional practice and community realities.

The current state of ethics education in the health care professions reflects a deep-rooted asymmetry in epistemic power. While Western bioethical frameworks continue to dominate, indigenous and African ethical traditions remain marginalized or misrepresented. This imbalance not only affects the legitimacy of the curriculum but also shapes how future professionals perceive ethical responsibility and relate to the communities they serve.

As we turn toward constructing solutions in the next section, it becomes clear that transforming ethics education will require more than content revisions. It demands a fundamental rethinking of whose knowledge counts, how we teach ethics, and what it means to be an ethical health professional in a globally interconnected yet locally grounded world.

Toward decolonized bioethics education: Recommendations and approaches

Integrating African bioethics as equal and valid in curricula

A decolonized approach to bioethics education requires more than simply "adding" African perspectives into an already Western-centric curriculum—it calls for a paradigmatic shift. African bioethics, with its roots in a relational worldview, must be positioned not as a cultural supplement but as an equal and valid ethical framework alongside Western theories such as deontology (actions are inherently right or wrong regardless of their outcomes, i.e. morality is a matter of duty), utilitarianism (the morality of an action is judged by the outcome, i.e. the ends justify the means), and liberal theories of justice (Metz, 2021; Tangwa, 1996).

This integration starts by recognizing that African ethical systems—such as Ubuntu—offer coherent and philosophically rigorous moral reasoning grounded in community, interdependence, and spiritual continuity (Molefe, 2020). These are not merely "values" but form the basis of structured moral traditions with internal consistency and explanatory power that can meaningfully inform bioethics education (Metz, 2011; van Niekerk, 2007).

To accomplish this integration:

- Ethics syllabi must give African frameworks equal coverage and analytical depth.
- Assessment methods should evaluate students' ability to apply ethical reasoning from relational worldviews.
- Accreditation bodies and professional councils should endorse the legitimacy of diverse ethical frameworks in competency guidelines.

This shift is not about replacing Western bioethics but about expanding the ethical repertoire of future health care professionals so they can navigate pluralistic societies ethically, respectfully, and skillfully (Stoeklé and Hervé, 2024).

Co-creating curricula with local communities and indigenous scholars

Curriculum transformation should not be the domain of academics alone. One of the most profound strategies in decolonizing bioethics education is co-creation, where communities and indigenous scholars participate meaningfully in the development, teaching, and evaluation of ethics content (Eichbaum et al., 2021).

This approach addresses two critical gaps: epistemic exclusion and practical relevance. indigenous communities possess embodied ethical knowledge rooted in lived experiences, oral traditions, rituals, and relational structures. By involving elders, traditional healers, and indigenous ethicists in curriculum development, universities ensure that ethical education is grounded in the contexts where students will practice (Karsgaard et al., 2025).

Practical strategies include:

- Community-embedded ethics modules, where students learn directly from community leaders.
- Advisory boards composed of local stakeholders who review and inform curriculum content, delivery, and assessment.
- Reciprocal knowledge exchanges that value indigenous voices not as "informants on the sideline" but as fully recognized co-authors of ethical knowledge. The inclusion of these community and indigenous voices promotes ethical literacy that is not only theoretical but also situated, responsive, and accountable to the communities that health professionals serve (Rashid et al., 2025).

Methods of learning and teaching relational ethics: Storytelling, dialog circles, and case-based reflection

Relational ethics thrives in pedagogical environments that mirror its principles (Bergum, 2003). Traditional lecture formats, which focus on abstract theorizing and hierarchical knowledge transfer, are poorly suited to the education of ethics grounded in relationships, empathy, and mutual recognition (Campbell and Coker, 2025).

Innovative and culturally congruent learning and teaching methods include:

Storytelling

A foundational method in many African traditions, storytelling conveys moral lessons, ethical dilemmas, and shared values in ways that are emotionally resonant and contextually rich. Ethics education can use narratives drawn from local contexts to foster critical thinking and moral reflection (Osei-Tutu, 2022).

Dialog circles

Inspired by African communal practices, dialog circles create spaces of horizontal engagement where all voices are heard and valued. These settings encourage collaborative meaning-making and cultivate ethical sensitivity grounded in empathy, listening, and mutual respect (Barkaskas and Gladwin, 2021).

Case-based reflection

Rather than applying ethical principles mechanically and in an abstract sense, case-based reflection encourages students to explore diverse perspectives—including spiritual, communal, and sociopolitical dimensions—of ethical dilemmas.

Facilitators can guide students to consider how different frameworks, such as Ubuntu or Western individualism, can be used to interpret the same scenario (Forster, 2024; Ngubane-Mokiwa, 2016).

By shifting the pedagogical emphasis from "knowing," which focuses on foundational understanding, to "relating," which demands higher-order cognitive skills such as synthesis, perspective-taking, and ethical reasoning, these methods nurture ethically attuned professionals equipped for the moral complexity of real-world health care.

Training educators to teach diverse ethical frameworks with cultural humility

Educator readiness is pivotal to curriculum transformation. Many lecturers in ethics courses have themselves been trained in Western paradigms and may be unfamiliar—or uncomfortable—with teaching indigenous ethical theories. Some educators may doubt the philosophical rigor of these approaches or worry about applying them incorrectly (Thorpe et al., 2021; Riley et al., 2024).

To overcome this, ethics educators require structured training in cultural humility, a lifelong process of self-reflection and critical engagement with difference, to teach ethically in diverse contexts (Konidaris and Petrakis, 2025; Ogunyemi et al., 2024). Cultural humility differs from cultural competence in that it does not assume mastery over another's worldview but instead promotes respectful engagement and ongoing learning (Lambor et al., 2024; Schiavo, 2023).

Training programs for educators should include:

- Exposure to indigenous philosophies and their application to health care ethics.
- Workshops co-facilitated by community and academic ethicists.
- Reflective practice sessions that explore positionality, privilege, and bias, and how these may influence learning and teaching.

Ultimately, educators must see themselves not as gatekeepers of knowledge but as life-long learners and facilitators of ethical inquiry across epistemic traditions.

The path toward decolonized bioethics education is both a challenge and an opportunity. It calls for courage to critique entrenched systems, imagination to envision alternatives, and humility to share power and knowledge. By integrating African bioethics, co-creating curricula with local communities, adopting relational teaching methods, and supporting educators in this journey, health sciences education can become a space of ethical inclusivity, epistemic justice, and transformative learning.

In preparing future health care professionals, we are not merely transmitting ethical theories—we are shaping moral agents who will stand at the intersections of culture, power, and care. Let us give them the tools to stand wisely.

Case studies and practical applications

While theoretical frameworks provide the scaffolding for ethical reasoning, their true test lies in practice. This section explores how African bioethics—especially frameworks grounded in Ubuntu—offers alternative, often more culturally congruent, interpretations and resolutions to real-world ethical challenges in health care. These examples are not merely academic but are grounded in the lived moral realities of patients, communities, and practitioners in African contexts.

Conflicting ethical interpretations: Western vs. Ubuntu-based ethics

To illustrate the practical implications of differing ethical paradigms, this section examines how Western and Ubuntu-based frameworks approach key bioethical dilemmas, beginning with debates around organ donation and bodily integrity.

Organ donation and bodily integrity

In Western bioethics, decisions about organ donation are typically governed by the principles of individual autonomy and informed consent. The decision belongs to the individual, and upon death, the legal declaration of donor status is considered final (Ehtuish, 2024).

However, in many African communities, the body is viewed not merely as a biological entity but as integrally connected to family, ancestry, and spiritual continuity (Kunle, 2020). Decisions around organ donation are often collective, with family and even community elders playing a role. In some traditions, disrupting the bodily wholeness of the deceased is believed to impact their journey into the ancestral realm (Monareng Moropa et al., 2022).

Ubuntu-based approach: Ethical deliberation would center on relational consultation—what would bring harmony and dignity to the family and community? Consensus and communal flourishing, rather than individual autonomy alone, becomes the central moral ideal considerations (Chuwa, 2014; Sambala et al., 2020; Ewuoso, 2021); Ethical teaching in such a context would emphasize negotiation, respect for spiritual beliefs, and community dialog over unilateral decision-making (Sambala et al., 2020).

Human immunodeficiency virus (HIV) disclosure and relational autonomy

Western principles often prioritize confidentiality and respect for patient autonomy when disclosing an HIV-positive status (Souleymanov et al., 2025; Wilkinson and Shah Arora, 2015). A health care provider may be legally and ethically constrained from informing a patient's sexual partner without explicit consent.

In an Ubuntu-informed approach, relational autonomy redefines this obligation. The individual is seen as inherently embedded within a network of moral relationships, and ethical action considers the welfare of others in this network. Withholding information that endangers others would be morally questionable (Ewuoso, 2021).

Ubuntu-based approach: Disclosure may be ethically encouraged or mandated through a process of communal accountability and restorative justice, not punitive action. The patient could be supported through dialog with family or community mediators to take responsible action that preserves relationships and promotes collective wellbeing.

Community consent in research and clinical practice

Informed consent is a cornerstone of Western ethics, emphasizing the individual's right to accept or refuse participation in research or treatment (National Health and Medical Research Council, 2018). However, in many indigenous contexts, health and decision-making are embedded in collective structures. Individual consent may be insufficient or even inappropriate when decisions affect shared land, cultural resources, or communal bodies of knowledge (Peltier et al., 2024).

Ubuntu-based approach: Ethical legitimacy arises through community consent, where elders or community leaders provide collective guidance (Peltier et al., 2024). Researchers and practitioners must engage in participatory processes that prioritize respect, relationship-building, and accountability to the whole.

This approach is particularly relevant in genomics, traditional medicine studies, or health interventions conducted in small or closely-knit communities (Amoakoh Coleman et al., 2023).

Justice and distributive fairness

In Western bioethics, justice often centers on the fair distribution of resources, guided by frameworks such as Rawlsian "justice as fairness," utilitarian benefit maximization, or equal treatment (Cohen-Almagor, 2017). Decisions about access are typically framed in terms of individual rights and impartial principles.

In contrast, Ubuntu-based ethics prioritizes *relational justice*—seeking communal harmony and moral solidarity rather than strict equality. Chuwa (2014) notes that Ubuntu-informed decision-making emphasizes contextual, relationship-based responsibilities, with a view toward restoring social equilibrium rather than distributing resources by abstract criteria.

Ubuntu-based approach: Resource allocation and ethical decisions involve deliberative community engagement, drawing on elders' wisdom and communal consensus to determine which action will best preserve collective dignity and relationships (Chuwa, 2014; Ewuoso, 2021).

Community engagement and participatory ethics initiatives

True decolonization of ethics education requires ongoing, reciprocal engagement with the communities from which ethical knowledge emerges and to whom it is applied. Several community-participatory models have shown promise in redefining how ethics is learned, taught, and practiced.

Example: Community Ethics Forums

In Limpopo, South Africa, a rural clinic collaborated with traditional healers and local leaders to create monthly community ethics forums. These gatherings provided a platform for health care workers, patients, and elders to discuss ethical dilemmas grounded in local realities. Issues ranged from refusals of biomedical treatment due to spiritual practices to decisions about resource allocation. Through open dialog, the community co-developed responses guided by Ubuntu principles, emphasizing harmony, collective care, and mutual respect. This model illustrates how ethical deliberation can be decentralized and locally grounded while still aligned with professional values (Ramavhoya et al., 2024).

Example: Participatory Curriculum Development

At Lilitha Nursing College in the Eastern Cape, South Africa, a participatory initiative invited community members—including traditional leaders, elders, and health service users—to co-develop aspects of the ethics curriculum. This collaboration ensured that the training incorporated cultural values such as intergenerational respect, ancestral knowledge, and collective responsibility. Community representatives helped review case studies, adapt terminology to local dialects, and reframe learning outcomes to include ethical principles rooted in Ubuntu. This process not only improved the cultural relevance of ethics education but also deepened student engagement and understanding of moral complexity in real-world contexts (Mbatha and Mtshali, 2013).

These examples demonstrate that ethical reasoning and curriculum development become more contextually relevant when rooted in community engagement. They also show that Ubuntu-based ethics is not simply conceptual but can be institutionally implemented through inclusive pedagogical practices.

The San Code of Research Ethics: A community-led ethical framework

One of the most powerful illustrations of indigenous ethical agency in Africa is the San Code of Research Ethics—a landmark document launched in 2017 by the South African San Institute in partnership with the San Council. This code emerged in response to generations of research exploitation, misrepresentation, and ethical marginalization of the San peoples, one of the world's oldest continuous cultures. Historically, researchers have conducted studies on San communities without obtaining proper consent, sharing benefits, or respecting cultural values, often treating participants as mere subjects rather than ethical partners.

In a historic reversal of this power dynamic, the San Code asserts a community-led ethical framework rooted in the values and worldviews of the San people (South African San Institute, 2017). The code identifies four central ethical values as the foundation for all research and engagement:

1. Respect—for San culture, elders, social structures, and spiritual beliefs.
2. Honesty—transparency in research aims, methods, and expected outcomes.
3. Justice and fairness—equitable benefit-sharing and no exploitation.
4. Care—attention to community wellbeing, dignity, and legacy.

These principles are not presented as abstract universals but are grounded in the lived ethical practices of the San people. For instance, "respect" in the San Code includes honoring oral traditions, nonwritten knowledge systems, and the authority of elders as moral guides. "Care" encompasses not only individual welfare but also the wellbeing of the entire community and future generations.

The San Code of Research Ethics serves as a model for ethical engagement, prioritizing community leadership, culturally appropriate values, and equitable partnerships in research involving indigenous populations (South African San Institute, 2017).

Implications for bioethics education

The San Code represents a decolonial turn in ethics—not simply critiquing Western dominance, but constructively articulating a local, relational, and communal ethical system. Its integration into health professions education offers several key pedagogical benefits:

- Case-based learning: Students can explore how applying the San Code alters decision-making in clinical trials, informed consent processes, and public health initiatives.
- Comparative ethics reflection: Learners can analyze tensions between the San Code and Western research ethics guidelines, such as the Declaration of Helsinki or the Belmont Report, particularly pertaining to autonomy, consent, and benefit-sharing.
- Community-partnered pedagogy: The code models how ethics curricula can be co-developed with communities to reflect their priorities, values, and historical experiences.

As an educational tool, the San Code encourages students to shift from a compliance-based to a relational and justice-oriented mindset regarding ethics. It also fosters humility and responsiveness—core components of cultural safety and ethical sensitivity in African contexts.

A model for indigenous ethical sovereignty

The San Code is not only an ethics guideline; it is also a political statement of ethical sovereignty. By developing their own code, the San have reclaimed agency over how they are engaged and represented in research. It stands as a model for other indigenous and marginalized communities to define ethical terms on their own grounds and to engage with researchers—and health care practitioners—as equals (South African San Institute, 2017).

As such, it represents a living expression of Ubuntu ethics, emphasizing community, relationship, dignity, and mutual respect. The San Code does not reject international ethical standards; instead, it demands that such standards meet local values halfway, in dialog, not domination.

The case studies and examples presented affirm that African bioethics is not a theoretical abstraction but a living, contextually grounded framework capable of addressing complex ethical tensions in health care. African ethical systems emphasize interdependence, harmony, and collective responsibility, offering nuanced alternatives to Western individualistic paradigms.

Ultimately, the question is no longer whether we can teach African bioethics, but whether we have the courage and commitment to revise entrenched power structures, recognize the legitimacy of diverse moral worldviews, and embrace ethical plurality as a core feature of professional formation. In doing so, ethics education extends beyond the classroom—and beyond the individual—to communities and relationships founded on respect, reciprocity, and shared humanity.

Comparative Indigenous ethical systems beyond Africa

While this chapter centers on African moral philosophies, particularly Ubuntu, many indigenous traditions across the world articulate comparable relational worldviews that likewise challenge Western individualism. In the Americas, for instance, First Nations and Native American philosophies emphasize kincentric ethics—the idea that moral life arises from kinship with all living beings, human and nonhuman alike (Cajete, 2000). Such a reconceptualization of human–nature ethical relations underscores the extent to which indigenous knowledge systems provide indispensable insights, in this case for advancing ecologically responsible, mutually sustaining modes of coexistence with the Earth (Gauthier et al., 2025; Hird, 2025). These perspectives carry significant health implications: by affirming the interdependence of ecological integrity and human wellbeing, kincentric ethics highlight how clean air and water, biodiversity, stable climates, and reciprocal relationships with land constitute foundational determinants of health. Moreover, the emphasis on relationality and stewardship expands conventional biomedical models by placing individual and community health squarely within broader socioecological systems, potentially influencing health policies and practices that promote resilience, equity, and sustainable futures.

The Navajo concept of *hózhó* embodies balance, harmony, and beauty as moral ideals, guiding right action through the restoration of relationships rather than adherence to abstract principles (Kahn-John and Koithan, 2015). In Latin America, the Andean-Amazonian indigenous notion of *sumak kawsay*, or *buen vivir* ("Good Living"), similarly expresses a communal ethic of reciprocity and equilibrium between people and nature (Richter, 2025) and has now been enshrined in the national constitutions of Ecuador and Bolivia (Gudynas, 2011). As with kincentric ethics, *sumak kawsay/buen vivir* centers ethics in relationality, ecological limits, and collective wellbeing, thereby challenging biomedical ethics to move beyond individual autonomy toward a conceptualization that emphasizes interdependence and cultural plurality, similar to Ubuntu.

In the Asia-Pacific region, similar relational moral systems appear in indigenous and Eastern traditions. The Māori principle of *whanaungatanga* underscores interconnectedness and collective responsibility within the community, while *manaakitanga* foregrounds care, hospitality, and respect as moral imperatives (Wilson et al., 2021). Together, these concepts speak to relational ethics, closely aligned with an ethics of care, urging researchers to consider how their actions shape the lived experiences of those with whom they work and to foster relationships grounded in respect, responsibility, and connection (Fraser, 2025). In East and South Asia, Confucian *ren* (humaneness) and the Buddhist ideal of *interbeing* (as espoused by Thích Nhất Hạnh, 2016) resonate with Ubuntu's vision of personhood as fundamentally co-constituted through relationships. These systems, like African bioethics, situate moral worth not in isolated autonomy but in the capacity to sustain harmony, reciprocity, and balance among persons, ancestors, and the natural world.

Recognizing these convergences does not erase cultural particularity; instead, it demonstrates that relational ethics is a global phenomenon, expressed through diverse cosmologies and practices. Such cross-cultural dialog enriches global bioethics by revealing that the moral imagination of community, care, and interdependence extends well beyond any single continent.

Conclusion

The story of bioethics in Africa is not merely a critique of the past—it is an invitation to imagine a different future. This chapter has traced how Western paradigms came to dominate ethical thought and education, often silencing or sidelining African and other indigenous worldviews and epistemologies. Yet, it has also shown that this dominance is neither inevitable nor unchangeable. By reclaiming and centering relational moral traditions such as Ubuntu, we can reframe bioethics not as a set of abstract, impartial principles but as a profoundly human practice rooted in reciprocity, solidarity, and shared responsibility (refer to Fig. 13.1).

Such a shift demands more than superficial curricular reform. It requires us to confront epistemic injustice head-on, to challenge the mistaken universality of Western moral frameworks, and to embrace ethical pluralism as a strength. It calls on educators to teach beyond the textbook and in collaboration with communities, on practitioners to care and act beyond bureaucratic individual consent, and on institutions to

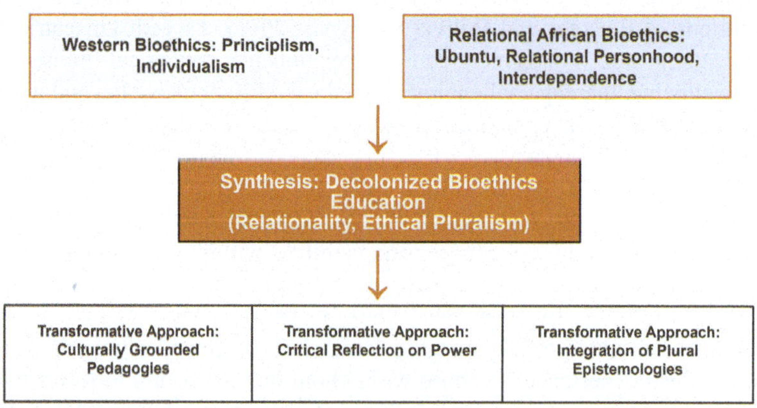

FIG. 13.1

Framework for decolonized bioethics education.

value knowledge born from lived experience. Decolonizing bioethics is, at its core, a project of emancipation—of knowledge, of pedagogy, and of practice. It is about humanizing and contextualizing health care by acknowledging that personhood is relational and that ethical responsibility is often shared.

African health sciences curricula are enriched through the integration of Ubuntu and other indigenous philosophies. In this way, we prepare health care professionals to navigate the convergence of cultures and histories with humility and courage. Yet, Ubuntu is not alone in its vision of relational personhood and collective responsibility. Across the world, indigenous and Eastern ethical traditions echo similar moral insights: from the Native American concept of *hózhó* (harmony and balance) and the Andean-Amazonian ideal of *sumak kawsay/buen vivir* (living well in community), to the Māori value of *whanaungatanga* (kinship and relational belonging) and the Confucian and Buddhist emphasis on *ren* and *interbeing*.

These parallel traditions reveal that the moral imagination of interdependence, care, and harmony is not exclusive to Africa but is woven through the world's diverse fibers of ethical heritage. Recognizing these resonances strengthens the call for a genuinely global, relational bioethics—one that honors the plurality of moral worlds while affirming our shared humanity. Through such an inclusive vision, we move beyond the Western focus on the autonomous individual toward a bioethics that restores connection, reciprocity, and dignity at every level of human life.

Glossary

African bioethics Ethical frameworks grounded in African worldviews, values, and traditions, often emphasizing community and spirituality.

Autonomy The ethical principle that individuals have the right to make decisions for themselves, central to Western bioethics.

Belmont Report A foundational 1979 US report on research ethics, outlining the principles of respect for persons, beneficence, and justice. It was written in response to ethical failures such as the Tuskegee Syphilis Study and it heavily influenced principlism.

Bioethics The study of ethical issues and decision-making in health care, medical research, and clinical practice.

Case-based reflection A teaching strategy that uses real or hypothetical cases to explore ethical dilemmas from multiple perspectives.

Cognitive dissonance Psychological discomfort experienced when one's beliefs or values conflict with newly presented information.

Common morality A set of moral norms or principles that are considered universal and shared across different individuals and societies.

Community consent An ethical practice where decisions affecting groups are made with the approval and participation of community representatives.

Communitarianism The view that individuals are constituted by their social ties and communities, and this perspective on the person should shape our approaches to ethics, politics, and public policy.

Cultural humility A lifelong process of self-reflection and learning that emphasizes openness to others' cultural identities and power imbalances.

Cultural safety A practice that addresses power imbalances and historical inequities by creating environments where patients and students feel respected and empowered—with safety defined by those receiving care or education.

Curricular colonization The dominance of Western knowledge systems in education that marginalize local or indigenous perspectives.

Declaration of Helsinki An international ethical framework adopted by the World Medical Association in 1964, providing guidance on medical research involving human subjects.

Decolonization (of Education) The process of critically dismantling colonial epistemologies and restoring diverse, local, and indigenous knowledge systems within curricula.

Dialog circles A participatory teaching method inspired by communal practices, where all voices are heard and valued equally.

Enlightenment (Western) An intellectual movement of the 17th and 18th centuries that emphasized reason, science, and individual autonomy. It profoundly shaped Western bioethics but has been critiqued for its colonial, individualistic, and Eurocentric biases.

Epistemic injustice The marginalization or devaluation of certain knowledge systems, particularly those of non-Western or indigenous origin.

Epistemologies Ways of knowing and systems that define what counts as valid knowledge. Epistemologies include the assumptions, methods, and perspectives through which people understand reality. In bioethics, this refers to whether knowledge is grounded in Western analytic philosophy, African relational worldviews (e.g., Ubuntu), indigenous oral traditions, or other traditions of thought.

Ethical imperialism The imposition of one ethical framework (often Western) as universal, disregarding the validity of alternative or indigenous moral systems.

Ethical pluralism The recognition and acceptance of multiple ethical frameworks as valid, particularly across diverse cultural contexts.

Fifth Industrial Revolution The current era marked by artificial intelligence, biotechnology, and digital innovation, raising new ethical questions about health care, personhood, and human dignity.

Hermeneutical injustice A form of epistemic injustice that occurs when people lack the shared concepts, language, or interpretive tools needed to understand and communicate their own experiences within dominant cultural frameworks.

Indigenous ethical sovereignty The right of indigenous communities to define and apply their own ethical norms and practices, particularly in research and health care.

Informed consent A process in which individuals voluntarily agree to participate in research or receive treatment after being adequately informed. It is central to Western bioethics but may conflict with communal decision-making traditions.

Kantian or Utilitarian theory Two dominant traditions in Western moral philosophy. Kantian (deontological) theory emphasizes duty, universal moral rules, and respect for persons as ends in themselves, regardless of consequences. Utilitarian (consequentialist) theory, associated with Jeremy Bentham and John Stuart Mill, judges actions by their outcomes, seeking to maximize overall happiness or wellbeing. Both frameworks strongly shape Western bioethics, particularly in debates around autonomy, beneficence, and justice.

Moral relativism The view that right and wrong are not absolute but depend on the moral standards of particular people or groups, and that serious moral disagreements exist across cultures.

Nuremberg Code A set of ten ethical principles established after the Nuremberg Trials (1947) to protect human subjects in research, emphasizing voluntary consent.

Principlism A dominant Western framework in bioethics that includes four principles: autonomy, beneficence, nonmaleficence, and justice.

Reflective equilibrium A method of ethical reasoning that involves balancing principles and intuitions in a coherent system.

Relational autonomy A concept of autonomy that acknowledges individuals as embedded within social and relational contexts.

Relational ethics An ethical approach that prioritizes relationships, context, and interdependence over individual autonomy.

Restorative justice An approach that prioritizes repairing harm, rebuilding trust, and restoring relationships in response to ethical breaches or historical injustices. In health care ethics and education, it involves acknowledging past harms, fostering dialog, and promoting accountability to create more equitable and inclusive practices.

San Code of Research Ethics A community-led ethical framework developed by the San people of South Africa to guide respectful, just, and culturally grounded research practices.

Solidarity An ethical principle emphasizing social cohesion, mutual support, and collective responsibility. Some African scholars propose it as a more fundamental principle than autonomy in African bioethics.

Testimonial injustice A form of epistemic injustice that occurs when someone's credibility as a speaker or knowledge-holder is unfairly diminished because of prejudice regarding their social identity.

Ubuntu An African ethical philosophy emphasizing community, relational personhood, and mutual care, often summarized as "I am because we are."

References

Amoakoh Coleman, M., Vieira, D., Abugri, J., 2023. Ethical considerations for biobanking and use of genomics data in Africa: a narrative review. BMC Med. Ethics 24, 108. doi:10.1186/s12910-023-00985-y.

Ashby, M., 2019. All across the universe: bioethics across cultures. Bioeth. Inq. 16, 299–300. doi:10.1007/s11673-019-09941-1.

Barkaskas, P., Gladwin, D., 2021. Pedagogical talking circles: decolonizing education through relational Indigenous frameworks. J. Teach. Learn. 15 (1), 20–38. doi:10.22329/jtl.v15i1.6519.

Bayuo, J., 2025. Indigenous African philosophy as a paradigm for health and social care research: a philosophical and methodological discussion. Nurs. Inq. 32 (2), e70002. doi:10.1111/nin.70002.

Beauchamp, T.L., Childress, J.F., 2019. Principles of Biomedical Ethics, eighth ed., Oxford University Press, Oxford.

Behrens, K.G., 2017. Hearing sub-Saharan African voices in bioethics. Theor. Med. Bioeth. 38, 95–99. doi:10.1007/s11017-017-9406-z.

Bell, D., 2024. Communitarianism. In: Zalta, E.N., Nodelman, U. (Eds.), The Stanford Encyclopedia of Philosophy. Available from: https://plato.stanford.edu/archives/sum2024/entries/communitarianism/ (Accessed 23 September 2025).

Bergum, V., 2003. Relational pedagogy: embodiment, improvisation, and interdependence. Nurs. Philos. 4 (2), 121–128. doi:10.1046/j.1466-769X.2003.00128.x.

Brandt, A.M., 1978. Racism and research: the case of the Tuskegee Syphilis study. Hastings Cent. Rep. 8 (6), 21–29. doi:10.2307/3561468.

Buyondo, J.T., 2024. The Critique of Bioethical Principlism in Contrast to an African Approach to Bioethics. Wipf and Stock Publishers, Eugene, OR.

Cajete, G., 2000. Native Science: Natural Laws of Interdependence. Clear Light Publishers, Santa Fe, NM.

Campbell, L., Coker, H., 2025. Community, culture and care: mapping relational pedagogy in online higher education. Discov. Educ. 4, 143. doi:10.1007/s44217-025-00570-y.

Chuwa, L.T., 2014. African Indigenous Ethics in Global Bioethics: Interpreting Ubuntu. Springer, Dordrecht.

Cohen-Almagor, R., 2017. On the philosophical foundations of medical ethics: Aristotle, Kant, J.S. Mill, and Rawls. Ethics, Medicine and Public Health 3 (4), 436–444. doi:10.1016/j.jemep.2017.09.009.

Ehtuish, E.F.A., 2024. Ethics of organ donation and transplantation: toward self-sufficiency. Exp. Clin. Transplant. 22 (Suppl 5), 30–34. doi:10.6002/ect.pedsymp2024.O1.

Eichbaum, Q.G., Adams, L.V., Evert, J., Ho, M.-J., Semali, I.A., van Schalkwyk, S.C., 2021. Decolonizing global health education: rethinking institutional partnerships and approaches. Acad. Med. 96 (3), 329–335. doi:10.1097/ACM.0000000000003473.

Ewuoso, C., 2021. An African relational approach to healthcare and big data challenges. Sci. Eng. Ethics 27 (3), 34. doi:10.1007/s11948-021-00313-w.

Fayemi, A.K., 2021. Reconsidering solidarity in an African modified Principlism. S. Afr. J. Bioeth. Law 14 (2), 50–54.

Fayemi, A.K., Macaulay-Adeyelure, O.C., 2016. Decolonizing bioethics in Africa. BEOnline 3 (4), 68–90. doi:10.20541/beonline.2016.0009.

Fischer, B.A., 2006. A summary of important documents in the field of research ethics. Schizophr. Bull. 32 (1), 69–80. doi:10.1093/schbul/sbj005.

Forster, D.J., 2024. Normative case studies, reflective equilibrium, and the ethics of belief in teacher education. Educ. Theory 74 (3), 340–349. doi:10.1111/edth.12635.

Fraser-Burgess, S., Heybach, J., Metro-Roland, D., 2024. Traditions in Ethics and Education. Cambridge University Press, Cambridge.

Fraser, G., 2025. Relational ethics between researchers: an indigenous qualitative perspective. Methods Psychol. 13, 100196. doi:10.1016/j.metip.2025.100196.

Fricker, M., 2007. Epistemic Injustice: Power and the Ethics of Knowing. Oxford University Press, Oxford.

Gauthier, P.E., Chungyalpa, D., Goldman, R.I., Davidson, R.J., Wilson-Mendenhall, C.D., 2025. Mother Earth kinship: centering Indigenous worldviews to address the Anthropocene and rethink the ethics of human-to-nature connectedness. Curr. Opin. Psychol. 64, 102042. doi:10.1016/j.copsyc.2025.102042.

Gudynas, E., 2011. Buen Vivir: today's tomorrow. Development 54 (4), 441–447. doi:10.1057/dev.2011.86.

Haggerty, K.D., 2004. Ethics creep: governing social science research in the name of ethics. Qual. Sociol. 27, 391–414. doi:10.1023/B:QUAS.0000049239.15922.a3.

Halse, C., 2011. Confessions of an ethics committee chair. Ethics Educ. 6 (3), 239–251. doi:10.1080/17449642.2011.632718.

Hird, C., 2025. Indigenous theory as epistemic ground for environmental scholarship and practice. People Nat. 7, 2262–2277. doi:10.1002/pan3.70137.

Iyioke, I.V., 2019. Bioethics via Africology. In: Ed. (Keengwe, J.), *Handbook of Research on Indigenous Knowledge and Bi-Culturalism in a Global Context*. Advances in Religious and Cultural Studies. IGI Global, Hershey, PA, pp. 21–37. doi:10.4018/978-1-5225-6061-6.ch003.

Jecker, N.S., Atuire, C., 2022. Bioethics in Africa: a contextually enlightened analysis of three cases. Dev. World Bioeth. 22 (2), 112–122. doi:10.1111/dewb.12324.

Kahn-John, M., Koithan, M., 2015. Living in health, harmony, and beauty: the Diné (Navajo) Hózhó wellness philosophy. Glob. Adv. Health Med. 4 (3), 24–30. doi:10.7453/gahmj.2015.044.

Kamwangamalu, N.M., 1999. Ubuntu in South Africa: a sociolinguistic perspective to a pan-African concept. Crit. Arts 13 (2), 24–41. doi:10.1080/02560049985310111.

Karsgaard, C., Parlee, B., Catholique, A., MacKay, M., 2025. The "gift" of Indigenous knowledge and critical, place-based curriculum development through ethical relationality. Curric. Inq. 54 (3), 208–230. doi:10.1080/03626784.2025.2456811.

Konidaris, M., Petrakis, M., 2025. Cultural humility training in mental health service provision: a scoping review of the foundational and conceptual literature. Healthcare 13 (11), 1342. doi:10.3390/healthcare13111342.

Kunle, A.J., 2020. Organ donation in an African culture. Bangladesh J. Bioeth. 10 (1), 21–25. Available at: https://www.doii.org/10.3329/bioethics.v10i1.48973.

Kurakina, O.D., 2023. The ethics of Enlightenment in the foundations of modern science. 12 Kant-Readings Conference, "Kant and the Ethics of Enlightenment: historical Roots and Contemporary Relevance". Web Conf. 161, 07005. doi:10.1051/shsconf/202316107005.

Lambor, P.D., Bhageerathy, R., Devi, E.S., Gudi, N., R, V.L., 2024. Cultural humility training interventions for healthcare students to improve care for gender and sexual minority groups: protocol for a systematic review and a proposed meta-analysis. MethodsX 13, 102963. doi:10.1016/j.mex.2024.102963.

Le Grange, L., 2020. Research ethics: examining the tension between Principlism and rational self-interest in a neoliberal university context. Transform. High. Educ. 5 (0), a88. doi:10.4102/the.v5i0.88.

MacIntyre, A., 1988. Whose Justice? Which Rationality? Notre Dame. Notre Dame University Press, IN.

Makanjee, C.R., Thambura, M.J., Engel-Hills, P., 2018. Ethics for healthcare professionals in radiography – an African perspective. In: Nortjé, N., Jongh, J.C.D., Hoffmann, W. (Eds.), *African Perspectives on Ethics for Healthcare Professionals*. In: Advancing Global Bioethics, vol. 13. Springer, Cham, pp. 189–199. doi:10.1007/978-3-319-93230-9_14.

Makola, Z., Ntoyanto-Tyatyantsi, N., 2023. Post graduate students' experiences with research ethics: a South African perspective. J. Empir. Res. Hum. Res. Ethics 18 (4), 208–217. doi:10.1177/15562646231188004.

Maqutu, L., 2023. African Principlism: A Southern African Approach to Bioethics. University of Johannesburg. [online]. Available from: https://ujcontent.uj.ac.za/esploro/outputs/graduate/African-Principlism–a-Southern-African/9950209007691#file-0.

Marks, D., 1999. Disability: Controversial Debates and Psychosocial Perspectives. Routledge, London.

Matahela, V.E., Ngwenya, N., 2025. A meta-synthesis of studies on ubuntu philosophy in nursing: implications for nursing education. Curationis 48 (1), e1–e15. doi:10.4102/curationis.v48i1.2652.

Mbatha, A.N., Mtshali, G.N., 2013. Conceptualization of responsiveness of nursing education programme by stakeholders at Lilitha Nursing College in Eastern Cape, South Africa. Afr. J. Phys. Act. Health Sci. 19 (Suppl 1), 1–10. Available at: https://www.ajol.info/index.php/ajpherd/article/view/95572.

Metz, T., 2011. Ubuntu as a moral theory and human rights in South Africa. Afr. Hum. Rights Law J. 11 (2), 532–559. Available at: https://www.ahrlj.up.ac.za/metz-t.

Metz, T., 2021. A Relational Moral Theory: African Ethics in and beyond the Continent. Oxford University Press, Oxford.
Molefe, M., 2020. An African Ethics of Personhood and Bioethics: A Reflection on Abortion and Euthanasia. Palgrave Macmillan, Cham.
Moll, T., 2021. Medical mistrust and enduring racism in South Africa. J. Bioeth. Inq. 18 (1), 117–120. doi:10.1007/s11673-020-10072-1.
Monareng Moropa, J., Mokwena, J., Makgahlela, M., Sodi, T., 2022. Older black rural South African adults' perceptions on organ donation: an exploratory study. J. Psychol. Afr. 32 (3), 1–5. doi:10.1080/14330237.2021.2017597.
Moreno, J.D., Schmidt, U., Joffe, S., 2017. The Nuremberg Code 70 years later. JAMA 318 (9), 795–796.
Musitha, M.E., Mafukata, M.A., 2018. Decolonising higher education in Africa: arriving at a global solution. S. Afr. J. Educ. 38 (5), 1–9. https://www.scielo.org.za/scielo.php?pid=S0256-01002018000500003&script=sci_arttext.
National Health and Medical Research Council (NHMRC), 2018. National Statement on Ethical Conduct in Human Research (2007, Updated 2018). NHMRC, Canberra. Available at: https://nhmrc.gov.au/about-us/publications/national-statement-ethical-conduct-human-research-2007-updated-2018.
Ngubane Mokiwa, S.A., 2016. Delivering open distance e-learning through Ubuntu values. In: Letseka, M., Ed., Open Distance Learning (ODL) through the Philosophy of Ubuntu. Nova Science Publishers, New York, pp. 147–162.
Nhất Hạnh, T., 2016. *The Miracle of Mindfulness: An Introduction to the Practice of Meditation* (Gift Ed.). Beacon Press, Boston, MA.
Nielsen, K.M., Nordgaard, J., Henriksen, M.G., 2025. Fundamental issues in epistemic injustice in healthcare. Med. Health Care Philos. 28, 291–301. doi:10.1007/s11019-025-10259-6.
Ogunyemi, D., Thind, B.S., Teixeira, A., Sams, C.M., Ojo, M., Dinkins, G.A.E., Serseni, D., 2024. Integrating cultural humility into medical education using a structured and interactive workshop. Adv. Med. Educ. Pract. 15, 575–583. doi:10.2147/AMEP.S460970.
Onuoha, C., 2007. Bioethics across borders: an African perspective. In: Doctoral Dissertation. Uppsala University. Uppsala Studies in Social Ethics, Vol. 34. Acta Universitatis Upsaliensis, Uppsala. ISBN 978-91-554-6869-9.
Osei-Tutu, A.A.Z., 2022. Developing African oral traditional storytelling as a framework for studying with African peoples. Qual. Res. 23 (6), 1497–1514. doi:10.1177/14687941221082263.
Peltier, C., Dickson, S., Grandpierre, V., Oltean, I., McGregor, L., Hageltorn, E., Young, N.J., 2024. Culturally appropriate consent processes for community driven indigenous child health research: a scoping review. BMC Med. Ethics 25, 3. doi:10.1186/s12910-023-00996-9.
Ramavhoya, T.I., Kgatla, M.N., Rasweswe, M.M., Mulaudzi, F.M., 2024. Empowering a Limpopo rural community to integrate Ubuntu in caring for PLWHIV/AIDS or TB. Curationis 47 (2), e1–e7. doi:10.4102/curationis.v47i2.2628.
Rashid, M., Clark, W., Foulds, J.L., Nguyen, J., John, I., Forgie, S.E., 2025. Exploring approaches to teaching Indigenous health curricula from the perspectives of faculty and residents. Med. Educ. Online 30 (1), 2516673. doi:10.1080/10872981.2025.2516673.
Richter, K., 2025. Cosmological limits to growth, affective abundance, and Rights of Nature: insights from Buen Vivir/sumak kawsay for the cultural politics of degrowth. Ecol. Econ. 228, 108442. doi:10.1016/j.ecolecon.2024.108442.

Riley, T., Meston, T., Cutler, C., Low-Choy, S., McCormack, B.A., Kim, E.-J.A., Nakar, S., Vasco, D., 2024. Weaving stories of strength: ethically integrating Indigenous content in teacher education and professional development programmes. Teach. Teach. Educ. 142, 104513. doi:10.1016/j.tate.2024.104513.

Sambala, E.Z., Cooper, S., Manderson, L., 2020. Ubuntu as a framework for ethical decision making in Africa: responding to epidemics. Ethics Behav. 30 (1), 1–13. doi:10.1080/10508422.2019.1583565.

Schiavo, R., 2023. Embracing cultural humility in clinical and public health settings: a prescription to bridge inequities. J. Commun. Healthc. 16 (2), 123–125. doi:10.1080/17538068.2023.2221556.

Schrepel, C., Jauregui, J., Brown, A., Shandro, J., Strote, J., 2019. Navigating cognitive dissonance: a qualitative content analysis exploring medical students' experiences of moral distress in the emergency department. AEM Educ. Train. 3 (4), 331–339. doi:10.1002/aet2.10380.

Shih, Y.-H., 2024. Culturally responsive curriculum: a systematic overview. Edelweiss Appl. Sci. Technol. 8 (6). doi:10.55214/25768484.v8i6.3493.

Souleymanov, R., Akinyele-Akanbi, B., Njeze, C., Ukoli, P., Migliardi, P., Larcombe, L., Restall, G., Ringaert, L., Payne, M., Kim, J., 2025. Navigating ethics in HIV data and biomaterial management within Black, African, and Caribbean communities in Canada. BMC Med. Ethics 26, 5. doi:10.1186/s12910-025-01161-0.

South African San Institute (SASI), 2017. San Code of Research Ethics. SASI, Cape Town. Available at: https://trust-project.eu/wp-content/uploads/2017/03/San-Code-of-RESEARCH-Ethics-Booklet-final.pdf.

Stanfield, M.A., 2019. Autonomy, Suffering, and the Practice of Medicine: A Relational Approach, ProQuest Dissertations & Theses.

Stoeklé, H.-C., Hervé, C., 2024. Global bioethics and moral pluralism: towards a criteriology at the hospital. J. Transl. Med. 22, 625. doi:10.1186/s12967-024-05440-z.

Sutrop, M., 2024. Vilhjálmur Árnason's call for expanding bioethical discourse: "The Personal is Political". Etikk I Praksis 18 (2), 11–21. doi:10.5324/eip.v18i2.6193.

Tangwa, G.B., 1996. Bioethics: an African perspective. Bioethics 10 (3), 183–200. doi:10.1111/j.1467-8519.1996.tb00118.x.

Ten Have, H., Patrão Neves, M.D.C., 2021. Nuremberg code. In: Dictionary of Global Bioethics. Springer International Publishing, Cham, p. 47. doi:10.1007/978-3-030-54161-3_22.

Thorpe, K., Burgess, C., Egan, S., 2021. Aboriginal community-led preservice teacher education: learning from country in the city. Aust. J. Teach. Educ. 46 (1). doi:10.14221/ajte.202v46n1.4.

Tobin, M.J., 2022. Fiftieth anniversary of uncovering the Tuskegee Syphilis study: the story and timeless Lessons. Am. J. Respir. Crit. Care Med. 205 (10). doi:10.1164/rccm.202201-0136SO.

Tosam, M.J., 2020. Global bioethics and respect for cultural diversity: how do we avoid moral relativism and moral imperialism? Medicine, Health Care and Philosophy 23 (4), 611–620. doi:10.1007/s11019-020-09972-1.

Trimble, M., 2024. Ethics - A matter of principle? Ulster Med. J. 93 (2), 83–86.

Ujewe, S.J., 2023. Personal autonomy and shared-value in bioethics: an African communal ethics outlook. In: Imafidon, E., Tshivashe, M., Freter, B. (Eds.), *Handbook of African Philosophy*. Springer International Publishing, Cham, pp. 181–196. doi:10.1007/978-3-030-77898-9_7-1.

van Niekerk, J., 2007. In Defence of an autocentric account of Ubuntu. S. Afr. J. PhiloS. 26 (4), 364–368. doi:10.4314/sajpem.v26i4.31494.

van Zyl, S., Kruger, W.H., Walsh, C.M., 2023. Chronic diseases of lifestyle curriculum: students' perceptions in primary health care settings. Afr. J. Prim. Health Care Fam. Med. 15 (1), a3775. doi:10.4102/phcfm.v15i1.3775.

Vest, M., 2025. Bioethics as a language game: probing the quality of moral guidance in Principlism. Theor. Med. Bioeth. 46 (1), 51–65. doi:10.1007/s11017-025-09702-9.

Wallace, J., 1999. Women in Love: Introduction and Notes, Wordsworth Editions Limited, Hertfordshire.

Wilkinson, B., Shah Arora, K., 2015. A systematic review on confidentiality, disclosure, and stigma in the United States: lessons for HIV care in pregnancy from reproductive genetics. New Bioeth. 21 (2), 142–154. doi:10.1179/2050287715Z.00000000067.

Wilson, D., Moloney, E., Parr, J.M., Aspinall, C., Slark, J., 2021. Creating an indigenous Māori-centred model of relational health: a literature review of Māori models of health. J. Clin. Nurs. 30 (23-24), 3539–3555. doi:10.1111/jocn.15859.

Ziatdinov, R., Atteraya, M.S., Nabiyev, R., 2024. The Fifth Industrial Revolution as a transformative step towards Society 5.0. Societies 14 (2), 19. doi:10.3390/soc14020019.

Further reading

Makerere University, n.d. CEB 8103 Bioethics: Year 2 Course Outline, Master of Science in Clinical Epidemiology & Biostatistics. College of Health Sciences. Available at: https://courses.mak.ac.ug/courses/bioethics.

University of the Western Cape, 2024. Road to Ubuntu Initiative and Community-engaged Ethics Teaching. School of Public Health. Available at: https://www.uwc.ac.za/events/road-to-ubuntu.

CHAPTER

14

Reclaiming wholeness: Ayurveda, colonization, and the healing of medicine

Sheila Patel[a,b]

[a]*University of California San Diego School of Medicine, San Diego, CA, United States*

[b]*Gaja Collective, San Diego, CA, United States*

Medicine as a discipline has continuously evolved, shaped by diverse cultural and philosophical paradigms throughout history. To understand the contemporary state of medical systems and education globally, and to continue to evolve, it is essential to examine traditional and indigenous healing systems, including their potential role in shaping future health care models. Ayurveda is an ancient system of medicine originating from the Indian subcontinent, practiced for millennia, and remains widely used, especially in this region. Ayurveda is often considered the oldest healing science, with its roots deeply embedded in ancient Indian philosophy, and is believed to have originated >5,000 years ago (Pondomatti et al., 2024). A substantial portion of the Indian population still uses Ayurveda and other indigenous systems for the prevention or treatment of ailments, with it being the most commonly used system in rural and urban India (Survey on Ayush Fact Sheet, 2023, p. 9). Since its introduction to the West in the 1970s and 1980s, Ayurveda has been embraced by a growing number of people worldwide. Internationally, Ayurvedic products are increasingly utilized, with one estimate suggesting that their market value exceeded $18 billion globally in 2024 (Market.us, 2025). This reflects its growing relevance and accessibility.

To appreciate Ayurveda's potential role in evolving current medical systems, it is essential to understand the enduring effects of the colonial occupation of India and the challenges it presents to the integration of Ayurveda into modern medical practice. While Ayurvedic knowledge was suppressed for many centuries, in recent years, there has been a concerted effort to revitalize and integrate it more fully into India's health care system. This also includes efforts to incorporate it into health care systems around the world, as evidenced by the establishment of professional and scientific Ayurvedic organizations in the United States (US), Europe, and Latin America (Yeola et al., 2024). Moreover, such efforts include the establishment of the World Ayurveda Congress in 2002 (Tillu et al., 2024). Ayurveda's emphasis on balanced living, dietary modifications, and natural remedies could offer a much-needed complement to the modern medical approach (Hegde et al., 2024). Integration offers several

Decolonizing Medical Education, Research and Practice. https://doi.org/10.1016/B978-0-443-44998-7.00017-X

potential advantages, including lower health care costs, improved access in rural and underserved regions, and the promotion of preventive, holistic practices that can mitigate the burden of chronic diseases. Moreover, Ayurveda's emphasis on individualized care, lifestyle optimization, and disease prevention offers valuable insights for Western health care systems, which often focus primarily on acute interventions and symptom management. Integrating Ayurvedic principles into Western medicine may, therefore, enhance patient-centered care, promote wellness and resilience, and support broader strategies for chronic disease prevention and health promotion. Overall, integrating Ayurveda with modern medicine can enhance patient care by providing a more comprehensive approach toward health management (Verma et al., 2024). Therefore an understanding of its foundational principles and practices is critical, as Ayurveda could provide a complementary and sustainable framework to address the growing health care crisis facing many health systems today.

Medicine through the lens of colonization

To appreciate how contemporary medicine could benefit from the insights of traditional healing systems and to address the obstacles to their integration, it is essential to understand the history of colonialism and its enduring impact on medical systems and education. As stated by Marya and Patel (2021), colonialism and colonization are not simply the physical occupation of land. Instead, they represent a process where one paradigm is extinguished and then replaced with another to exert power and control over a population and "to replace one cosmology with another" (Marya and Patel, 2021, p. 14). Colonialism not only incorporated a core set of social, cultural, political, and economic hierarchies but also intellectual ones (Demir, 2022, p. 2). As part of colonialism, there is typically a systematic conditioning that the beliefs, ideas, and practices of the colonizing country are superior to those of the country being colonized, as occurred in India during colonization. In the context of social systems such as health care or medical education, the term "decolonization" refers to the process of changing something, such as a curriculum, in a way that considers the cultural beliefs behind it. For example, this includes the belief that European writers, artists, or ideas are superior and more important than ones from countries that were colonized (or controlled) by Europe (Cambridge Dictionary, 2025).

The effects of colonization on medicine led to the active suppression of a large body of knowledge across the world. When applied to medicine, decolonization involves recognizing and rectifying the long-standing suppression of indigenous healing systems, which, in many cases, have been disregarded or erased. By changing the prevailing paradigms and knowledge systems of indigenous populations to colonial frameworks, traditional systems no longer possessed context or meaning and, hence, could more easily be devalued. The suppression of indigenous healing systems has a long and complex history that we are only now beginning to fully realize.

Because of colonization, the Western world cultivated a dominant perspective that marginalizes indigenous wisdom, often deeming it inferior while remaining largely uninformed about the depth and richness of indigenous cultures. Many people in mainstream American culture—from my direct experience growing up in the United States as a child of Indian immigrants—were taught significant misunderstandings about Indian culture and beliefs in general. The lens of colonization, through generations of systematized conditioning, has shaped the landscape of medical education and practice as well, leading many medical professionals to relegate Ayurveda to superstition or claim it is not scientific or valid, or that there is no evidence that it works. This was in direct contradiction to my personal experience at home, where Ayurvedic remedies were utilized and proved helpful in treating many health conditions. Furthermore, throughout my training to become an MD at a conventional medical school, indigenous medical systems such as Ayurveda were openly mocked, devalued, and deemed "not scientific."

By exploring the intersection of medicine and colonialism, we begin to uncover the profound implications of these historical forces, and this process confirms that true healing and wellness can only be achieved through a more inclusive, decolonized approach to health.

Ayurveda and the colonization of Indian medicine: A historical perspective

The history of India, including Ayurveda, was carried forward through a rich oral tradition. Knowledge was transmitted via Sanskrit chants, which could range from 40,000 to 200,000 sutras (verses) in length and were memorized and recited with extraordinary precision for thousands of years (Smith, 2013, p. 15). While written language did exist in ancient Indian culture, writing was considered a secondary or inferior method of preserving knowledge—a sentiment that exists in some contexts even today. Sanskrit is unique in that its structure, intonation, and meter are designed not only to convey semantic meaning, but also to encode subtle energetic and vibrational patterns. Each sound and syllable carry specific physiological and cognitive effects, engaging both the analytical mind and the "emotional brain," or limbic system, thereby facilitating memory, emotional resonance, and the embodiment of knowledge. In fact, research has shown that the verbalization of Sanskrit verses coordinately engages multiple brain regions and suggests cognitive and neural benefits (Chhandak et al., 2025). In Ayurveda, this oral tradition was central to teaching complex concepts. For example, the recitation of Sanskrit sutras allowed students to internalize knowledge not just intellectually, but experientially, aligning mind, body, and consciousness—a process that remains foundational to the Ayurvedic understanding of healing.

The first evidence of Ayurvedic practices can be found during the Early Vedic period from 1,500 to 1,000 BCE. The Vedas are India's ancient sacred texts that contain spiritual teachings, rituals, early knowledge of medicine, philosophy, and the natural world, and also mention medicinal herbs and guidelines for healthy living. Here, one of the texts (Atharva Veda) contains detailed descriptions of various diseases, their causes, symptoms, and treatments (Verma et al., 2024, pp. 2–3). During this period, medical practice was delineated into two distinct lineages. Firstly, there is the Atreya School, led by Atreya, the legendary Ayurvedic physician who taught the principles of internal medicine. Secondly, there is the Dhanvantari School, which originated from Dhanvantari (a mythological figure considered the originator of Ayurveda) and taught the principles of surgery and therapeutics.

During the Classical period (1,000–500 BCE), Ayurvedic teaching advanced and was codified into two texts. These texts are still used in contemporary Ayurvedic medical schools and include pictures, herbal medicinal recipes, and additional information in volumes known as *samhitas* (Charaka Samhita and Sushruta Samhita). Charaka was a student of Atreya and founded the Charaka School. His samhita is a foundational text on internal medicine, pharmacology, physiology, and preventive care. Sushruta is another Ayurvedic physician who is associated with a direct lineage of surgeons and physicians and developed the Sushruta School. He was the author of a foundational surgical text that described over 100 surgical instruments and procedures and was also the head of a lineage of surgical teachers. A third classical lineage was developed during this time by Vagbhata, a physician and scholar, who synthesized and systematized earlier Ayurvedic knowledge. His main work is the Ashtanga Hridaya, which is a concise, practical compendium combining the principles of Charaka (internal medicine) and Sushruta (surgery) into a single accessible text. Despite their differences, these schools shared a common belief in the importance of a balanced diet, herbal remedies, and lifestyle changes for maintaining health (Jaiswal and Williams, 2016).

From about 1,000 BCE, Ayurveda transitioned from being an oral tradition to a written teaching. Although Ayurveda had been practiced for millennia prior to this codification, it was during this time that it became more structured and formalized, i.e., organized into eight specialties, namely, internal medicine, ear/nose/throat, surgery, toxicology, psychiatry, pediatrics, rejuvenation (preventative medicine), and reproduction/fertility (Smith, 2013, pp. 17–21). This period is considered the golden age of Ayurveda, marked by the development of schools, hospitals, and dispensaries. However, from around 1,000 CE onwards, Ayurveda suffered due to multiple Mughal invasions, which were followed by colonization by the British Empire. While Ayurveda was not entirely suppressed during the Mughals' reign, it faced challenges, particularly in terms of its official recognition. The system was largely preserved in regional and rural areas, as well as within certain families and communities.

The Portuguese were the first to introduce Western medicine into India during the 16th century. During this time, medical officers who arrived with the British East India Company's first fleet of ships also brought Western medicine to India. Here, medical departments staffed with surgeons were established to provide medical relief to the troops and employees of the East India Company. Moreover, hospital boards comprising the Surgeon General and Physician General were established (Anshu and Supe, 2016). Ayurveda was more significantly suppressed during the British occupation and colonization period (1858–1947). This suppression occurred through various methods, including the imposition of Western medical practices, the establishment of colonial policies, and the creation of an educational system that marginalized indigenous knowledge. Of note, during the early 1800s, the prevailing colonial ideology was centered around the belief that Western biomedicine and education were the sole legitimate systems. For example, the British politician and historian Thomas Babington Macaulay—who served on the Supreme Council of India from 1834 to 1838—played a key role in introducing a Western-based educational system into India. Here, Macaulay's "Minute on Indian Education" advocated for English-based education in 1835, while regarding European-based science as superior and dismissing indigenous knowledge systems as inferior. Ancient texts on Ayurvedic knowledge systems and spiritual and metaphysical teachings were regarded as inferior to English texts. For example, Macaulay notes in his "Minute on Indian Education" that "It is, I believe, no exaggeration to say, that all the historical information which has been collected from all the books written in the Sanskrit language is less valuable than what may be found in the most paltry abridgements used at preparatory schools in England" (Internet Modern History Sourcebook, 1957). This laid the foundation for a narrative that European knowledge and sciences were more advanced than traditional systems such as Ayurveda. This contributed to the devaluation of Ayurvedic philosophy and Sanskrit terminology until contemporary times. The British colonial government actively promoted this agenda, systematically positioning Western medicine as the dominant system and relegating Ayurveda and other indigenous knowledge to the margins of society.

The Calcutta Medical College was, therefore, established in 1835 to teach only Western medicine, thereby creating a medical education system that excluded traditional knowledge. This British-style education now replaced the traditional *gurukul* system, where students learned medicine through a lineage of knowledge passed down by experienced Ayurvedic practitioners. This meant that indigenous practitioners (or *vaidyas*), who had been recognized as knowledgeable healers within their communities, were now systematically excluded from state-sponsored medical practice. This notion of "qualified" medical professionals being solely those trained in Western medicine remains entrenched in much of our contemporary society.

In this manner, empirical, "evidence-based," and rational medicine (based on reductionist philosophy) was positioned as superior due to its association with the colonial powers. The cultural impact of such steps was the internalization of an inferiority complex among Indians regarding their own traditions and knowledge. In fact, Ayurveda and other indigenous healing systems were now regarded as superstitious, unverified, and backward, and were considered inferior to Western medicine. To reinforce this superiority, the British hailed public health measures such as vaccination and sanitation campaigns as prime examples of Western medicine's progressiveness. Although beneficial, such initiatives were framed as gifts of modernity, designed to combat disease and improve public health. This further elevated the standing of Western medicine and reinforced the notion that colonial rule brought advanced scientific solutions to a supposedly "backward" India. Colonial officials promoted the idea that Western medicine represented rational, scientific progress, while indigenous practices like Ayurveda were inadequate, outdated, and incapable of addressing the pressing health issues of the time.

In contrast, Ayurveda and other indigenous systems were relegated to folk practices with no institutional support. They were framed as incompatible with modern scientific knowledge. Ayurveda came to be viewed as outdated and increasingly irrelevant. The resulting shift to the English medical discourse meant that Sanskrit and regional Ayurvedic texts lost ground in favor of Western-style medical education and literature. Ayurveda's role in formal medicine was progressively undermined, and the internalized inferiority led to a gradual loss of pride in Ayurvedic knowledge, contributing to a decline in its transmission. Such healing systems were further marginalized by Western pharmaceutical products (often manufactured in Britain), which were promoted as more effective than Ayurvedic remedies.

Many Ayurvedic institutions that were once central to the intergenerational transmission of knowledge were dismantled or ignored, and traditional *gurukul* systems were disrupted. With fewer students interested in Ayurveda and the loss of livelihood for many vaidyas, the teaching of Ayurveda began to fade. Sanskrit texts went untranslated or were critiqued through a Western framework without an appreciation for the richness of the language, leading to widespread misinterpretations or outright dismissals of key Ayurvedic concepts. The loss of intergenerational transmission of Ayurvedic knowledge further entrenched the decline of the system. As a result, Ayurveda became devalued by the educated elite, who increasingly adopted Western medical practices. The British colonial narrative not only framed Indian culture (including Ayurveda) as inferior, but also actively worked to strip it of its legitimacy in Indian society. This cultural devaluation, together with the internalized inferiority, played a significant role in undermining the value of Ayurveda, a sentiment that still persists in modern-day India. As Western medicine gained more prominence, Ayurveda was seen as obsolete and incapable of contributing to India's future.

However, Ayurveda never fully disappeared despite its extensive suppression during British rule. Despite the introduction of Western medical systems, Ayurveda remained intact as the primary medical system for Indians, especially in the rural areas (Jha, 2023). Many Ayurvedic practitioners continued to practice through word of mouth, and traditional knowledge was preserved through informal channels. As stated by Jha (2023), "It may be concluded that there is an unbroken continuity of Ayurvedic medicinal tradition in rural India. Ayurvedic treatment and prescriptions mentioned in classical literature are still in vogue in rural India in the same way. This leads us to the conclusion that Ayurveda was very effective in responding to the challenges posed by the introduction of colonial biomedicine and, as a result, succeeded in maintaining its relevance and contributing to medical pluralism." After India gained independence in 1947, the Indian government took steps to revive Ayurveda by recognizing it as a legitimate system of medicine. Such efforts included the establishment of Ayurvedic colleges, research institutes, and the translation of key Sanskrit texts into regional languages. These efforts helped reintegrate Ayurveda into Indian society, although the effects of colonial marginalization still exist.

The wisdom of Ayurveda: Principles rooted in consciousness and the nature of reality

Ayurveda is often translated as "The Science of Life" from two Sanskrit terms: *ayus*, meaning "life," and *veda*, meaning "knowledge, science, or wisdom." Based on knowledge from the Vedas, the foundational principles of Ayurveda were laid down during the second century BCE by the Vedic philosophical schools of Vaisheshika and Nyaya, alongside the Samkhya framework. These schools significantly shaped the diagnostic and treatment processes in Ayurveda and emphasized a logical and empirical approach to medicine (Pondomatti et al., 2024, pp. 1–2). These philosophies are rooted in the Vedas, which laid the foundation for understanding life, health, and the entire cosmos, emphasizing harmony between humans and their environment. Building on this concept, the Vaisheshika school categorized the material world and its properties, helping Ayurveda understand the body, tissues, and medicinal substances. Nyaya, the school of logic and reasoning, provided methods to observe, analyze, and infer, shaping Ayurvedic diagnosis and treatment. Samkhya, with its dualist cosmology of consciousness (*Purusha*) and nature (*Prakriti*), underpins the understanding of the mind–body connection, the principles that influence an individual's health (*tridoshas*) and that shape the interplay between mental and physical health. Together, these philosophies provide Ayurveda with a profound framework for considering human health as an interconnected balance of body, mind, and consciousness (spirit).

Although some religions are based on similar philosophies, Ayurveda is not based on any religion, and individuals from all religions (or nonreligious) can benefit from Ayurveda. The fusion of Hindu mythology and ancient Indian medicine (particularly Ayurveda) is a fascinating synthesis of cultural heritage and scientific endeavor. Ayurveda encompasses a wide range of practices, including pharmacology, anatomy, physiology, surgery, and obstetrics, and integrates the rich tapestry of mythology to provide a comprehensive understanding of health and disease that can easily be transmitted to the masses through story telling. The inclusion of mythological figures and narratives in the discourse of ancient Indian medicine offers a unique perspective on the integration of spiritual and empirical knowledge (Pondomatti et al., 2024). By integrating philosophy, observation, and narrative, Ayurveda provides a holistic approach to health that encompasses body, mind, and spirit, although its depth and reliance on traditional concepts can make it less familiar to a Western audience.

The following sections explore how Ayurveda can be reintroduced as a holistic and scientifically relevant medical system, offering unique perspectives on diagnosis, treatment, and prevention that complement and enrich modern medical paradigms.

Demystifying Ayurveda

The contemporary teaching of Ayurveda in the West was pioneered by a small but influential group of physicians and scholars who sought to bridge ancient wisdom with modern scientific frameworks. Here, personalities like Vasant Lad (2002), Deepak

Chopra (1997), David Simon (1997), Marc Halpern (1995), and David Frawley (1989) translated classical concepts into terminology more familiar to Western audiences, emphasizing physiology, nutrition, and psychology alongside traditional Ayurvedic principles. Their efforts laid the groundwork for a growing network of Ayurvedic training programs, with more than a dozen schools now offering certification and professional courses in North America. This includes, for example, the Ayurvedic Institute, the California College of Ayurveda, Kerala Ayurveda Academy, and the Gaja Collective. In addition, the Maharishi Ayurveda University was established (under the guidance of Maharishi Mahesh Yogi) to provide formalized training, research, and standardization of Ayurvedic knowledge for practitioners in the West. These institutions have helped establish Ayurveda as a legitimate, structured system of knowledge in contemporary Western contexts. They have also provided information on both the traditional principles and the scientifically recognized benefits and applications of Ayurvedic herbs (Simon and Chopra, 2000; Frawley and Lad, 1986).

As previously discussed, one of the enduring effects of colonization was the marginalization and misinterpretation of traditional medical terminology, including key concepts expressed in Sanskrit. Terms such as *agni*, *ama*, and *doshas* were often reduced to literal translations and dismissed as nonsensical within a Western biomedical framework. However, a deeper understanding of Sanskrit reveals that these terms function as metaphors, describing complex biological and physiological processes. Recognizing this metaphorical and conceptual depth highlights Ayurveda's sophisticated grasp of the interconnectedness and complexity inherent in human anatomy, biology, and physiology. While preserving key Sanskrit terms is important when teaching Ayurveda, they can also be translated into modern terminology to make the teachings more accessible to a wider audience without compromising their richness and authenticity.

Advancement in the understanding of Ayurveda in a modern context can be facilitated by studying the mechanisms of action of Ayurvedic practices and treatments, and hence, there has been a rapid modernization of the ways traditional medicine is being studied. Whole systems research is now a topic of conversation in research areas, while artificial intelligence is being used to map evidence and trends in traditional medicine and also to screen natural products for pharmacokinetic properties. Moreover, functional magnetic resonance imaging is now employed to study brain activity and the relaxation response for traditional medicine therapies such as meditation, yoga, and Ayurvedic practices, which are increasingly being employed for mental health and well-being in stressful times (WHO, 2022). Therefore several Ayurvedic concepts and therapies can now be explained using modern scientific methodologies and terminologies.

Ayurveda—A complete medical system

Ayurveda is a complete medical system that focuses on providing tools and practices to create health on a daily basis by utilizing the innate healing ability of the human mind–body system. Such an approach allows individuals to thrive physically, mentally, emotionally, and spiritually. However, reintroducing Ayurveda as a complete medical system in the postcolonial era requires more than just a revival of ancient practices; it requires a re-examination of the profound wisdom that was once disregarded under colonial rule.

Ayurveda is rooted in several foundational principles that explain its effectiveness. Of note, such principles are increasingly being validated by the modern scientific method. Firstly, it is a holistic mind–body–spirit approach to health and wellbeing, offering tools for thriving in all aspects of life. Research confirms that when all such aspects of life are addressed, wellbeing is maximized (Bożek et al., 2020). Secondly, it is a whole-systems medical approach, as it aims to address the entire system at once and not just one part of the whole body. Thus overall harmony and wellbeing can be maintained by collectively addressing particular imbalances within the entire mind–body system. This contrasts with some Western medical treatments which, while treating one condition, can create side effects elsewhere in the system.

Self-healing is also a fundamental principle in Ayurveda, i.e., the recognition that the body and mind possess innate and powerful self-regulatory mechanisms that help to maintain homeostasis. Pathological complications and diseases arise when such mechanisms are disrupted by internal or external factors. Consequently, Ayurvedic treatment focuses on addressing the root cause of disease rather than merely managing symptoms and hence guides individuals to reverse the disease process by correcting the underlying imbalances. Key concepts in this framework include *ama* and *agni*. Ama, often translated as "toxicity," broadly refers to anything that interferes with the body's natural capacity for self-regulation. Agni is commonly understood as "digestive fire," which encompasses not only the ability to digest food but also all experiences and stimuli encountered throughout daily life. As individuals are constantly "digesting" physical, emotional, and sensory inputs, sustaining a strong agni through daily practices is essential for maintaining balance. In practice, Ayurveda addresses diseases by either removing obstacles that impair self-healing or by providing the support needed to strengthen the body's innate restorative processes, or both being simultaneously applied.

Another unique aspect of Ayurvedic medicine is the concept of personalization of treatments based on an individual's natural constitution, or *dosha*. Ayurveda recognizes that there is no one-size-fits-all treatment or lifestyle and that unique choices need to be made based on an individual's specific constitution, state of balance, environment, stage of life, and even the season of the year. This concept is known as "bio-individuality," a term coined by Joshua Rosenthal, who is the founder of the Institute for Integrative Nutrition. In agreement, earlier pioneering work by the biochemist Roger Williams (Williams, 1956) demonstrated that individual differences in biochemical makeup and variability in metabolic characteristics hold certain implications in terms of various treatment protocols (Eckhart, 2011). Ayurveda also stretches beyond the physical by describing differences in mental and emotional tendencies, a concept that is being validated by the modern scientific method (Sanchez-Roige et al., 2018).

Finally, Ayurveda emphasizes that both individual wellbeing and the health of society are deeply interconnected. Thus our actions toward ourselves through daily practices, as well as behaviors toward others, play a crucial role in maintaining balance and vitality. Ayurvedic science teaches that a healthy life is attainable for anyone who follows its guidelines, which are organized into four key categories. These include *Sadvritta*, a code of conduct for ethical living and mental harmony; *Ritucharya*, practices aligned with the seasonal cycles; *Ratricharya*, a nightly routine; and *Dinacharya*, daily practices that support physical and mental health. These behaviors are foundational for the prevention of illness and the promotion of overall wellbeing (Chhandak et al., 2025, p. 1912). Beyond their practical benefits, these guidelines reflect Ayurveda's spiritual understanding of interconnectedness. It recognizes that the health of the individual, community, and natural environment is inseparable and that living in alignment with this principle fosters harmony across all levels of existence.

Core Ayurvedic concepts and their complementarity with modern science and medicine

In this section, we delve deeper into foundational Ayurvedic concepts while introducing additional principles essential for creating and sustaining health. There is also an examination of how such time-honored teachings are supported by modern scientific insights.

Layers of life: Body, mind, and spirit

Unlike the fragmented approach to physical, mental, and spiritual health often found in modern medicine, Ayurveda emphasizes the fundamental interconnectedness of all aspects of human experience. In this system, life comprises three essential layers: the physical body (*sthula sharira*), the mind (*sukshma sharira*), and the spirit (*karana sharira*). Physical experiences are mediated through the body and senses, mental experiences through the mind, and spiritual experiences transcend both body and mind. These layers are deeply interdependent, such that an imbalance in one inevitably influences the others. Ayurveda, therefore, provides therapeutic approaches tailored to restore balance across all three domains. Treatment is inherently multidimensional, aiming to address the body, mind, and spirit through diet, lifestyle practices, meditation, subtle energy work, and other modalities. In agreement, contemporary medical science (particularly through the biopsychosocial model) increasingly recognizes the integration of biological, psychological, and social factors in health. Moreover, such research also underscores the profound impact of mental and emotional states on physical wellbeing (Galvano, 2024), thus corroborating Ayurvedic medicine's philosophy that the mind and spirit are not separate from, but integral, to overall health.

Building on this understanding, Ayurveda places particular emphasis on the spiritual dimension of human experience. Here, consciousness (or spirit) is considered the true and unchanging nature of humans, while the body and mind are regarded as transient expressions occurring within this field of consciousness. Hence, by cultivating a connection to such fundamental consciousness (the foundation of individual existence), an individual can access the deepest levels of healing when applying Ayurvedic teachings. Thus the outcome of spiritual practices such as meditation, mindfulness, and yoga is linked to improved physical and mental health.

Panchamahabhutas: The five elements

The concept of the *Panchamahabhutas*, or the five elements—space (*akasha)*, air (*vayu*), fire (*tejas*), water (*jala*), and earth (*prithvi*)—forms a central foundation of Ayurvedic philosophy. The elements are regarded as the fundamental building blocks of the known universe, encompassing all of nature, including the human body and mind. As previously discussed, the Sanskrit terms for the elements function as metaphors for universal principles of nature, providing a lens through which to understand physiological and psychological processes. Space (or ether) represents the principle of potential, i.e., the capacity for anything to arise. This can manifest as the space around human beings, the cellular spaces within the body, or the pauses between thoughts in the mind. Air (or vayu) embodies movement within such spaces, observed in nature as wind, within the body as circulation, nerve impulses, and the flow of lymph, as well as the movement of thoughts in the mind. Fire (or tejas) represents transformation, exemplified in nature by the sun, flames, and light, while within the body it governs digestion, metabolism, and detoxification. It also supports the cognitive transformation of ideas and perceptions in the mind. Water (or jala) provides cohesion and lubrication, manifesting as water in nature, and in the body through the fluids that sustain tissues. In the mind, it represents the cohesion of thoughts. Earth (or prithvi) signifies structure and stability, expressed physically in bones, muscles, and organs, and mentally as the organization and consolidation of thoughts and memory. Together, maintaining balance among these five elements is essential for the optimal functioning of both body and mind. By mapping physiological and psychological processes through the lens of the Panchamahabhutas, Ayurveda offers a comprehensive foundation for implementing treatments that restore and sustain health.

Tridosha theory: Psychophysiological principles

At the heart of Ayurvedic medicine lies the theory of the Tridosha—*Vata*, *Pitta*, and *Kapha*—which describes the "psychophysiological" tendencies that shape both body and mind. The five elements interact within the body to form these three doshas, or principles, which govern patterns of movement, metabolism, and structure within everyone, as well as in nature. These concepts do not have direct equivalents in the reductionist biomedical framework and are best understood through the lens of Ayurveda. Vata is composed of the qualities of space and air (lightness, dryness, cold, irregular) and governs movement, communication, and nervous system function. Pitta is a combination of fire and water and their qualities (hot, light, intense, penetrating), which regulate digestion, metabolism, and transformative processes. Kapha, with qualities of water and earth (heavy, slow, cold, solid), provides structure, cohesion, and lubrication. Each dosha reflects specific aspects of physiological and psychological function. All three are present in every individual and shape how the body and mind respond to internal and external stimuli. Depending on lifestyle, diet, environment, and other influences, these principles can become imbalanced, thereby disrupting the harmonious functioning of the system.

In Ayurveda, an imbalance is created when a particular dosha or doshic quality builds up (or accumulates) due to internal or external influences. A core principle called *Samanya-Vishesha Siddhanta* teaches that "like increases like, and opposites bring balance." Thus restoring harmony by reducing what has accumulated and introducing opposing qualities is central to both preventing illness and supporting effective treatment.

Prakruti and vikruti: Nature and nurture

The concepts of *Prakruti* (innate constitution) and *Vikruti* (current imbalance) provide a framework for understanding individual health, closely paralleling modern insights from genetics and epigenetics. Prakruti reflects each person's unique proportion of the five elements and three doshas and thus their qualities. Prakruti is established at conception and influenced by parental constitution, gamete quality, uterine environment, and the timing of conception. This can be affected by diet, behavior, season, and emotions. This innate constitution determines lifelong physical and psychological traits, much as DNA determines inherent traits. Ayurvedic practitioners can, therefore, assess an individual's prakruti and recommend specific practices that maintain daily health.

Vikruti represents the dynamic fluctuations in dosha expression caused by lifestyle, environment, and mental–emotional factors that can lead to imbalance and disease. A deviation from a person's prakruti is an acquired, disturbed state known as vikruti. Modern research in epigenetics demonstrates how diet, exercise, stress management, and mindfulness influence gene expression and overall health (Ostaiza-Cardenas et al., 2025), highlighting the interaction between innate constitution and environmental influences. Ayurveda's emphasis on restoring balance in the doshas underscores that health is an individualized, dynamic process and that any imbalance can be rectified by identifying the root cause of such imbalances and implementing appropriate changes. Both ancient and modern perspectives converge on the principle that optimal health arises from honoring each person's unique constitution rather than applying uniform interventions. In support, studies demonstrate the significant impact of genetic variation, which can influence everything from drug metabolism to the degree of susceptibility to illnesses (Schärfe et al., 2017). This reinforces the relevance of prakruti in terms of guiding individualized care. The concept of dosha balance forms the foundation for personalized medicine and can be taught so that individuals can understand their own nature and ultimately recognize the most suitable choices depending on the context.

Each person possesses one (or two) of the doshas that are dominant in their physiology and can be observed through their natural tendencies in terms of body and mind. Vata types, dominated by the qualities of space and air (cold, light, dry, mobile, irregular), are typically slender and agile, with bursts of energy that can quickly give way to sudden fatigue. Their digestion is naturally irregular and often varies from day to day. Psychologically, they thrive on excitement and change and display kindness and a desire to please, with quick emotional shifts. When in balance, Vata supports creativity, alertness, light but restful sleep, and an agile mind and body. However, when out of balance, Vata can manifest as anxiety, restlessness, weight loss, degenerative conditions, constipation, dietary and chemical sensitivities, poor sleep, and digestive disturbances such as irritable bowel syndrome.

Pitta individuals, with primary elements of fire and water qualities (hot, light, penetrating, sharp), are usually of medium build with good muscle tone, strong digestion, and robust energy. They may possess thinner and oily hair and oily skin. They are mentally sharp, goal-oriented, and outspoken, with robust concentration spans and natural leadership tendencies. When balanced, Pitta produces efficient digestion, a healthy appetite, sharp intellect, and overall contentment. However, when not in balance, it may lead to excessive heat, irritability, skin inflammation, disturbed sleep, acid reflux, peptic ulcers, and/or cardiovascular and liver disorders.

Kapha constitutions, with water and earth qualities (cold, heavy, slow, steady), are characterized by sturdy and larger frames, stamina, and resilience, with a tendency toward easy weight gain. They often possess thick skin and hair. They are calm, loving, tolerant, and steady, with excellent memory retention. Balanced Kapha, therefore, ensures mental stability, patience, physical strength, resilience, and good immunity. However, when not balanced, Kapha may present as lethargy, excessive sleep, procrastination, depression, weight gain, sluggish digestion, fluid retention, respiratory congestion, and/or asthma.

An Ayurvedic practitioner, therefore, considers both an individual's inherent constitution (prakruti) and any current dosha imbalances (vikruti) to design a personalized treatment plan. This approach aims not only to alleviate symptoms at presentation but also to address underlying causes. Thus the ultimate aim is to restore individuals to their natural state of balance and health.

Agni: Digestion and metabolism

The concept of agni (or digestive fire) is central to Ayurvedic medicine. While agni traditionally refers to the body's ability to break down food, assimilate nutrients, and eliminate waste, it also encompasses the broader capacity to "digest" all life experiences. Agni, therefore, parallels the modern concept of resilience; i.e., strong agni supports energy, mental clarity, and adaptability, whereas weak agni allows the accumulation of toxins (ama), contributing to illness. Beyond digestion, agni governs metabolism and sustains physiological function.

Because gut health is foundational in Ayurveda, all treatment plans prioritize the support of agni along with other areas of balance. The classic Ayurvedic axiom that "all disease begins in the gut" has found increasing validation in modern medicine through research on the gut microbiome and its influence on systemic health (Afzaal et al., 2022). Within the gastrointestinal tract, the doshas perform distinct functions, and imbalances here can trigger disturbances that propagate throughout the body. In support, contemporary studies consistently highlight the gut's crucial role in immunity, mental health, and chronic conditions (Shao et al., 2023; Singh et al., 2024). Ayurvedic practices such as aligning diet with one's dosha, ingesting nutrition seasonally, and adopting appropriate lifestyle routines therefore promote optimal digestion and metabolic function, complementing modern strategies for gut health and microbiome balance.

Many Ayurvedic tools and practices help maintain and strengthen agni, including consuming appropriate amounts of nutrition, using spices to support digestive function, following seasonal dietary guidelines, ensuring restorative sleep, managing stress through meditation or mindfulness, and engaging in yoga or other forms of mindful movement. Commonly recommended practices, such as drinking warm water with lemon in the morning, sipping ginger tea before meals, and using digestive spices and herbs, are also employed to stimulate and balance agni. Interestingly, Peterson et al. (2019) found that compounds present in Ayurvedic culinary spices can beneficially alter gut microbial communities, thereby influencing their collective metabolism and supporting digestive efficiency. Together, these practices help optimize digestion, metabolism, and support overall health.

Prana, Tejas, and Ojas: Subtle energies and vital forces

Prana, Tejas, and Ojas are considered the subtle energies in Ayurveda that guide the vital functions of the body, mind, and consciousness. *Prana* is the life force or energy that animates the body, governing processes such as breathing, circulation, and the flow of energy through the nervous system. *Tejas* represents the energy of transformation, overseeing metabolic processes, digestion, and the refinement of cognition, emotion, and mental clarity. *Ojas* is the essence of vitality, immunity, and the overall capacity to maintain health.

These subtle energies are sometimes described as the more refined, energetic, and subtle expressions of the three doshas: Prana corresponds most closely with Vata, governing movement and communication; Tejas aligns with Pitta, representing metabolism and clarity; and Ojas reflects Kapha, providing stability and resilience. While these concepts do not have direct correlates in Western medicine, they can be understood through modern physiological systems. Here, Prana is analogous to the autonomic nervous system, i.e., through the movement of energy orchestrating the balance between sympathetic and parasympathetic functions to regulate energetic metabolism. Tejas parallels processes of metabolism and cognitive-emotional regulation, like the integrated functioning of the endocrine system and the central nervous system. Ojas mirrors immune function and overall vitality, reflecting the body's capacity for regeneration and longevity. From an Ayurvedic perspective, these vital forces serve as the bridge between physiology and consciousness, describing how subtle energy flows through the body to regulate bodily functions, metabolism, immunity, mental clarity, and emotional resilience.

Dhatus and malas: Building healthy tissue and eliminating waste

In Ayurveda, the *dhatus* are the seven fundamental tissues—plasma (*rasa*), blood (*rakta*), muscle (*mamsa*), fat (*meda*), bone (*asthi*), marrow/nerve (*majja*), and reproductive tissue (*shukra/artava*)—that form the structural and functional foundation of the body. Proper nourishment of each dhatu by the consumption of healthy foodstuffs is essential for maintaining strength, vitality, and overall health, as each tissue relies on the support and transformation of the one preceding it. Each tissue possesses its own agni to digest incoming nutrients. This highlights the sophisticated Ayurvedic understanding of the interconnectedness and interdependence of all bodily systems.

By contrast, *malas* are the waste products of metabolism, with the major ones including urine, feces, and sweat. These must be efficiently eliminated to prevent the accumulation of toxins (ama) and to maintain homeostasis. Ayurvedic detoxification processes aim to cleanse the body by removing accumulated toxins through these channels and thus prevent disease. In agreement, emerging medical paradigms (e.g., functional and integrative medicine) are increasingly recognizing the importance of detoxification by highlighting the role of the liver, kidneys, lymphatic system, and gut in such waste removal. However, the broader Ayurvedic approach to systemic detoxification has not been fully embraced by Western medicine. This is partly due to concerns about the lack of evidence concerning benefits and safety aspects, as well as misunderstandings regarding traditional *panchakarma* (or Ayurvedic detoxification therapies). Moreover, such acceptance is further limited by contemporary wellness trends and practices often turning such "detox" into a fad, while omitting many essential components of complete Ayurvedic practices. However, when these procedures (when clinically indicated) are performed by experienced Ayurvedic practitioners, such detoxification processes can be highly beneficial. This applies especially to individuals who are not improving despite following healthy lifestyle practices. Accumulated toxicity often interferes with the body's innate self-regulation, thereby reducing the effectiveness of both lifestyle interventions and medications. Here, Ayurveda's seasonal detoxification practices can support the elimination of metabolic waste and help restore the digestive tract on a regular basis. Hence, the careful attention to dhatu nourishment and mala elimination reflects Ayurveda's understanding of how nutrition, metabolism, and waste management are interconnected and essential for overall health.

Nutrition as medicine: An Ayurvedic perspective

Nutrition occupies a central place in Ayurveda, not merely as sustenance but also as a medicinal form, with a balanced diet being essential for maintaining optimal health. Ayurvedic dietary guidelines are grounded in the principle of *ahara*—proper nourishment—which encompasses all nutrition, liquids, and anything ingested orally via the oral cavity. Here, the emphasis is placed on eating in accordance with an individual's dosha, the seasons, and digestive capacity (agni). Ayurvedic philosophy centers around a more precise adage, namely, "we are what we digest," compared to the more popular notion that "we are what we eat." This outlook highlights the pivotal role of a healthy gut and strong digestive fire in digesting and absorbing nutrients. The foundational role of nutrition in maintaining health and preventing disease is emphasized in classical Ayurvedic texts, for example, "When diet is proper, medicine is of no need; when diet is poor, medicine is of no use."

Nutrition in Ayurveda functions not only as nourishment but also as a primary tool of medicine, capable of restoring balance between the elements and doshas within the body. As all matter contains the five elements and displays specific qualities, nutrition can be used to correct imbalances by introducing opposite elemental qualities. Ayurvedic texts, therefore, classify foods into six tastes (or *rasas*)—sweet, sour, salty, pungent, bitter, and astringent—each performing specific functions within the body. This was described long before the biochemical classification of foodstuffs. Each taste has a predominance of particular elements and associated qualities that provide foodstuffs with their characteristic actions (Chopra, 1997; Lad, 2002). By incorporating a variety of tastes, all the elements are supported, thereby optimizing the full range of bodily functions. Thus through such a deep understanding of the primary elements and qualities of foodstuffs, as well as how preparation can affect these qualities, an Ayurvedic practitioner can recommend personalized dietary plans tailored to an individual's unique constitution and current state of balance.

Modern nutritional science and lifestyle medicine increasingly recognize the therapeutic potential of nutrition, particularly in reducing inflammation, regulating metabolism, and supporting immunity. Hence, most of the spices used in Ayurvedic cooking have been shown to possess antiinflammatory and antioxidant properties and can prove helpful in treating certain health issues. For example, a metaanalysis by Garza et al. (2024) revealed that several Ayurveda-recommended spices are effective at controlling blood sugar and insulin levels. Moreover, a review by Verma et al. (2024) showed that several Ayurvedic herbs and spices (e.g., ginger and black pepper) are beneficial for various ailments.

Ayurvedic dietary practices also align closely with the principles of functional and lifestyle medicine, with an emphasis on a whole-food, plant-based diet that minimizes processed foods. Specific dietary recommendations are tailored to an individual's dosha constitution and current state of health, reflecting a time-tested approach to personalized nutrition.

Five senses therapy: Using the senses as tools for healing

In Ayurveda, the five senses—sound, touch, sight, taste, and smell—are viewed as vital gateways to engage with the world, through which the external environment can influence an individual's internal state. While modern science describes this process in terms of sensory receptors converting stimuli into electrical signals that are processed by the brain, Ayurveda goes further by recognizing that the senses are direct pathways to the mind and body, capable of influencing both physical and psychological balance.

Ayurvedic therapies uniquely leverage the senses to promote healing and restore equilibrium. Sound and music therapy are, therefore, used to harmonize mental states, reduce stress, and even influence heart rate and breathing patterns. Moreover, touch (applied through Ayurvedic massage [*abhyanga*]) or other somatic therapies, such as acupressure, engages skin mechanoreceptors to modulate circulation, lymphatic flow, and nervous system activity. The depth, pressure, and rhythm of such strokes can elicit different physiological effects, from relaxation to stimulation, thereby demonstrating the specificity of sensory input in terms of therapeutic outcomes (Kopf, 2021). Visual stimuli, such as color therapy and/or exposure to images of nature, can also soothe the nervous system and uplift moods, while aromas from essential oils or medicinal herbs act on the limbic system, regulating emotions and hormonal responses, as supported by studies on aromatherapy and olfaction (Cui et al., 2022).

Thus by consciously engaging the senses, practitioners can influence mood, cognition, state of arousal, and metabolism, thereby providing a holistic approach that is difficult to replicate through sole pharmaceutical therapeutic approaches. Modern neuroscientific research increasingly supports this view by showing that sensory experiences can trigger neuroplastic changes in the brain (Karawani et al., 2022; Kollndorfer et al., 2014), improve emotional resilience, and restore homeostasis. This underscores the timeless insights of Ayurveda in harnessing the healing potential of the senses.

Circadian rhythms and Ayurvedic routines

As Ayurveda is a nature-based indigenous system, it places a profound emphasis on living in harmony with the environment and viewing it as an intrinsic part of human health. The environment is recognized not merely as a backdrop to human life, but instead as an active determinant of wellbeing. Ayurveda, therefore, highlights that human health is deeply intertwined with the rhythms of nature, including daily cycles, lunar phases, and seasonal changes (Kshirsagar, 2018). Hence, the alignment of an individual's life with these natural rhythms is proposed to promote balance and prevent disease, echoing contemporary science's views regarding the crucial role of circadian rhythms in regulating sleep, metabolism, and overall health (Foster, 2020).

The Ayurvedic view that the environment serves as an extension of the body underscores a holistic understanding. This means that every aspect of an individual's surroundings (e.g., nutritional intake, inspiration of life-giving air, exposure to energizing sunlight) can profoundly influence physical, mental, and emotional states. Hence, human beings are equally molded by the external environment as they are shaped by their internal constitution and imbalances (prakruti and vikruti). This Ayurvedic emphasis on harmony with nature resonates strongly with modern ecological and planetary health principles, which highlight the intimate connection between an individual and the external world. This also instills a desire to be good stewards of nature and the environment. The daily and seasonal routines (dinacharya and ritucharya), which synchronize human activity with natural cycles, are therefore central to Ayurveda. In fact, modern medicine increasingly supports such practices through research on circadian biology that is increasingly demonstrating the importance of regular sleep-wake patterns and timed meals (Boege et al., 2021) together with an alignment with environmental cues for metabolic regulation and longevity. Ayurveda's guidance to rise with the sun, eat at consistent times, and follow lifestyle rhythms in concert with nature closely aligns with evidence showing that regularity in daily activities supports health and longevity (Sato et al., 2022).

Mind–body connection and emotional root causes

Perhaps the most important principle in Ayurveda is the acknowledgment of the profound connection between the mind and body. It posits that mental and emotional imbalances can manifest as physical illness, and conversely, that physical disease can affect emotional wellbeing. The principle that the mind creates the body, i.e., energy precedes matter, and that mental and emotional states shape physiological processes is central to Ayurvedic philosophy. Contemporary research in behavioral science and trauma supports this view by showing that emotions such as chronic stress and anxiety can alter immune function, metabolism, and gene expression (Musaelyan et al., 2020; Alotiby, 2024). Moreover, individuals diagnosed with anxiety, depression, and posttraumatic stress disorder are at increased risk for a range of acute and chronic health conditions (Bobo et al., 2022).

In contrast, contemporary medicine often addresses mental health primarily through a pharmaceutical lens. Although this approach can alleviate the symptoms, it does not always cultivate a deep understanding of how emotional states can contribute to physical illness. Ayurveda's holistic approach, therefore, emphasizes treating the emotional root causes of disease, recognizing that unresolved emotional patterns can perpetuate physiological imbalances. Hence, practices such as meditation, mindfulness, yoga, and other techniques for emotional release are central to this approach, and their benefits have now been shown in clinical studies (Hoge et al., 2013; Park and Slattery, 2021). Thus Ayurveda provides a model for integrating mental, emotional, and physical care by focusing on the mind as a foundational aspect of human health.

Rasayana: Lifestyle for longevity and healthy aging

Rasayana is often translated as the "science of rejuvenation" and is one of the eight specialties in Ayurveda. It emphasizes the creation and preservation of health rather than simply treating disease. It focuses on nourishing the body, mind, and spirit to enhance vitality, immunity, and overall wellbeing. Ayurvedic texts describe rasayana as essential for maintaining balance in the doshas, strengthening dhatus, and supporting the body's natural regenerative processes to ultimately promote longevity and an improved quality of life. This approach underscores the Ayurvedic principle that true health is proactive. This means that, through the correct routines and practices, individuals can prevent disease, enhance resilience, and cultivate optimal function throughout their lifespans. Here, rasayana therapies extend beyond herbal supplementation and encompass a broad range of practices, including meditation, mindful breathing, yoga, sensory therapies, seasonal routines, and dietary patterns that support digestion and tissue nourishment.

Contemporary interest in longevity and healthy aging has drawn attention to the scientific parallels of rasayana; for example, areas such as telomere biology, oxidative stress, and cellular repair align closely with Ayurvedic principles, highlighting the importance of supporting the body's intrinsic regenerative capacity. In support, Ayurvedic herbs (classified as rasayanas) such as Ashwagandha (*Withania somnifera*), Amalaki (*Phyllanthus emblica*), Guduchi (*Tinospora cordifolia*), and Shatavari (*Asparagus racemosus*) are rich in antioxidants and adaptogenic compounds that help mitigate the physiological effects of stress, enhance immunity, and support cognitive function (Shivani et al., 2025). The integration of these ancient principles with contemporary findings underscores Ayurveda's unique contribution to the science of creating and sustaining health and longevity.

Herbal medicine

Although Ayurveda places a primary emphasis on lifestyle and behavioral practices as the foundation of health, contemporary perceptions often highlight herbal therapies as its central focus. This reflects a reductionist mindset which oversimplifies Ayurveda's nuanced, systemic approach to healing. Indeed, Ayurvedic herbal medicine is highly effective for a wide range of health concerns. For example, the vast Ayurvedic pharmacopoeia encompasses ~1,200–1,800 medicinal plants. However, the adoption of classical herbal formulations faces practical challenges that include variability in quality, complexities in harvesting and preparation, and a relative scarcity of rigorous clinical trials (Sen and Chakraborty, 2016). A challenge in studying Ayurvedic pharmacology is that such treatments usually involve a combination of herbs that are often formulated in proportions and combinations tailored for a particular individual. This makes it challenging to investigate their efficacy using current, reductive research models that can only compare one variable at a time. Some of the reported side effects of Ayurvedic herbs stem from their "allopathic" use, where a single herb is taken at relatively high doses to achieve a specific effect without considering an individual's constitution or dosha balance. For example, there have been some reports of liver damage associated with the herb ashwagandha. Although these cases are rare with relatively mild presentations, they may be more severe in individuals with preexisting liver conditions (Philips et al., 2023). From an Ayurvedic perspective, ashwagandha possesses heating qualities and should therefore be used more cautiously and at lower doses in individuals with a Pitta imbalance—with liver complications being one possible manifestation. As herbs are widely available and often self-prescribed, many commence treatments such as ashwagandha based on scientific studies highlighting its benefits for sleep and fatigue. However, this is often done without accounting for Ayurvedic principles in a holistic sense. This underscores the importance of using Ayurvedic herbs within their traditional context to ensure both safety and effectiveness. Certain Ayurvedic herbs, such as triphala (a blend of three dried fruits), are considered useful for balancing all doshas. Due to such broad effects, triphala displays a relatively low side effect profile and can be safely recommended for a range of concerns, particularly digestive health. Studies have also shown that triphala supports the gut microbiome by encouraging the growth of beneficial bacteria while inhibiting harmful species, as well as by enhancing gut antioxidant activity (Peterson et al., 2017).

Research interest in Ayurvedic herbs is growing, with recent work by Verma et al. (2024) highlighting studies on numerous medical applications. Ongoing research is essential for helping practitioners navigate the modern context, especially in cases where patients also receive pharmaceutical therapeutic interventions. This is an important consideration to ensure that herbal therapies can be safely and effectively used. Thus a deep and thoughtful use of the Ayurvedic framework, based on the evaluation of doshas and traditional formulations, can successfully complement conventional care.

Ayurveda and modern lifestyle practices as treatment

Lifestyle and daily routines form the cornerstone of Ayurvedic medicine, where health is cultivated by addressing several components such as nutrition, movement, sleep, and emotional regulation. Daily routines should, therefore, include all these practices so that the body and mind can function optimally. A few key practices include:

- Digestive care: drinking warm water in the morning and using digestive spices (e.g., ginger, black pepper, cumin) to optimize agni and eating meals with a wide variety of foods that meet all the needs of the body [Ayurveda describes the six "tastes" of food and provides specific guidelines on nutritional balance (Chopra, 1997, pp. 284–293)].

- Sensory support: ear and nose drops to keep eyes and nose moist and lubricated, self-massage with oil for skin health, eye exercises to keep eye muscles healthy, oil pulling (an Ayurvedic oral cleansing practice using oil), tongue cleaning, and neti (nasal cleansing with a saline solution to maintain clear respiratory passages), among others.
- Physical activity: daily movement tailored to dosha balance.
- Sleep and circadian alignment: regular waking and sleeping times, eating the largest meal at midday, and reducing evening activity to support restorative rest.
- Mind–body practices: meditation, restorative yoga, breathing exercises, and massage to regulate stress and enhance emotional balance.
- Nature: regularly spending time in nature.
- Seasonal and stress-related detoxification: practices that clear accumulated toxins and support digestive and metabolic functions.

Ayurveda thus remains both accessible and practical by emphasizing small and consistent daily adjustments rather than costly or complex interventions. Treatments often begin with a few carefully chosen practices, allowing the body to gradually adapt. Hence, healing can be naturally supported by employing such relatively simple tools.

The cost of separation: The limits of modern medicine

When we examine the holistic framework Ayurveda offers for healing the body, mind, and spirit, the limitations of modern medicine become evident. The conventional allopathic model often focuses on alleviating symptoms or addressing objective laboratory values, rather than understanding the root causes of disease and/or treating the whole person. Health conditions are frequently managed using standardized algorithms that are applied broadly to all patients, rather than through truly individualized care. Modern medicine is grounded in reductionist, materialistic science that primarily treats the body as a physical machine. While there is a growing recognition of the mind–body connection, this philosophy still largely remains peripheral in terms of conventional medical training and practice. Hence, the social, environmental, and emotional contexts that shape health are given insufficient attention during clinical encounters and in terms of curricula. Such a materialistic outlook means that spiritual domains are often not addressed or well understood by physicians.

The need for proper integration is urgent, especially within the context of rising chronic illnesses globally, where patients increasingly seek care outside of conventional medical options. This can result in their accessing wellness influencers and/or using supplements and herbal therapies without professional guidance. Although modern medicine excels in acute care, emergency interventions, surgical procedures, vaccinations, and infection control, it often falls short in addressing chronic diseases, preventive care, emotional wellbeing, and spiritual health. In addition, patients frequently present with complex, multifactorial challenges (e.g., fatigue, digestive disturbances, hormonal imbalances, persistent emotional stress) that do not always neatly align with a particular diagnosis based on laboratory results and/or imaging studies. Our reliance on medications and procedures—while lifesaving in certain contexts—can also inadvertently neglect the dynamic interplay between mind, body, and spirit. Clinical encounters, therefore, risk becoming increasingly transactional, where the practitioner is primarily trained to treat symptoms instead of fostering holistic health as a more transformative process.

A further limitation lies in the framework of scientific validation. Reductionist research methods are designed to isolate single variables and are thus poorly suited to evaluate whole systems approaches like Ayurveda, where treatments are inherently multifactorial and synergistic. Focusing on one variable only will miss the broader, systemic effects of Ayurvedic interventions and has resulted in the delayed integration of such transformative healing tools into conventional medical practice. However, newer scientific paradigms are beginning to bridge this gap. Here, multiomics research—integrating genomics, proteomics, metabolomics, and other systems-level analyses—allows for a more holistic understanding of biological processes and their interconnections. These tools align with Ayurvedic principles, which describe the interdependence of tissue layers, physiological processes, and subtle energies. Hence, by embracing such methodological approaches, the mechanisms by which Ayurvedic lifestyle practices and interventions support health and wellbeing can be explained.

To fully address the limitations of modern medicine and the growing complexity of patient needs, it is essential to also consider the wellbeing of physicians themselves. The re-humanizing of health care requires the restoration of empathy, the cultivation of practitioner presence, and the valuing of intuition together with intellect. Modern medicine's separation from the spiritual nature of healing and its sacredness can create a sense of disconnect between the purpose and meaning of work—a factor that contributes to physician burnout (West et al., 2018, p. 522). Ayurveda offers an alternative model by emphasizing that a good physician is not only knowledgeable and skilled, but also ethically grounded, present, compassionate, and attuned to the patient's physical, emotional, and spiritual state. The cultivation of self-care, mindfulness, and intuition allows the practitioner to maintain their own wellbeing while fostering a relationship with patients that is transformative rather than procedural. In addition, it is taught that "a well-versed Ayurvedic physician is a good teacher in the client-practitioner relationship" (Frawley and Kshirsagar, 2016, p. 13) and empowers the patient with skills to help implement suitable lifestyle changes. The integration of such principles and the incorporation of Ayurvedic concepts of interconnectedness into medical training and clinical practice can result in beneficial outcomes for both the physician and the patient. For example, physicians can reconnect with the sacredness of healing, foster patient-centered care, and cultivate a model of medicine that nurtures the whole person (both healer and patient) rather than simply treating the presenting symptoms.

Bridging two worlds: A journey of integration

Although modern medicine has provided remarkable life-saving tools, and ongoing research has helped to develop more effective therapeutic interventions and procedures, it is also becoming evident that a sole symptom- and disease-focused approach is insufficient. In addition, the impacts of chronic stress, poor nutrition, environmental exposures, and lifestyle imbalances accumulate over time in ways that pharmacologic treatments cannot fully address. Hence, patients who recover from acute illnesses often return with ongoing health concerns or side effects from medications, underscoring the crucial need for preventive strategies that support long-term health and resilience—an area where Ayurveda offers extensive guidance.

Over the last 15 years of practicing and teaching Ayurveda globally, I have found that its foundational principles are universally applicable, grounded in the laws of nature, and require only contextual modifications for geography, climate, and culture. Integrating such principles into my own practice as a physician has expanded the tools and guidance that I can offer patients. Thus Ayurveda and Western medicine can complement each other and are not mutually exclusive, as each system offers unique insights and interventions that, together, can provide more holistic care.

As a result of my experience in integrating both conventional and Ayurvedic medicine, I have collaborated with local medical schools (with established integrative medicine departments) to educate students, residents, and physicians on Ayurveda. This includes providing opportunities for medical students and residents to shadow me during Ayurvedic consultations and conducting didactic sessions. It also includes presentations to physicians and medical audiences on practical strategies for incorporating Ayurvedic principles into clinical care. I have also trained other physicians to perform Ayurvedic consultations in a manner that complements conventional lifestyle medicine frameworks. Initiatives aimed at integrating lifestyle and nutritional medicine into medical school curricula are gaining momentum, though they often face resistance due to the entrenched structures of conventional medical education.

In addition to working with medical students, I also do the converse, i.e., teach Ayurvedic practitioner students about Western medicine, anatomy, and clinical practice, helping them to better understand the perspectives and experiences that their patients may bring. This bidirectional approach, i.e., educating both conventional and Ayurvedic practitioners while fostering openness to collaboration, is essential as we continue to deepen our understanding and integration of these two complementary systems.

India provides a model for integrative practices where Ayurveda is woven into the framework of Western medicine. Here, hospitals and clinics often offer both systems under one roof. This means using Ayurvedic remedies alongside modern diagnostics, fostering interdisciplinary education and research, and applying Ayurvedic principles to chronic disease management. National policies supporting traditional systems in primary care, collaborative research leveraging advanced analytics to validate Ayurvedic practices, and educational initiatives training scientists who understand both traditions exemplify the potential of integration. These models offer valuable lessons on how Ayurveda can be responsibly and effectively incorporated into contemporary medical practice globally.

Case studies

While waiting for this paradigm shift to occur at a systemic level, there are many ways a physician can, in the meantime, integrate Ayurveda into conventional medical care. This ranges from addressing root causes that allow patients to safely reduce or discontinue medications that may no longer be necessary to applying a more holistic body–mind–spirit approach that supports deeper healing and resilience. The following case studies highlight examples from my own clinical practice, where I utilized such integrative methods, and demonstrate how Ayurvedic principles can be woven into modern care to restore balance and promote lasting health.

Case 1: Integrative approach for anxiety, palpitations, and sleep disturbances

A 55-year-old female presented to me as a new patient with palpitations, longstanding anxiety, and sleep issues, which were worsening over time. She also had headaches that were more frequent when she did not sleep well. She reported a history of anxiety since her teenage years, though it had generally been manageable. However, over the past 2 years, her symptoms worsened following her husband's heart surgery, as she developed a significant fear that he might die in his sleep. This ongoing fear further disrupted her sleep, as she would lie awake to make sure that her husband was breathing. The lack of sleep then heightened her anxiety and headaches. She also had a history of irritable bowel syndrome with constipation, a multinodular goiter, and osteopenia. At the time of her initial visit, she was taking levothyroxine and diltiazem to reduce palpitations, Fiorinal (a combination of butalbital, aspirin, and caffeine) as needed for headaches, and Ambien (zolpidem) for sleep. These medications had been prescribed over time by her previous physician to help her manage the symptoms she displayed. Several years earlier, she had been started on a low-dose thyroid medication because of a goiter, some fatigue, and thyroid laboratory results that were "slightly out of range," though she did not recall further details. Her palpitations had been worked up and identified as benign premature ventricular contractions. She was slightly anxious but otherwise normal upon examination. Her blood pressure was 104/64 mmHg, her pulse was 80 beats per minute, and she displayed a body mass index of 22 kg/m^2. She exhibited no palpable nodules on her thyroid. Laboratory evaluations showed thyroid-stimulating hormone (TSH) levels of 2.3 mIU/L and low-density lipoprotein cholesterol (LDL) levels of 160 mg/dL (4.1 mmol/L). All other values were within normal limits, including negative thyroid autoantibodies.

From an Ayurvedic perspective, I assessed her prakruti as predominantly Vata, as she exhibited classic signs of Vata imbalance, including anxiety, insomnia, palpitations, and irregular digestion with constipation. An integrative treatment plan was therefore initiated that included guided meditation, breathing exercises, yoga, and counseling, along with the introduction of ashwagandha, which supports sleep, mood, energy, and thyroid hormone secretion and conversion. I determined that her thyroid function could be carefully monitored as we lowered her medication, as her Vata nature likely made her more sensitive to the stimulating effects of the thyroid medication. She also started on triphala to help maintain regular bowel movements, was prescribed a Vata-balancing diet with regular warm, cooked foods and teas, and was encouraged to spend more time in nature daily, especially in the morning sun to synchronize her circadian rhythm.

She was followed closely, and after 6 months, she reported sleeping well without the need for Ambien, was able to discontinue diltiazem with resolution of palpitations, and described improved energy and mood. With careful monitoring, she was also weaned off the levothyroxine, maintaining stable thyroid function without medication and with TSH levels ranging from 2.7 to 4.3 mIU/L. Overall, she felt happier, more balanced, and confident in her ability to manage stress using lifestyle and Ayurvedic practices.

Clinical pearls

- Assessing prakruti and vikruti provided important guidance—the Vata imbalance explained her insomnia, anxiety, and irregular digestion.
- Ashwagandha was chosen for its grounding and adaptogenic properties. Its heating qualities make it particularly balancing for Vata, while also supporting sleep, reducing anxiety, and calming palpitations.
- Ashwagandha supports the natural conversion of thyroxine (T4) to triiodothyronine (T3), providing thyroid support. This justifies a careful plan to monitor TSH levels while gradually weaning her off the levothyroxine.
- Mind–body interventions, such as meditation, breathing, and yoga, addressed the root cause of her insomnia—fear and anxiety—while providing her with lasting tools for resilience.
- With an integrative approach, she was able to discontinue three medications (Ambien, diltiazem, levothyroxine) and reduce the use of Fiorinal, while improving her sleep, mood, energy, and overall wellbeing.

Case 2: Integrative approach for hyperlipidemia and overweight

A 45-year-old male presented with elevated cholesterol and excess weight. His initial laboratory results revealed the following: total cholesterol of 266 mg/dL (6.9 mmol/L), LDL of 172 mg/dL (4.4 mmol/L), triglycerides of 168 mg/dL (1.9 mmol/L), and high-density lipoprotein cholesterol (HDL) of 60 mg/dL (1.6 mmol/L). He was reluctant to initiate statin treatment and was instead motivated to use Ayurveda and lifestyle modification. He admitted to not regularly exercising, was overweight, and displayed habits of overeating and late-night snacking. He was a regular meditator for many years, had no significant family history of premature heart disease, nor did he exhibit other risk factors for heart disease. He was generally quite healthy, slept well, was in a long-term and healthy relationship, and enjoyed his work.

From an Ayurvedic perspective, I determined his prakruti to be Kapha, with its current imbalance contributing to his overweight condition and hyperlipidemia. This imbalance reflects an accumulation of Kapha dosha in the body, as well as diminished agni and accumulated ama (toxic by-products of impaired digestion). Treatment, therefore, focused on restoring digestive balance, stimulating metabolism, and clearing ama (cholesterol) while supporting healthier lipid metabolism. To accomplish this, I utilized a combination of guggulu (*Commiphora mukul*) with its heating qualities, and triphala. Of note, guggulu (through its active compounds known as guggulsterones) can lower LDL and triglycerides, raise HDL, modulate bile acid metabolism, and enhance thyroid function via modulation of gene expression (Shishodia et al., 2008), thereby improving cholesterol clearance in some individuals. Although such findings are not replicated in all studies, and while there are some known side effects due to guggulu's heating qualities (Ulbricht et al., 2005), it was still thought to be a good choice due to this patient's Kapha prakruti and vikruti. Triphala is a classical blend of amalaki (*Emblica officinalis*), haritaki (*Terminalia chebula*), and bibhitaki (*Terminalia bellirica*), and it complemented guggulu's actions by supporting digestion, promoting elimination, and preventing reabsorption of bile acids and cholesterol. In addition, its mild lipid-lowering, antioxidant, and antiinflammatory effects synergize with guggulu to enhance metabolic balance in the gut. Due to the combination of such herbs, I was able to use a smaller dose (versus some single-herb studies) with good effect, thereby lowering the chance of side effects.

After the initiation of a Kapha-balancing diet (a plant-based diet with all calories consumed before 6:00 p.m.) that was combined with Ayurvedic herbs, the repeat laboratory tests 6 weeks later demonstrated a significant improvement. Here, total cholesterol decreased to 191 mg/dL (4.9 mmol/L), LDL to 116 mg/dL (3 mmol/L), triglycerides to 79 mg/dL (0.9 mmol/L), with HDL remaining stable at 59 mg/dL (1.5 mmol/L). He lost approximately 9 kg over several months and felt more energetic.

Together, guggulu and triphala acted synergistically and provided a comprehensive strategy, where the former acted at the hepatic and metabolic levels, while triphala improved gut health and detoxification through effective elimination. This integrative plan, therefore, contributed to his rapid and clinically significant improvement in lipid levels, together with suitable dietary and lifestyle modifications. The guggulu treatment was withdrawn after 6 months, as it is not recommended for long-term use in Ayurveda, and long-term safety studies are currently lacking. However, he continued with occasional triphala for gut support, together with lifestyle changes that emphasized regular exercise and lighter meals with spices to support agni. He was able to maintain his weight and clinical biomarker levels in the desirable range for several years of follow-up.

Clinical pearls

- Assessing prakruti and vikruti provided important guidance: his Kapha imbalance explained his relatively high cholesterol levels and weight issues.
- Hyperlipidemia in Ayurveda reflects impaired agni and ama accumulation, which is often (but not always) linked to a Kapha constitution with Kapha aggravation. In cases where there is a Pitta prakruti or vikruti with more inflammation as a root cause, a herb with cooling qualities (e.g., amalaka) can be used to lower cholesterol levels and reduce cardiovascular risk (Gopa et al., 2012) instead of using guggulu.
- Guggulu and triphala act synergistically: guggulu enhances lipid metabolism and clearance, while triphala improves digestion, aids elimination, and reduces systemic inflammation.
- The combination of Ayurvedic formulations with lifestyle interventions resulted in significant improvements in his lipid profile within 6 weeks, which were maintained in the long term with suitable lifestyle modifications that included mind–body practices.

Case 3: Ayurvedic framework for treating Type 1 diabetes, emotional healing, and rediscovering purpose

A 28-year-old woman presented to me as a new patient with poorly controlled Type 1 diabetes after running out of insulin supplies several days prior. On arrival at the clinic, her examination and vital signs were within normal limits, apart from her appearing withdrawn and depressed. She was well hydrated and did not report any nausea or vomiting. Although her fingertip blood glucose test was in the 400s mg/dL (≥22 mmol/L), laboratory analyses showed no evidence of acidosis or electrolyte abnormalities. Her electrocardiogram was normal. Upon inquiry, she explained that she had been too busy to collect her prescription and was tired of poking herself all the time. She denied any suicidal intent. Although I recognized there were deeper issues at play, the immediate priority was to address her acute physical risk from uncontrolled diabetes. I refilled her insulin and asked her to return daily for several days so we could monitor her progress. Her blood sugars gradually stabilized with consistent dosing of insulin, and a trusting patient–physician relationship was established through an empathetic approach rather than being judgmental or reprimanding her.

After addressing the acute danger, I next explored the underlying emotional and social factors contributing to the decline in her health. She had a prior history of depression, with symptoms fluctuating over time. She was in a toxic relationship and had been drinking daily to cope, while describing her job as highly stressful and unfulfilling. As I got to know her better, she disclosed that she had experienced childhood trauma, which she came to recognize had shaped many of her relationship choices and contributed to her current struggles. At this stage, the focus of care expanded to her mental and emotional health. I offered compassion, recommended counseling, and encouraged her to create space for honest self-reflection. She agreed to begin counseling and returned regularly for follow-up visits.

Of note, meaningful shifts began to occur over the next few months. Here, she ended the unhealthy relationship, which was instrumental in lowering her dependence on alcohol. She also remained consistent with her insulin therapy, leading to stable blood sugar control. As she regained a sense of emotional stability, she began to engage in practices that supported deeper self-awareness, including meditation and journaling. These practices, along with counseling, helped her recognize that much of her unhappiness stemmed from living a life out of alignment with her values and passions.

She gradually explored what truly inspired her and what kind of career would bring her fulfillment. Over the course of the following year, she found the courage to leave her stressful job and began working with an organization devoted to empowering young women. This new role provided her not only with meaningful work but also with a profound sense of purpose. She expressed gratitude for the support and guidance she received, sharing that she felt as though I had "saved her life."

Clinical pearls

- This case highlights how integrative care can move in stages: first addressing the body (acute stabilization of blood sugars), then the mind (counseling, meditation, self-inquiry, processing trauma), and ultimately the spirit (reconnecting with purpose or *dharma*).
- Modern medicine provided the immediate tools necessary to prevent life-threatening complications, while Ayurvedic principles and practices supported emotional healing and spiritual growth.
- Recognizing and addressing the influence of childhood trauma as a root cause were key to breaking cycles of unhealthy relationships and empowering her to make new choices.
- True healing extends beyond physical stabilization to include rediscovering meaning, joy, and alignment with life's purpose, which sustains ongoing health and well-being.

In each of these clinical cases, the patient was offered a framework for becoming more aware of their own state of balance in body, mind, and spirit and was provided with tools to understand how to care for themselves. This empowered them to take an active role in their healing journeys and helped to build resilience and foster a long-term transformation.

These case summaries demonstrate the multitude of ways an integrative approach can support patients by combining the diagnostic tools and therapies of modern medicine with the holistic framework and practices of Ayurveda. By addressing root causes, honoring the connection between body, mind, and spirit, and engaging the patient in their own health, care becomes patient-centered, personalized, and effective, guiding patients toward balance, resilience, and healing of the whole person.

Toward a decolonized medical future

Decolonizing medicine extends far beyond simply offering "alternative" therapies. As the previous cases demonstrate, it requires a profound transformation in terms of *how* we value and engage with indigenous healing systems and philosophies. At its core, this process demands cultural humility, inclusivity, and recognition that health is deeply shaped by context, tradition, and community. Effective care must honor this complexity, moving beyond a purely reductionist or disease-focused model toward one that embraces the interconnectedness of body, mind, spirit, and environment (Marya and Patel, 2021, pp. 23–24).

Although individual practitioners can provide traditional practices through direct patient care, the decolonization of medical practice also requires structural support and policy initiatives. For example, in India, efforts to reclaim cultural and medical heritage included the establishment of the Ministry of AYUSH (Ayurveda, Yoga & Naturopathy, Unani, Siddha, and Homeopathy) in 2014. Its mission is "reviving the profound knowledge of our ancient systems of medicine and ensuring the optimal development and propagation of the AYUSH systems of health care" (Survey on Ayush Fact Sheet, 2023).

There is also a growing global interest among health care professionals to learn from traditional healing systems like Ayurveda, and hence, an increasing number of Ayurvedic practitioners have been trained over the last few decades. Some of these are physicians who are now providing holistic care in their communities. Moreover, in the US, organizations such as the National Ayurvedic Medical Association are creating standardized frameworks for training, credentialing, and clinical practice, as well as offering recognition of Ayurvedic education completed in India.

However, the decolonization and integration of traditional medical systems into conventional allopathic medicine still face a complex set of challenges. As described by Mehjabeen et al. (2025), some of these challenges include creating an agreed-upon definition of decolonizing health and also defining the aims of decolonizing global health. Furthermore, urbanization and the disconnection from natural environments make it increasingly difficult to implement traditional principles of balance and harmony with nature. This makes it challenging to heal even when the proper Ayurvedic therapies are employed. For example, environmental pollutants that act as endocrine disruptors, air and water contaminants, and persistent organic compounds from industrial pollution can contribute to chronic inflammation, impair digestive function, and disrupt metabolic and hormonal homeostasis (Shetty et al., 2023; Fuller et al., 2022). This creates physiological complexities that historical practitioners did not previously encounter. Beyond environmental considerations, the globalization and commercialization of traditional and complementary medicine present complex ethical and cultural dilemmas. The widespread use of "imported" traditional therapies often fails to recognize the communities and cultural knowledge systems from which these practices originate, raising issues around intellectual property, equitable benefit sharing, and the preservation of indigenous medical heritage. Here, social media is contributing to wider awareness of Ayurveda, though often emphasizing isolated interventions, e.g., herbal remedies, rather than the comprehensive lifestyle, dietary, and mind–body practices that are central to authentic Ayurvedic care.

Policy and funding challenges can further complicate the integration of indigenous knowledge systems into conventional medical practice. During 2022, the World Health Organization (WHO) and the Indian government established the WHO Global Center for Traditional Medicine with an initial investment of $250 million "to harness the potential of traditional medicine from around the world through modern science and technology" (WHO, 2022). However, while the WHO has provided leadership and guidance, many countries—including those with growing interest in traditional, complementary, and alternative medicine (TCAM)—have only allocated limited resources to research and implementation in this regard. In Western countries, research efforts have disproportionately focused on the quality, safety, and efficacy of select modalities. This especially includes herbal medicines due to both perceived risks and the potential for drug discovery, while the broader lifestyle, dietary, and mind–body Ayurvedic principles remain underinvestigated. This narrow focus risks reducing Ayurveda to isolated pharmacological interventions rather than a holistic system. Addressing these challenges requires not only rigorous scientific evaluation but also critical reflection on the structural, cultural, and systemic biases that have historically marginalized indigenous medical knowledge. Such an approach should help ensure that the integration process preserves both the therapeutic integrity and the cultural richness of such systems (Xue, 2008).

In Xue's review (Xue, 2008) of Bodeker and Burford's comprehensive book volume *Traditional, Complementary, and Alternative Medicine* (2007), he emphasized that education is a vital means of ensuring the safe and effective practice of TCAM. However, levels of education vary substantially from country to country, and hence some curricula may be inadequate to ensure that the potential benefits of various TCAM therapies are fully realized. In addition, Xue (2008) recounts that the health benefits of TCAM have not yet been fully assessed in ways that can be universally accepted as "evidence-based." This is partly due to the philosophical differences and cultural barriers between places where such therapies originated and Western countries where they are also now being practiced. Many TCAM research studies are of questionable quality, as they often do not employ modern scientific methodological approaches and techniques. However, there is also an emerging body of scientific research supporting the efficacy of Ayurvedic herbs, therapies, and formulations (Verma et al., 2024). In support, clinical studies demonstrate the effectiveness of Ayurveda in managing various complications such as arthritis, diabetes, hypertension, and anxiety (Gupta, 2024). To gain wider acceptance, there should be increasing efforts fostering the scientific basis of the tenets of Ayurveda and integrating modern analytical techniques with the broader Ayurveda principles (Jaiswal and Williams, 2016, p. 53).

Another challenge regarding the relevance and benefits of Ayurveda and other indigenous knowledge systems revolves around the concept of the credibility of ways of knowing. Here, it is essential that leading journals publishing impactful global health research also provide a platform for original research articles that explore the world from alternate paradigms or frameworks of knowing. Gradual formalization and mainstreaming of alternate paradigms and forms of analysis can start with the publication of methodological articles or research studies that were conceptualized and originated from their respective locales (as opposed to only from Western-based regions). Highlighting such methodologies in leading journals should help to demystify their workings for interested practitioners that were differently schooled and trained. This should result in broader dissemination, awareness, and cross-fertilization of ideas and knowledge. If progress is to be made toward justice in global health, there should be room for alternate ways of knowing, as mainstream journals often function as gatekeepers for what qualifies as "credible" or "high-quality" research and knowledge. Here, peer-reviewed journals often judge the validity and value of research work and knowledge through a one-dimensional Eurocentric lens, thereby perpetuating the injustice against knowledge-producers expounding alternate, indigenous ways of knowing. This leads to the exclusion of testimony that originates from alternate interpretations of how the world works (Khan, 2022, pp. 4–5). This is especially applicable to Ayurveda as a consciousness-based healing system, where this concept is not regarded as valid in contemporary medicine or medical research, thereby excluding the validation of such foundational principles.

The vision of a decolonized medical future is not a return to premodern practice nor a rejection of Western science; rather, it is the integration of modern empirical tools with the paradigms of traditional medicine, which offer a more holistic and complete view of nature. Empirical tools of modern science could validate many Ayurvedic practices, allowing practitioners to critically assess both the benefits and risks of therapies while maintaining rigorous standards of care. This will, at the same time, allow for the appreciation and understanding of such practices within the context of Ayurveda's foundational philosophy. To achieve this, we must critically examine the colonial influences and reductionist assumptions that have shaped conventional medicine.

The decolonization of medical research and practice is both a responsibility and an opportunity. It is a call to action for practitioners, educators, institutions, and policymakers. By reclaiming indigenous knowledge systems, reframing what it means to heal, and cultivating an integrative, holistic approach to care, this effort should enable the establishment of a unique health care paradigm that honors not only human health but also the wellbeing of society as a whole.

Reclaiming our roots, redefining health

As India emerges from the shadows of colonial oppression, there is an urgent need to reclaim and honor Ayurveda not as an alternative or adjunct but as a fully integrated and scientifically validated system of medicine. The core principles of Ayurveda reveal a deeply integrative approach to health. Ayurveda provides a comprehensive framework for individualized, holistic healing that addresses not only the body but also the mind, spirit, and environment. Central to this approach is the recognition of human health as being intimately connected with the natural world.

Ayurveda inherently links personal wellbeing with ecological health, teaching that balance within the individual mirrors balance within the environment. By observing natural cycles, Ayurvedic practices cultivate harmony with the earth, supporting vitality while promoting sustainability. Principles such as *ahimsa* (nonharming), mindful consumption, and seasonal dietary adjustments not only nourish the body but also reduce strain on natural resources, conserve biodiversity, and minimize waste. In this framework, nurturing personal health is inseparable from caring for the planet, and hence the cultivation of vitality and resilience within the individual reflects the flourishing of the natural world. Conversely, environmental restoration supports human wellbeing.

At its philosophical core, Ayurveda views consciousness as the ground state of reality—the eternal, unchanging essence from which body, mind, and life emerge. Healing is, therefore, not merely the restoration of physical function but a reconnection to this deeper, unified consciousness. Health is understood as a dynamic state of balance encompassing the individual, their community, and the larger universe. True healing is a return to the wholeness of who we are. By reclaiming these principles, we not only honor an ancient tradition but also create a pathway for modern medicine that integrates indigenous wisdom with empirical science. As stated by Hegde et al. (2024, p. 52): "The future of health care lies in a more inclusive and integrated approach that combines the best aspects of both ancient and conventional healing traditions. Rather than seeing Ayurveda and modern medicine as competing paradigms, they should be viewed as complementary systems of care that together address the full spectrum of human health. For this vision to become a reality, both systems must evolve. Conventional medicine could benefit from incorporating holistic approaches that account for the emotional and spiritual aspects of health, while Ayurveda could gain from modern research techniques that confirm and expand its therapeutic repertoire."

Ultimately, redefining health through an Ayurvedic lens invites all health care professionals to honor the interconnectedness of body, mind, spirit, and environment. The most meaningful work in medicine is found by supporting rigorous scientific practice while embracing the profound insights of traditional healing systems, guiding both patients and practitioners toward wellbeing. Through this reawakening, we not only recover lost traditions but also create a space for a truly integrated health care system that respects both the past and the future of human, societal, and planetary health.

Glossary

Abhyanga Ayurvedic oil massage used for nourishment and balance.
Agni Digestive fire; metabolic and transformative energy.
Akasha Space or ether; principle of potential and expansion.
Ama Toxicity; impurities that obstruct the body's natural healing processes.
Dhatus The seven bodily tissues that sustain structure and function.
Dinacharya Daily routine aligned with natural rhythms.
Dosha Mind–body principle governing psychophysiological function.
Gurukul Traditional lineage-based school or mentorship system.
Jala Water; principle of cohesion and fluidity.
Kapha Dosha that represents structure, stability, and lubrication.
Malas The body's waste products, which are eliminated to maintain balance.
Ojas Essence of vitality, immunity, and resilience.
Panchakarma Five purification therapies performed for detoxification.
Panchamahabhuta The five great elements: space, air, fire, water, earth.
Pitta Dosha of transformation, digestion, and metabolism.
Prakruti One's innate constitution or natural balance.
Prana Life force or vital energy.
Prithvi Earth; principle of matter and solidity.
Rasa The "taste" of food that influences balance in the body and mind.
Rasayana Rejuvenation therapy; science of longevity.
Ratricharya Nightly routine promoting rest and restoration.
Ritucharya Seasonal routine aligned with natural seasons.
Tejas Subtle form of fire that sustains cellular activity, radiance, intelligence.
Vaidya Ayurvedic physician or healer.
Vata Dosha of movement, change, and communication.
Vayu Air; principle of motion and mobility.
Vikruti Current state of imbalance or deviation from prakruti (natural state).

References

Afzaal, M., et al., 2022. Human gut microbiota in health and disease: unveiling the relationship. Front. Microbiol. 13, 999001. Available from: https://www.frontiersin.org/journals/microbiology/articles/10.3389/fmicb.2022.999001/full. doi:10.3389/fmicb.2022.999001.

Alotiby, A., 2024. Immunology of stress: a review article. J. Clin. Med. 13 (21), 6394. https://pmc.ncbi.nlm.nih.gov/articles/PMC11546738/. doi:10.3390/jcm13216394.

Anshu,, Supe, A., 2016. Evolution of medical education in India: the impact of colonialism. J. Postgrad. Med. 62 (4), 255–259. doi:10.4103/0022-3859.191011.

Bobo, W.V., et al, 2022. Association of depression and anxiety with the accumulation of chronic conditions. JAMA Netw. Open 5 (5), e229817. Available from: https://jamanetwork.com/journals/jamanetworkopen/fullarticle/2791768. doi:10.1001/jamanetworkopen.2022.9817.

Boege, H.L., Bhatti, M.Z., St-Onge, M.P., 2021. Circadian rhythms and meal timing: Impact on energy balance and body weight. Curr. Opin. Biotechnol. 70, 1–6. doi:10.1016/j.copbio.2020.08.009.

Bożek, A., Nowak, P.F., Blukacz, M., 2020. The relationship between spirituality, health-related behavior, and psychological well-being. Front. Psychol. 11, 1997. doi:10.3389/fpsyg.2020.01997.

Cambridge Dictionary, 2025. *Decolonization*. Available from: https://dictionary.cambridge.org/us/dictionary/english/decolonization. (Accessed 30 August 2025).

Chhandak, V.K., Dabi, N., Barulia, K.C., 2025. Ayurveda's role in promoting health and sustainability: a one health perspective. Int. J. Innov. Sci. Res. Technol. 1911–1913. doi:10.38124/ijisrt/25mar1397.

Chopra, D., 1997. Perfect Health. Three Rivers Press, New York.

Cui, J., et al, 2022. Inhalation aromatherapy via brain-targeted nasal delivery: natural volatiles or essential oils on mood disorders. Front. Pharmacol. 13. Available from: https://www.frontiersin.org/journals/pharmacology/articles/10.3389/fphar.2022.860043/full.

Demir, I., 2022. How and why should we decolonize global health education and research? Ann. Glob. Health 88 (1), 30. doi:10.5334/aogh.3787.

Eckhart, R., 2011. Genetic research and nutritional individuality. In: *American Society for Nutritional Sciences Symposium: Nutritional and Metabolic Diversity: Understanding the Basis of Biologic Variance in the Obesity/Diabetes/Cardiovascular Disease Connection*. Available from: https://jn.nutrition.org/article/S0022-3166(22)14634-1/pdf (Accessed 15 July 2025).

Foster, R.G., 2020. Sleep, circadian rhythms and health. Interface Focus. 20190098. 10 (3), Available from: https://pmc.ncbi.nlm.nih.gov/articles/PMC7202392/. doi:10.1098/rsfs.2019.0098.

Frawley, D., 1989. Ayurvedic Healing: A Comprehensive Guide. Lotus Press, Twin Lakes, WI.

Frawley, D., Kshirsagar, S., 2016. The Art and Science of Vedic Counseling. Lotus Press, Twin Lakes, WI.

Frawley, D., Lad, V., 1986. The Yoga of Herbs. Lotus Press, Twin Lakes, WI.

Fuller, R., et al, 2022. Pollution and health: a progress update. Lancet Planet. Health 6 (6), e535–e547. doi:10.1016/S2542-5196(22)00090-0.

Galvano, F., 2024. *Physiological changes associated with emotions*. Research Gate. Available from: https://www.researchgate.net/publication/381311629_Physiological_Changes_Associated_with_Emotions.

Garza, M.C., et al, 2024. Effect of aromatic herbs and spices present in the Mediterranean diet on the glycemic profile in type 2 diabetes subjects: a systematic review and meta-analysis. Nutrients 16 (6), 756. doi:10.3390/nu16060756.

Gopa, B., Bhatt, J., Hemavathi, K.G., 2012. A comparative clinical study of hypolipidemic efficacy of Amla (*Emblica officinalis*) with 3-hydroxy-3-methylglutaryl-coenzyme—a reductase inhibitor simvastatin. Indian J. Pharmacol. 44 (2), 238–242. doi:10.4103/0253-7613.93857.

Gupta, R., 2024. Integrating Ayurveda with modern medicine for enhanced patient care—analysis of realities. Physician 9 (1), 1–6. doi:10.38192/1.9.1.3.

Halpern, M., 1995. Principles of Ayurvedic Medicine. Harmony is Health Press, Nevada City, CA.

Hegde, H., Barvaliya, M., Roy, S., 2024. Integrating Ayurveda and conventional medicine: a bridge to be built for holistic health. Indian J. Ayurveda Integr. Med. 5 (2), 51–53. doi:10.4103/ijaim.ijaim_48_24.

Hoge, E.A., et al, 2013. Randomized controlled trial of mindfulness meditation for generalized anxiety disorder: effects on anxiety and stress reactivity. J. Clin. Psychiatry 74 (8), 786–792. doi:10.4088/JCP.12m08083.

Internet Modern History Sourcebook, 1957. Thomas Babington Macaulay (1800–1859) on Empire and Education. In: Macaulay, T.B., Prose and Poetry. Selected by G.M. Young. Cambridge, MA: Harvard University Press, pp. 716–718. Available from: https://sourcebooks.fordham.edu/mod/1833macaulay-india.asp.

Jaiswal, Y.S., Williams, L.L., 2016. A glimpse of Ayurveda—the forgotten history and principles of Indian traditional medicine. J. Tradit. Complement. Med. 7 (1), 50–53. doi:10.1016/j.jtcme.2016.02.002.

Jha, S.K., 2023. Resistance and response to colonial biomedicine: a critical study of survival of Ayurvedic tradition in rural India. J. Ayurveda 17 (4), 323–328. doi:10.4103/joa.joa_365_21.

Karawani, H., Jenkins, K., Anderson, S., 2022. Neural plasticity induced by hearing aid use. Front. Aging Neurosci. 14, 884917. doi:10.3389/fnagi.2022.884917.

Khan, S.A., 2022. Decolonising global health by decolonising academic publishing. BMJ Glob. Health 7 (3), e007811. doi:10.1136/bmjgh-2021-007811.

Kollndorfer, K., Kowalczyk, K., Hoche, E., Mueller, C.A., Pollak, M., Trattnig, S., Schöpf, V., 2014. Recovery of olfactory function induces neuroplasticity effects in patients with smell loss. Neural Plast. 2014, 140419. doi:10.1155/2014/140419.

Kopf, D., 2021. Massage and touch-based therapy: Clinical evidence, neurobiology and applications in older patients with psychiatric symptoms. Z. Gerontol. Geriatr. 54 (8), 753–758. doi:10.1007/s00391-021-01995-4.

Kshirsagar, S., 2018. Change Your Schedule, Change Your Life. Harper Wave, New York.

Kumar, U., et al, 2025. Neural correlates of auditory comprehension and integration of Sanskrit verse: a functional MRI study. Brain Struct. Funct. 230 (1), 28. doi:10.1007/s00429-025-02892-x.

Lad, V., 2002. *Textbook of Ayurveda, Fundamental Principles*, Vol. 1. The Ayurvedic Press, Albuquerque, NM.

Market.us, 2025. *Ayurveda market*. Available from: https://market.us/report/ayurveda-market/. (Accessed 30 August 2025).

Marya, R., Patel, R., 2021. Inflamed. Farrar, Straus, and Giroux, New York.

Mehjabeen, D., Patel, K., Jindal, R.M., 2025. Decolonizing global health: a scoping review. BMC Health Serv. Res. 25 (1), 828. doi:10.1186/s12913-025-12890-8.

Musaelyan, K., et al, 2020. Chronic stress induces significant gene expression changes in the prefrontal cortex alongside alterations in adult hippocampal neurogenesis. Brain Commun. 2 (2), fcaa153. doi:10.1093/braincomms/fcaa153.

Ostaiza-Cardenas, J., et al, 2025. Epigenetic modulation by lifestyle: advances in diet, exercise, and mindfulness for disease prevention and health optimization. Front. Nutr. 12, 1632999. doi:10.3389/fnut.2025.1632999.

Park, C.L., Slattery, J.M., 2021. Yoga as an integrative therapy for mental health concerns: an overview of current research evidence. Psychiatry Int. 2 (4), 386–401. doi:10.3390/psychiatryint2040030.

Peterson, C.T., et al, 2019. Prebiotic potential of culinary spices used to support digestion and bioabsorption. Evid. Based Complement. Alternat. Med. 2019, 8973704. doi:10.1155/2019/8973704.

Peterson, C.T., Denniston, K., Chopra, D., 2017. Therapeutic uses of triphala in ayurvedic medicine. J. Altern. Complement. Med. 23 (8), 607–614. doi:10.1089/acm.2017.0083.

Philips, C.A., et al, 2023. A case series from India and literature review. Hepatol. Commun. 7 (10), e0270. doi:10.1097/HC9.0000000000000270.

Pondomatti, S.C., Tyagi, I., Shrivastava, K.K., et al, 2024. A literature review of the integration of ancient Indian mythology in clinical medicine: a holistic approach to health and healing. Cureus 16 (7), e63779. doi:10.7759/cureus.63779.

Sanchez-Roige, S., et al., 2018. The genetics of human personality. Genes Brain Behav. 17 (3), e12439. doi:10.1111/gbb.12439.

Sato, S., Solanas, G., Sassone-Corsi, P., Benitah, S., 2022. Tuning up an aged clock: circadian clock regulation in metabolism and aging. Transl. Med. Aging 6, 1–13. doi:10.1016/j.tma.2021.11.003.

Schärfe, C.P.I., Tremmel, R., Schwab, M., et al, 2017. Genetic variation in human drug-related genes. Genome Med. 9, 117. doi:10.1186/s13073-017-0502-5.

Sen, S., Chakraborty, R., 2016. Revival, modernization and integration of Indian traditional herbal medicine in clinical practice: importance, challenges and future. J. Tradit. Complement. Med. 7 (2), 234–244. doi:10.1016/j.jtcme.2016.05.006.

Shao, T., et al, 2023. The gut ecosystem and immune tolerance. J. Autoimmun. 141, 103114. doi:10.1016/j.jaut.2023.103114.

Shetty, S.S., et al, 2023. Environmental pollutants and their effects on human health. Heliyon 9 (9), e19496. doi:10.1016/j.heliyon.2023.e19496.

Shishodia, S., Harikumar, K.B., Dass, S., Ramawat, K.G., Aggarwal, B.B., 2008. The guggul for chronic diseases: ancient medicine, modern targets. Anticancer Res. 28 (6A), 3647–3664. Available from: https://ar.iiarjournals.org/content/anticanres/28/6A/3647.full.pdf.

Shivani, D., et al, 2025. Rasayana in the era of anti-ageing medicine: insights from Rasayana and modern gero-therapeutics. JETIR 12 (4). Available from: https://www.jetir.org/papers/JETIR2504755.pdf.

Simon, D., 1997. The Wisdom of Healing. Three Rivers Press, New York.

Simon, D., Chopra, D., 2000. The Chopra Center Herbal Handbook. Three Rivers Press, New York.

Singh, J., et al, 2024. Microbiota-brain axis: exploring the role of gut microbiota in psychiatric disorders—a comprehensive review. Asian J. Psychiatry 97, 104068. doi:10.1016/j.ajp.2024.104068.

Smith, A., 2013. Ayurvedic Medicine for Westerners, Vol. 1: Anatomy and Physiology in Ayurveda, EIVS GmbH.

Survey on Ayush Fact Sheet, 2023. *2022–2023*. Government of India, Ministry of Statistics and Programme Implementation. National Statistical Office. Available from: https://www.mospi.gov.in/sites/default/files/publication_reports/Fact_sheet-Survey_on-Ayush.pdf.

Tillu, G., Bhalerao, S., Godatwar, P., 2024. The World Ayurveda Congress—a movement for advancement of Ayurveda. J. Ayurveda Integr. Med. 15 (6), 101123. doi:10.1016/j.jaim.2024.101123.

Ulbricht, C., et al, 2005. Guggul for hyperlipidemia: a review by the Natural Standard Research Collaboration. Complement. Ther. Med. 13 (4), 279–290. Available from: https://www.ncbi.nlm.nih.gov/books/NBK72258/.

Verma, S.K., et al, 2024. Exploring ayurveda: principles and their application in modern medicine. Bull. Natl. Res. Cent. 48, 77. doi:10.1186/s42269-024-01231-0.

West, C.P., Dyrbye, L.N., Shanafelt, T.D., 2018. Physician burnout: contributors, consequences and solutions (Review). J. Intern. Med. 283, 516–529.

WHO, 2022. WHO establishes the global centre for traditional medicine in India'. Available from: https://www.who.int/news/item/25-03-2022-who-establishes-the-global-centre-for-traditional-medicine-in-india#:~:text=Around%2080%25%20of%20the%20world's, traditional%20medicine%20practices%20and%20products.

Williams, R.J., 1956. Biochemical Individuality. Wiley, New York. Available at: https://archive.org/details/in.ernet.dli.2015.547843/page/n17/mode/2up.

Xue, C.C., 2008. Traditional, complementary and alternative medicine: policy and public health perspectives. Bull. World Health Organ. 86 (1), 77–78. doi:10.2471/BLT.07.046458.

Yeola, G., et al, 2024. International Delegate Assembly: an overview of two decades. J. Ayurveda Integr. Med. 15 (2), 100892. doi:10.1016/j.jaim.2024.100892.

CHAPTER

Harnessing the health-promoting potential of indigenous foods: Enhancing public health and advancing environmental sustainability

15

Xikombiso Mbhenyane[a], Nomakhushe Nxusani[a], Vanessa Mbhatsani[b], Sefora Makuse[c], and Mthokozisi Zuma[a,d]

[a]*Division of Human Nutrition, Department of Global Health, Faculty of Medicine and Health Sciences, Stellenbosch University, Cape Town, South Africa*

[b]*Department of Nutrition, Faculty of Health Sciences, University of Venda, Thohoyandou, South Africa*

[c]*Department of Human Nutrition and Dietetics, School of Health Care Sciences, Faculty of Health Sciences, University of Limpopo, Polokwane, South Africa*

[d]*Research Support and Coordination Division, Agricultural Research Council, Hatfield, Pretoria, South Africa*

Introduction

According to research, consuming crop products rich in essential nutrients is associated with a lower risk of viral infections and chronic diseases (Panghal et al., 2021; Kaur and Kapoor, 2019). This is because plant-based bioactivities do not promote resistance; instead, they are reliable and effective against pathogens (Mahendranathan and Abhayarathne, 2021). Consuming fruits and vegetables can positively impact overall health and quality of life, as they may help prevent the onset of diseases. Indigenous fruits and vegetables could, therefore, serve as a solution, an alternative, or a complementary solution to hunger and ongoing nutritional deficiencies.

Indigenous foods include those that are native to a specific region and that have been traditionally consumed by a particular cultural group, representing a rich and underutilized resource for improving human health, enhancing dietary diversity, and promoting sustainable food systems. Across Africa, and particularly in South Africa, a wide variety of indigenous foods are either grown or foraged from the wild, utilizing indigenous knowledge systems; these foods are deeply embedded in the ecological, spiritual, and cultural fabric of communities (Nxusani et al., 2023).

Decolonizing Medical Education, Research and Practice. https://doi.org/10.1016/B978-0-443-44998-7.00007-7

In this context, African rural communities have incorporated these crops into their cultural heritage and daily staple diet for generations. Such crops are, therefore, deeply ingrained in cultural heritage systems and are vital to communities in terms of food security, dietary diversity, and environmental sustainability. They provide a significant source of nutrition for rural communities and may act as a potential safety net during drought and famine (Kunene et al., 2020; Mabhaudhi et al., 2017; Mbhenyane, 2017).

Across the globe, indigenous foods similarly underpin health, resilience, and cultural identity across diverse regions. In the Mediterranean, traditional diets in countries such as Italy, Greece, and Spain, typically enriched by olives, pulses, whole grains, fruits, and fish, are now recognized for their cardioprotective effects and longevity benefits (Willett et al., 2019). Moreover, rural and forest-dependent communities in India, Thailand, and the Philippines continue to rely on wild edibles and traditional crops, harnessing deep-rooted, place-based knowledge systems to safeguard food security and ecological sustainability (NTFP-EP, 2025). Across South America, indigenous agronomic wisdom in Peru, Bolivia, and Brazil, from the Andes to the Amazon, nurtures high-biodiversity agroecosystems that buffer families from hunger, uphold food sovereignty, and resist the threats of industrial monoculture (IDRC, 2024). Likewise, First Nations in Canada and Native American tribes in the United States are revitalizing ancestral foods such as salmon, corn, beans, squash, and berries to combat diet-related illnesses like diabetes and obesity, and to reassert food sovereignty over land and seed systems (Kuhnlein et al., 2020). Finally, in Europe, ethnobotanical practices and heirloom varieties in France, Germany, and the United Kingdom, once relegated to "forgotten" status and indigenous foods, are being revalorized for their resilience, nutritional virtues, and cultural value, feeding into local food heritage and conservation efforts (Burlingame and Dernini, 2019). These global examples reinforce the universality of indigenous foods in nurturing nutrition, cultural identity, and environmental sustainability.

Indigenous foods encompass grains, legumes, vegetables, fruits, roots, and tubers, as well as nuts and seeds. Indigenous grain crops yield seeds rich in protein and high in starch. Cereals include pearl millet (*Pennisetum glaucum/nwahuva*[a]) and grain sorghum (*Sorghum bicolor/amazimba*), while pulses consist of cowpea (*Vigna unguiculata/munawa*), Bambara groundnuts (*Vigna subterranea/tindluwa*), and mung bean (*Vigna radiata/ditlhodi*). Indigenous vegetable crops suitable for food preparation typically include tender leaves, stems, and stalks. These include leafy vegetables such as cleome (*Cleome gynandra/lerotho*), amaranth (*Amaranthus/imbuya*), blackjack (*Bidens pilosa* L./*mushiji*), Jew's mallow (*Corchorus olitorius* L./*guxe*), and pumpkin (*Cucurbita pepo/xiphaswo*), as well as roots and tubers like cassava (*Manihot esculenta/ntsumbula*) and *amadumbe* (*Colocasia esculenta*). South African indigenous nuts and seeds include the

[a] The second name in brackets is from one of the indigenous South African languages, including Afrikaans, while some are named in the original language

nutritious marula nut (*Sclerocarya caffra/makanyi*), baobab seeds (*Adansonia digitata/muvuyu*), mongongo nut (*Schinziophyton rautanenii/manketti*), wild almond seeds (*Brabejum stellatifolium/keurboom*), sesame seeds (*Sesamum* spp./*umhlwathi*), *Leucadendron seeds* (*knokkers*), and *Strelitzia* seeds (Natal wild banana/*iSinga*). These nuts and seeds can provide utility as oils, flours, and snacks, showcasing diverse uses from traditional cooking to commercial products and enriching the region's food heritage.

Moreover, indigenous fruits encompass fleshy, seed-associated structures, which are palatable and delicious when uncooked. These fruit crops comprise marula (*S. caffra/makanyi*), red milkwood (*Mimusops zeyheri/nhlatswa*), mobola plum (*Parinari curatellifolia/mbulwa*), wild medlar (*Vangueria infausta/umvilo*), num-num (*Carissa macrocarpa/umthungulu*), and kei apple (*Dovyalis caffra/umqokolo*), among others (Takaidza, 2023). Given this context, potential solutions for food and nutritional security can be explored, as such crops have the potential to enhance and strengthen the existing food system and are adapted to and tolerant of marginal and harsh conditions. Thus they can help promote sustainable and resilient agricultural and food systems (refer to Fig. 15.1).

Considering increasing concerns regarding malnutrition, noncommunicable diseases, and environmental sustainability, the call for revalorizing indigenous foods has become more urgent than ever. This extends beyond modern food systems, public health, and academic research to help unlock the potential of indigenous foods as sources of health, resilience, and cultural continuity. Mbhenyane (2016) asserts that a dynamic food and nutrition system is self-reliant, locally controlled, accessible, safe, sustainable, resilient, and free from external influence. Such a system reflects the principles of food sovereignty, which advocate for the rights of communities to define their own food systems, prioritizing culturally appropriate and ecologically sound practices. Within this framework, the promotion and utilization of indigenous food crops play a central role, strengthening the links between sustainable natural resource management, local food production, dietary practices, nutritional status, and overall wellbeing. Indigenous foods are rooted in traditional knowledge and biodiversity and are vital to building resilient food systems that support both health and cultural identity.

This chapter examines the nutritional and therapeutic properties, as well as the diverse health benefits, of indigenous foods, drawing on both scientific evidence and indigenous knowledge. It highlights their bioactive properties, roles in food sovereignty for development, and their significance in addressing current health challenges, with the opportunity to both revalue and reintroduce indigenous products into contemporary diets, as well as their inclusion in sustainable food and nutrition policies. We argue that there is significant potential to enhance public health, protect cultural heritage, and promote environmental sustainability.

FIG. 15.1 Collage of indigenous foods.

Source: Mbhatsani, H.V., Maluleke, S.D.C., Tambe, B.A., Zuma, M.K., Mbhenyane, X.G., 2023. Innovative dishes and micronutrient composition of traditionally preserved Amaranthus, Vigna unguiculata, and Bidens pilosa. S. Afr. J. Bot. 163, 478–487. https://doi.org/10.1016/j.sajb.2023.08.019; Nxusani, Z.N., Zuma, M.K., Mbhenyane, X.G., 2023. A systematic review of indigenous food plant usage in Southern Africa. Sustainability 15 (11), 8799. https://doi.org/10.3390/su15118799.

Nutraceutical properties in indigenous food crops

Beyond basic nourishment, indigenous foods possess significant nutraceutical potential, offering bioactive compounds that contribute to disease prevention, therapeutic effects, and the promotion of overall health. Nutraceuticals are defined as food-derived substances with health-enhancing properties beyond basic nutrition, including phytochemicals, antioxidants, vitamins, minerals, probiotics, dietary fiber, and other functional nutrients. Indigenous food crops are often rich in such compounds due to their natural resilience and adaptation to local agroecological conditions. Recognizing and harnessing the nutraceutical value of indigenous foods can support sustainable health interventions, strengthen food and nutrition security, and advance community-based approaches to improved wellbeing.

Phytochemicals and indigenous food crops

Indigenous foods contain phytochemicals, which are natural bioactive compounds that contribute to human health by providing antioxidant, antiinflammatory, and disease-preventive benefits. The connection between indigenous foods and phytochemicals helps promote healthy diets while preserving biodiversity and cultural heritage (Hammoudi Halat et al., 2023; Mbhenyane, 2017). Phytochemicals, also known as plant chemicals, are secondary metabolites that serve nonessential functions, yet they are beneficial to human physiology due to their antioxidant properties (Janarthanan and Kumar, 2013). Although plants produce such chemicals to protect themselves from herbivores and pathogens, evidence suggests they can also protect humans from diseases. Phytochemicals are naturally occurring compounds found in plants that possess protective effects against various diseases and include, for example, polyphenols, flavonoids, carotenoids, and alkaloids. As indigenous foods often contain high concentrations of such phytochemicals, they offer value in terms of promoting human health by possessing disease-preventive properties (Sivarkumar et al., 2018).

Nutritional content of indigenous food crops

Many indigenous foods are celebrated for their health benefits. They are often more nutrient-dense than commercially grown crops, as they have been selectively bred or naturally adapted to their environment over generations. These foods are typically rich in vitamins, minerals, antioxidants, fiber, and healthy fats, contributing to overall health and wellbeing. Their nutritional composition varies slightly from one location to another, depending on the climate, soil, and crop variety. Selected indigenous food crops and their nutritional composition have been documented as commonly consumed in South Africa (Mbhenyane et al., 2013; Makuse and Mbhenyane, 2011). Refer to Table 15.1 for the nutritional composition of selected food crops, followed by a narrative for some of them.

Amadumbe

Amadumbe (*C. esculenta*), also known as taro, is a tropical tuber crop cultivated for its underground corms (short, fleshy stems) in various African regions (Kaur et al., 2013). In South Africa, it is primarily cultivated in the KwaZulu-Natal, Limpopo, and Eastern Cape provinces, where agroecological conditions favor its growth and traditional use (Agricultural Research Council (ARC), 2023). The Agricultural Research Council (ARC) of South Africa has identified these regions as suitable for promoting the production of amadumbe due to its potential to contribute to food security,

Table 15.1 Nutritional value of selected indigenous foods.

Scientific name	Common name	Indigenous name[a]	Moisture (%)	Energy (kJ)	Protein (g)	Vit A (µg)	Vit C (mg)	Iron (mg)	Zinc (mg)
Acacia erioloba	Camel thorn	*Umwhohlo*	13.9	1308	6.8	–	–	10.4	0.3
Amaranthus tricolor	Amaranth	*Imbuya*	91.7	96	2.5	292	43.3	2.3	0.9
Cajanus cajan	Pigeon pea	*Lethlodi*	71.8	464	6.0	2	28.1	1.6	0.8
Chenopodium album	Fat hen	*Imbilikicane*	88.9	134	3.2	391	37	0.7	0.3
Colocasia esculenta	Taro root	*Amadumbe*	66.8	1231	1.96	3	5	0.68	3.2
Corchorus olitorius	Jute	*Guxe*	87.2	155	3.4	156	33	3.1	0.8
Dioscorea dumetorum	African bitter yam	*Ingwevu*	67	520	3.2	–	–	–	–
Ficus sycomorus	Sycamore fig	*Makuwa*	82.7	210	1.4	–	7.3	1.7	0.4
Manihot esculenta	Cassava	*Ntsumbula*	62.8	625	1.4	30	15	0.23	0.48
Momordica charantia	Bitter cucumber	*Tshibavhe*	94	79	0.8	11	33	0.4	0.8
Pennisetum glaucum	Millet	*Unyawothi*	11.6	1442	10.5	–	–	6.5	1.7
Pseudolachnostylis maprouneifolia	Kudu-berry	*Mutondowa*	60	147	3.3	–	–	2.7	2.6
Sechium edule	Choko	*Chayote*	93.4	100	0.6	5	8	0.2	0.3
Sorghum bicolor	Sorghum	*Amazimba*	–	1459	11.1	0	–	–	–
Vigna subterranea	Bambara groundnut	*Tindluwa*	7.3	1572	18.4	–	–	4.6	2.2
Vigna unguiculata subsp. *unguiculata*	Cowpea	*Munawa*	11.2	1189	23.5	–	1.5	6.4	–

%, percentage; µg, microgram; g, gram; kJ, kilojoule; mg, milligram; Vit, vitamin.

[a] *The indigenous names of these foods are provided in one of South Africa's nine official indigenous languages.*

Source: Mbhatsani, H.V., Maluleke, S.D.C., Tambe, B.A., Zuma, M.K., Mbhenyane, X.G., 2023. Innovative dishes and micronutrient composition of traditionally preserved Amaranthus, Vigna unguiculata and Bidens pilosa. S. Afr. J. Bot. 163, 478–487. https://doi.org/10.1016/j.sajb.2023.08.019; Food Plants Solution Field Guide, 2013. Potentially Important Food Plants of South Africa. Available from: www.foodplantsolutions.org. (Accessed 10 July 2025); Makuse, S.N., Mbhenyane, X.G., 2011. Health benefits and omega-3 fatty acid content of selected indigenous foods in the Limpopo Province, South Africa. Indilinga: Afr. J. Indig. Knowl. Syst. 1683-0296. 10 (2), 182–194. https://hdl.handle.net/10520/EJC61398.

nutrition, and income generation, especially among smallholder farmers and rural communities. Amadumbe is also widely cultivated by smallholder farmers across Africa as a vital subsistence and food security crop. For example, Nigeria, Cameroon, and Ghana account for over two-thirds of the continent's production (Oladimeji et al., 2022; Grimaldi et al., 2018). In East Africa, it is grown in Ethiopia, Kenya, and Tanzania, while in North Africa, its cultivation is pursued in Egypt, Algeria, Morocco, and Tunisia—often in irrigated or riparian zones (Grimaldi et al., 2018). In addition, amadumbe thrives on Indian Ocean islands such as Madagascar, Mauritius, and Réunion (Oladimeji et al., 2022). Although large-scale commercial production remains centered in West Africa, crops are broadly utilized across the continent for its capacity to support food resilience under diverse environmental conditions.

Amadumbe is a valuable staple food containing high moisture levels, ranging from 63.6% to 72.4% in fresh corms (Chandrakar et al., 2022; Alinnor and Akalezi, 2010). It is a good source of carbohydrates, protein, and fat, as well as insoluble and soluble dietary fiber, which may positively affect glycemic control. Significant vitamins (per 100 g of dry weight) include β-carotene (up to 93.6 mg) and ascorbic acid (12–21 mg), while others such as niacin (0.64–2.7 mg), riboflavin (0.05–0.1 mg), and thiamine (0.05–0.07 mg) are also present (Alinnor and Akalezi, 2010). Amadumbe corm is also high in minerals, including potassium, sodium, magnesium, calcium, phosphorus, iron, zinc, and copper (Alinnor and Akalezi, 2010).

Amaranthus

The *Amaranthus* (*imbuya*) species is indigenous to South Africa and belongs to the Amaranthaceae family, which includes *Amaranthus dubius*, *Amaranthus hybridus*, *Amaranthus esculentus*, *Amaranthus viridis*, *Amaranthus tricolor*, and *Amaranthus thunbergii*, with *A. dubius* and *A. viridis* (Department of Agriculture, Forestry and Fisheries (DAFF), 2010). These species are well adapted across sub-Saharan Africa, growing abundantly in the wild, while also being cultivated in the Limpopo, KwaZulu-Natal, Northwest, and Mpumalanga provinces of South Africa (Agricultural Research Council (ARC), 2023). Amaranthus is a cosmopolitan genus also cultivated and consumed across many regions of the world; for example, species such as *A. tricolor* and *A. viridis* are widely grown in India, Nepal, Bangladesh, China, and the Philippines as leafy vegetables and grain crops (Shukla and Singh, 2010). Moreover, amaranth grains were historically cultivated by the Incas in Peru and Bolivia and remain important in countries such as Brazil for their nutritional and cultural value (National Research Council, 1989). In North America, native species occur in Mexico and the southwestern United States where they are grown for grain and as ornamentals and are increasingly promoted as part of indigenous food sovereignty initiatives (Kuhnlein et al., 2020). In Europe, countries such as Italy and Greece cultivate amaranth primarily as a leafy vegetable and for specialty grains, often linked to sustainable diets and biodiversity conservation (Burlingame and Dernini, 2019). The genus is also adaptable to diverse climates and soils and is, for example, present in tropical and subtropical regions across the globe, including the

Caribbean islands and parts of Australia (Rastogi and Shukla, 2013). This global distribution underscores amaranth's versatility and its potential role in promoting food security and dietary diversity.

According to Beswa et al. (2016), amaranthus leaves are highly nutritious, containing elevated levels of protein, calcium, iron, magnesium, potassium, and β-carotene, together with vitamins A, C, and E, and dietary fiber. Mbhatsani et al. (2023) further highlight the richness of amaranthus in crude protein and calcium, noting that nutrient concentrations vary with environmental conditions such as water availability and harvesting frequency. These nutritional properties make amaranthus a valuable component of traditional diets and a promising crop for addressing micronutrient deficiencies and enhancing food and nutrition security in resource-limited settings.

Bambara groundnut

The Bambara groundnut (*V. subterranea/tindluwa*), also known as the Jugo nut, is a drought-tolerant legume crop classified as a neglected and underutilized species in South Africa. It is primarily cultivated by smallholder farmers in the northern and eastern provinces, including Limpopo, Mpumalanga, and KwaZulu-Natal, often as inter- or sole crops (Agricultural Research Council (ARC), 2023). The crop is well adapted to low-input agricultural systems and thrives in semiarid environments, making it suitable for regions affected by climate variability. Similarly, it is primarily cultivated by smallholder farmers across West Africa, notably in Nigeria, Ghana, and Burkina Faso, and in parts of East Africa (including Tanzania and Kenya), as well as Southern Africa, where it is grown in Malawi, Zambia, and Zimbabwe (Hillocks et al., 2012; Azam-Ali et al., 2001).

Although traditionally grown for household consumption, the Bambara groundnut is increasingly recognized for its high protein content and resilience, offering potential for food security and climate-smart agriculture in sub-Saharan Africa (Mabhaudhi et al., 2019). The Bambara groundnut is considered a complete food that contains balanced proportions of proteins, carbohydrates, lipids, and essential amino acids. It is particularly rich in iron and zinc, making it a valuable source of plant-based protein for low-income populations with limited access to animal protein. In addition, it contributes to soil fertility through nitrogen fixation, while its resilience under harsh conditions makes it a strategic crop for improving food and nutrition security (Agricultural Research Council (ARC), 2023).

Bidens pilosa

B. pilosa (*mushiji*) is an annual plant native to tropical America that is commonly found growing wild and foraged in various regions of South Africa, particularly in the Limpopo, KwaZulu-Natal, Northwest, and Mpumalanga provinces (Agricultural Research Council (ARC), 2023). In Africa, it thrives widely in tropical and subtropical zones, especially in Cameroon, Ethiopia, and Zimbabwe (Shackleton et al.,

2017). Similarly, it is abundant in Brazil, Colombia, and Peru Souza-Filho and Takaki (2018). It is also common in China, India, and Thailand, where it colonizes agricultural fields and disturbed soils (Wu et al., 2016). In the Mediterranean region, it is well established in Spain and Italy (Greuter, 2006), while in the Oceania region (e.g., Australia and Fiji), it is considered an invasive weed due to its dense growth (Knope et al., 2020). This global spread demonstrates its ecological adaptability and resilience in diverse environments (Miller, 2022).

As noted by Mavengahama et al. (2013), it contains appreciable levels of iron, zinc, manganese, and copper, with iron concentrations particularly notable in indigenous leafy vegetables. It also provides vitamins A and C, which support immune function and antioxidant defenses. These micronutrients are essential in combating hidden hunger and improving nutritional status, especially in vulnerable populations. However, nutrient concentrations may vary depending on environmental conditions, plant maturity, and postharvest handling, underscoring the need for further research to optimize its dietary use.

Cassava

Cassava (*M. esculenta/ntsumbual*) is a tropical root vegetable from the *Euphorbiaceous* family that is primarily cultivated in the KwaZulu-Natal, Mpumalanga, and Limpopo provinces of South Africa. These provinces offer favorable climatic and soil conditions (as identified by the Agricultural Research Council (ARC), 2023) for mechanized plantation development. Cassava is increasingly recognized as a strategic crop for food security in subtropical and drought-prone regions. Cassava is not only cultivated in South Africa but is also abundant in other tropical regions. For example, it is native to South America (particularly Brazil and Paraguay), where it remains a staple food (Burns et al., 2010). Moreover, Nigeria, the Democratic Republic of Congo, and Ghana are leading producers, establishing cassava as a cornerstone of food security (FAO, 2023). In Asia, countries such as Thailand, Indonesia, and Vietnam dominate production for both food and industrial uses (Howeler, 2022). The Oceania region also grows cassava extensively in Fiji and Papua New Guinea for subsistence and commercial purposes (Lebot, 2019). Its adaptability to poor soils and drought conditions underscores its role as a strategic crop in climate-stressed regions globally (Sakadzo et al., 2025).

Cassava leaves are nutritionally rich, particularly in vitamins and minerals. They contain high levels of thiamine, riboflavin, niacin, and vitamin A, often surpassing those found in legumes, leafy vegetables, and some animal-based foods. For example, thiamine levels exceed those in eggs, cheese, and milk, while riboflavin and niacin contents are higher than in cereals, legumes, and dairy products. Vitamin A levels are comparable to those in carrots and superior to most leafy vegetables. Dietary fiber content ranges from 1 to 10 g per 100 g fresh weight, establishing cassava leaves as a superior source compared to legumes and leafy greens (Montagnac et al., 2009; Chavez et al., 2000). Cassava leaves also offer substantial protein, with concentrations ranging from 14% to 40% dry matter depending on the variety. This is comparable to fresh eggs (10.9 g/100 g), while their amino acid profile is well-balanced and often exceeds the Food and Agriculture Organization (FAO) recommended intake (e.g., surpassing soybeans in this instance) (Rawel and Kroll, 2003; Gil and Buitrago, 2002). Carbohydrate levels are similar to those in green beans, carrots, and soybeans, and higher than in leafy vegetables like lettuce. Mineral content, including iron, zinc, manganese, magnesium, and calcium, is notably high. Iron and zinc levels are comparable to those in sweet potato and peanut leaves, respectively, while calcium and magnesium concentrations rival those in peanuts and broccoli (Montagnac et al., 2009; Chavez et al., 2000).

Cowpea

Cowpea (*V. unguiculata* L. Walp/*munawa*) is a nutrient-rich, affordable legume widely cultivated in tropical regions, including South Africa. It thrives in dryland farming systems, with major production areas in Limpopo, Mpumalanga, Northwest, and KwaZulu-Natal (Agricultural Research Council (ARC), 2023). Likewise, cowpea is a major crop cultivated across sub-Saharan Africa, with Nigeria, Niger, and Burkina Faso among the largest producers (Boukar et al., 2019). In Asia, it is grown extensively in India and Myanmar for food and fodder (Singh et al., 2021), while in Brazil, it serves as an important source of protein for rural communities (Freire Filho et al., 2012). In North America, production occurs mainly in the southern United States (particularly Texas and California) for both fresh and dry markets (Ehlers and Hall, 1997). Its adaptability to poor soils and drought conditions makes cowpea a strategic crop for food security in semiarid regions globally (Boukar et al., 2019).

Cowpea seeds are rich in protein (23–32%), carbohydrates (50–60%), and low in fat (1%), offering two to four times more protein than cereals and root crops due to their high lysine content (Gonçalves et al., 2016; Liyanage et al., 2014). They also contain digestible and nondigestible carbohydrates, essential amino acids, and polyphenols with antioxidant properties (Affrifah et al., 2022). The seeds provide potassium, calcium, magnesium, phosphorus, and modest amounts of iron, sodium, zinc, and selenium. Cowpea is a good source of thiamine, niacin, riboflavin, and vitamins A, C, B6, and pantothenic acid, though folate levels are low. Dietary fiber content varies from 2.5% to 32% (dry matter), with dehulled cowpea containing 14.8% insoluble and 3.3% soluble fiber. The total fiber includes cellulose (6%), hemicellulose (3.9%), lignin (2%), and pectin (1.8%), amounting to 13.60–15.99 g per 100 g of seeds (Mbhatsani et al., 2023; Eshwarage et al., 2017; Burridge et al., 2016; Asare et al., 2013).

Momordica balsamina

Momordica balsamina (*tshibavhe*) is commonly known as balsam apple and is a climbing annual plant from the Cucurbitaceae family that is indigenous to tropical and subtropical Africa. In South Africa, it grows wild in the Eastern Cape, Northern

Cape, Limpopo, and KwaZulu-Natal provinces (Thiaw et al., 2023). It is also abundant in regions such as West Africa (Senegal, Mali, and Nigeria) and East Africa (Kenya and Tanzania), where it grows in savannas and fallow lands (Thiaw et al., 2023). It has also naturalized in parts of Asia (including India and Pakistan), often in dry, open habitats (Schmelzer and Gurib-Fakim, 2008). In the Mediterranean, it occurs sporadically in southern Spain and Italy as an introduced species (Greuter, 2006). In South America, it is reported in Brazil and Paraguay, but primarily in disturbed areas (Burkill, 1985). Its resilience to drought and poor soils makes it an important traditional food and medicinal plant in semiarid regions globally (Thiaw et al., 2023).

Nutritionally, the leaves of *M. balsamina* are rich in micronutrients, including iron, calcium, magnesium, zinc, and phosphorus, as well as proteins and essential amino acids. These attributes make it a valuable indigenous leafy vegetable for addressing micronutrient deficiencies and enhancing dietary diversity in rural communities (Thiaw et al., 2023).

Sorghum

Sorghum (*S. bicolor* [L.] *Moench/amazimba*) is a drought-tolerant cereal crop indigenous to Africa and widely cultivated in South Africa, particularly in the Free State, Mpumalanga, and Limpopo provinces. It thrives in semiarid regions and is well adapted to low-input agricultural systems, making it a strategic crop for smallholder farmers facing climate variability and water scarcity (ARC, 2021a). Agronomically, sorghum is resilient to heat and drought, with a deep root system and waxy leaf surfaces that reduce transpiration. Similarly, sorghum is widely cultivated in semiarid regions worldwide. It is a staple in West and East Africa, with Nigeria, Sudan, and Ethiopia among the largest producers (FAO, 2023). In Asia, it is grown extensively in India and China, where it serves as food, fodder, and for brewing (Reddy et al., 2006). In the United States, Kansas and Texas dominate global sorghum exports for feed and biofuel (USDA, 2022), while in South America, Brazil and Argentina cultivate sorghum for livestock feed and ethanol production (de Morais Cardoso et al., 2015). Sorghum's resilience to heat and drought, supported by its deep root system and waxy leaf surfaces that reduce transpiration, makes it a strategic crop for food security under climate variability (FAO, 2023; Reddy et al., 2006).

Nutritionally, sorghum is a rich source of carbohydrates, proteins, dietary fiber, and essential minerals such as iron, calcium, and potassium. It is naturally gluten-free, making it suitable for individuals with celiac disease or gluten sensitivity. Sorghum also contains antioxidants and polyphenols, which contribute to its health benefits, including improved digestion, reduced cholesterol, and better blood sugar regulation (ARC, 2021b). From a food security perspective, sorghum plays a vital role in household nutrition and livelihoods, especially in drought-prone areas (Fig. 15.2) (ARC, 2021c).

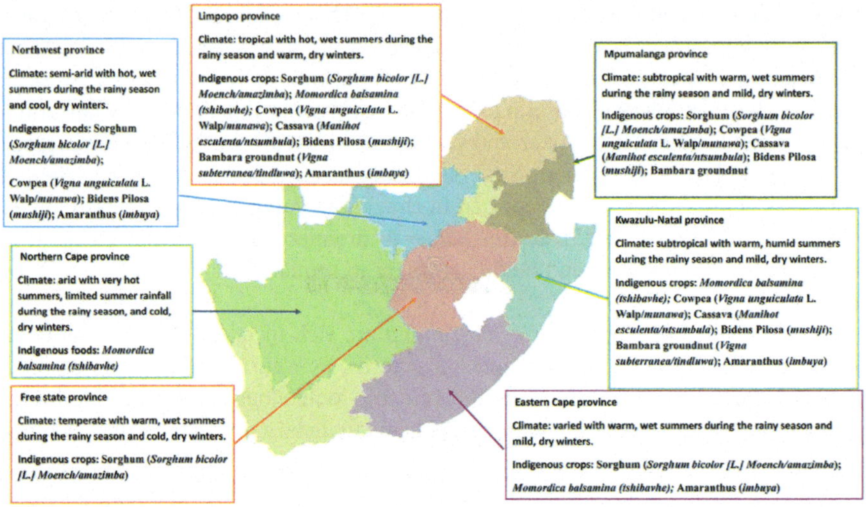

FIG. 15.2

Distribution of selected indigenous vegetables across various localities in South Africa, with a description of the prevailing climatic conditions in each region.

Antioxidant activities of selected indigenous food crops

The antioxidant activities of selected indigenous food crops are explored below, highlighting their potential to combat oxidative stress and contribute to the prevention of chronic diseases. These crops are rich in bioactive compounds such as phenolics, flavonoids, and carotenoids, which play a crucial role in neutralizing free radicals, supporting immune function, and promoting overall health. Their inclusion in local diets not only enhances nutritional quality but also offers therapeutic benefits rooted in traditional knowledge and supported by emerging scientific evidence.

Amaranthus cruentus L.

Amaranth leaves (*Amaranthus cruentus* L./*imbuya*) possess a sweet flavor and are known to be rich in carotenoids. These leaves display a relatively higher carotenoid content than white cabbage (0.42–0.009 mg 100 g^{-1} fresh weight), red cabbage (0.016–1.09 mg 100 g^{-1} fresh weight), and Brussels sprouts (1.160–0.051 mg 100 g^{-1} fresh weight). Amaranth leaves are an excellent source of rutin (flavonoid) content, which offers protection against stroke, hemorrhoids, atherosclerosis, and osteoarthritis, while also exhibiting anticarcinogenic and antiinflammatory properties (Mampholo et al., 2016; Van Eylen et al., 2007). Furthermore, Ogie-Odia et al. (2022) reported that the dried and powdered forms contained alkaloids, flavonoids, saponins, phenols, steroids, and glycosides. Here, *Amaranthus spinosus* var. green exhibited the highest concentrations of 161.1, 214.11, 38.11, 21.03, and 89.16 mg/g for alkaloids, flavonoids, saponins, steroids, and glycosides, respectively. In contrast, *A. dubius* possessed the highest concentration of phenols at 18.03 GAE (GAE refers to gallic acid equivalents, which are used to express total phenolic content). The antioxidant potential of these extracts is advantageous for preventing cardiovascular and neurodegenerative diseases, as well as bone diseases and cancer (Ogie-Odia et al., 2022). Moreover, a South African study by Makuse and Mbhenyane, 2011 found that *Amaranth* contained omega-3 fatty acids (particularly α-linolenic acid), and it is well known that such fatty acids are linked to increased high-density lipoprotein cholesterol levels and a lowered risk of coronary heart disease. Sattara et al. (2024) also highlighted the nutritional and therapeutic properties of *Amaranthus* species from regions, including Syria and Pakistan, noting their richness in proteins, polyphenols, tocopherols, and bioactive compounds with antioxidant, antihypertensive, and antidiabetic effects. Similarly, in India, Gupta et al. (2025) reported high levels of phenolic compounds, essential fatty acids, and minerals in *A. cruentus*, reinforcing its potential as a functional food. These findings align with studies by Makuse and Mbhenyane, 2011 and Ogie-Odia et al. (2022), which demonstrated the presence of omega-3 fatty acids, flavonoids such as rutin, and significant antioxidant activity, supporting its role in reducing cardiovascular and metabolic disease risks.

Bambara groundnut

The Bambara groundnut (*V. subterranea/tindluwa*) is a legume belonging to the *Fabaceae* family and the *Faboideae* subfamily. It is generally regarded as the second most significant indigenous legume after cowpea. The plants serve as a source of polyphenolic compounds, primarily flavonoids such as medioresinol, catechin, catechin dimer, anthocyanin, epicatechin, and phenolic acids, including quinic acid, chlorogenic acid, caffeic acid, and ellagic acid (Okafor et al., 2021). These compounds are recognized for their health benefits and can aid in strengthening the immune system against infectious diseases. Bambara groundnut also exhibits antioxidant properties, and plant extracts have effectively inhibited lipid oxidation in food products by limiting the complementary reaction of transition metals and oxygen with food components. Its bioactive phytochemicals have been linked to the prevention of various chronic conditions, such as diabetes, stroke, atherosclerosis, heart disease, cancer, Alzheimer's disease, and other cardiovascular disorders (Harris et al., 2018; Nyau et al., 2015).

Bidens pilosa

B. pilosa (*mushiji*) is native to South America but is widely distributed across tropical and subtropical regions. It is valued for both its medicinal and nutritional properties, containing ascorbic acid, vitamin A, protein, iron, and a range of phytochemicals with antioxidant activity (Lee et al., 2013a; Xu et al., 2017). It is rich in flavonoids, polyacetylenes, and polyphenols, which contribute to its health-promoting effects, including antioxidant, antiinflammatory, antihypertensive, and antidiabetic properties (Lee et al., 2013b; Falowo et al., 2017a). At a molecular level, active compounds such as ethyl acetate and tetraene inhibit endothelial cell proliferation and migration, showing potential in cancer treatment. Moreover, cytopiloyne (a polyacetylenic glucoside found in the aerial parts) has displayed hypoglycemic, antipyretic, and anticancer effects (Chung et al., 2016). *B. pilosa* also contains hydrophilic antioxidants (phenolics, flavonoids, anthocyanins, stilbene, lignan) and liposoluble antioxidants (β-carotene, α-carotene, lycopene, lutein, zeaxanthin), together with essential minerals and vitamins (Xu et al., 2017; Lee et al., 2013a). Its traditional uses include the treatment of hepatitis, laryngitis, headaches, digestive disorders, and liver protection through the inhibition of NF-kappa B activation (a key transcriptional regulator mediating increased immune activation) (Chung et al., 2016). Furthermore, in South Africa, the leaves can be consumed or applied to wounds, used as a deworming agent for children, and used to promote satiety during famine (Mbhatsani et al., 2011). These practices highlight its dual role in terms of nutrition and health.

Cleome gynandra

C. gynandra (*lerotho*) is a member of the *Cleomaceae* family, widely consumed due to its health-promoting properties. In addition to its high nutrient content, *Cleome* contains a diverse array of antioxidant phytochemicals that enhance its overall health benefits. *Cleome* leaves are highly nutritious, rich in minerals and vitamins, and possess antiinflammatory and antioxidant effects. Due to its active phytochemicals, the leaves are known to help lower blood pressure and are suggested to function as antidiabetic and anticancer agents (Mishra et al., 2011; Makuse and Mbhenyane, 2011). Its leaves contain important mineral constituents and dietary vitamins, specifically provitamin A carotenoids, E, and C (ascorbic acid). The prominent macronutrients in *Cleome* are potassium, calcium, magnesium, phosphorus, and sulfur, while the salient microminerals include iron, zinc, and manganese (Dehayem-Kamadjeu and Okonda, 2019; Omondi et al., 2017). The main carotenoids identified are α-carotene, β-carotene, lutein, violaxanthin, and β-cryptoxanthin. Furthermore, the principal forms of vitamin E include α-tocopherol, β-tocopherol, and γ-tocopherol (Moyo et al., 2018).

Jew's mallow

Jew's mallow (*C. olitorius* L./*guxe*) is a tropical, fast-growing leafy green vegetable from the *Tiliaceae* family. It is commonly referred to by this name but also as bush okra, jute, or jute mallow and is widely recognized as a highly nutritious vegetable. Jew's mallow contains abundant bioactive compounds such as α-linolenic acid, β-sitosterol, chlorogenic acid, 3,5-dicaffeoylquinic acid, flavone, quercetin, tannins, diacylglycerol, and uronic acid, which can provide therapeutic effects (Kaboré et al., 2021; Oyewole et al., 2015; Ijioma et al., 2014). It is well documented for its numerous health benefits, including antitumor, antioxidant, antibacterial, antidiabetic, antihypertensive, and cholesterol-lowering effects, as well as anticancer, antiulcer, and antimicrobial activities, which contribute to its proposed ability to treat or prevent diseases. In Egypt, Ali et al. (2024) examined the chemical composition, total phenolic compounds, chlorophyll *a*, chlorophyll *b*, β-carotene, and antioxidant activity of dried Jew's mallow extract. The authors observed high antioxidant potential in the extract, which reached 67%, with 276.37 mg/100 g of total phenolic compounds. They identified the phenolic constituents as ellagic acid (4,905.26 μg/kg), naringin (4,296.94 μg/kg), luteolin-7-O-glucoside (4,314.48 μg/kg), apigenin-6-rhamnose-8-glucoside (3,078.87 μg/kg), chlorogenic acid (3,757.08 μg/kg), and ferulic acid (3,628.29 μg/kg) (Ali et al., 2024).

Mung beans

Mung beans (*V. radiata*//*ditlhodi*) are a source of flavonoids, phenolics, and antioxidants. They are also important pulses worldwide, with historical roots in Asia, where they have been used as a staple in traditional medicine for the treatment of various chronic diseases. In China, mung beans have been recognized as a medicinal herb since at least the 1050s, known for alleviating fever and aiding detoxification (Zhu et al., 2012). The nutrients in mung beans—rich in protein, dietary fiber, minerals, and vitamins, together with bioactive compounds such as polyphenols, polysaccharides, and peptides—classify them as functional foods that promote health. Mung beans have shown potential in managing hyperglycemia, hyperlipidemia, and hypertension while protecting against cancer and melanogenesis. Furthermore, the active compounds in mung beans offer hepatoprotective and immunomodulatory benefits. Active compounds in mung beans primarily include polyphenols (vicenin-2, vitexin, and isovitexin), and the preferred peptides consist of hydrophobic amino acids, exhibiting relatively low molecular weights that enhance bioactivity (Hou et al., 2019).

Momordica balsamina

M. balsamina L. (*tshibavhe*) is commonly known as balsam apple and Southern balsam pear and is a climbing herbaceous plant from the Cucurbitaceae family. It is indigenous to sub-Saharan Africa and widely used both as a food and medicinal plant in rural communities across southern Africa, including Limpopo, KwaZulu-Natal, and the Eastern Cape (Ramalhete et al., 2022). The plant is traditionally consumed in various ways. For example, the young fruits and leaves are cooked and eaten, while the bright red pulp of mature fruits is consumed (e.g., in Namibia). However, mature fruits may cause gastrointestinal distress due to the presence of alkaloids and saponins. The leaves are often used in soups and stews, while in Botswana, it is served with porridge (Ramalhete et al., 2022). Medicinally, the plant exhibits a wide range of bioactivities, including antidiabetic, antimalarial, antibacterial, antiinflammatory, antiviral, analgesic, and antioxidant effects. These properties are attributed to its high content of cucurbitane-type triterpenoids (particularly cucurbitacins), which have shown potential in reversing multidrug resistance in cancer cells and bacteria

(Ramalhete et al., 2022; Qadir et al., 2019). Advanced extraction techniques, such as enzyme-assisted extraction, have enhanced the recovery of phenolic compounds from the fruit, including ferulic acid, quercetin, benzoic acid, and gallic acid, which contribute to its antioxidant potential (Qadir et al., 2019). Importantly, Mbhenyane et al. (2001) evaluated the glycemic index (GI) of indigenous South African foods and found that stiff mealie meal porridge consumed with *M. balsamina* revealed a GI of 105, which was lower than white bread. When combined with dried green bean leaf stew, the GI dropped further to 87, indicating its potential role in managing blood glucose levels (Mbhenyane et al., 2001). Singh and Devi (2018) reported that methanolic fruit extracts of *M. balsamina* (from Haryana, India) contained exceptionally high levels of polyphenols (201 mg GAE/g) and flavonoids (103 mg quercetin equivalents/g), demonstrating strong antioxidant activity and antimicrobial effects against *Pseudomonas aeruginosa*, *Staphylococcus aureus*, and *Candida albicans*. A Nigerian study by Baso et al. (2025) also showed that ethanol leaf extracts contained alkaloids, flavonoids, glycosides, steroids, saponins, and tannins, and exhibited significant antibacterial activity, particularly against *Salmonella typhi* and *Escherichia coli*, with inhibition zones up to ~29 mm and minimum inhibitory concentrations (MICs) of 12.5–25 mg/mL. These findings align closely with studies reporting high cucurbitacin and phenolic content, antioxidant and antidiabetic effects, and antimicrobial properties, thereby reinforcing *M. balsamina*'s consistent bioactive profile and functional potential across India, Nigeria, and southern Africa.

Sorghum

Sorghum (*S. bicolor/amazimba*) is a significant food source that is widely consumed as a staple in many parts of the world, including South Africa. Its gluten-free nature and rich nutritional profile make sorghum an appealing option for health-conscious individuals. It is an exceptional source of phenolic acids and fiber, which help to regulate blood sugar levels and aid digestion. Various sorghum varieties are also rich in phytochemicals, particularly phenolic compounds with robust antioxidant properties and numerous health benefits. Some sorghum grains are high in bran, exhibiting beneficial antioxidant activity, like all fractions of cereal crops. The phenolic profile of sorghum offers health-promoting and medicinal properties, including the management of oxidative stress and antitumor efficacy. Sorghum grains contain a range of phenolic compounds with a total phenolic content ranging from 445 to 2850 μg, including carotenoid xanthophylls (lutein and zeaxanthin) (Girard and Awika, 2018; González-Montilla et al., 2012). Sorghum exhibits significant antioxidant properties due to its inherent polyphenolic compounds, including procyanidins (condensed tannins), 3-deoxyanthocyanidins, and phenolic acids. Not only does sorghum exhibit antioxidant activity and help manage the risk of oxidative stress-related diseases, but it also acts as a cancer-preventive agent due to its antiproliferative effects, possesses antimicrobial efficacy, and enhances glucose metabolism, particularly in the context of diabetes (Mohammed et al., 2022). Punia et al. (2021) analyzed five pigmented sorghum genotypes in India using high-performance liquid chromatography and antioxidant assays, finding red and brown pericarp varieties particularly rich in proanthocyanidins and phenolic acids such as caffeic acid, taxifolin, luteolinidin, and apigeninidin. These compounds correlated strongly ($r > 0.95$) with high antioxidant activity as measured by the 2,2-diphenyl-1-picrylhydrazyl radical scavenging assay (DPPH), the 2,2′-azino-bis(3-ethylbenzothiazoline-6-sulfonic acid radical cation decolorization assay, and the Ferric Reducing Antioxidant Power assay. Moreover, Ajiboye et al. (2014) assessed traditional beverages made from sorghum grains ("Obiolor" and "Pito") in Nigeria, and they demonstrated its potent antioxidant effects by scavenging 81–87% of DPPH radicals, 59–65% of superoxide ions, and 76–85% of hydrogen peroxide levels, as well as showing significant inhibition of lipid peroxidation and protein carbonylation.

Global perspective

While a significant portion of literature concentrates on South African contexts, similar evidence exists in other low- and middle-income countries, emphasizing the universality of indigenous foods as contributors to health, resilience, and food sovereignty. Indigenous foods are increasingly being rediscovered across different continents for their dense nutrient profiles, rich phytochemical content, and health-promoting (nutraceutical) properties. In support, a study by Xiang et al. (2019), analyzing finger millet (*Eleusine coracana* L.) harvested in Malawi, showed that the total phenolics, flavonoids, and condensed tannins in the free fractions ranged from 114.43 to 179.19 mg ferulic acid equivalent (FAE)/100 g, 90.24–202.94 mg catechin equivalent (CE)/100 g, and 31.76–83.59 mg CE/100 g, respectively. The total phenolic contents of the bound fractions ranged from 58.27 to 123.23 mg FAE/100 g. Moreover, Lee et al. (2011) in South Korea reported that mung bean (*V. radiata*) extract contained significantly high total phenol at 2.03 GAE g-1 and flavonoid contents at 1.49 GAE g-1. Hou et al. (2019) in China found that these beans contain vitexin and isovitexin, flavonoids with hepatoprotective and antiinflammatory properties. Furthermore, a South American study by Repo-Carrasco-Valencia (2020) analyzed the bioactive compounds and nutritional compositions of different varieties of quinoa (*Chenopodium quinoa*), *kañiwa* (*Chenopodium pallidicaule*), *kiwicha* (*Amaranthus caudatus*), and *tarwi* (*Lupinus mutabilis*), which showed protein contents ranging from 13% to 20%, with very high lysine content and 27–58 mg gallic acid per 100 g of phenolic compounds. The authors reported that these grains (as ancient crops) contain nutritional and bioactive components. This includes proteins, starch, dietary fiber, fat, polyphenols, and saponins, and they are associated with antioxidant, antibacterial, antiinflammatory, anticancer, hypoglycemic, hypolipidemic, and other biological properties. This broader view emphasizes the value of indigenous food systems, which extend beyond a single nation and constitute a viable strategy for enhancing diet quality, food security, and resilience in many regions, particularly those susceptible to nutritional vulnerabilities and climate change.

Potential health benefits of indigenous foods

Indigenous foods offer diverse health benefits in both traditional healing and modern dietary science. Their therapeutic roles are especially significant in preventing and managing infections and chronic diseases. Indigenous grains, vegetables, and legumes are rich in dietary fiber, resistant starch, and plant-based phenolics, which support blood glucose regulation, insulin sensitivity, and weight management. Antioxidants and phytochemicals (particularly flavonoids and tannins) exert anticancer and antitumor effects by inhibiting cell growth and reducing oxidative stress. Many indigenous herbs and plants also possess antimicrobial and antibacterial properties due to bioactive compounds such as alkaloids, saponins, and essential oils. Furocoumarins and antiinflammatory agents help mitigate chronic inflammation linked to cardiovascular diseases, rheumatoid arthritis, and metabolic syndrome.

While modern science increasingly validates these benefits, indigenous knowledge holders have long understood the healing potential of such foods. Their use of indigenous grains, legumes, vegetables, and herbs for nourishment and disease prevention is rooted in lived experience and oral tradition. Here, Makuse and Mbhenyane, 2011 documented the omega-3 content and health benefits of indigenous foods in Limpopo through community engagement. Moreover, Mbhenyane et al. (2013) found that over 90% of rural households in Limpopo consume wild edible herbs and vegetables seasonally, relying on them for nutrition and survival, while Nxusani et al. (2023) highlighted the erosion of indigenous food knowledge among youth and called for its preservation. Ndhlovu et al. (2023) emphasized the role of indigenous knowledge holders in managing childhood diseases using medicinal plants. In addition, Masuku and Bhengu (2023) showed how indigenous foods improve food security in KwaZulu-Natal, while Takaidza, 2023 stressed their underappreciated nutritional and medicinal value in combating noncommunicable diseases.

Indigenous knowledge holders worldwide have long understood the healing potential of traditional foods. For example, in Africa, Adebowale et al. (2005) and Akinola et al. (2020) emphasize the role of indigenous cropping systems and wild foods in sustaining nutrition and resilience. Similarly, in Asia, Aweya et al. (2025) highlight ancient grains such as millet and sorghum for their antioxidant, antidiabetic, and cardioprotective properties, while FAO and Bioversity International (2023) show how Arctic and Himalayan communities maintain biodiverse food systems that support health and cultural identity. In the Americas, Jernigan et al. (2023) and Elm-Hill et al. (2023) discuss Native food sovereignty movements as pathways to address chronic disease and restore cultural heritage. Finally, in Oceania, Lopes et al. (2024) report that programs promoting native foods among First Nations communities improve micronutrient intake and strengthen food security.

Antidiabetic and antiobesity effects

Indigenous food crops show promising potential in managing type 2 diabetes mellitus (T2DM) through their hypoglycemic properties. In a Taiwanese study involving 14 T2DM patients, consumption of *B. Pilosa* (*mushiji*) significantly reduced fasting blood glucose from 11.2 to 6.9 mmol/L and glycated hemoglobin from 9.1 ± 1.7% to 7.2 ± 0.7%, indicating its value in dietary diabetes management (Xin et al., 2021; Yang et al., 2020). Similarly, an Indian study observed that aqueous extracts of *Amaranthus* spp. demonstrated antiglycation, anti-α-amylase, and anti-α-glucosidase activities, along with increased serum total proteins, suggesting protective effects against diabetic nephropathy (Balasubramanian and Karthikeyan, 2015).

Opuntia basilaris (prickly pear/*itolofia*) also exhibits antihyperglycemic properties by inhibiting α-amylase and α-glucosidase, as revealed by a study conducted in Mexico and Spain (Gómez-Maqueo et al., 2019). Moreover, in Brazil, *A. digitata* (baobab/*muvuyu*) fruit extract has shown efficacy in reducing capillary blood glucose levels in humans and improving carbohydrate metabolism in diabetic Wistar rats, nearly eliminating dysglycemia caused by streptozotocin (Silva et al., 2023).

In addition, peptides derived from *V. unguiculata* (cowpea/*munawa*) demonstrated insulin activation through phosphorylation in skeletal muscle cell cultures, supporting glucose homeostasis (Jayathilake et al., 2018). Moreover, both cowpea and *V. subterranea* (Bambara groundnut/*tindluwa*) contain high levels of soluble and insoluble fiber, which help lower postprandial blood glucose and cholesterol levels (Oladosu & Njoku, 2021; Liyanage et al., 2018).

Sorghum is another valuable crop for diabetes and obesity management due to its resistant starch and tannin content. Resistant starch delays carbohydrate digestion through protein-starch cross-linking and disulfide bonding, promoting satiety and stabilizing blood glucose. Tannins further lower digestibility and contribute to a relatively low glycemic response. Sorghum's phenolics inhibit digestive enzymes, with some varieties outperforming acarbose in attenuating blood glucose levels—as shown by studies completed in Australia (Stefoska-Needham, 2024), Australia and India (Khoddami et al., 2021) and the United States (Leboe et al., 2019).

Marula fruit and bark (*Sclerocarya birrea/nkanyi*) have long been used in traditional medicine for diabetes. Scientific studies confirm that marula bark extract significantly lowers blood glucose levels in streptozotocin-induced diabetic rats (Dimo et al., 2007). Chronic administration over 5 weeks yielded glycemic reductions comparable to metformin (Gondwe et al., 2008). Marula fruit also contains phenolic compounds like epicatechin-3-galloyl ester, which can enhance glucose tolerance in diabetic mice and humans (Mashau et al., 2022).

M. balsamina (*tshibavhe*) is traditionally used for treating diabetes, malaria, inflammation, and digestive disorders. It contains polyphenols, flavonoids, terpenes, carotenoids, alkaloids, saponins, and glycosides, which contribute to its antioxidant and antidiabetic properties (Thiaw et al., 2023). Research by Mbhenyane et al. (2001) showed that *M. balsamina* has a low GI, supporting its use in blood sugar regulation and nutritionally sensitive diets. In support, studies in India (Singh and Devi, 2018) and Nigeria (Baso et al., 2025) confirmed the antidiabetic health benefits of *M. balsamina*.

Anticancer and antitumor-promoting activities

Recent epidemiological evidence links a high intake of plant-based foods, especially leafy greens and fruits, with reduced cancer risk, particularly gastrointestinal cancers. *C. olitorius* (Jew's mallow/*guxe*), which is rich in phytochemicals like genistein (34.96 μg/kg), has demonstrated anticancer potential across multiple cancer types in studies conducted in the Ivory Coast and Nigeria (Kaboré et al., 2021; Oyewole et al., 2015). Its edible stems, affordability, and accessibility make it a promising candidate for pharmaceutical development (Biswas et al., 2023).

Sour fig (*Carpobrotus edulis/Igcukuma*) is traditionally used to treat tuberculosis, dysentery, throat infections, burns, and ulcers, and it possesses confirmed pharmacological properties, including antifungal, antioxidant, antiplasmodial, anticancer, and antibacterial activities (Otang-Mbeng et al., 2019; Omoruyi et al., 2012, 2014). For example, oleanolic acid from *C. edulis* shows potent antibacterial effects against pathogens such as *Enterococcus faecalis* (MIC 6.25 mg/L), methicillin-resistant *S. aureus* (MIC 25–50 mg/L), and *Mycobacterium tuberculosis* H37Rv (MIC 100 mg/L) (Martins et al., 2011).

Olufunso et al. (2025) also highlight the anticancer potential of indigenous Nigerian plants, including *Spondias mombin*, *Xanthosoma sagittifolium*, *Elaeis guineensis*, *Irvingia gabonensis*, *Allium cepa*, *Blighia sapida*, *Dioscorea dumetorum*, *Psidium guajava*, and *Talinum triangulare*. These plants act through apoptosis induction, cell cycle arrest, angiogenesis inhibition, and inflammation modulation. Their efficacy is attributed to flavonoids, tannins, terpenoids, alkaloids, and organosulfur compounds. For instance, *S. mombin* suppresses tumor formation via flavonoids, *X. sagittifolium* shows cytotoxicity against leukemia cells, and *E. guineensis* counters oxidative stress through antioxidant activity. In South Africa, such indigenous foods are found in specific regions: *X. sagittifolium* (cocoyam) is grown by small-scale farmers (particularly in KwaZulu-Natal), where it is locally referred to as "amadumbe," while *D. dumetorum* (bitter yam/*nkansa*) likely grows wild and/or in subsistence contexts in regions like Limpopo or Mpumalanga, though it is not widely cultivated commercially.

B. pilosa (*mushiji*) is widely used in traditional medicine and has gained formal recognition in Brazil's national public health system (Gilbert et al., 2022). Its roots and aerial parts are used to treat malaria, cancer, pain, fever, diabetes, and infections. Scientific studies confirm its broad bioactivity, including antihyperglycemic, antihypertensive, antiinflammatory, antibacterial, antimalarial, antioxidant, and anticancer effects (Falowo et al., 2017b; Sanoussi et al., 2015; Mavengahama et al., 2013). In Brazil and China, *Bidens* is consumed as teas and extracts for ailments like diarrhea, dysentery, colic, and pharyngitis (Silva et al., 2011; Lee et al., 2013b), while in South Africa, its use is widespread among Zulu communities for gastrointestinal disorders and wound care (Arthur et al., 2012). Adetunji et al. (2025a) in Nigeria and Njume et al. (2016) in South Africa further confirm its popularity in rural areas, where it is consumed both as a vegetable and medicinal plant, reflecting its integration into indigenous knowledge systems and community health practices.

Antimicrobial and antibacterial properties

Marula (*nkanyi*) seed oil has been shown to possess antibacterial activity. For example, Elisha et al. (2017) reported that marula oil demonstrated antibacterial effects against several bacterial strains, including *S. aureus*, *E. coli*, *Bacillus subtilis*, *P. aeruginosa*, and *Salmonella choleraesuis*. This information supports the potential use of marula seed oil as a natural antimicrobial, which can be applied in food preservation, pharmaceutical formulations, and cosmetic products. It has also been reported that marula bark is used to treat fever, gastrointestinal conditions, and ulcers, as well as to cure sore eyes (Mashau et al., 2022). In addition, the root is used raw in alcoholic medicine to treat an internal disease known as *kati*, while the bark is used to treat digestive disorders (Mashau et al., 2022).

Tamarind (*T. indica/musika*) ethanolic extracts exhibited antibacterial effects against *S. mutans* and *E. coli*, with inhibition zones ranging from 11.41 mm to 7.04 mm and from 6.88 mm to 10.40 mm, respectively (Elisha et al., 2017). Indigenous fruits such as baobab (*A. digitata/muvuyu*), brown ivory (*Berchemia discolor/nyiri*), and white ivory (*Berchemia albitrunca/muthobi*) were also identified as possessing the highest antimicrobial activity against *S. typhi*, *Streptococcus pyogenes*, and *Bacillus cereus*. Moreover, brown ivory was reported to display antimicrobial properties, including activity against drug-resistant *M. tuberculosis*, while white ivory showed positive results against *B. subtilis* and *Klebsiella pneumoniae* (Shai et al., 2013).

Antinutritional factors and toxicological concerns

Antinutritional factors are naturally occurring agents present in many indigenous foods that hinder or interfere with the absorption, digestion, or utilization of nutrients. Although these compounds play crucial protective roles in plants, such as pest deterrence and pathogen inhibition, when consumed in large amounts, there are concerns regarding their potential effects.

Understanding antinutritional factors, their nature, and their impact on health, along with strategies to mitigate them, is essential to optimize the health benefits of indigenous foods when they are incorporated into modern dietary patterns.

Phytates (or phytic acid) are among the most common antinutritional compounds found in indigenous legumes such as cowpea (*V. unguiculata/munawa*), Bambara groundnut (*V. subterranea/tindluwa*), and mung beans (*V. radiata/dithlodi*), as well as in cereals like pearl millet (*P. glaucum/unyawothi*) and sorghum (*S. bicolor/amazimba*). Phytates serve as potent chelating agents by binding essential minerals such as iron, calcium, zinc, and magnesium, and forming insoluble complexes that reduce absorption in the gastrointestinal tract. The interaction between phytates and minerals can adversely affect mineral bioavailability, potentially leading to micronutrient deficiencies, particularly among vulnerable groups such as children, pregnant women, and individuals with limited dietary variety. Health concerns linked to this include iron deficiency anemia and impaired bone mineralization. However, traditional food processing and cooking methods, including soaking, fermentation, germination, and boiling, can reduce phytate content. In addition, combining high-phytate foods with vitamin C-rich items can enhance iron absorption in such meals (Reddy et al., 1989; Goyal et al., 2023; Botanical Online Editorial Team, 2025).

Oxalates are a specific type of antinutritional factor present in various leafy indigenous vegetables, including *Amaranth* (*Amaranthus* spp./imbuya), *B. pilosa* (*mushiji*), and *C. gynandra* (*lerotho*). These compounds bind dietary calcium to form insoluble compounds that contain calcium oxalate crystals and have poor absorption in the gut. Excessive dietary oxalate intake may lead to calcium deficiency and an increased risk of developing kidney stones in individuals predisposed to nephrolithiasis. A simple way to reduce oxalates in vegetables is to boil them and then discard the cooking water. Moreover, consuming high-oxalate foods in conjunction with calcium-rich foods may reduce oxalate absorption in the gut (Lee et al., 2023; Mwanri et al., 2018).

Tannins are polyphenolic compounds prevalent in sorghum, pearl millet, legumes, and some leafy greens. These compounds can bind dietary proteins and digestive enzymes, impairing protein digestion and mineral absorption. This can sometimes affect iron absorption and the bioavailability of proteins; as a result, some researchers have suggested that bioactivity, growth, and immune response in children may be compromised. Currently, several indigenous and modern methods of preparation have been proposed to reduce tannin levels and increase nutrient availability, including decortication, dehulling, sun drying, and fermentation (Gwanzura et al., 2012; Mahmood et al., 2014).

Saponins are a group of glycosides that can foam and disrupt the intestinal mucosa and are found in *Amaranth*, mung beans, cowpeas, and other legumes. At high doses, they may interfere with nutrient absorption and irritate the gastrointestinal tract. Although hemolytic effects have been observed in vitro (Francis et al., 2002), such effects rarely occur at dietary levels in vivo (Timilsena et al., 2023; Kareem et al., 2022). It has been established that traditional food preparation methods (such as soaking, boiling, and roasting) can lower saponin levels, thereby improving food safety and palatability (Timilsena et al., 2023; Kareem et al., 2022).

Lectins are carbohydrate-binding proteins found in legumes such as cowpeas and mung beans. They can attach to the epithelial cells lining the gut, potentially limiting or disrupting nutrient absorption and damaging the intestinal lining. Lectins are, therefore, likely to cause gastrointestinal discomfort if raw or undercooked beans are consumed. Although it is well established that proper cooking inactivates lectins, it is advisable and best practice not to eat raw or undercooked legumes (Katoch and Tripathi, 2021; Gautam et al., 2018).

Considering recommendations to enhance indigenous foods for their health benefits, it is essential to recognize that antinutritional factors are present in significant quantities in such foods. However, the presence of antinutritional factors in such foods largely depends on dietary context, processing methods, and individual nutritional status. Indigenous food preparation techniques—many of which have been employed for generations—can play a vital role in reducing their antinutritional effects. To maximize the health benefits of indigenous foods while reducing adverse effects, it is suitable to combine indigenous knowledge with modern nutritional science. This approach will also support food security, dietary diversity, and sustainable nutrition within these culturally and ecologically valuable food systems.

Table 15.2 provides a summary of the potential health benefits of selected crops, together with their chemical composition and associated modulatory mechanisms. The data are compiled from the sources discussed in this chapter.

Table 15.2 Selected indigenous crops indicating phytochemicals reported and the associated potential health benefits (collated from the sources discussed in this chapter and to be read alongside the narrative).

Food crop	Phenolic compounds	Modulation activities/ potential health benefits	Sources
Amarathus/ imbuya	• Alkaloids • Carotenoids • Flavonoids (rutin) • Glycosides • Phenols • Saponins • Steroids	• Antioxidant • Anticarcinogenic • Antiinflammatory • Cardiovascular and stroke • Antihypertensive • Antidiabetic effects (antiglycation, anti-α-amylase, and anti-α-glucosidase activities) • Neurodegenerative diseases • Bone diseases • Hemorrhoids • Atherosclerosis • Osteoarthritis	Makuse and Mbhenyane, 2011 Balasubramanian and Karthikeyan (2015) Yang et al. (2020) Xin et al. (2021) Ogie-Odia et al. (2022) Sattara et al. (2024) Gupta et al. (2025)
Bambara groundnut (*Vigna subterranea/ tindluwa*)	• Flavonoids (medioresinol, catechin, catechin dimer, anthocyanin, epicatechin) • Phenolic acids (including quinic acid, chlorogenic acid, caffeic acid, and ellagic acid)	• Antiimmune • Antioxidant • Diabetes • Stroke • Atherosclerosis • Heart disease • Cancer • Alzheimer's disease • Cardiovascular disorders	Nyau et al. (2015) Harris et al. (2018) Okafor et al. (2021)

Continued

Table 15.2 Selected indigenous crops indicating phytochemicals reported and the associated potential health benefits (collated from the sources discussed in this chapter and to be read alongside the narrative)—cont'd

Food crop	Phenolic compounds	Modulation activities/ potential health benefits	Sources
Bidens pilosa (*mushiji*)	• Polyacetylenes • Polyphenols • Hydrophilic antioxidants (phenolics, flavonoids, anthocyanins, stilbenes, lignans) • Liposoluble antioxidants (β-carotene, α-carotene, lycopene, lutein, zeaxanthin)	• Antioxidant • Antiinflammatory • Antihypertensive • Antidiabetic properties • Hypoglycemic, antipyretic, and anticancer effects • Hepatitis, laryngitis, headaches, digestive disorders, and liver protection • Applied to wounds • Used as a deworming agent for children	Chung et al. (2016) Mbhatsani et al. (2011) Lee et al. (2013a) Falowo et al. (2017a,b) Xu et al. (2017)
Cleome gynandra (*lerotho*)	• Provitamin A carotenoids (α-carotene, β-carotene, lutein, violaxanthin, and β-cryptoxanthin)	• Antiinflammatory • Antioxidant effects • Lower blood pressure • Antidiabetic • Anticancer agent	Omondi et al. (2017) Moyo et al. (2018) Dehayem-Kamadjeu and Okonda (2019)
Jew's mallow (*Corchorus olitorius* L./ *guxe*)	• Bioactive compounds, such as α-linolenic acid, β-sitosterol, chlorogenic acid, 3,5-dicaffeoylquinic acid, flavone, quercetin, tannins, diacylglycerol, and uronic acid • Phenolic constituents (ellagic acid, naringin, luteolin-7-O-glucoside, apigenin-6-rhamnose-8-glucoside, chlorogenic acid, and ferulic acid)	• Antioxidant • Antitumor • Antibacterial • Antidiabetic • Antihypertensive • Cholesterol-lowering effects • Anticancer • Antiulcer • Antimicrobial activities	Ijioma et al. (2014) Oyewole et al. (2015) Kaboré et al. (2021) Biswas et al. (2023) Ali et al. (2024)

Food crop	Phenolic compounds	Modulation activities/ potential health benefits	Sources
Mung beans (*Vigna radiata*// *ditlhodi*)	• Flavonoids • Phenolics • Bioactive compounds (polyphenols, polysaccharides), • Peptides • Polyphenols (vicenin-2, vitexin, and isovitexin) • Preferred peptides (hydrophobic amino acids)	• Antioxidants • Antiinflammatory • Hyperglycemia • Hyperlipidemia • Hypertension • Protecting against cancer and melanogenesis • Hepatoprotective and immunomodulatory benefits • Alleviating fever and aiding detoxification	Zhu et al. (2012) Hou et al. (2019)
Momordica balsamina L., (*tshibavhe*)	• Alkaloids • Saponins • Cucurbitane-type triterpenoids, particularly cucurbitacins • Phenolic compounds (ferulic acid, quercetin, benzoic acid, and gallic acid) • Flavonoids • Glycosides • Steroids • Tannins	• Antioxidant • Antidiabetic • Antimalarial • Antibacterial • Antiinflammatory • Antiviral • Analgesic effects • Multidrug resistance in cancer cells and bacteria • Antimicrobial effects	Mbhenyane et al. (2001) Singh and Devi (2018) Qadir et al. (2019) Ramalhete et al. (2022) Thiaw et al. (2023) Baso et al. (2025)
Sorghum (*Sorghum bicolor*/ *amazimba*)	• Phenolic compounds include carotenoid xanthophylls (lutein and zeaxanthin) • Procyanidins (condensed tannins) • 3-Deoxyanthocyanidins • Phenolic acids (caffeic acid, taxifolin, luteolinidin, and apigeninidin)	• Antioxidant • Antiproliferative effects • Enhances glucose metabolism • Manage the risk of oxidative stress-related diseases • Cancer preventive agent • Antimicrobial efficacy • Diabetes mellitus	González-Montilla et al. (2012) Ajiboye et al. (2014) Girard and Awika (2018) Leboe et al. (2019) Khoddami et al. (2021) Punia et al. (2021) Mohammed et al. (2022) Stefoska-Needham (2024)

Promotion of the utilization of indigenous food crops

The advocacy of indigenous foods is being recognized as a critical pathway toward food security, preserving biodiversity and culture, and promoting public health. However, even with a growing body of evidence supporting their nutritional, health, environmental, and socioeconomic importance, indigenous foods are persistently excluded from dominant food systems.

Contribution to food and nutrition security

Indigenous foods are characterized by their excellent micronutrient density, drought resistance, and ecological resilience. They can be utilized to address malnutrition and food insecurity caused by climate change. Crops such as sorghum, *Amaranth*, and various varieties of Bambara groundnuts can be utilized as future climate-resilient foods to diversify food options and serve as climate-smart adaptations that enhance resilience at the household level (Lara-Arevalo et al., 2024; Lugo-Morin, 2020; Mbhenyane, 2017).

Revoking cultural identity

Promoting indigenous foods also helps communities reclaim their identity and foster a sense of cultural pride. Festivals, storytelling, culinary traditions, and passing knowledge to different generations enable communities to connect with their ancestral practices, thereby strengthening social cohesion and identity (Fisher and Du Rand, 2022; Makuse and Mbhenyane, 2011).

Emergence of niche and health-focused markets

There is a rising demand among consumers for organic, gluten-free, and heritage foods, which creates market opportunities for indigenous crops. Health-conscious consumers in urban areas, existing organic food markets that focus on heritage foods, and expanding global markets present opportunities for indigenous foods as consumers become aware of the nutritional and functional benefits of traditional foods (Akinola et al., 2020).

Integration into climate change adaptation strategies

Indigenous food crops are resilient to adverse conditions, and climate change can present a valuable opportunity to incorporate these foods into climate-smart agriculture. Similarly, promoting indigenous foods within food systems can enhance agro-ecological diversification and reduce individual households' reliance on import-dependent food systems that are susceptible to socio-political-economic "shocks" and global fluctuations (Zavaleta-Cortijo et al., 2024; Rankoana, 2020). Furthermore, the Lancet Planetary Health Diet approach (Willett et al., 2019) advocates for a predominantly plant-based diet that is environmentally sustainable and nutritionally adequate, aligning closely with the inclusion of indigenous foods in global health and sustainability strategies.

Policy shifts and institutional recognition

This shift will be achievable due to recent policies promoted by countries and the recognition by institutions of various indigenous foods, such as academic research institutions, national food-based dietary guidelines, and biodiversity conventions. This acceptance indicates openness to incorporating indigenous foods into regional and national development agendas (Madzivhandila, 2024; Akinola et al., 2020).

Policy implications

How can we bring this type of information and knowledge into the lecture halls at medical schools and universities?

Although indigenous food systems are deeply connected to nutrition, health, culture, and sustainability, they remain largely absent from curricula in medical and allied health sciences. This omission perpetuates a biomedical model that overlooks food sovereignty, cultural relevance, and ecological resilience. To address this gap, we propose the following:

- Curriculum reform: Medical and health sciences curricula should integrate modules on indigenous food systems, their nutritional and therapeutic properties, and their role in addressing malnutrition and noncommunicable diseases. This aligns with global health priorities and the Lancet Planetary Health Diet framework.
- Interdisciplinary teaching: Encourage collaboration between nutrition, agriculture, anthropology, and public health departments to deliver holistic education on indigenous foods.
- Case-based learning: Use real-world examples of indigenous crops (e.g., sorghum, *Amaranth*, Bambara groundnut) to teach clinical nutrition, food security, and climate adaptation strategies.
- Community engagement: Incorporate experiential learning through partnerships with local communities, enabling students to understand cultural practices and traditional food preparation methods.
- Research Integration: Promote student-led research on indigenous foods' bioactive compounds, antinutritional factors, and health benefits, fostering innovation in functional foods and nutraceuticals.

Recommendations for decision-makers

The following is a clear and succinct list of actionable policy recommendations that we propose:

1. Curriculum reform: Integrate indigenous food systems and their health implications into medical, nutrition, and allied health sciences curricula.
2. Institutional recognition: Mandate the inclusion of indigenous foods in national food-based dietary guidelines, the national school feeding program, public facility meals, and public health nutrition programs.
3. Research investment: Allocate funding for research on nutrient composition, bioactive compounds, and safe processing methods of indigenous foods.
4. Community-driven interventions: Develop culturally appropriate food system strategies with the full participation of local communities to ensure food sovereignty.
5. Market development: Support niche markets for indigenous foods through incentives for smallholder farmers and entrepreneurs.
6. Climate-smart agriculture: Promote indigenous crops as part of national climate adaptation strategies to enhance resilience and biodiversity.
7. Public awareness campaigns: Launch education programs to inform consumers about the nutritional and cultural value of indigenous foods.
8. Policy harmonization: Align agricultural, health, and trade policies to reduce barriers to indigenous food production and commercialization.
9. Monitoring and evaluation: Establish indicators to track the integration of indigenous foods into food systems and their impact on nutrition and health outcomes.

Conclusion

Indigenous food crops play a vital yet underutilized role in the broader food system, offering solutions to persistent challenges such as malnutrition, noncommunicable diseases, and food insecurity. As demonstrated throughout this chapter, these crops are not only rich in essential nutrients but also contain a diverse range of bioactive compounds with health-promoting and disease-preventive properties. Examples include nutrient dense grains like sorghum and pearl millet, leafy vegetables such as *Amaranth* and cleome, and legumes like cowpea and Bambara groundnut. Such foods possess antioxidant, antidiabetic, antiinflammatory, antimicrobial, and anticancer activities, supported by both scientific evidence and generations of indigenous knowledge.

Despite their benefits, the presence of antinutritional factors and potential toxicities calls for further research, alongside consumer education and the promotion of safe food processing methods (both traditional and modern) to improve nutrient bioavailability and ensure safety. Beyond their physiological value, indigenous foods are deeply embedded in cultural identity, ecological knowledge systems, and the principles of food sovereignty. Their integration into food systems supports global efforts to eliminate poverty and hunger, improve health and wellbeing, promote responsible consumption, and address climate change, aligning directly with Sustainable Development Goals 1 (No Poverty), 2 (Zero Hunger), 3 (Good Health and Wellbeing), 12 (Responsible Consumption and Production), and 13 (Climate Action).

However, the continued exclusion of indigenous foods from mainstream food systems and policy frameworks limits their potential impact. This chapter calls for a paradigm shift, namely one that recognizes and supports indigenous food systems through inclusive policies, community-driven initiatives, public health nutrition interventions, and sustained research investment. Only through such a comprehensive and inclusive approach can indigenous foods be repositioned from the margins to the center of food system transformation, contributing to healthier populations, resilient communities, and more equitable and sustainable food futures.

Colonial interference resulted in the marginalization of indigenous foods, evident in their minimal production, marketing, preparation, and consumption (Mbhenyane, 2016). These foods were not considered valuable contributors to formal food systems. For example, this is demonstrated by the high-value, naturally growing *Amaranth* crop that was historically (and still at present) considered a weed, especially in commercial farm areas, and hence often weeded and destroyed or fed to animals. Through the adoption of a Westernized diet, colonialism displaced indigenous diets with processed, industrial, and often imported foods, leading to a form of cultural erosion from within. To avoid a new wave of recolonization, attention to local food systems must center on the needs and agency of local people. The concept of food sovereignty, first articulated by the transnational agrarian movement La Via Campesina in 1996, is central to this shift. It asserts the right of people to define their own food systems, prioritizing ecological sustainability, cultural relevance, and democratic participation (Pimbert and Claeys, 2024; Wittman, 2011; Windfuhr and Jonsén, 2005). Appropriate and culturally acceptable food system interventions should be developed with the full participation of local communities.

Glossary

Indigenous food crops Crops native to a region and traditionally cultivated by local communities, often resilient and nutrient-rich (Chen and Byrne, 2023).

Nutraceutical A food-derived product offering health benefits beyond basic nutrition, often used in disease prevention (Bhuvaneshwari et al., 2024).

Phytochemicals Bioactive compounds in plants that contribute to health through antioxidant and protective effects (Thiede and Zidenberg-Cherr, 2016).

References

Adebowale, A.A., Adebayo, A., Akinola, R., Ajayi, O., Alabi, O., 2005. Role of indigenous knowledge systems in food security in Africa. Afr. J. Food Sci. Technol. 15 (7), 1–2. Available from https://www.interesjournals.org. doi:10.14303/ajfst.2024.092.

Adetunji, T.L., Ogunro, O.B., Ogunremi, O.O., Andrew, O.R., 2025a. The genus Bidens L. in African traditional medicine. In: Bibi, Y., Zahara, K., Qayyum, A., Jenks, M.A. Eds.. The Genus Bidens, Springer, Singapore. pp. 65–76. doi:10.1007/978-981-96-4257-1_5.

Affrifah, N.S., Phillips, R.D., Saalia, F.K., 2022. Cowpeas: nutritional profile, processing methods and products—a review. Legume Sci. 4 (3), e131. doi:10.1002/leg3.131.

Agricultural Research Council (ARC), 2023. Cowpea Production in South Africa: participatory Variety Trials and Demonstration Sites. Available from: https://www.arc.agric.za (Accessed 27 August 2025).

Ajiboye, T.O., Iliasu, G.A., Adeleye, A.O., Abdussalam, F.A., Akinpelu, S.A., Ogunbode, S.M., Jimoh, S.O., 2014. Nutritional and antioxidant dispositions of sorghum/millet based beverages indigenous to Nigeria. Food Sci. Nutr. 2 (5), 597–604. doi:10.1002/fsn3.139.

Akinola, R., Pereira, L.M., Mabhaudhi, T., De Bruin, F.-M., Rusch, L., 2020. A review of indigenous food crops in Africa and the implications for more sustainable and healthy food systems. Sustainability 12 (8), 3493. doi:10.3390/su12083493.

Ali, M.R., Ibrahim, H.H., Salah-Eldin, A.A., 2024. Unveiling the chemical composition, bioactive profile and antioxidant capacity of dried Egyptian Jew's mallow stems as a promising anticancer agent. Molecules 29 (6), 1377. doi:10.3390/molecules29061377.

Alinnor, I.J., Akalezi, C.O., 2010. Proximate and mineral compositions of *Dioscorea rotundata (White Yam)* and *Colocasia esculenta (White Cocoyam/amadumbe*). Pak. J. Nutr. 9 (10), 998–1001. doi:10.3923/pjn.2010.998.100.

ARC, 2021a. Sorghum Production. Agricultural Research Council –Grain Crops Institute. https://www.arc.agric.za/arc-gci/Fact%20Sheets%20Library/Sorghum%20Production.pdf.

ARC, 2021b. SA Graan – october 2021. ARC-Grain Crops, Potchefstroom. https://arc.agric.za/arc-gci/News%20Articles%20Library/ARC%20evaluates%20sorghum%20cultivars.pdf

ARC, 2021c. Sorghum Trial Results – These Cultivars Can Adapt. ARC-Grain Crops & University of the Free State. https://www.arc.agric.za/Agricultural%20Sector%20News/Trial%20results%20-%20these%20cultivars%20can%20adapt.pdf.

Arthur, G.D., Naidoo, K.K., Coopoosamy, R.M., 2012. *Bidens pilosa* L.: agricultural and pharmaceutical importance. J. Med. Plants Res. 6 (17), 3282–3287. doi:10.5897/JMPR012.195.

Asare, A.T., Agbemafle, R., Adukpo, G.E., Diabor, E., Adamtey, K.A., 2013. Assessment of functional properties and nutritional composition of some cowpea (*Vigna unguiculata* L.) genotypes in Ghana. J. Agric. Biol. Sci. 8, 465–469. http://hdl.handle.net/123456789/5929.

Aweya, J.J., Sharma, D., Bajwa, R.K., Earnest, B., Krache, H., Moghadasian, M.H., 2025. Ancient grains as functional foods: integrating traditional knowledge with contemporary nutritional science. Foods 14, 2529. doi:10.3390/foods14252529.

Azam-Ali, S.N., Sesay, A., Karikari, S.K., Massawe, F.J., Aguilar-Manjarrez, J., Bannayan, M., Hampson, K.J., 2001. *Assessing the potential of an underutilized crop – bambara groundnut*. FAO Agricultural Services Bulletin No. 155. FAO, Rome. Available from: https://www.fao.org/3/y2779e/y2779e00.htm.

Balasubramanian, T., Karthikeyan, M., 2015. Therapeutic effect of *Amaranthus hybridus* on diabetic nephropathy. J. Dev. Drugs 5 (1), 147. doi:10.4172/2329-6631.1000147.

Baso, A.A., Sule, S.Y., Sheshe, F.A., Rasheed, H.I., 2025. Phytochemical screening and antimicrobial properties of *Momordica balsamina* L. leaves extracts. IJLTEMAS. 14 (6), 318–325. Available from: https://www.ijltemas.in/DigitalLibrary/Vol.14Issue6/318-325.pdf.

Beswa, D., Dlamini, N.R., Amonsou, E.O., Siwela, M., Derera, J., 2016. Effects of amaranth addition on the pro-vitamin A content, and physical and antioxidant properties of extruded pro-vitamin A-biofortified maize snacks. J. Sci. Food Agric. 96 (1), 287–294. doi:10.1002/jsfa.7092.

Bhuvaneshwari, G., Ganiger, V.M., Narayanpur, V.B., Allolli, T.B., 2024. Nutraceuticals: classification, sources and relation with medicine. In: Eds., Yaradoddi, J.S., Meti, B.S., Mudgulkar, S.B., Agsar, D.. Frontiers in Food Biotechnology, Springer Nature, Singapore, pp. 29–55. doi:10.1007/978-981-97-3261-6_3.

Biswas, P., Bibi, S., Yousafi, Q., Mehmood, A., Saleem, S., Ihsan, A., Dey, D., Hasan Zilani, M.N., Hasan, M.N., Saleem, R., Awaji, A.A., 2023. Study of MDM2 as prognostic biomarker in Brain-LGG cancer and bioactive phytochemicals inhibit the p53-MDM2 pathway: a computational drug development approach. Molecules 28 (7), 2977. doi:10.3390/molecules28072977.

Botanical Online Editorial Team, 2025. Iron-rich plant-based foods and dietary strategies to improve iron absorption. *Botanical Online*. Available from: https://www.botanical-online.com/en/food/iron-food (Accessed 10 June 2025).

Boukar, O., Fatokun, C., Huynh, B.L., Roberts, P.A., Close, T.J., 2019. Genomic tools in cowpea breeding programs: status and perspectives. Front. Plant Sci. 10, 1346. doi:10.3389/fpls.2019.01346.

Burkill, H.M., 1985. *The Useful Plants of West Tropical Africa*, Vol. 1. Royal Botanic Gardens, Kew. Available from: https://www.kew.org/science/projects/useful-plants-of-west-tropical-africa.

Burlingame, B., Dernini, S., 2019. Sustainable Diets: linking Nutrition and Food Systems. FAO. Available from: https://www.fao.org/documents/card/en/c/ca6640en.

Burns, A., Gleadow, R.M., Cliff, J., Zacarias, A., Cavagnaro, T.R., 2010. Cassava: the drought, war and famine crop in a changing world. Sustainability 2 (11), 3572–3607. doi:10.3390/su2113572.

Burridge, J., Jochua, C.N., Bucksch, A., Lynch, J.P., 2016. Legume shovelomics: high-throughput phenotyping of common bean (*Phaseolus vulgaris* L.) and cowpea (*Vigna unguiculata* subsp. *unguiculata*) root architecture in the field. Field Crops Res. 192, 21–32. doi:10.1016/j.fcr.2016.04.008.

Chandrakar, S., Pisalkar, P.S., Patel, S., Khokhar, D., Mishra, N.K., 2022. Physico-chemical properties of the taro (*Colocasia esculenta*) root. Pharma Innov. J. 11 (9), 98–100. https://www.thepharmajournal.com/archives/2022/vol11issue9/PartB/11-8-92-741.pdf.

Chavez, A.L., Bedoya, J.M., Sánchez, T., Iglesias, C., Ceballos, H., Roca, W., 2000. Iron, carotene, and ascorbic acid in cassava roots and leaves. Food Nutr. Bull. 21 (4), 410–413. doi:10.1177/156482650002100413.

Chen, K., Byrne, P., 2023. The importance of indigenous crops. In: Chen, K., Byrne, P. (Eds.), Understudied Indigenous Crops. Colorado State University, Fort Collins, Colorado. Available from: https://colostate.pressbooks.pub/understudiedindigenouscrops/chapter/importance/.

Chung, C.Y., Yang, W.C., Liang, C.L., Liu, H.Y., Lai, S.K., Chang, C.L.T., 2016. Cytopiloyne, a polyacetylenic glucoside from *Bidens pilosa*, acts as a novel anticandidal agent via regulation of macrophages. J. Ethnopharmacol. 184, 72–80. doi:10.1016/j.jep.2016.02.036.

de Morais Cardoso, L., Pinheiro Sant'Ana, H.M., Stringheta, P.C., Chaves, J.B.P., de Cássia Gonçalves Alfenas, R., 2015. Sorghum (*Sorghum bicolor* L.): nutrients, bioactive compounds, and potential impact on human health. Crit. Rev. Food Sci. Nutr. 57 (2), 372–390. doi:10.1080/10408398.2014.887057.

Dehayem-Kamadjeu, A., Okonda, J., 2019. X-ray fluorescence analysis of selected micronutrients in ten African indigenous leafy vegetables cultivated in Nairobi, Kenya. Pan Afr. Med. J. 33, 296. doi:10.11604/pamj.2019.33.296.19501.

Department of Agriculture, Forestry and Fisheries (DAFF) 2010. Amaranthus production guideline. Pretoria, South Africa: Directorate Plant Production. 1–15. https://www.agri4africa.com/wp-content/uploads/2020/06/amaranthus.pdf.

Dimo, T., Rakotonirina, S.V., Tan, P.V., Azay, J., Dongo, E., Kamtchouing, P., Cros, G., 2007. Effect of *Sclerocarya birrea* stem bark methylene chloride/methanol extract on streptozotocin-diabetic rats. J. Ethnopharmacol. 110 (3), 434–438. doi:10.1016/j.jep.2006.10.020.

Ehlers, J.D., Hall, A.E., 1997. Cowpea (*Vigna unguiculata* L. Walp.). Field Crops Res. 53 (1–3), 187–204. doi:10.1016/S0378-4290(97)00031-2.

Elisha, I.L., Botha, F.S., McGaw, L.J., Eloff, J.N., 2017. The antibacterial activity of extracts of nine plant species with good activity against *Escherichia coli*, against five other bacteria and cytotoxicity of extracts. BMC Complement. Altern. Med. 17, 183. doi:10.1186/s12906-017-1683-y.

Elm-Hill, R., Webster, R.M., Allen, A., 2023. Perspectives on Native Food Sovereignty & Health Equity. First Nations Development Institute. Available from: https://www.firstnations.org/publications/perspectives-on-native-food-sovereignty-health-equity/.

Eshwarage, I.S., Herath, T., Gunathilake, T., 2017. Dietary fiber, resistant starch and in-vitro starch digestibility of selected eleven commonly consumed legumes (mung bean, cowpea, soybean and horse gram) in Sri Lanka. Res. J. Chem. Sci. 7, 27–33. http://www.isca.in/.

Falowo, A.B., Muchenje, V., Hugo, A., Aiyegoro, O.A., Fayemi, P.O., 2017b. Antioxidant activities of *Moringa oleifera* L. and *Bidens pilosa* L. leaf extracts and their effects on oxidative stability of ground raw beef during refrigeration storage. CyTA J. Food 15 (2), 249–256. doi:10.1080/19476337.2016.1245296.

Falowo, A.B., Mukumbo, F.E., Idamokoro, E.M., Afolayan, A.J., Muchenje, V., 2017a. Phytochemical constituents and antioxidant activity of *Bidens pilosa* and Moringa oleifera extracts. J. Food Biochem. 41 (6), e12481. doi:10.1111/jfbc.12481.

FAO & Alliance of Bioversity International and CIAT, 2023. In Brief: Indigenous Peoples' Food systems—Insights on Sustainability and Resilience from the Front Line of Climate Change. FAO, Rome. Available from: doi:10.4060/cc5122en.

FAO, 2023. FAOSTAT: Cassava Production Statistics. Food and Agriculture Organization of the United Nations, Rome. Available from: https://www.fao.org/faostat/en/#data/QCL.

Fisher, H., Du Rand, G., 2022. Promoting South African indigenous and cultural foods through contemporary menu adaptations—a case study. Afr. J. Hosp. Tour. Leis. 11 (SE2), 1685–1697. doi:10.46222/ajhtl.19770720.317.

Food Plants Solution Field Guide, 2013. Potentially Important Food Plants of South Africa. Available from: www.foodplantsolutions.org. (Accessed 10 July 2025).

Francis, G., Kerem, Z., Makkar, H.P.S., Becker, K., 2002. The biological action of saponins in animal systems: a review. Br. J. Nutr. 88 (6), 587–605. doi:10.1079/BJN2002725.

Freire Filho, F.R., Ribeiro, V.Q., Rocha, M.M., Lopes, Â.C.A., Silva, K.J.D., 2012. Feijão-caupi No Brasil: Produção, Melhoramento Genético, Avanços E Desafios. Embrapa, Brasília. Available from: https://www.embrapa.br/busca-de-publicacoes/-/publicacao/937095/feijao-caupi-no-brasil-producao-melhoramento-genetico-avancos-e-desafios.

Gautam, A.K., Shrivastava, N., Sharma, B., Bhagyawant, S.S., 2018. Current scenario of legume lectins and their practical applications. J. Crop Sci. Biotechnol. 21 (3), 217–227. doi:10.1007/s12892-018-0002-0.

Gil, J.L., Buitrago, A.J.A., 2002. La yuca en la alimentación animal. In: Ospina, B., Ceballos, H. (Eds.), La Yuca En El Tercer Milenio: Sistemas Modernos de Producción, Procesamiento, Utilización Y Comercialización. Centro Internacional de Agricultura Tropical, Cali, Colombia, pp. 527–569. Available from: http://www.clayuca.org/PDF/libro_yuca/capitulo28.pdf.

Gilbert, B., Alves, L.F., Favoreto, R.F., 2022. Bidens pilosa. In: Monografias de Plantas Medicinais Brasileiras E Aclimatadas: Vol. II. Abifisa, Rio de Janeiro; Editora FIOCRUZ, pp. 73–96. doi:10.7476/9786557081778.0005.

Girard, A.L., Awika, J.M., 2018. Sorghum polyphenols and other bioactive components as functional and health promoting food ingredients. J. Cereal Sci. 84, 112–124. doi:10.1016/j.jcs.2018.10.009.

Gómez-Maqueo, A., García-Cayuela, T., Fernández-López, R., Welti-Chanes, J., Cano, M.P., 2019. Inhibitory potential of prickly pears and their isolated bioactives against digestive enzymes linked to type 2 diabetes and inflammatory response. J. Sci. Food Agric. 99 (14), 6380–6391. doi:10.1002/jsfa.9917.

Gonçalves, P., Goufo, P., Trindade, H., Rosa, E.A., Ferreira, L., Dominguez-Peris, R., et al, 2016. Cowpea (*Vigna unguiculata* L. Walp), a renewed multipurpose crop for a more sustainable agri-food system: nutritional advantages and constraints. J. Sci. Food Agric. 96 (10), 2941–2951. doi:10.1002/jsfa.7644.

Gondwe, M., Kamadyaapa, D.R., Tufts, M., Chuturgoon, A.A., Musabayane, C.T., 2008. *Sclerocarya birrea* [(A. Rich.) Hochst.] [Anacardiaceae] stem-bark ethanolic extract modulates blood glucose, glomerular filtration rate and mean arterial blood pressure of STZ-induced diabetic rats. Phytomedicine 15 (19), 699–709. doi:10.1016/j.phymed.2008.02.004.

González-Montilla, F.M., Chávez-Santoscoy, R.A., Gutiérrez-Uribe, J.A., Serna-Saldivar, S.O., 2012. Isolation and identification of phase II enzyme inductors obtained from black shawaya sorghum (*Sorghum bicolor* L. Moench) bran. J. Cereal Sci. 55 (2), 126–131. doi:10.1016/j.jcs.2011.10.009.

Goyal, S., Nagar, S., Mallesh, G., Kumari, K., Sonu, Kharor, N., 2023. A review on the role of phytic acid and phytase in food and feed. Chem. Sci. Rev. Lett. 11 (44). chesci.com https://chesci.com/wp-content/uploads/2023/01/v11i44_20_CS205311544_Completd.pdf.

Greuter, W., 2006. Mediterranean Plant Invasions: Checklist and Status. Euro+Med Plantbase, Berlin. Available from: https://www.europlusmed.org.

Grimaldi, I.M., Leke, W.N., Borokini, I., Wanjama, D., Van Andel, T., 2018. From landraces to modern cultivars: field observations on taro (*Colocasia esculenta*) in sub-Saharan Africa. Genet. Resour. Crop Evol. 65, 1809–1828. doi:10.1007/s10722-018-0663-4.

Gupta, V., Gupta, S.K., Sharma, A., Javeria, S., Amin, Z., Sharma, M., 2025. Metabolomic and nutritional profiling of Amaranthus cruentus landraces from Jammu and Kashmir: insights into phytochemical diversity and functional potential. Eur. Food Res. Technol. 251, 3935–3951. doi:10.1007/s00217-025-04789-3.

Gwanzura, T., Ng'ambi, J.W., Norris, D., 2012. Nutrient composition and tannin contents of forage sorghum, cowpea, lablab and mucuna hays grown in Limpopo Province of South Africa. Asian J. Anim. Sci. 6 (5), 256–262. doi:10.3923/ajas.2012.256.262.

Hammoudi Halat, D., Hallit, S., Younes, S., AlFikany, M., Khaled, S., Krayem, M., El Khatib, S., Rahal, M., 2023. Exploring the effects of health behaviors and mental health on students' academic achievement: a cross-sectional study on Lebanese university students. BMC Public Health 23 (1), 1–11. doi:10.1186/s12889-023-16184-8.

Harris, T., Jideani, V., Le Roes-Hill, M., 2018. Flavonoids and tannin composition of Bambara groundnut (*Vigna subterranea*) of Mpumalanga, South Africa. Heliyon 4, e00833. doi:10.1016/j.heliyon.2018.e00833.

Hillocks, R.J., Bennett, C., Mponda, O.M., 2012. Bambara nut: a review of utilisation, market potential and crop improvement. Afr. Crop Sci. J. 20 (1), 1–16. Available from: https://www.ajol.info/index.php/acsj/article/view/78609.

Hou, D., Yousaf, L., Xue, Y., Hu, J., Wu, J., Hu, X., Feng, N., Shen, Q., 2019. Mung bean (*Vigna radiata* L.): bioactive polyphenols, polysaccharides, peptides, and health benefits. Nutrients 11 (6), 1238. Available from: https://www.mdpi.com. doi:10.3390/nu11061238.

Howeler, R., 2022. Cassava in Asia: Trends and Challenges. CIAT, Bangkok. Available from: https://cgspace.cgiar.org/handle/10568/119711.

IDRC, 2024. Indigenous cosmovision at the heart of a food security project in Latin America. International Development Research Centre. Available from: https://idrc.ca/en/news/indigenous-cosmovision-heart-food-security-project-latin-america.

Ijioma, N., Madubuike, K., Nwankudu, O., Nwosu, C., Emelike, C., 2014. Uterine relaxation potential of ethanol leaf extract of *Moringa oleifera* Lam. via the muscarinic receptor pathway. Br. J. Pharm. Res. 4, 2455–2462. doi:10.9734/BJPR/2014/12140.

Janarthanan, M., Kumar, M.S., 2013. Qualitative and quantitative analysis of phytochemical studies on selected seaweeds *Acanthopora spicifera* and *Sargassum wightii*. Int. J. Eng. Res. 2278-800X. 7 (3), 11–15. 2278-067X www.ijerd.com.

Jayathilake, C., Visvanathan, R., Deen, A., Bangamuwage, R., Jayawardana, B.C., Nammi, S., Liyanage, R., 2018. Cowpea: an overview of its nutritional facts and health benefits. J. Sci. Food Agric. 98 (13), 4793–4806. doi:10.1002/jsfa.9074.

Jernigan, V.B.B., Salvatore, A.L., Styne, D.M., Wetherill, M.S., 2023. Food sovereignty and health: a conceptual framework to advance research and practice. Health Promot. Pract. 24 (6), 1070–1074. doi:10.1177/15248399231173191.

Kaboré, S., Touré, A., Kouamé, B.K., Bamba, A.E., Kabran, A.F., Kadio, B.D., Konan, D.J., Attioua, K.B., Coulibaly, W.K., Kablan, A.L., et al., 2021. Phytochemical screening and antioxidant activity of leaves of *Amaranthus hybridus* L., *Corchorus olitorius* L. and *Hibiscus sabdariffa* L. grown in northern Côte d'Ivoire. GSC Biol. Pharm. Sci. 16 (2), 182–189. doi:10.30574/gscbps.2021.16.2.0227.

Kareem, I., Elisha, I.L., Botha, F.S., McGaw, L.J., Eloff, J.N., 2022. Blood hemolytic activity and acute toxicity of saponin extract from *Lepidium aucheri* Boiss. Int. J. Drug Deliv. Technol. 12 (2), 478–483. https://impactfactor.org/PDF/IJDDT/12/IJDDT,Vol12,Issue2,Article73.pdf.

Katoch, R., Tripathi, A., 2021. Research advances and prospects of legume lectins. J. Biosci. 46, 104. doi:10.1007/s12038-021-00225-8.

Kaur, C., Kapoor, H.C., 2019. Antioxidants in fruits and vegetables—the millennium's health. Int. J. Food Sci. Technol. 36 (7), 703–725. doi:10.1046/j.1365-2621.2001.00513.

Kaur, M., McEwan, M., Naidoo, P., Alcantara, T.P., 2013. Nutritional and functional properties of amadumbe (*Colocasia esculenta*) cultivated in South Africa. Food Res. 4 (4), 1333–1343. https://www.myfoodresearch.com/uploads/8/4/8/5/84855864/ 48 fr-2020-069_mokhele_1.pdf.

Khoddami, A., Messina, V., Vadabalija Venkata, K., Farahnaky, A., Blanchard, C.L., Roberts, T.H., 2021. Sorghum in foods: functionality and potential in innovative products. Crit. Rev. Food Sci. Nutr. 63 (9), 1170–1186. doi:10.1080/10408398.2021.1960793.

Knope, M.L., Funk, V.A., Johnson, M.A., Wagner, W.L., Datlof, E.M., Johnson, G., Crawford, D.J., Bonifacino, J.M., Morden, C.W., et al., 2020. Dispersal and adaptive radiation of Bidens (Compositae) across the remote archipelagoes of Polynesia. J. Syst. Evol. doi:10.1111/jse.12704.

Kuhnlein, H.V., Erasmus, B., Spigelski, D., 2020. Indigenous peoples' food systems & well-being: interventions & policies for healthy communities. FAO. Available from: https://www.fao.org/documents/card/en/c/ca7868en.

Kunene, E.N., Nxumalo, K.A., Ngwenya, M.P., Masarirambi, M.T., 2020. Domesticating and commercialisation of indigenous fruit and nut tree crops for food security and income generation in the Kingdom of Eswatini. Curr. J. Appl. Sci. Technol. 39 (16), 37–52. doi:10.9734/CJAST/2020/v39i1630733.

Lara-Arevalo, J., Laar, A., Chaparro, M.P., Drewnowski, A., 2024. Nutrient-dense African indigenous vegetables and grains in the FAO Food Composition Table for Western Africa (WAFCT) identified using Nutrient-Rich Food (NRF) scores. Nutrients 16 (17), 2985. doi:10.3390/nu16172985.

Leboe, D.W., Dhuha, N.S., Wahyuddin, M., Blum, A., Kim, J., Park, S., Ratnasari, R., 2019. The potential of *Sorghum bicolor* L. as a blood glucose lowering agent: a review. J. Kesehat. 1 (1), 1–10. Available from: https://journal3.uin-alauddin.ac.id/index.php/kesehatan/article/download/18181/10642.

Lebot, V., 2019. Tropical Root and Tuber Crops: Cassava, Sweet Potato, Yams and Aroids. CABI, Wallingford. Available from: doi:10.1079/9781789243364.0000.

Lee, J.H., Jeon, J.K., Kim, S.G., Kim, S.H., Chun, T., Imm, J.Y., 2011. Comparative analyses of total phenols, flavonoids, saponins and antioxidant activity in yellow soy beans and mung beans. Int. J. Food Sci. Technol. 46 (12), 2513–2519. Available from: https://www.academic.oup.com. doi:10.1111/j.1365-2621.2011.02775.x.

Lee, J.H., Kim, Y.S., Kim, J.H., Lee, S.H., 2013b. Anti-inflammatory and antimicrobial activities of *Bidens Pilosa* extracts. J. Ethnopharmacol. 148 (2), 701–709. Retrieved from: https://www.sciencedirect.com/science/article/pii/S0378874112005843.

Lee, M., Bargagli, M., Tio, M.C., Waikar, S.S., Ferraro, P.M., 2023. Dietary oxalate intake and kidney outcomes: a review. Nutrients 12 (9), 2673. doi:10.3390/nu12092673.

Lee, W.C., Peng, C.C., Chang, C.H., Huang, S.H., Chyau, C.C., 2013a. Extraction of antioxidant components from *Bidens Pilosa* flowers and their uptake by human intestinal Caco-2 cells. Molecules 18 (2), 1582–1601. https://www.mdpi.com/1420-3049/18/2/1582.

Liyanage, R., Perera, O., Lakmini, G.W.A.S., Weththasinghe, P., Visvanathan, R., Jayathilake, C., Jayawardana, B.C., Vidanarachchi, J., Sivakanesan, R., 2018. Boiled, sprouted, and raw cowpea-incorporated diets modulate high-fat diet-induced hypercholesterolemia in rats. Food Sci. Nutr. 6 (6), 1762–1769. doi:10.1002/fsn3.727.

Liyanage, R., Perera, O.S., Wethasinghe, P., Jayawardana, B.C., Vidanaarachchi, J.K., Sivaganesan, R., 2014. Nutritional properties and antioxidant content of commonly consumed cowpea cultivars in Sri Lanka. J. Food Legumes/Indian J. Pulses Res. 27, 215–217. https://www.researchgate.net/publication/272996965.

Lopes, C.V.A., Neri, J.L.S.A., Hunter, J., Ronto, R., Mihrshahi, S., 2024. Interventions and programs using native foods to promote health: a scoping review. Nutrients 16 (23), 4222. doi:10.3390/nu16234222.

Lugo-Morin, D., 2020. Indigenous communities and their food systems: a contribution to the current debate. J. Ethn. Foods 7 (6). doi:10.1186/s42779-019-0043-1.

Mabhaudhi, T., Chibarabada, T.P., Modi, A.T., 2019. Water-food-nutrition-health nexùs: linking water to improving food, nutrition and health in sub-Saharan Africa. Int. J. Environ. Res. Public Health 16 (3), 297. doi:10.3390/ijerph16030297.

Mabhaudhi, T., Chimonyo, V.G.P., Chibarabada, T.P., Modi, A.T., 2017. Developing a roadmap for improving neglected and underutilized crops: a case study of South Africa. Front. Plant Sci. 8, 2143. doi:10.3389/fpls.2017.02143.

Madzivhandila, T.S., 2024. Integrating indigenous knowledge systems (IKS) to climate change adaptation in South Africa: lessons from the "past". S. Afr. J. High. Educ. 38 (5), 159–168. doi:10.20853/38-5-6417.

Mahendranathan, C., Abhayarathne, A., 2021. Antibacterial activities of some medicinal plants and green leafy vegetables: a review. Int. J. Eng. Appl. Sci. Technol. 6 (4), 78–84. doi:10.33564/ijeast.2021.v06i04.011.

Mahmood, S., Ali, H., Ahmad, F., Iqbal, Z., 2014. Estimation of tannins in different sorghum varieties and their effects on nutrient digestibility and absorption of some minerals in caged White Leghorn layers. Int. J. Agric. Biol. 16 (1), 217–221. https://fspublishers.org/published_papers/28312.

Makuse, S.M., Mbhenyane, X.G., 2011. Health benefits and omega-3-fatty acid content of selected indigenous foods in the Limpopo Province, South Africa. Indilinga: Afr. J. Indig. Knowl. Syst. 1683-0296. 10 (2), 182–194. https://hdl.handle.net/10520/EJC61398.

Mampholo, B.M., Sivakumar, D., Thompson, A.K., 2016. Maintaining overall quality of fresh traditional leafy vegetables of Southern Africa during the postharvest chain. Food Rev. Int. 32 (4), 400–416. https://www.researchgate.net/publication/283199519.

Martins, A., Vasas, A., Viveiros, M., Molnár, J., Hohmann, J., Amaral, L., 2011. Antibacterial properties of compounds isolated from *Carpobrotus edulis*. Int. J. Antimicrob. Agents 37 (5), 438–444. doi:10.1016/j.ijantimicag.2011.01.016.

Mashau, M.E., Kgatla, T.E., Makhado, M.V., Mikasi, M.S., Ramashia, S.E., 2022. Nutritional composition, polyphenolic compounds and biological activities of marula fruit (*Sclerocarya birrea*) with its potential food applications: a review. Int. J. Food Prop. 25 (1), 1549–1575. doi:10.1080/10942912.2022.2064491.

Masuku, M.M., Bhengu, A.S., 2023. The value of indigenous foods in improving food security in *Emaphephetheni rural* setting. Indilinga: Afr. J. Indig. Knowl. Syst. 20 (1), 22–35. https://catalogue.leidenuniv.nl/permalink/31UKB_LEU/s5ab2f/alma9940578360102711.

Mavengahama, S., McLachlan, M., De Clercq, W., 2013. The role of wild vegetable species in household food security in maize-based subsistence cropping systems. Food Secur. 5 (2), 227–233. doi:10.1007/s12571-013-0243-2.

Mbhatsani, H.V., Maluleke, S.D.C., Tambe, B.A., Zuma, M.K., Mbhenyane, X.G., 2023. Innovative dishes and micronutrient composition of traditionally preserved *Amaranthus*, *Vigna unguiculata* and *Bidens Pilosa*. S. Afr. J. Bot. 163, 478–487. doi:10.1016/j.sajb.2023.08.019.

Mbhatsani, H.V., Mbhenyane, X.G., Makuse, S.H.M., 2011. Knowledge and consumption of indigenous food by primary school children in Vhembe District in Limpopo Province. Indilinga: Afr. J. Indig. Knowl. Syst. 10 (2), 210–227. https://www.africabib.org/htp.php?RID=370410548.

Mbhenyane, X.G., 2016. The Contribution of 'Indigenous Foods' to the Elimination of Hidden Hunger and Food Insecurity: An Illusion or Innovation? Inaugural lecture, Stellenbosch University, Tygerberg Campus. https://www.su.ac.za/en/node/1838.

Mbhenyane, X.G., 2017. Indigenous foods and their contribution to nutrient requirements. S. Afr. J. Clin. Nutr. 30 (4), 5–7. http://www.sajcn.co.za/index.php/SAJCN/article/view/1299.

Mbhenyane, X.G., Mushaphi, L.F., Mabapa, N.S., Makuse, S.H., Amey, A.K.A., Nemathaga, L.H., Lebese, R.T., 2013. The consumption of indigenous fruits and vegetables and health risk in rural subjects of Limpopo province, South Africa. Indilinga: Afr. J. Indig. Knowl. Syst. 12 (1), 160–168. https://hdl.handle.net/10520/EJC141610.

Mbhenyane, X.G., Venter, C.S., Vorster, H.H., Steyn, H.H., 2001. The glycemic index of South African indigenous foods. S. Afr. J. Clin. Nutr. 14 (3), 91–95. https://www.sajcn.co.za/index.php/SAJCN/article/view/1619.

Miller, J., 2022. GBIF's (Global Biodiversity Information Facility) new strategic framework to address biodiversity data needs. Biodivers. Inf. Sci. Stand. 6. doi:10.3897/biss.6.91093.

Mishra, A., Chweya, J.A., Mnzava, N.A., 2011. Cleome Gynandra: A Promising Leafy Vegetable for Nutritional and Medicinal Applications. Bioversity International, Rome. https://www.bioversityinternational.org/e-library/publications/detail/cleome-gynandra-a-promising-leafy-vegetable/.

Mohammed, A., Li, Z., Zhao, X., Zhang, X., Liu, H., 2022. Bioactive compounds and biological activities of sorghum grains. Foods 10 (11), 2868. doi:10.3390/foods10112868.

Montagnac, J.A., Davis, C.R., Tanumihardjo, S.A., 2009. Nutritional value of cassava for use as a staple food and recent advances for improvement. Compr. Rev. Food Sci. Food Saf. 8 (3), 181–194. doi:10.1111/j.1541-4337.2009.00077.x.

Moyo, M., Amoo, S.O., Aremu, A.O., Gruz, J., Šubrtová, M., Jarošová, M., Tarkowski, P., Doležal, K., 2018. Determination of mineral constituents, phytochemicals and antioxidant qualities of *Cleome gynandra*, compared to *Brassica oleracea* and *Beta vulgaris*. Front. Chem. 5, 128. doi:10.3389/fchem.2017.00128.

Mwanri, L., Worsley, A., Ryan, P., 2018. Dietary strategies to reduce oxalate absorption and prevent calcium deficiency: a public health perspective. Afr. J. Food Agric. Nutr. Dev. 18 (2), 13545–13560. https://www.ajfand.net/Volume18/No2/Mwanri182.pdf.

National Research Council, 1989. Lost Crops of the Incas: Little-Known Plants of the Andes with Promise for Worldwide Cultivation. The National Academies Press, Washington, DC. Available from: doi:10.17226/1398.

Ndhlovu, P.T., Asong, J.A., Omotayo, A.O., Otang-Mbeng, W., Aremu, A.O., 2023. Ethnobotanical survey of medicinal plants used by indigenous knowledge holders to manage healthcare needs in children. PLoS One 18 (3), e0282113. doi:10.1371/journal.pone.0282113.

Njume, C., Gqaza, B.M., Rozani, C., Goduka, N.I., 2016. Studies on bioactivity and secondary metabolites of crude extracts of *Bidens pilosa* L.(Asteraceae): a medicinal plant used in the Transkei region of South Africa. Pak. J. Pharm. Sci. 29 (3), 877–885.

NTFP-EP, 2025. *Wild Foods in Asia: Sustaining Biodiversity and Livelihoods*. Non-Timber Forest Products Exchange Programme. Available from: https://ntfp.org/publications/wild-foods-in-asia.

Nxusani, Z.N., Zuma, M.K., Mbhenyane, X.G., 2023. A systematic review of indigenous food plant usage in Southern Africa. Sustainability 15 (11), 8799. doi:10.3390/su15118799.

Nyau, V., Prakash, S., Rodrigues, J., Farrant, J., 2015. Antioxidant activities of Bambara groundnuts as assessed by FRAP and DPPH assays. J. Food Nutr. Res. 3 (1), 7–11. doi:10.12691/ajfn-3-1-2.

Ogie-Odia, E.A., Mensah, J.K., Ehilen, O.E., Eseigbe, D.A., 2022. Nutrient, mineral and phytochemical properties of selected underutilized Amaranthu*s* vegetable species in Ekpoma, Edo State, Nigeria. Niger. J. Biotechnol. 38 (1), 1–12. doi:10.4314/njb.v38i.1S.

Okafor, J.N., Rautenbauch, F., Meyer, M., Le Roes-Hill, M., Harris, T., Jideani, V.A., 2021. Phenolic content, antioxidant, cytotoxic and anti-proliferative effects of fractions of *Vigna subterranea* (L.) Verdc from Mpumalanga, South Africa. Heliyon 7 (11), e08397. doi:10.1016/j.heliyon.2022.e09024.

Oladimeji, J.J., Kumar, P.L., Abe, A., Vetukuri, R.R., Bhattacharjee, R., 2022. Taro in West Africa: status, challenges and opportunities. Agronomy 12 (9), 2094. doi:10.3390/agronomy12092094.

Oladosu, N.O., Njoku, O.C., 2021. Haematopoietic potential of Bambara nut (*Vigna subterranea* (L.) Verdc.) in Wistar rats. Sci. Res. J. Agric. Life Sci. 1 (1), 32–37. doi:10.47310/srjals.2021.v01i01.004.

Olufunso, A.B., et al, 2025. Anticancer mechanisms of indigenous food plants in Nigeria. Future Integr. Med. 2 (1), 1–25. doi:10.14218/FIM.2024.00042.

Omondi, E.O., Engels, C., Nambafu, G., Schreiner, M., Neugart, S., Abukutsa-Onyango, M., Winkelmann, T., 2017. Nutritional compound analysis and morphological characterization of spider plant *(Cleome gynandra)*—an African indigenous leafy vegetable. Food Res. Int. 100 (Pt 1), 284–295. doi:10.1016/j.foodres.2017.06.050.

Omoruyi, B.E., Afolayan, A.J., Bradley, G., 2014. Chemical composition profiling and antifungal activity of the essential oil and plant extracts of *Mesembryanthemum edule* (L.) Bolus leaves. Afr. J. Tradit. Complement. Altern. Med. 11 (1), 19–30. doi:10.4314/ajtcam.v11i4.4.

Omoruyi, B.E., Bradley, G., Afolayan, A.J., 2012. Antioxidant and phytochemical properties of *Carpobrotus edulis* (L.) Bolus leaf used for the management of common infections in HIV/AIDS patients in Eastern Cape Province. BMC Complement. Altern. Med. 12, 215. doi:10.1186/1472-6882-12-215.

Otang-Mbeng, F.N.N., Materechera, S.A., Mbeng, W., 2019. Carpobrotus edulis L. (Sour Fig): phytochemistry, pharmacology, and toxicology. In: The Therapeutic Properties of Medicinal Plants, Apple Academic Press, Palm Bay, FL. pp. 1–12.

Oyewole, A.L., Owoyele, B.V., Alimi, M.L., Sanni, S.A., Oyeleke, S.A., 2015. Anti-inflammatory and antipyretic properties of *Corchorus olitorius* aqueous root extract in Wistar rats. J. Basic Clin. Physiol. Pharmacol. 26, 363–368. doi:10.1515/jbcpp-2013-0166.

Panghal, A., Gusain, R., Chhikara, N., Singh, A., 2021. Role of plant-based functional foods in boosting immunity and preventing viral infections: a review. J. Funct. Foods 87, 104732. doi:10.1016/j.jff.2021.104732.

Pimbert, M.P., Claeys, P., 2024. Food Sovereignty. Oxf. Res. Encycl. Anthropol. doi:10.1093/acrefore/9780190854584.013.297.

Punia, H., Tokas, J., Malik, A., Satpal, Sangwan, S., 2021. Characterization of phenolic compounds and antioxidant activity in sorghum (*Sorghum bicolor* L. Moench) grains. Cereal Res. Commun. 49 (3), 343–353. doi:10.1007/s42976-020-00118-w.

Qadir, M.I., Abbas, R.N., Rasool, S., Javid, A., 2019. Enzyme-assisted extraction of *Momordica balsamina* L. fruit phenolics: process optimized by response surface methodology. Food Meas. 13, 697–706. doi:10.1007/s11694-018-9982-2.

Ramalhete, C., Mulhovo, S., Ferreira, M.J.U., Duarte, N., 2022. *Momordica balsamina*: a review of its nutritional, phytochemical and pharmacological properties. J. Ethnopharmacol. 284, 114799. doi:10.1016/j.jep.2021.114799.

Rankoana, S.A., 2020. Climate change impacts on water resources in a rural community in Limpopo province, South Africa: a community-based adaptation to water insecurity. Int. J. Clim. Change Strateg. Manag. 12 (5), 587–598. doi:10.1108/IJCCSM-03-2019-0019.

Rastogi, A., Shukla, S., 2013. Amaranth: a new millennium crop of nutraceutical values. Crit. Rev. Food Sci. Nutr. 53 (2), 109–125. doi:10.1080/10408398.2010.517876.

Rawel, H.M., Kroll, J., 2003. Die Bedeutung von Cassava (*Manihot esculenta*, Crantz) als Hauptnahrungsmittel in tropischen Ländern. Dtsch. Lebensm.-Rundsch. 99, 102–110. https://www.researchgate.net/publication/259600158.

Reddy, B.V.S., Ramesh, S., Sanjana Reddy, P., Ramaiah, B., Salimath, P.M., Kachapur, R., 2006. Sorghum breeding research at ICRISAT—goals, strategies, and impacts. Int. Sorghum Millets Newsl. 47, 1–10. Available from https://www.icrisat.org.

Reddy, N.R., Pierson, M.D., Sathe, S.K., Salunkhe, D.K., 1989. Phytates in Cereals and Legumes. CRC Press, Boca Raton, FL.

Repo-Carrasco-Valencia, R., 2020. Nutritional value and bioactive compounds in Andean ancient grains. Proceedings 53 (1), 1. Available from: https://www.mdpi.com. doi:10.3390/proceedings2020053001.

Sakadzo, N., Kugedera, A.T., Rangani, N., Kokera, L.K., 2025. Cassava: practices and technologies to improve food security in sub-Saharan Africa. Cogent Food Agric. 11 (1), 2518758. doi:10.1080/23311932.2024.2518758.

Sanoussi, F., Ahissou, H., Dansi, M., Hounkonnou, B., Agre, P., Dansi, A., 2015. Ethnobotanical investigation of three traditional leafy vegetables [*Alternanthera sessilis* (L.) DC., *Bidens pilosa* L., *Launaea taraxacifolia* Willd.] widely consumed in southern and central Benin. J. Biodivers. Environ. Sci. 6 (2), 187–198. http://www.innspub.net/.

Sattara, M., Saeed, F., Afzaal, M., Rasheed, A., Asifa, A., Sharifa, S., Hussain, M., Rehman, H.A.U., Raza, M.A., Munir, H., Al Jbawi, E., 2024. An overview of the nutritional and therapeutic properties of amaranth. Int. J. Food Prop. 27 (1), 263–272. doi:10.1080/10942912.2023.2297134.

Schmelzer, G.H., Gurib-Fakim, A., 2008. Plant Resources of Tropical Africa 11(1): Medicinal Plants. PROTA Foundation, Wageningen. Available from https://www.prota4u.org.

Shackleton, R.T., Witt, A.B., Aool, W., Pratt, C.F., 2017. Distribution of the invasive alien weed, Lantana camara, and its ecological and livelihood impacts in eastern Africa. Afr. J. Range Forage Sci. 34 (1), 1–11. doi:10.2989/10220119.2017.1301551.

Shai, L.J., Chauke, M.A., Magano, S.R., Mogale, A.M., Eloff, J.N., 2013. Antibacterial activity of sixteen plant species from Phalaborwa, Limpopo Province, South Africa. J. Med. Plants Res. 7 (26), 1899–1906. https://www.researchgate.net/publication/309532269.

Shukla, S., Singh, S.P., 2010. Studies on genetic variability for foliage yield and its contributing traits in vegetable amaranth. Indian J. Hortic. 67 (1), 124–128. Available from: https://www.indianjournals.com/ijor.aspx?target=ijor:ijh&volume=67&issue=1&article=024.

Silva, F.L., Fischer, D.C.H., Tavares, J.F., Silva, M.S., de Athayde-filho, P.F., Barbosa-Filho, J.M., 2011. Compilation of secondary metabolites from *Bidens pilosa* L. Molecules 16 (2), 1070–1102. doi:10.3390/molecules16021070.

Silva, M.L., Oliveira, A.C., Santos, R.M., Pereira, L.C., 2023. Nutritional and pharmacological properties of *Adansonia digitata*: a narrative review. J. Ethnopharmacol. 305, 116093. doi:10.1016/j.jep.2023.116093.

Singh, B.B., Ajeigbe, H.A., Tarawali, S.A., Fernandez-Rivera, S., Abubakar, M., 2021. Improving the production and utilization of cowpea as food and fodder. Field Crops Res. 82 (2), 169–177. doi:10.1016/S0378-4290(03)00039-6.

Singh, S., Devi, B., 2018. Phytopharmacological evaluation of *Momordica balsamina* Linn. from southern Haryana, India. J. Pharm. Pract. Health Care 4, 17–34. Available from: https://www.researchgate.net/publication/Phytopharmacological_evaluation_of_Momordica_balsamina.

Sivarkumar, D., Soundy, P., Sultanbawa, Y., 2018. Phytochemicals in indigenous leafy vegetables of Southern Africa: a review. Food Res. Int. 104, 110–123. doi:10.1016/j.foodres.2017.10.039.

Souza-Filho, P.R.M., Takaki, M., 2018. Germination constraints of dicarpic cypselae of Bidens pilosa L. Braz. J. Biol. 79 (3), 383–394. doi:10.1590/1519-6984.178222.

Stefoska-Needham, A., 2024. Sorghum and health: an overview of potential protective health effects. J. Food Sci. 89 (S1), A30–A41. doi:10.1111/1750-3841.16978.

Takaidza, S., 2023. Indigenous South African Food: nutrition and Health Benefits. In: Phytochemicals in Agriculture and Food. IntechOpen. pp.1–20. https://www.intechopen.com/chapters/86715.

Thiaw, C., Diop, A., Diatta, C., Sarr, A., Gueye, M., 2023. Ethnobotanical and nutritional importance of *Momordica balsamina* in West Africa. J. Ethnobiol. Ethnomed. 19 (1), 1–12. doi:10.1186/s13002-023-00591-8.

Thiede, A.A., Zidenberg-Cherr, S., 2016. Phytochemicals. In: Nutrition and Health Info Sheet for Health Professionals. University of California. pp. 1–5, Davis. Available from: https://nutrition.ucdavis.edu/sites/g/files/dgvnsk426/files/content/infosheets/factsheets/fact-pro-phytochemical.pdf.

Timilsena, Y.P., Phosanam, A., Stockmann, R., 2023. Perspectives on saponins: food functionality and applications. Int. J. Mol. Sci. 24 (17), 13538. doi:10.3390/ijms241713538.

USDA, 2022. Sorghum: World Markets and Trade. United States Department of Agriculture, Washington, DC. Available from: https://www.fas.usda.gov/data/sorghum-world-markets-and-trade.

Van Eylen, D., Hendrick, O.M., Van Loey, A., 2007. Kinetics of the stability of broccoli (*Brassica oleracea* Cv. Italica) myrosinase and isothiocyanates in broccoli juice during pressure/temperature treatments. J. Agric. Food Chem. 55, 2163–2170. doi:10.1021/jf0630631.

Willett, W., Rockström, J., Loken, B., Springmann, M., Lang, T., Vermeulen, S., Garnett, T., Tilman, D., De Clerck, F., Wood, A., et al, 2019. Food in the Anthropocene: the EAT–Lancet Commission on healthy diets from sustainable food systems. Lancet 393 (10170), 447–492. doi:10.1016/S0140-6736(18)31788-4.

Windfuhr, M., Jonsén, J., 2005. Food Sovereignty: Towards Democracy in Localised Food Systems, FIAN International.

Wittman, H., 2011. Food sovereignty: a new rights framework for food and nature? Environ. Soc. Adv. Res. 2, 87–105. doi:10.3167/ares.2011.020106.

Wu, Y., Liu, J., Lu, H., Wu, C., Kerr, P., 2016. Periphyton: an important regulator in optimizing soil phosphorus bioavailability in paddy fields. Environ. Sci. Pollut. Res. Int. 23 (21), 21377–21384.

Xiang, J., Apea-Bah, F.B., Ndolo, V.U., Katundu, M.C., Beta, T., 2019. Profile of phenolic compounds and antioxidant activity of finger millet varieties. Food Chem. 275, 361–368. doi:10.1016/j.foodchem.2018.09.120.

Xin, Y.J., Liu, H., Wang, Y., Zhang, Q., Li, Y., Zhang, Y., Liu, J., 2021. Anti-inflammatory activity and mechanism of isookanin, isolated by bioassay-guided fractionation from *Bidens pilosa* L. Molecules 26 (2), 255. doi:10.3390/molecules26020255.

Xu, D.P., Li, Y., Meng, X., Zhou, T., Zhou, Y., Zheng, J., Zhang, J.J., Li, H., 2017. Natural antioxidants in foods and medicinal plants: extraction, assessment and resources. Int. J. Mol. Sci. 18. doi:10.3390/IJMS18010096.

Yang, Y.C., Mong, M.C., Wu, W.T., Wang, Z.H., Yin, M.C., 2020. Phytochemical profiles and anti-diabetic benefits of two edible *Amaranthus* species. CyTA J. Food 18 (1), 94–101. doi:10.1080/19476337.2020.1716850.

Zavaleta-Cortijo, C., Silvera-Ccallo, R., Lancha-Rucoba, G., Chanchari, J., Inuma, N., Pizango, M., et al, 2024. Indigenous knowledge and leadership for climate change adaptation in nutrition. PLOS Glob. Public Health 4 (11), e0003917. doi:10.1371/journal.pgph.0003917.

Zhu, F., Du, B., Xu, B., 2012. Mung bean (*Vigna radiata*) as a functional food: a review. Food Sci. Hum. Wellness 1 (1), 11–18. doi:10.1016/j.fshw.2012.07.001.

PART

Future horizons III

CHAPTER

Conclusion 16

The decolonization of medicine: reflections on the past, understanding the present, and contemplating the future

Faadiel Essop

Centre for Cardio-metabolic Research in Africa (CARMA), Division of Medical Physiology, Faculty of Medicine and Health Sciences, Stellenbosch University, Cape Town, South Africa

This timely and comprehensive volume addresses an existing knowledge gap by providing an in-depth and substantive overview of the decolonization of medical teaching, research, and practice. The exciting array of chapters provides significant insights into the decolonization of medicine, while also reminding readers that it is a complex process by revealing the multiple layers constituting this challenge. When adopting a more reflective and integrative vantage point of the book's contents, it is clear that three important cross-cutting themes emerge:

Contemporary medicine is not a neutral discipline but was forged by power, culture, and political contexts

It is evident throughout the volume that contemporary medicine (as a discipline) has been profoundly shaped by its past, which includes, for example, the notion of scientific racism, power asymmetries that impacted the nature of health care, education, and institutions, the suppression of indigenous knowledge systems, and the exploitation of native populations, with lesser value and dignity assigned to Black bodies. In contrast, contemporary medicine is generally considered a value-neutral discipline and/or a purely technical endeavor. Such amnesia represents a form of scientific elitism where medical and allied health sciences training is largely devoid of historical contexts impacting current practices. This mindset is reflected in multiple ways, including the nature of medical research design and endeavors, power dynamics underlying collaborative research projects between the Global North and the Global South, and existing medical curricula and clinical practices. Some may argue that the colonizers have long departed and that we should forget the past, instead focusing on the present and then charting the way forward. However, this volume reveals that the phenomenon of coloniality endures, with the persistence of structural patterns and frameworks established during the colonial era. Here,

Decolonizing Medical Education, Research and Practice. https://doi.org/10.1016/B978-0-443-44998-7.00005-3

many of the chapters converge on the idea that colonialist-derived structural and systemic patterns in contemporary medicine remain a problem. For example, this was underscored by the recent COVID-19 pandemic in terms of vaccine inequality (rich versus poor countries), with the Global North in control as knowledge producers and the Global South as passive recipients. Other examples include the effects of epistemological and race-entrenched medicine-based tools and frameworks (e.g., the development of the BiDil drug for heart failure for Black patients and race-based genetic tests), which can result in ethical harm and injustices. Likewise, global research collaborations are generally skewed in favor of the Global North (often under the guise of benevolence), with clear power imbalances in terms of *who* leads such projects, *who* is allocated the bulk of the resources, and limited acknowledgments to research team members based in the Global South. Collectively, the authors of this volume demonstrate that the myth of ahistoricity in medicine should be abandoned and that there should be a strong impetus for the implementation of more inclusive approaches.

Medicine should embrace epistemic plurality, humility, and restorative justice

This volume also focuses on the dominance of the Western biomedical sciences paradigm that determines *who* are recognized as the knowers and *whose* knowledge is considered legitimate and worthy of forming part of evidence-based medicine. Here, the phenomenon of scientism means that Western scientific and medical knowledge is regarded as superior to other forms of knowing and, hence, the only legitimate way to obtain true knowledge. This philosophical framework also enabled the triumph of reductionism in science and medicine (versus the holistic emphasis of most indigenous knowledge systems), with values and humanness condensed to the biological.

Several chapters in this book reveal that this framework assisted in the conquering of indigenous populations, while their knowledge systems were suppressed and/or outlawed. Subsequently, the dominant Western paradigm of knowing was enforced on subjugated populations, propelling the colonial project to entrench power and domination, while ensuring economic exploitation. Significant resources were made available for such purposes, while financial support for indigenous knowledge systems was curtailed or stopped. In parallel, indigenous knowledge systems were relegated to the status of myths and irrational beliefs. Such subordination of other knowledge systems reveals that medical knowledge production was closely linked to the power of the dominant forces. Ironically, the colonizers also stole knowledge from indigenous populations for commercial gain, without acknowledging their contributions in this regard. Such epistemic exclusion was deliberate and part of a grand strategy to control, dominate, and subjugate the Other. These structures sadly persist in contemporary times, such as the setting of global research agendas and the concentration of publishing houses and prestigious peer-reviewed medical journals in the Global North.

Some of the chapters consider ways to navigate such structural challenges by calling for the inclusion of indigenous knowledge systems into clinical practice as a complementary strategy and not as a replacement. For example, the argument is made that biomedical ethics frameworks should adopt a more pluralistic approach to include ideas derived from indigenous knowledge systems. Here, some authors propose an African bioethics relational framework (Ubuntu) to rehumanize contemporary health care. This should enable a context-sensitive approach to bioethics and a centering of the notion of dignity in modern medicine, forming part of restorative justice and a healing process for wounds inflicted during the colonial and postcolonial eras. Such an inclusive bioethics framework would also be expected to lead to enhanced moral credibility and increased trust in and legitimacy of the medical system by marginalized communities and those residing in the Global South. However, some authors push back more firmly by arguing against current global collaborative frameworks for research projects and calling for more radical and structural changes. Such tensions are understandable when considering the relatively early stage of the debate regarding the future of medicine and health care, and when weighing up the best options in terms of integration or establishing parallel systems.

This highlights that modern medicine should acknowledge its epistemic dominance and earnestly deliberate on how to create a more inclusive discipline as part of a restorative justice process. Thus we argue that medical research evidence should also be considered through this historical lens and should aim to be more culturally and morally grounded compared to current frameworks, procedures, and applications. Such an approach will also require serious remodeling of curricula, as discussed, while being mindful to pursue genuine structural changes rather than mere tokenistic gestures.

Rethinking medical and allied health sciences curricula and clinical training

Some of the chapters in this volume argue for the decolonization of medical, allied health sciences, and biomedical sciences curricula at both undergraduate and postgraduate levels. Moreover, it is clear that such initiatives should also extend to include the clinical training of health care practitioners. Reaching this goal will require curriculum redesign in terms of both the content and the nature of engagement with the students and trainees (active versus passive learning strategies). Such a shift will likely result in discomfort for educators, students, and trainees when considering the nature of content to be included as part of curriculum change. The case studies included in this volume reveal that such attempts included classes and student engagements that focused, for example, on medical racism and coloniality, and their damaging and lingering effects on health care. Of note, the authors show that such curriculum redesign can indeed foster an environment of open discussion and learning and thereby promote greater inclusivity. It also emerges that the decolonization of the medical curriculum is an iterative, fragmented, and ongoing process, as part of the continual striving toward more equitable practices in medicine. In this volume, some explore the integration of indigenous knowledge systems into health sciences curricula as part of the broader decolonization agenda. By employing appropriate case studies, one of the chapters showcases innovative practices and partnerships that can be utilized to bridge the gap between indigenous and biomedical paradigms. Here, the argument is made that the integration of indigenous knowledge systems into health sciences curricula is not only a matter of inclusion, but rather a necessary step toward justice, healing, and the creation of a more contextually responsive health care system.

Some authors call for a rethinking of clinical reasoning, evidence-based medicine, and associated guidelines that are rooted in statistical abstraction. They argue that evidence-based guidelines should be expanded to include, for example, causal explanations and lived experiences. Moreover, they propose that clinical reasoning should become a more normative process rather than diligently following a set of guidelines. Others focus on how the embedding of decolonial implementation science into medical training can equip practitioners to engage with complexity, interrogate power, and design responses that honor local priorities. The authors show that implementation science and decolonial practice are not parallel agendas but are deeply interconnected, with each strengthening the other. They propose that training future practitioners in such an integrative manner is necessary if health systems are to be reimagined as equitable, responsive, and genuinely transformative. Such initiatives reveal that graduates should be trained not only to diagnose pathological complications and diseases, but also to interrogate systems and structural frameworks that impact the health and well-being of their patients. The cultivation of a mindset where practitioners are aware of such complexities should also help to establish a more just and inclusive health care system.

Taken together, this volume issues a challenge for a reconfiguration regarding the pursuit of medical research, education, and clinical practice. It advances a cogent and persuasive appeal for transformative change that extends beyond the inclusion of the Other to institutional accountability regarding the ongoing harms of coloniality in contemporary medicine and health care. As we navigate the complex and wicked problems facing modern medicine, it is clear that such a recalibration requires a discipline that is grounded in epistemic humility, that recognizes and values the co-existence of various knowledge systems, and that pursues more inclusive pedagogies that are not only focused on technical expertise but also actively engage the historical, ethical, and moral challenges that have shaped contemporary medicine over the last few centuries. Collectively, such a realignment would form part of a restorative justice process and result in curricula, research agendas, and clinical practices that genuinely benefit all populations in terms of their health care and well-being.

Index

Note: Page numbers followed by *f* indicate figures, *b* indicate boxes, *t* indicate tables.

F

G

H

Y

Printed and bound by CPI Group (UK) Ltd, Croydon, CR0 4YY
14/09/2020
02216408-0013